CRC Series
in
MARINE SCIENCE

Editor-in-Chief
J. Robert Moore
Director, Marine Science Institute
The University of Texas
Austin, Texas

CRC HANDBOOK OF MARINE SCIENCE
Editors
F. G. Walton Smith
Dean Emeritus, School of Marine and Atmospheric Science
University of Miami
President, International Oceanographic Foundation
Miami, Florida
Frederick A. Kalbler
President, Hydrobiological Services, Inc.
Naranja, Florida

CRC HANDBOOK OF GEOPHYSICAL EXPLORATION AT SEA
Editor
Richard A. Geyer
Professor Emeritus
Department of Oceanography
Texas A & M University
College Station, Texas

MAN AND THE MARINE ENVIRONMENT
Editor
Robert A. Ragotzkie
Director, Sea Grant Institute
University of Wisconsin
Madison, Wisconsin

CRC HANDBOOK OF MARINE SCIENCE: COMPOUNDS FROM MARINE ORGANISMS
Authors
Joseph T. Baker
Vreni Murphy
Roche Research Institute of Marine Pharmacology
Sydney, Australia

CRC HANDBOOK OF COASTAL PROCESSES AND EROSION
Editor
Paul D. Komar
School of Oceanography
Oregon State University
Corvallis, Oregon

CRC HANDBOOK OF MARICULTURE
Editor
James P. McVey
Fishery Biologist
(Aquaculture Specialist)
National Marine Fisheries Service
Office of International Fisheries
Jakarta, Indonesia

CRC Handbook of Mariculture

Volume I
Crustacean Aquaculture

Editor

James P. McVey, Ph.D.
Fishery Biologist (Aquaculture Specialist)
National Marine Fisheries Service
Office of International Fisheries
Jakarta, Indonesia

CRC Series in Marine Science

Editor-in-Chief

J. Robert Moore, Ph.D.
Director, Marine Science Institute
The University of Texas
Austin, Texas

CRC Press, Inc.
Boca Raton, Florida

Library of Congress Cataloging in Publication Data
Main entry under title:

Crustacean aquaculture.

(CRC handbook of mariculture ; v. 1) (CRC series in marine science)
Bibliography: p.
Includes index.
1. Shellfish culture. I. McVey, James P.
II. Series. III. Series: CRC series in marine science.
SH138.C73 1983 vol. 1 [SH370] 639s [639'.5] 83-6085

Direct all inquiries to CRC Press, Inc., 2000 Corporate Blvd., N.W., Boca Raton, Florida, 33431.

Second Printing, 1986
Third Printing, 1986
Fourth Printing, 1989

International Standard Book Number 0-8493-0220-X

Library of Congress Card Number 83-6085
Printed in the United States

PREFACE

Crustacean aquaculture is a relatively new area of endeavor that has tremendous appeal to the public and private industry. The Japanese may have been the first to succeed in spawning and rearing commercial shrimp in captivity in the 1940s and 1950s. However, it was not until the late 1960s and 1970s that technology had developed to the point that commercial culture seemed possible.

Several research groups around the world have made significant advances in the control of crustacean life history and in the development of systems and techniques needed to mass produce many species of crustaceans. Each specie and each culture location requires unique conditions and adaptations in order for the culture process to be successful. In addition, no two crustacean aquaculture researchers will tackle the same problem in exactly the same way.

This Handbook attempts to bring the various techniques currently in use (1982) for the culture of commercially important crustaceans together, under one cover, so that individuals interested in culturing crustaceans, either commercially or for research purposes, can choose the methods that best apply to their particular situation. The authors are recognized as world leaders in their field and they have been asked to be as specific as possible about the techniques they are presently using.

The Handbook is divided into four sections: Section I: Larval Foods for Crustaceans; Section II: Maturation, Spawning Hatchery and Grow-Out Techniques for Crustaceans; Section III: Pathology and Disease Treatments for Crustaceans; Section IV: Crustacean Nutrition. Section I contains specific methods for the culture and preparation of algae, rotifers, *Artemia,* and other foodstuffs for use in crustacean hatcheries. Section II provides detailed descriptions of all aspects of culture techniques for the three main groups of crustaceans covered in this volume: *Macrobrachium* shrimps, penaeid shrimps, and lobsters. Section III describes the various diseases and the treatments and methods used in combating disease problems experienced in crustacean culture. Section IV provides the most recent information available on crustacean nutritional requirements.

There is a small degree of overlap on the descriptions of techniques by the different authors. These areas of overlap were not edited out so that the reader could fully understand the author's perspective and the importance of certain techniques.

The Table of Contents is ordered so that papers dealing with *Macrobrachium* shrimp, or penaeid shrimp, or lobster are grouped together. Hopefully, this will help to make quick comparisons so that the reader can decide the preferred approach for his/her situation.

The techniques for the experimental production of crustaceans are well established, but commercial culture has only been successful for specific situations and species. The prospective investor is cautioned to research any project very carefully before making substantial investments.

James P. McVey

EDITOR-IN-CHIEF

The Editor-in-Chief of the CRC Marine Science Series, **Professor J. Robert Moore,** received his B.S. degree (honors) from the University of Houston, in geology, his M.A. from Harvard University, also in geology, and his Ph.D. from the University of Wales (Aberystwyth), in geology and oceanography. From 1956 to 1966, he was a Senior Scientist at the Texaco Research Laboratories, Houston, Texas, engaged in research on marine sedimentation and geo-resources. From 1966 to 1977, Dr. Moore was Professor of Geology at the University of Wisconsin, Madison, where he was also Director of the Marine Research Laboratory and the Marine Mining Program. In 1977, he became Director, Institute of Marine Science, University of Alaska, and in 1979, assumed his present affiliation as Professor of Marine Studies, at the University of Texas, at Austin. Professor Moore's principal research interests are in seafloor mineral resources, marine geology, and global oceanographic processes and resources. His research has involved studies in the Bering Sea, Irish Sea, Western Pacific, Gulf of Mexico, Gulf of Alaska, Atlantic shelf and several coastal areas in the United States and abroad. He is a member of the Marine Board of the National Academy, the A.A.P.G. Marine Geology Committee, the N.A.C.O.A. Marine Mining Panel, Organizing Chairman of the CHEMRAWN IV Conference, and member of The Geochemical Society, Soc. of Economic Paleontologists and Mineralogists, Marine Technology Society, Am. Assoc. of Petroleum Geologists, Challenger Society, and a Fellow of the Geological Society of London. He is Editor of the *Marine Mining Journal,* and Exec. Secretary of the International Marine Mining Association. Since 1966, he has been an active consultant to major international corporations, chiefly on matters related to ocean resources, utilization, exploration, and development. He is currently conducting two major marine research programs, one in Palauan waters of the Western Pacific and the other in the Bering Sea off Alaska. He resides in Austin, Texas.

THE EDITOR

James P. McVey, Ph.D., is presently with The National Marine Fisheries Service, Office of International Fisheries, and is on a two-year special assignment to Indonesia where he will be working on improvement of *Macrobrachium* shrimp production.

Dr. McVey received his B.S. degree from the University of Miami in 1965 and his M.S. and Ph.D. degrees from the University of Hawaii in 1967 and 1970, respectively. He then spent 7 years as Fisheries Biologist for the U.S. Trust Territory in Palau, Micronesia. In this capacity he was responsible for the design and construction of the Micronesian Mariculture Demonstration Center (MMDC) and served as its Director from 1974 to 1977. During this time the MMDC introduced brackishwater pond culture to the principal island groups of Micronesia; successfully reared the popular rabbitfish, *Siganus,* through several generations; operated a freshwater shrimp hatchery; and worked with other commercial marine resources such as giant clams, oysters, saltwater shrimp, and tuna baitfish.

From 1977 to 1982 Dr. McVey served as Aquaculture Division Chief for the National Marine Fisheries Service, Southeast Fisheries Center, Galveston Laboratory. He was involved in developing the technology for mating and spawning of saltwater shrimp in captivity and was also the main coordinator for the Kemp's ridley turtle headstart program.

Dr. McVey is a member of the World Mariculture Society, The Western Society of Naturalists, and the American Fisheries Society. He has published numerous scientific and popular articles on aquaculture and contributed to the preparation of the National Aquaculture Plan.

ADVISORY BOARD

CONTRIBUTORS

Donald W. Anderson, Ph.D.
Oceanic Institute
Makapuu Point
Waimanalo, Hawaii

AQUACOP
Aquaculture Team of the Centre Océanologique du Pacifique
D. Amaru
O. Avalle
M. Autrand
A. Bennett
J. P. Blancheton
G. Breuil
C. Calinie
J. Calvas
D. Coatanea
G. Cuzon
Ch. De La Pomelie
F. Fallourd
A. Febvre
P. Garen
J. Goguenheim
J. M. Griessinger
P. J. Hatt
M. Jarillo
D. La Croix
J. J. Laine
J. P. Landret
J. F. Le Bitoux
J. L. Martin
A. Michel
O. Millous
J. Moriceau
Y. Normant
J. M. Peignon
S. Robert
D. Sanford
M. Tauhiro
V. Vanaa
J. F. Virmaux
V. Vonau
Centre National pour l'Exploitation des Oceans
Taravao, Tahiti, French Polynesia

Etienne Bossuyt
Agricultural Engineer
Sanitary Engineer
Research Assistant
Artemia Reference Center
State University of Ghent
Ghent, Belgium

James A. Brock, D.V.M.
Aquaculture Disease Specialist
Aquaculture Development Program
Department of Land and Natural Resources
State of Hawaii
Honolulu, Hawaii

Ernest S. Chang, Ph.D.
Bodega Marine Laboratory
Bodega Bay, California
and
Assistant Professor
Department of Animal Science
University of California, Davis
Davis, California

Nai-Hsien Chao
Senior Specialist
Tungkang Marine Laboratory
Taiwan Fisheries Research Institute
Tungkang, Taiwan

Yi-Peng Chen
Specialist
Tungkang Marine Laboratory
Taiwan Fisheries Research Institute
Tungkang, Taiwan

Douglas E. Conklin, Ph.D.
Assistant Nutritionist
Bodega Marine Laboratory
Bodega Bay, California
and
Department of Animal Science
University of California, Davis
Davis, California

John S. Corbin, Ph.D.
Manager
Aquaculture Development Program
Department of Land and Natural Resources
State of Hawaii
Honolulu, Hawaii

Louis R. D'Abramo, Ph.D.
Postgraduate Research Nutritionist
Bodega Marine Laboratory
Bodega Bay, California

Joe M. Fox
Research Associate
Texas A & M University
Agricultural Experiment Station
College Station, Texas

Michael Fujimoto
Aquatic Biologist
Anuenue Fisheries Research Center
Department of Land and Natural Resources
State of Hawaii
Honolulu, Hawaii

Dennis Hedgecock, Ph.D.
Assistant Geneticist
Bodega Marine Laboratory
Bodega Bay, California
and
Department of Animal Science
University of California, Davis
Davis, California

Thomas Y. Iwai, Jr.
Aquatic Biologist
Anuenue Fisheries Research Center
Department of Land and Natural Resources
State of Hawaii
Honolulu, Hawaii

Wallace E. Jenkins
Biologist
Marine Resources Research Institute
South Carolina Wildlife and Marine Resources Department
Marine Resources Division
Charleston, South Carolina

Patrick Lavens, M.Sc.
Research Assistant
Artemia Reference Center
State University of Ghent
Ghent, Belgium

Philippe Leger
Pharmacologist
Specialist in Environmental Sciences
Research Assistant at the Belgian National Science Foundation
Artemia Reference Center
State University of Ghent
Ghent, Belgium

Jorge K. Leong, Ph.D.
Microbiologist and Leader
Pathology Research Group
Aquaculture Research and Technology Division
Galveston Laboratory
Southeast Fisheries Center
National Marine Fisheries Service
Department of Commerce
National Oceanographic and Atmospheric Administration
Galveston, Texas

I-Chiu Liao, Ph.D.
Director and Senior Specialist
Tungkang Marine Laboratory
Taiwan Fisheries Research Institute
Tungkang, Taiwan

Donald V. Lightner, Ph.D.
Research Associate
Aquaculture Program
Environmental Research Laboratory
University of Arizona
Tucson, Arizona

Jaw-Hua Lin
Specialist
Tungkang Marine Laboratory
Taiwan Fisheries Research Institute
Tungkang, Taiwan

Richard Pretto Malca, Ph.D.
National Director of Aquaculture
Ministry of Agricultural and Livestock Development
Santiago de Veraguas
Republic of Panama
Adjunct Professor
International Center of Aquaculture and Fisheries
Auburn University
Auburn, Alabama

Spencer R. Malecha, Ph.D.
Principal Investigator
Prawn Aquaculture Research Program
Professor
Department of Animal Sciences
University of Hawaii
Honolulu, Hawaii

Mark R. Millikin
Research Fishery Biologist
National Marine Fisheries Service
Charleston Laboratory
Charleston, South Carolina

Karen Norman-Boudreau, Ph.D.
Bodega Marine Laboratory
Bodega Bay, California

Renée Rosemark, Ph.D.
Staff Research Associate
Bodega Marine Laboratory
Bodega Bay, California

Paul A. Sandifer, Ph.D.
Assistant Director
Marine Resources Research Institute
South Carolina Wildlife and Marine Resources Department
Charleston, South Carolina

Lowell V. Sick, Ph.D.
Supervising Research Physiologist
National Marine Fisheries Service
Charleston Laboratory
Associate Professor of Biochemistry
Medical University of South Carolina
Charleston, South Carolina

Noel Smith
Oceanic Institute
Makapuu Point
Waimanalo, Hawaii

Theodore I. J. Smith, Ph.D.
Associate Marine Scientist
Marine Resources Research Institute
South Carolina Wildlife and Marine Resources Department
Charleston, South Carolina

Patrick Sorgeloos, Ph.D.
Senior Scientist
Belgian National Science Foundation
Director
Artemia Reference Center
State University of Ghent
Ghent, Belgium

Alvin D. Stokes
Biologist
Marine Resources Research Institute
South Carolina Wildlife and Marine Resources Department
Charleston, South Carolina

Huei-Meei Su
Specialist
Tungkang Marine Laboratory
Taiwan Fisheries Research Institute
Tungkang, Taiwan

Paul Vanhaecke
Agricultural Engineer
Sanitary Engineer
Research Assistant
Artemia Reference Center
State University of Ghent
Ghent, Belgium

Danny Versichele
Agricultural Engineer
Sanitary Engineer
Research Assistant
Artemia Reference Center
State University of Ghent
Ghent, Belgium

TABLE OF CONTENTS

Section I
Larval Foods for Crustaceans

ALGAL FOOD CULTURES AT THE CENTRE OCÉANOLOGIQUE DU PACIFIQUE

Aquacop
Centre Océanologique du Pacifique*

The Centre Océanologique du Pacifique (COP), a laboratory of the French agency Centre National Pour l'Exploitation des Oceans (CNEXO), is located at Vairao, Tahiti, French Polynesia. Most of the island is surrounded with a coral reef delineating a lagoon. Around the COP, the lagoon forms a NW-SE basin with a maximum depth of 40 m and it is fed by oceanic water which results from sea swells coming over the barrier reef. The renewal rate of the lagoon water is once every day or every second day according to the strength of the waves. Under those conditions, the seawater pumped at the COP offers similar features to the superficial oceanic water: annual temperature range 25.5 to 30.1°C; annual salinity range 34.2 to 36.5 ppt; low quantities of nutrients (Table 1). It appears that rains are the main factor of variation of salinity and nutrient concentrations, the enrichment for the latter corresponding to the leaching of basaltic soils of the island. The COP aquaculture program was initiated in 1972. Work has been carried out on the freshwater prawn, *Macrobrachium rosenbergii*, on penaeid shrimp, and bivalve molluscs. For the latter two programs, the production of algal larval food was needed and thus unicellular algae production facilities were set up to supply penaeid and mollusc hatcheries, spat pregrowth tanks, and bivalve broodstock conditioning tanks.

INDOOR CULTURES

Cultured Species

Several temperate species have been cultured in the controlled algae culture rooms of the COP: *Isochrysis galbana, Pavlova lutheri, Platymonas tetrathele, Platymonas suesica, Chaetoceros calcitrans, Cylindrotheca* sp. Most of these species have been progressively replaced by tropical strains which have the advantage of proliferating in hatchery rearing tanks at a temperature of 27 to 30°C. Now the main cultured species are

1. *Chaetoceros gracilis* — origin Costa Rica (Maricultura penaeid farm)
2. *Isochrysis aff. galbana* [3] — isolated at the COP
3. *Platymonas* sp. — isolated at the COP
4. *Platymonas suesica* — isolated at The Culture Center of Algae and Protozoa, Cambridge, U.K.
5. *Chlorella* sp. — isolated at the COP
6. *Spirulina platensis* — isolated at The Culture Center of Algae and Protozoa, Cambridge, U.K.

General Facilities

Arrangement of the Algae Production Unit

The different parts of the algae unit are separated by doors or twin-door systems which

* The Aquaculture Team of the Centre Océanologique du Pacifique: Algal Cultures — D. Coatanea, J. L. Martin, Y. Normant; Nutrition — A. Febvre, G. Cuzon, J. J. Laine, J. M. Peignon; Water Quality Control and Treatment — J. P. Blancheton, J. Calvas, V. Vonau; Pathology — G. Breuil, J. F. Le Bitoux, S. Robert; Technology — J. F. Virmaux; Crustacean Cultures — P. J. Hatt, J. M. Griessinger, M. Jarillo, J. Goguenheim, M. Autrand, J. Mazurie, C. Calinie, O. Millous, J. P. Landret, P. Garen, D. LaCroix, F. Fallourd, J. Moriceau, O. Avalle, Ch. De La Pomelie, V. Vanaa, D. Amaru, A. Bennett, D. Sanford, and M. Tauhiro.

Table 1
NUTRIENT CONTENT[1-2] IN μ atg/ℓ OF SEAWATER AS PUMPED IN THE VAIRAO LAGOON

	N-NO_3	N-NO_2	N-NH_4	P-PO_4	Si-SiO_3
COP seawater in dry season	0.3	0.1	0	0.4	4.5
COP seawater after heavy rains	5.0	0.1	0.1	17.1	112.1

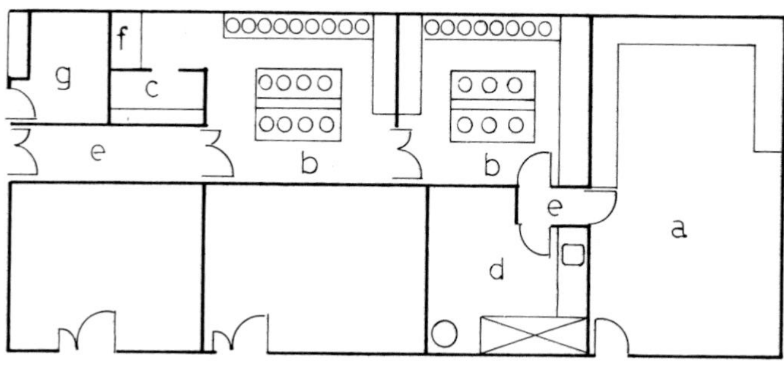

FIGURE 1. Algae production unit. a. Laboratory; b. culture rooms; c. test tube culture room; d. washing and storage room; e. air locks; f. filtration; g. security blower and electric control.

work as air-locks (Figure 1). The floors are laid with tiles and the whole building is air-conditioned.

Laboratory

The laboratory is equipped with benches supporting microscopes, a precise balance, and miscellaneous devices. The monitoring of algal cultures, the determination of cell densities by means of a Malassez hemacytometer, the preparation of nutrient stock solutions and weekly reinoculation of strain test tubes are performed in the laboratory.

Culture Rooms

Two separate 21 m^2 rooms are used routinely. Their capacity is 14 300-ℓ tanks each, plus the corresponding inoculation vessels of 20 ℓ, 5 ℓ, 2 ℓ, and 250 mℓ. In the first room, a bench has been fitted up to receive a set of the test tube cultures of the different strains.

Test Tube Culture Room

A special room has been set up to receive a second safety set of test tube cultures. This room has its own air conditioning and lighting distinct from the general systems.

Washing and Storage Room

This room is used for washing the 300-ℓ production tanks and the glassware with hot tap water. It contains a vertical autoclave and a steam generator to sterilize glassware which is thereafter stored in wall cupboards.

Temperature Regulation

The temperature control of the culture rooms is ensured by the general air conditioning system of the COP. Care must be taken for the proper mixing of cool air in the rooms; a safety switch automatically turns off the lights to avoid damage to the cultures in case of excessive increase of temperature in the culture rooms. Routinely the temperature of the rooms is regulated to 23°C so that the temperature inside the culture tanks stays within the range of 25 to 27°C.

Illumination of Cultures

The light was formerly produced by 40 W Sylvania® Gro-lux® fluorescent bulbs. In Tahiti, this type of bulb is hard to find; so it has been replaced by the more common 40 W Mazdafluor Daylight De Luxe TFRS 40 LJL bulb, without any consequence to the quality of cultures.

Test tube and 250-mℓ flask shelves are illuminated with one 40 W bulb providing 3000 lux of light. Other culture vessels receive 5000 lux of light measured at the center of empty vessels, provided by two or four 40 W bulbs according to the size of the containers.

Seawater System

The seawater pumped from the Vairao lagoon shows oceanic characteristics and its content in silt, detritus, organic matter, or live plankton is very low. The pumped water is directly delivered to the diverse equipment of the COP at 1 bar pressure.

In algae culture rooms, sea water is filtered through a set of four AMF CUNO® IM Type filters containing successively a 50 μ, a 10 μ, a 5 μ, and a 1 μ Super Micro Wynd cartridge. Cartridges are changed and the filter housings cleaned every two weeks. The 1 μ filtered seawater is directly used to fill up the 300-ℓ production tanks and its distribution is made through 16 bar resistant nontoxic PVC pipes.[4] The distribution system is built so that it can be drained and dried thoroughly every night. Once a week 60°C tap water is circulated through this system.

The seawater used for all the smaller culture vessels is previously filtered on a 0.22 μ stainless steel Millipore® filter and autoclaved for test tubes and 250-mℓ flasks.

Aeration System

Air is supplied at a 0.3 bar pressure by a general Roots blower. A second emergency Roots blower is also available for the algae rooms. Before use the air is filtered on a 1 μ Cuno® filter and then enriched with 0.2% carbon dioxide. PVC distribution pipes are fixed with a slope down so that condensing water can be easily purged. The distribution of gas mixture into the culture vessels is made through 6 mm diameter glass tubes.

Culture Vessels

Glassware (Figure 2)

For the different steps of culture, test tubes, 250-mℓ, 2-ℓ, and 5-ℓ Erlenmeyer flasks and 20-ℓ spherical carboys are used consecutively. Test tubes and 250-mℓ flasks are stopped with a cotton plug covered with an aluminum foil. Two-ℓ and 5-ℓ flasks have a rubber stopper pierced with two holes for the airpipe inlet and air outlet. Twenty-liter carboys are stopped with a glass cap fitted with two holes for aeration gas inlet and outlet. Glassware is autoclaved for 30 min at 125°C before use.

Fiberglass Tanks

Three hundred-liter tanks are made with translucent 1 mm fiberglass sheets. They are cylindrical with a conical bottom. At the base of the tank a rubber stopper equipped with a tap allows algal culture withdrawal. These tanks are washed and filled with chlorinated hot water during one night, then rinsed before use. They are covered with a fiberglass cap pierced with one hole for the aeration inlet.

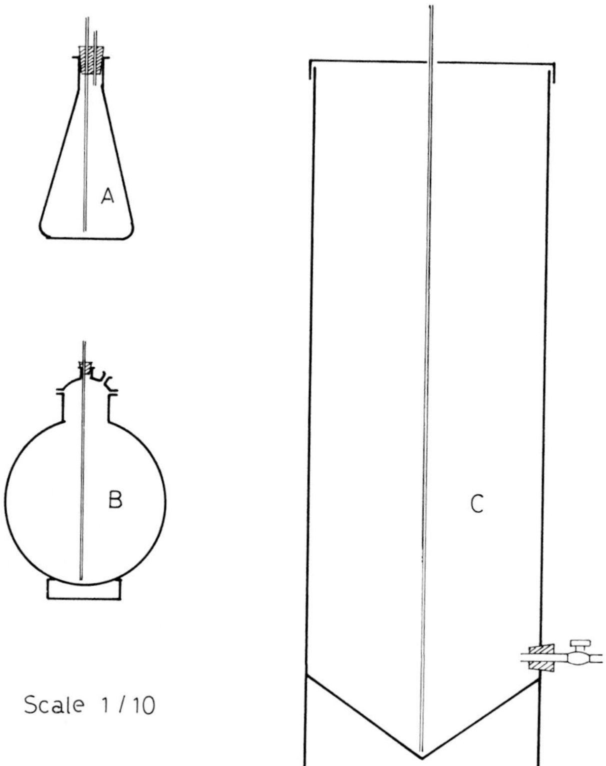

FIGURE 2. Culture vessels. A. 2-ℓ flask; B. 20-ℓ carboy; C. 300-ℓ tank.

Preparation of Culture Medium

Nutrient Stock Solutions

Walne Solution[5]

This solution is used for all species cultured.

Solution Number 1	
Na_2 EDTA	45.00 g
H_3 BO_3	33.60 g
Na NO_3 (KNO_3)	100.00 g (116 g)
NaH_2PO_4, $2H_2O$	20.00 g
Mn Cl_2, $4H_2O$	0.36 g
Fe Cl_3, $6H_2O$	1.30 g
Solution Number 2	1.00 mℓ

The above to be brought to 1 ℓ with distilled water.

Solution Number 2 (Trace Metal Solution)	
Zn Cl_2	2.1 g
Co Cl_2, $6H_2O$	2.0 g
$(NH_4)_6Mo_7O_{24}$,$4H_2O$	0.9 g
$CuSO_4$, $5H_2O$	2.0 g

The above to be brought to 100 mℓ with distilled water and HCl to get a good dissolution.

Solution Number 3 (Vitamin Mixture)	
Thiamin chlorhydrate	200 mg
Cyanocobalamin	10 mg

The above to be brought to 100 mℓ with sterilized distilled water.

Stock Solution for Diatom Culture

Solution Number 4	
Sodium metasilicate	20 g

The above to be brought to 1 ℓ with distilled water.

Solution Number 5	
KNO_3	100 g

The above to be brought to 1 ℓ with distilled water.

Except solution number 3, vitamin mixture, all the stock solutions are autoclaved 30 min at 125° C.

Algal Culture Mediums

Test Tubes

The test tubes are filled with 20 mℓ of 0.22 μ filtered seawater containing 1 mℓ/ℓ of solution number 1 and 0.1 mℓ/ℓ of solution number 3 and then sterilized. Test tubes destined for diatom culture are sterilized empty, and filled up while flaming with a Bunsen burner with 0.22 μ autoclaved seawater complemented after sterilization with 1 mℓ/ℓ of stock solutions number 1, number 4, and number 5 and 0.1 mℓ/ℓ of solution number 3.

250-mℓ Flasks

250-mℓ flasks are sterilized empty and then filled while flaming with a Bunsen burner with 200 mℓ of 0.22 μ sterilized seawater containing nutrient enrichment as described for test tubes.

2-, 5-, and 20-ℓ Vessels

2-, 5-, and 20-ℓ vessels are sterilized empty and filled with 0.22 μ filtered seawater. Enrichment is made at the time of inoculation with 1 mℓ/ℓ of solution number 1 and 0.1 mℓ/ℓ of solution number 3 for all species, plus 1 mℓ/ ℓ of solution numbers 4 and 5 for diatoms.

300-ℓ Fiberglass Tanks

These tanks as discussed before, are filled with 1 μ filtered seawater. Nutrient enrichment takes place at the time of inoculation as described for 2-, 5-, and 20-ℓ vessels.

Isolation and Maintenance of Strains

Isolation

The tropical species of *Platymonas* sp., *Chlorella* sp. and *Isochrysis aff. galbana* have been isolated from natural phytoplanktonic blooms occurring in the COP shrimp farming ponds. Two isolation techniques have been employed.

Gelose Medium

This medium is prepared by mixing 9 g agar-agar in 1 ℓ of seawater enriched with 1 mℓ of Walne's stock solution, number 1. Test tubes are filled with 10 mℓ of this medium, autoclaved and cooled in an inclined position. The tubes are thereafter inoculated by streaking the surface of the gelose with one drop of the wild mixed culture. The algae colonies will develop in 15 to 20 days. After that time, one colony of the wanted alga is picked up in the 15-day-old test tube and reinoculated in a new one. This operation is repeated as many times as needed to obtain a monospecific algal culture.

Dilution Method

The sample of wild mixed algal bloom is diluted to get a density of 1 cell per mℓ. Test tubes containing seawater enriched with Walne medium are inoculated with 1 mℓ of this diluted solution, so that each tube has a probability of containing one cell of one of the algae present in the natural bloom. The clones are allowed to develop two to three weeks and are thereafter observed to identify the species. By repeating this dilution from a two-week-old tube containing the wanted species, a monospecific culture will be obtained.

Maintenance of Strains

Every week, 7-day-old test tubes are reinoculated. This operation will be described with more details in the methodology of production.

Every month or every two months, a purification procedure is applied on the algae strains by using the dilution method described before. This allows us to work always with good quality strains.

For *Spirulina platensis* a particular procedure is used once a month to maintain strains in good shape. Because of their size, *S. platensis* are simply rinsed on an 85 μm sieve with 0.22 μm autoclaved seawater and then reinoculated in newly prepared test tubes.

Methodology of Production

All algae cultures are performed according to the batch technique. No continuous or semicontinuous cultures are made in the COP algae rooms.

Chaetoceros gracilis

Test tube

Every week, twelve test tubes containing 20 mℓ of culture medium are inoculated while flaming with a gas-operated Bunsen burner with 1 mℓ of the cultures coming from five, seven-day-old test tubes. The remaining seven-day-old tubes are kept till the next week's inoculation, in case of loss of the 12 newly inoculated tubes. Five of these 12 will serve as a source of the strain and the 7 remaining tubes as culture starters for the next weeks inoculation. These tubes are then placed on the shelf reserved for strains and exposed to a 3000 lux illumination.

A second safety set of 4 test tubes is inoculated in the same way and is placed into the safety strain room.

The average density on the seventh day is 2.8×10^6 cells per mℓ.

250-mℓ Flasks

Seven are inoculated each week while flaming with a burner from the seven culture starter test tubes inoculated the previous week. They will be used as starters for the 2 ℓ flasks from the fourth day after the inoculation day. The average density obtained on the seventh day is 3.2×10^6 cells per mℓ.

2-ℓ Flasks

One is inoculated every day from a 4 to 11-day-old 250 mℓ starter, and will remain in culture 3 days. The 2-ℓ flasks are the first in the culture procedure to be provided with aeration. The average density obtained on the third day is 1.9×10^6 cells per mℓ (Figure 3).

5-ℓ Flasks

Daily, two 5-ℓ flasks are inoculated from one 3-day-old 2-ℓ flask. The culture time is 3 days and the average density on the third day is 3.25×10^6 cells per mℓ (Figure 3).

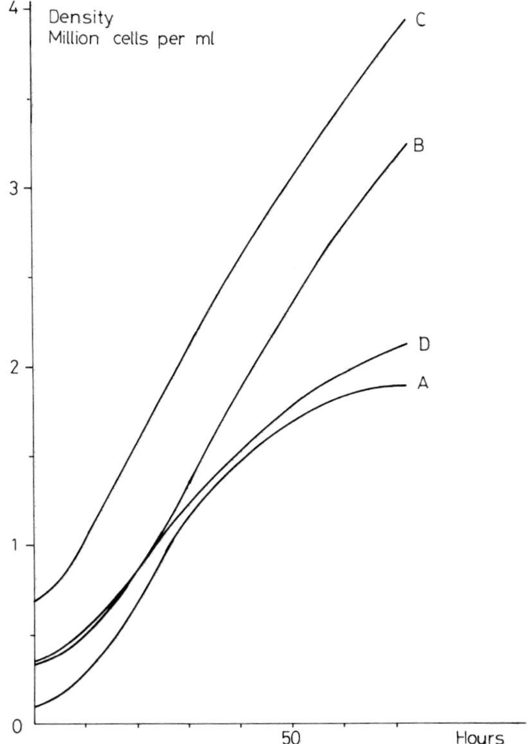

FIGURE 3. *Chaetoceros gracilis* growth in algae culture room. A. 2-ℓ flask; B. 5-ℓ flask; C. 20-ℓ carboy; D. 300-ℓ tank.

20-ℓ Carboys

Two of these are inoculated every day from two 3-day-old liter flasks. These spherical carboys give good cell division performance and the average cell density on the third day is 3.95×10^6 cells per mℓ (Figure 3).

300-ℓ Tanks

Two are inoculated every day from two 20-ℓ carboys. The 300-ℓ tanks are utilized for hatcheries on the third day at an average cell density of 2.1×10^6 cells per mℓ (Figure 3).

Other Species

Isochrysis aff. galbana

Almost the same methodology as for *Chaetoceros gracilis* is applied to *Isochrysis* and leads to the production of one 300-ℓ tank per day. There is no noticeable difference concerning the first two steps. One 2-ℓ flask is inoculated every day and will be used after four days of culture as starter for a 20-ℓ carboy. After four days, the latter will inoculate a 300-ℓ tank. The 5-ℓ flasks are not used for culturing this species. Growth curves are given in Figure 4.

Platymonas sp. and Chlorella sp.

Eight test tubes and eight 250-mℓ flasks of each species are inoculated once a week as described for *Chaetoceros gracilis*. Further inoculations take place every two days in 2, 5, 20, and 300-ℓ culture vessels. The culture duration between two consecutive steps is 4 days. For both species, a 300-ℓ tank is produced every two days. The average cell density obtained for *Platymonas* sp. on the fourth day in a 300-ℓ production tank is 1.5×10^6 per mℓ. The growth curves of *Chlorella* sp. are given in Figure 5.

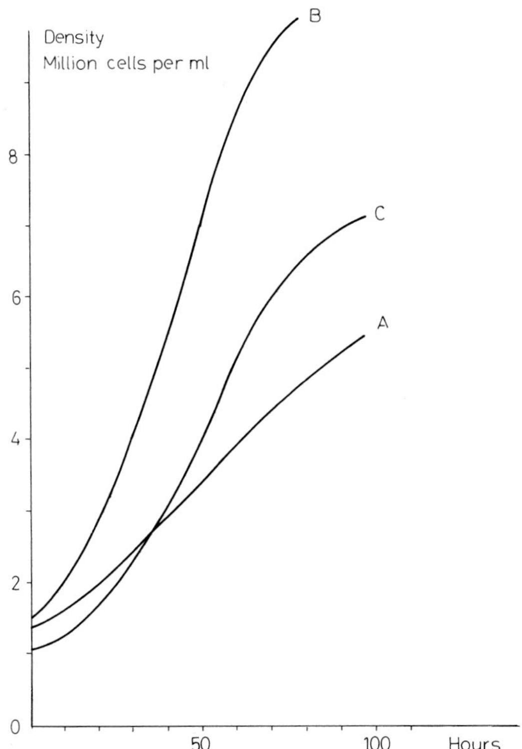

FIGURE 4. *Isochrysis aff. galbana* growth curves in algae culture room. A. 2-ℓ flask; B. 20-ℓ carboy; C. 300-ℓ tank.

Spirulina platensis

This species has recently been cultured as food for penaeid larvae and postlarvae. The culture methodology does not differ from *Platymonas* sp., except that the duration between two consecutive steps is 5 days instead of 4 for *Platymonas* sp.

Culture Problems

Few problems are encountered in running the indoor algae cultures. The short duration of algae blooms involve a rapid turnover of culture vessels. If not utilized on the planned day, cultures are only allowed to stay one more day and are then discarded. Special care is given to the test tubes and 250 mℓ flasks as they are the basis for the entire program. General cleanliness of the algae room will avoid contamination of the culture vessels. Once a year a sanitary stop is performed. All cultures are then stopped, except a set of strains in the safety strain room. Opportunity is taken at this stop to disassemble and thoroughly clean or change aeration and filtered seawater distribution pipes.

Some problems sometimes occur for unknown reasons on *Isochrysis aff. galbana* when an algal deposit appears on the walls of containers, mainly 300-ℓ tanks. Such cultures turn yellowish and foul within a few hours and must be discarded. There is some reason to suspect a fault in 1 μ seawater filtration, and in the future, a 0.22 μ filtration will be provided to all culture vessels, including 300-ℓ tanks.

OUTDOOR CULTURES

Mass quantities of good quality algae are required to support the growth of spat produced in a bivalve mollusc hatchery. These algae cannot be supplied by the lagoon seawater

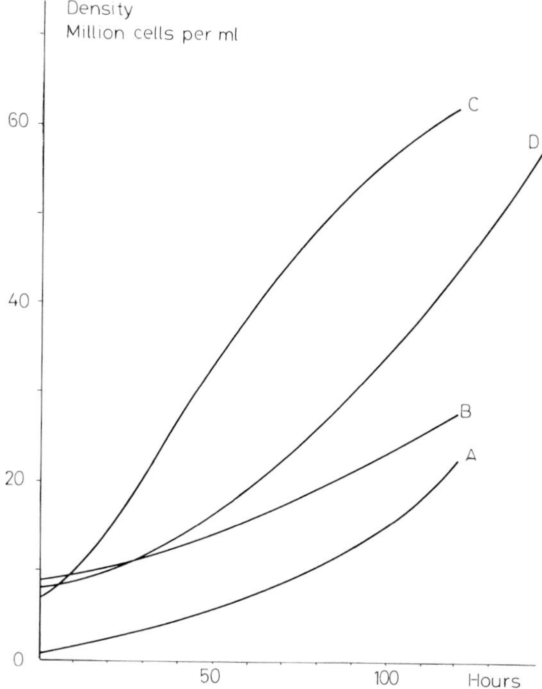

FIGURE 5. *Chlorella* sp. growth curves in algae culture room. A. 2-ℓ flask; B. 5-ℓ flask; C. 20-ℓ carboys; D. 300-ℓ tank.

characterized by its low phytoplankton content. It was thus decided to develop techniques for the mass production of algae in large outdoor tanks.

Cultured Species

Outdoor production is based upon the tropical species *Chaetoceros gracilis* which proved to induce good growth results when fed both to bivalve larvae and spat.

Description of Outdoor Tank Area

Culture Tanks

Four 8-ton and four 35-ton cylindrical tanks have been built. Walls and bottom are made of a 1 mm fiberglass sheet. The 35-ton tanks are girded with a 50 cm high concrete belt to reinforce walls. The slightly conical bottom rests on a bed of sand. Each tank is equipped with a central outflow PVC pipe. The 35-ton tanks are fitted with an extra lateral outflow PVC pipe at the base of the wall (Figure 6).

Water and Air Distribution Systems

All tanks have inlet pipes for freshwater, seawater, and blown air. No filtration is provided on the water distribution systems. Air is blown at a 0.3 bar pressure without carbon dioxide enrichment through a central pipe opening at the deepest part of the tank.

Algae Distribution System

Only 35-ton tanks have an algae distribution system, which is a lateral outflow PVC pipe. Through this lateral pipe, the algal culture feeds a Serseg "FLUXO" hydroejector which is also supplied with 1 bar pressure seawater coming from the general seawater distribution system. The mixture of algae culture and pumped seawater is then driven back through a

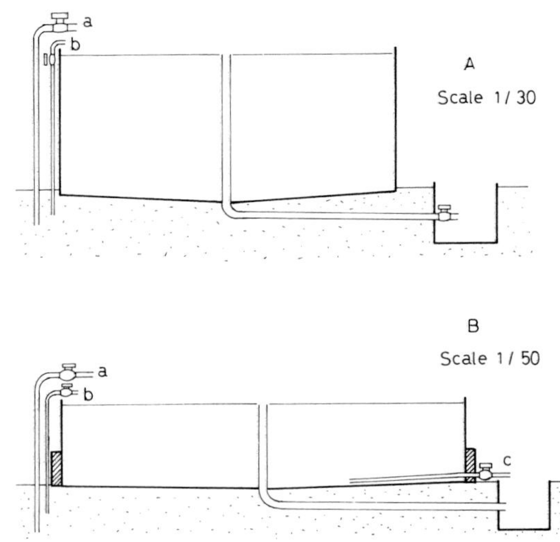

FIGURE 6. Outdoor tanks. A. 8-ton tank; B. 35-ton tank; a. Seawater inlet; b. Aeration inlet; c. Lateral outflow pipe.

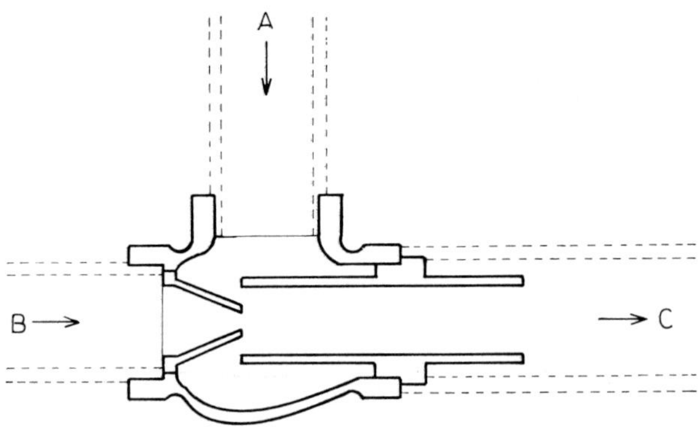

FIGURE 7. Serseg "FLUXO" hydroejector. A. Algal culture inlet; B. Seawater inlet; C. Mixture outflow.

PVC pipe to the spat culture basin (Figure 7). The algal cell density of the mixture is adjusted by two PVC valves, the first one regulating the outflow rate of algae culture and the second one regulating the seawater inflow. This device is regulated so that complete draining of the 35-ton tank takes about 24 hr.

Operation of the Outdoor Tanks

Culture Medium

Algal cultures are enriched with 20 g per cubic meter of "GROPLUS" (Yates, NZ) fertilizer, which is used as hydroponic fertilizer in agriculture for soilless cultures. The composition of "GROPLUS" is provided below.

N — KNO_3	5.26%
N — $(NH_4)_2$ SO_4	3.69%
P — Superphosphate	4.21%
Ca — Ca SO_4	2.63%
Mg — Mg SO_4	2.95%
S — SO_4	6.03%
Oligoelements	0.13%

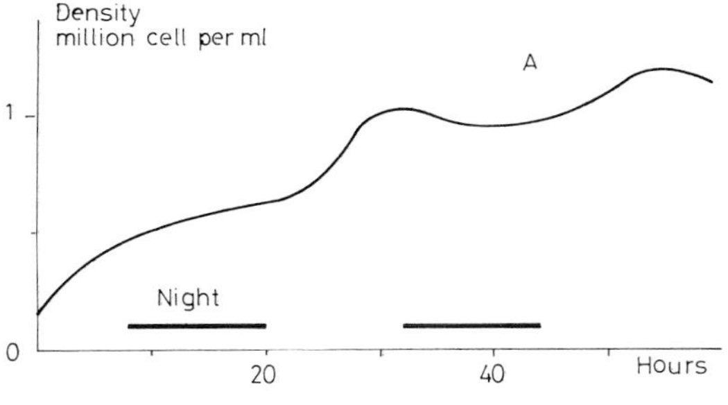

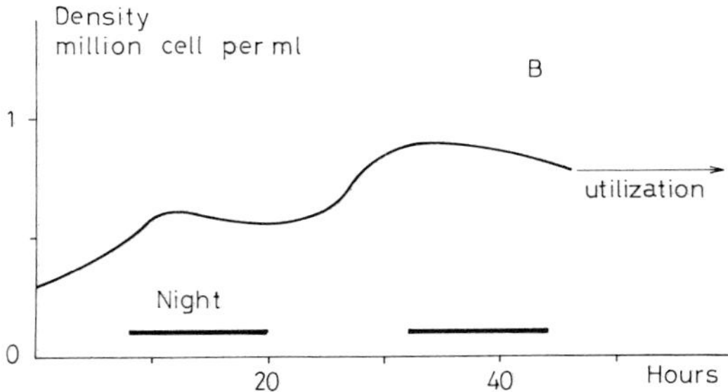

FIGURE 8. *Chaetoceros gracilis* growth curves in outdoor tanks. A. 8-ton tank; B. 35-ton tank.

The cultures are performed at a salinity of 26 ppt by adding one fourth of tap freshwater to the tanks. The freshwater is used for its silica content which is at least 5 mg/ℓ in dry season and up to 20 mg/ℓ in rainy season. The final content of silica in the culture tanks thus varies within the range 1.25 mg/ℓ to 5 mg/ℓ and appears not limiting for *Chaetoceros gracilis* culture.

8-Ton Tanks

Every day one 8-ton tank is inoculated either by pumping two 300-ℓ tanks from the algae culture room or by pumping an inoculum from a neighboring 8-ton tank, so that the initial density is about 150,000 to 200,000 cells per milliliter. The fertilizer is added at the inoculation time by placing the fertilizer into a bolting-cloth sock hung in the tank. On the third day, the average cell density is 1.2×10^6 cells per mℓ (Figure 8), and the 8-ton tank is then pumped into one 35-ton tank as the inoculant. On the fourth day, the tank is left empty for cleaning and drying purposes.

35-Ton Tanks

One 35-ton culture is inoculated every day using a 3-day-old culture of an 8-ton tank as starter. Fertilizer is added at the inoculation time. After two days of culture, the concentration reaches 700,000 to 800,000 cells per milliliter (Figure 8). The draining of the tank is then started and will last one more day. The fourth day is employed to clean the tank and let it dry before the next inoculation.

REFERENCES

1. **Anonymous,** Etude Hydrologique et Biologique du Lagon de Vairao et de ses Abords Exterieurs, Rapport de la Convention ORSTOM-CNEXO, Orstom, Noumea, 1976.
2. **Aquacop,** Données Hydrologiques et Météorologiques sur le Lagon de Vairao, Rapport CNEXO/COP, Taravao, Tahiti, 1976.
3. **Ewart, J. W. and Epifanio, C. E.,** A tropical flagellate food for larval and juvenile oysters *Crassostrea virginica* GMELIN, *Aquaculture,* 22, 297, 1981.
4. **Dupuy, J. L., Windsor, N. T., and Sutton, C. E.,** Manual for design and operation of an oyster seed hatchery for the American oyster *Crassostrea virginica, Special Report no. 142 in Applied Marine Science and Ocean Engineering of the Virginia Institute of Marine Science,* Hargis, W. J., Jr., Director, Gloucester Point, Va., 1977, 96.
5. **Walne, P. R.,** Experiments in the large-scale culture of the larvae of *Ostrea edulis* L., *Fish. Invest. Min. Agric. Fish.,* Ser. 2, 25, 1966, 53.

INTENSIVE ALGAL CULTURE TECHNIQUES

Joe M. Fox

INTRODUCTION

An area of major concern in most intensive aquaculture systems is the provision of a suitable food source. Unicellular microalgae have proven appropriate for a variety of culture situations.[1-5] According to Guillard,[6] six different classes of microalgae are cultured as food for aquaculturally reared animals (Table 1). These animals include shrimp (marine and freshwater), crabs, *Artemia* sp., bivalves, gastropods, lobsters, finfish, and even microzooplankton such as rotifers. The emphasis of this chapter will be on various techniques employed for mass culturing microalgae as food for penaeid larvae.

ENRICHMENT MEDIA

Surface seawater is most commonly used for preparation of media utilized in the mass cultivation of marine microalgae. Unfortunately, surface water seldom contains sufficient nutrients for mass cultivation to a density suitable for use in intensive culture systems. It is generally conceded that in most cases seawater must be enriched with essential growth nutrients to achieve high cell density (biomass).

Enrichment media can be classified as either natural or synthetic. Natural seawater media are those utilizing surface seawater to which nutrients have been added. Synthetic seawater media are those usually composed of distilled water, growth nutrients, and artificial sea salts.

Natural seawater media utilize a variety of inorganic macronutrients and both inorganic and organic micronutrients. One medium which has received extensive use and is suitable for the growth of most algae is Guillard's F medium.[6] This medium as well as other similar media, is described in Table 2.

Macronutrients found in Guillard's F medium include nitrate, phosphate, and silicate. Silicate is specifically used for the growth of diatoms which utilize this compound for production of an external shell or frustule. If diatoms are not being cultured, silicate is excluded from the medium.

Inorganic micronutrients include ferric chloride, the chelate EDTA, plus a variety of trace metals. Chelators serve to keep certain essential trace metals in solution thus insuring their availability to the cell. According to McLachlan,[7] EDTA is the most widely used chelate in marine media, and is not readily metabolized by microbes. Other chelators used in the preparation of seawater media are Versonal™, nitrilotriacetic acid (NTA), and citric acid.

Organic micronutrients used in the preparation of natural seawater media are the vitamins thiamine (B_1), cyanocobalamin (B_{12}), and biotin. Their use is not essential in mass cultivation of all algal species; however, this point should be established prior to the exclusion of any vitamin from a particular nutrient regime.

Synthetic seawater media, as previously mentioned, is composed of ''basal'' distilled water to which artificial salts and enrichment nutrients have been added. One synthetic medium which has proven suitable for growth of marine microalgae is Gates and Wilsons NH medium.[8] The composition of this medium is described in Table 3. Another suitable medium involves a combination of synthetic sea salts and Guillard's F medium. Instant Ocean™ is widely used as an additive and its formulation is described in Table 4. These salts are sold in a nonsterile powdered form and should be sterilized and homogenized (due to settling of constituents) prior to use. Various sources of synthetic sea salts are listed in Table 5.

Table 1
CLASSES AND SPECIES OF MICROALGAE PRESENTLY UNDER CULTURE

Class	Species
Bacillariophyceae	*Skeletonema costatum, Thalassiosira pseudomonas, Thalassiosira fluviatilis, Phaeodactylum tricornutum, Chaetoceros calcitrans, Chaetoceros curvisetus, Chaetoceros simplex, Ditylum brightwelli, Scenedesmus*
Haptophyceae	*Isochrysis galbana, Isochrysis* sp. (Tahiti), *Dicrateria inornata, Cricosphaera carterae, Coccolithus huxley*
Chrysophyceae	*Monochrysis* sp.
Prasinophyceae	*Pyraminimonas grossii, Tetraselmis suecica, Tetraselmis chuii, Micromonas pusilla*
Chlorophyceae	*Dunaliella tertiolecta, Chlorella autotrophica, Chlorococcum* sp., *Nannochloris atomus, Chlamydomonas coccoides, Brachiomonas submarina*
Chryptophyceae	*Chroomonas salina*
Cyanophyceae	*Spirulina*

Table 2
ENRICHED SEAWATER MEDIA

	Concentration μm/ℓ				
Additives	**F/2**	**H/2**	**F/2 beta**	**ES**	**SWM**
	Inorganic Macronutrients				
$NaNO_3$	880	—	880	660	500—2000
NH_4Cl	—	500	—	—	—
NaH_2PO_4	36.3	36.3	36.3	—	50—100
Na_2glycerophosphate	—	—	—	25.0	—
$Na_2SiO_3{\cdot}9H_2O$	54—107	54—107	54—107	—	200
	Inorganic Micronutrients				
Fe EDTA	—	—	—	7200	2.0
$FeCl_3{\cdot}6H_2O$	11.7	11.7	11.7	1.8	—
Na_2EDTA	11.7	11.7	11.7	26.9	48.0
$CuSO_4{\cdot}5H_2O$	0.04	0.04	0.04	—	0.3
$ZnSO_4{\cdot}5H_2O$	0.08	0.08	0.08	0.80	35.0
$CoCl_2{\cdot}6H_2O$	0.05	0.05	0.05	0.17	0.30
$MnCl_2{\cdot}4H_2O$	0.90	0.90	0.90	7.30	10.0
$Na_2MoO_4{\cdot}2H_2O$	0.03	0.03	0.03	—	5.0
Boron	—	—	—	185	400
	Organic Micronutrients				
Thiamine HCl(B_1)	100 μg	100 μg	100 μg	20 μg	—
Nicotinic acid	—	—	—	—	0.1 mg/ℓ
Ca. pantothenate	—	—	—	—	0.1 mg/ℓ
p-Aminobenzoic acid	—	—	—	—	10 μg/ℓ
Biotin	0.5 μg	0.05 μg	0.5 μg	0.8 μg	1.0 μg/ℓ
i-Inositol	—	—	—	—	5.0 mg/ℓ
Folic acid	—	—	—	—	2.0 μg/ℓ
Cyanocobalamin	0.5 μg	0.5 μg	0.5 μg	1.6 μg	1.0 μg/ℓ
Thymine	—	—	—	—	3.0 μg/ℓ
Tris	—	—	—	0.66	0—5000
Glycylglycine	—	—	—	—	5000
Soil extract	—	—	—	—	50 mℓ/ℓ
Liver extract	—	—	—	—	10 mg/ℓ

[a] Guillard, R. R. L., *Culture of Marine Invertebrate Animals*, Plenum Press, New York, 1975, 29.

Table 3
NH ARTIFICIAL SEAWATER MEDIUM

Additives	Amount / liter
NaCl	24.0 g
KCl	0.6 g
$MgCl_2 \cdot 6H_2O$	4.5 g
$MgSO_4 \cdot 7H_2O$	6.0 g
$CaCl_2$	0.7 g
K_2HPO_4	10.0 mg
KNO_3	10.0 mg
Vitamin B_{12}	1.0 μg
Thiamin HCl	10.0 mg
Biotin	0.5 μg
Sulfides[a]	1.0 mℓ
Vitamin mix 8[b]	0.1 mℓ
Metals T[c]	5.0 mℓ
Adenine sulfate	1.0 mg
Tris	0.1 g
NaEDTA	10.0 mg

[a] Sulfides:
NH_4Cl-0.2 g, KH_2PO_4-0.1 g, $MgCl_2 \cdot 6H_2O$-0.04 g, $NaHCO_3$-0.2 g, $Na_2SiO_3 \cdot 9H_2$)-0.15 g. Raise to 1.000 ℓ with distilled water.

[b] Vitamin mix:
Thiamin HCl-20 mg, biotin-50 μg, vitamin B_{12}-5 μg, folic acid-0.25 mg, PABA-1.0 mg, nicotine acid-10 mg, thymine-80 mg, choline-50 mg, inositol-100 mg, patrescine-0.8 mg, riboflavin-0.5 mg, pyridoxine-4.0 mg, pyridoxamine-2.0 mg, orotic acid-26 mg. Raise to 100 mℓ with distilled water (Provasoli, et al.; 1956).

[c] Metals T:
1% solutions; 2.5 mℓ Fe Tartrate (5 mg Fe), 3.0 mℓ H_3BO_3 (5.1 mg B), 0.1 mℓ H_2SeO_3 (1 mg Se), 0.12 mℓ NH_4VO_3 (0.5 mgV), 0.11 mℓ K_2CrO_4 (0.2 mg Cr), 0.37 mℓ $MnCl_2$ (1.0 mg Mn), 0.83 mℓ TiO_2 (5.0 mg Ti), 5.0 mℓ Na_2SiO_3 (5.0 mg Si), 0.4 mℓ $ZrOCl_2$ (2.0 mg Zr), 0.15 mℓ $BaCl_2$ (1.0 mg Ba). Raise to 100 mℓ with distilled water.[8]

OBTAINING CULTURES/ISOLATION

Algal species used in intensive aquaculture may be obtained from world-wide culture collections (commercial or nonprofit) or from the laboratory isolation of endemic species from local waters. Obtaining species from culture collections is a simple procedure requiring minimal effort. Stein lists several sources within the U.S. and world-wide.[9] Commercial outlets usually provide all items necessary for the shipping of algal cultures. Nonprofit sources almost always require the receipt of sterile stock culture tubes with medium prior to shipping. In this case, proper care should be taken to insure stock culture tubes (with medium) are sterile, securely sealed, and placed inside a protective shipping container.

Another means of obtaining unialgal cultures is laboratory isolation of endemic species. This method is especially helpful should stock cultures become contaminated. Many mariculture projects fail to investigate local species and find themselves totally dependent on external sources. Endemic species should be considered simply because of their ability to propagate in local waters.

Isolation of algal species is not a simple matter due to the nature of the cell itself. Particle units or individual cells are often quite small and usually associated with other nonsuitable epiphytic species. Isolation of a single unit is usually desired and accomplished by several proven techniques:

Table 4
INSTANT OCEAN™

Additive	Concentration ($\mu m/\ell$)
Cl	5.19×10^5
Na	4.44×10^5
So_4	2.60×10^4
Mg	4.94×10^4
K	9.50×10^3
Ca	9.20×10^3
HCO_3	2.30×10^3
H_3BO_3	4.04×10^2
Br	2.50×10^2
Sr	91.30
PO_4	10.50
Mn	18.00
MoO_4	4.40
S_2O_3	3.60
Li	28.80
Rb	1.20
I	0.55
EDTA	0.13
Al	1.50
Zn	0.31
V	0.39
Co	0.17
Fe	0.18
Cu	0.05

Table 5
COMMERCIAL SYNTHETIC SALTS

Dayno Sea Salt
Dayno Sales Company
678 Washington Street
Lynn, Massachusetts 01901

hw Marinemix
Hawaiian Marine Imports, Inc.
465 Town & Country Village
Houston, Texas 77024

Instant Ocean
Aquarium Systems Incorporated
1450 East 289th Street
Wyckliffe, Ohio 44092

Rila Marine Mix
Rila Products
Teaneck, New Jersey

Sea Salt Mix
Marine Research Associates
Box 7
Westport Point, Massachusetts 02791

Utility Seven Seas Mix #156
Utility Chemical Company
Manufacturing Chemists
145 Peel Street
Paterson, New Jersey

1. Simple capillary pipette isolation
2. Streak plating followed by capillary pipette isolation
3. Spray plating followed by capillary pipette isolation
4. Dilution followed by capillary pipette isolation
5. Treatment with antibiotics
6. Ultrasonic vibration (sonification)
7. Motile migration

Simple capillary pipette isolation involves the use of "pulled" pipettes of very small bore for separating larger single-celled and chain-forming species. Streak plating requires streaking of culture on solidified growth medium (agar plus Guillard's F medium or any other suitable liquid medium) followed by isolation with capillary pipettes. Spray plating is quite similar to streak plating in concept except the transfer vehicle is an atomizing jet of air. Simple dilution involves continued dilution of cells to a point where the subject species is well separated from any associated contamination. This method works well with motile species but is not suitable for isolation of nonmotile species (such as diatoms) which usually have some form of epiphytic contamination associated with them. If the desired phytoplankton species is contaminated with bacteria, antibiotics can be used to eliminate the contamination. This isolation technique is quite effective but requires strict monitoring. The usual procedure for isolation by antibiotic treatment requires the use of several concentrations of the antibiotic and frequent subculturing to an antibiotic-free medium. Algal cells are usually susceptible to antibiotics should exposure time and concentration of antibiotics be excessive. Each algal species has its own tolerance range; periodic evaluations and transfers must be made to insure maximum effect without harm. Ultrasonic vibration/isolation is accomplished by placing natural or contaminated cultures in a suitable ultrasonic device. This process physically removes epiphytic contaminants from the species in question. Further separation by centrifugation and capillary pipette is then recommended. Motile migration involves use of a light source to separate motile, positively phototactic cells from nonmotile cells.

For experimental purposes it may be necessary to prepare axenic clonal cultures. A culture termed "axenic" is one considered bacteria-free. A culture may be determined as bacteria-free if, when grown on bacteria-supporting agar, a negative bacterial response is shown. Peptose and sucrose agars are suitable for this purpose. However, one should be aware that certain bacteria will only grow on specific agar within a narrow range of experimental conditions. The "axenic" state is, therefore, somewhat easier to claim than to prove.

All the previously mentioned techniques should employ capillary pipette isolation. Usually variations and combinations of *all* methods are used and should definitely be considered. In all cases, after isolation has been performed, the isolation unit should be transferred to a suitable growth medium. As a final note on isolation, stringent aseptic conditions should be maintained at all times. The reader should refer to Stein for further details relevant to isolation.[9]

NUTRIENT STOCK SOLUTIONS

As exemplified by the section devoted to enrichment media, only small amounts of nutrients are necessary to achieve high cell densities. For instance, only 0.5 μg B_{12} is used to achieve F/2 conditions in 1ℓ of seawater. To avoid excessive repetition in weighing of similar small amounts, concentrated nutrient stock solutions are used. In some cases, the amount of nutrient to be added to even large volume cultures is difficult to weigh accurately. By using concentrated nutrient stock solutions less time is spent in weighing, reagents are handled less often, and, thus, the potential for contamination is reduced. With micronutrients it is even advisable to prepare not only primary stock solutions but also working stock solutions (prepared from the primary stock solution).

Different nutrients, in their reagent form, are not equally soluble in distilled water. Preparation of nutrient stock solutions must be made with consideration for each nutrient's solubility. These concentrates should also be prepared insuring all components will remain in solution. Nitrate and phosphate are fairly soluble in distilled water and can be stored together as a working stock solution. Silicate and trace metals are very soluble in distilled water. Since iron chloride and EDTA are used at somewhat higher concentrations than the trace metals, they can be added in crystal or powder form directly to the trace metals working stock. Iron chloride is highly soluble; however, EDTA is only slightly soluble. The addition of EDTA may be simplified, though, by addition in a hydrated form ($Na_2EDTA \cdot 2H_2O$). Vitamin solutions, as well as trace metal solutions, should be stored in the dark. If possible, they should also be refrigerated. A further precaution, enhancing consistency in results, requires maintenance of a stock solution notebook. By recording how and when each stock solution is prepared, one can maintain consistency in procedure and also help explain problems in production due to improper preparation of solution.

A final note on preparation of nutrient stock solutions regards sterility. In preparing nutrients for use in growing algae, from the stock culture tube to large tanks, it is advisable to maintain aseptic conditions in the initial smaller culture volumes. Nutrient media may be autoclaved in the tube or flask intended for growth of the culture.

GROWTH DYNAMICS

In order to achieve sufficient production of culture on a consistent basis, an understanding of algal growth dynamics is required. The goal of any algal production facility concerned with feeding shrimp (or any filter feeder) is to supply as much algae as required as efficiently as possible. A productive facility reflects compromise among several factors: growth characteristics of species used, the optimum nutrient regime, labor, equipment, production scheduling, etc., in order to reach its goal.

The growth of axenic cultures can be expressed in terms of cell division (doublings per day) or growth (relative growth rate). With suitable nutrient enrichment and favorable physical conditions, algal cultures will exhibit growth as shown in Figure 1.[10] This growth is characterized as having at least four separate phases. The first phase is referred to as the lag phase. This phase is sometimes only an apparent lag due to introduction of dead cells. Usually the lag is "real" and could possibly be attributed to several factors.

1. Deactivation of enzymes in inoculum
2. Decreased metabolite level in inoculum
3. Cell size increase, but no cell division
4. Some diffusible factors produced by the cells themselves is necessary for carbon fixation (CO_2, glycolic acid)
5. Metabolic activity of cells inactivates some toxic factor present in the medium
6. Introduction of inoculum into a medium containing high concentrations of some particular substance (phosphates, antibiotics, etc.)[10]

The second phase, or exponential phase, is characterized by constant rapid cell division. This relative growth rate is usually constant in the exponential phase and its value is dependent on cell size (surface area), light intensity (below saturation level), and temperature.

The third phase, or phase of declining relative growth can be caused by several factors.[11]

1. Depletion of a particular nutrient
2. Rate of supply of CO_2 or O_2
3. Change in pH due to preferential absorption
4. Light limitation due to shading (Beer's Law)
5. Autoinhibition by production of a toxic substance

This phase of growth occurs quite rapidly with a balance being formed between growth rate and the limiting factor. This phenomenon is known as the stationary phase (fourth stage).

The final stage is known as the death or culture collapse stage. This is usually the result of depletion of nutrients to a level incapable of sustaining growth and/or build-up of metabolites to a toxic level.

The key to successful and efficient algal production is maintaining all cultures in the exponential phase of growth. For a more detailed explanation of algal growth dynamics one should refer to Fogg's excellent treatise of this subject.[10]

CULTURE TECHNIQUES

Maintaining cultures in exponential growth may be accomplished several ways.

1. Transfer of algae to larger, enriched, culture volumes prior to stationary phase, using the transferred algae as inoculum; the larger culture volumes are then brought to a maximum density and fed to the filter feeding organisms being cultured
2. True continuous culture — a chemostat-like arrangement
3. Semicontinuous culture — maintenance of large volume cultures by periodic batch removal of culture followed by topping up and addition of nutrients to achieve original nutrient concentration

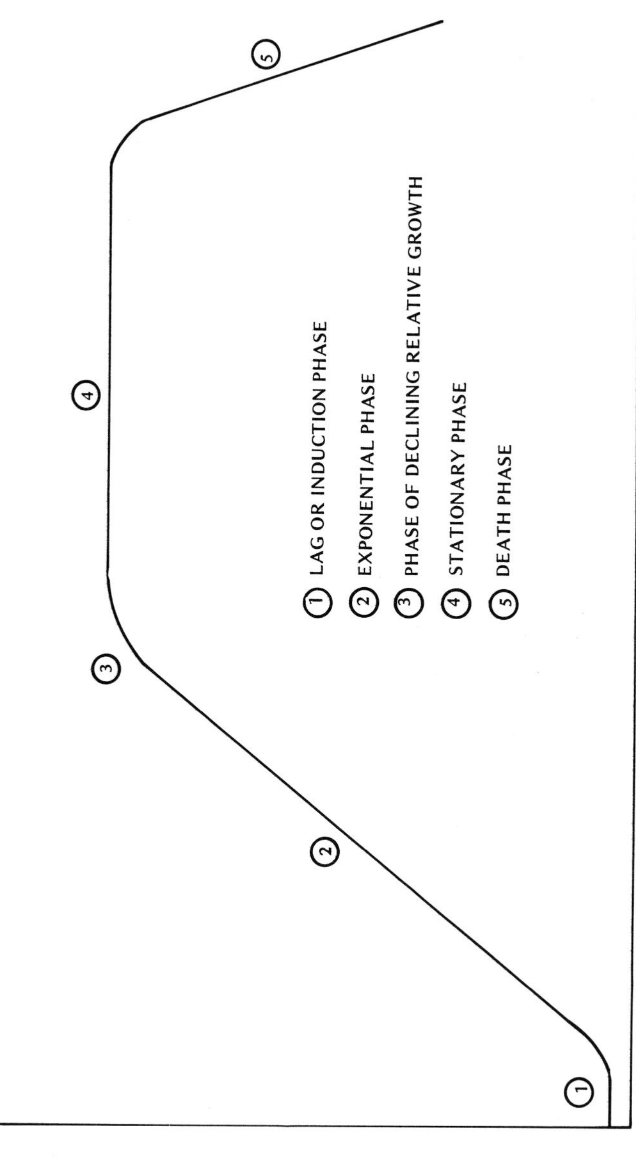

FIGURE 1. Typical culture growth.

The first method utilizes a stock culture from which periodic transfers are made to larger volume cultures. Usually, an intermediate volume is continuously utilized and stock cultures are maintained as the starter culture. The final culture volume can then be batch-fed to filter-feeding animals.

The second method involves maintenance of continuous mass cultures of algae at a suitable volume for feeding. This is a simple chemostatic arrangement whereby nutrients and filtered seawater are continuously added to a tank at a flow rate equalling the effluent or feeding flow rate. Given consistent illumination and nutrient enrichment, a working cell density may be obtained for long periods of time. Such a system utilizing natural sunlight and nutrient-rich Antarctic Intermediate Water was developed by pioneers at the St. Croix Artificial Upwelling Station and used in feeding brine shrimp (*Artemia salina*) and bivalved molluscs.[11]

The third method utilizes a semicontinuous approach to large tank culture. This technique prolongs use of large tank cultures by requiring partial periodic feeding from tanks. Immediately following feeding, the tank is "topped up" to its original volume and supplemented with nutrients to achieve the original level of enrichment. This method has proven quite satisfactory for production of algae at the Galveston hatchery.

ALGAL CULTURES

Stock cultures, as previously mentioned, are used as starter culture for the production of larger volumes of algal foods. Stock cultures can be maintained in small (20 mm × 160 mm) screw-top Pyrex™ test tubes. These tubes are made of high quality autoclaveable glass. Screw-tops simplify not only aseptic transfer of culture, but also the handling of tubes during and after the sterilization process (refer to the section on sterilization). As shown in Table 6, the enrichment level for stock cultures is F/8 (Guillard's F medium). This level restricts growth to a maintenance level, thus reducing the potential for build-up of metabolites or related contamination. Constant illumination is suitable for maintenance of flagellate stock cultures while a 12-hour photoperiod is sufficient for diatoms. Continuous illumination of diatom cultures can result in decreased cell size due to inherent reproductive characteristics of such species.[12] According to Guillard,[6] an incident light level of 750 to 1000 lux (measured horizontally) is sufficient for stock cultures. This may be provided by two 30 to 40 W cool white fluorescent bulbs placed directly in front of stock culture tubes. Stock cultures should be maintained under climate-controlled conditions at a temperature of approximately 24°C. This temperature is a good maintenance level for most cultured algal species. Frequent handling of stock cultures should be avoided; furthermore, they should be kept in a separate room in the laboratory. This room should be entered only for quick, visual examination of cultures or for removal of cultures for inoculation purposes. Stock cultures should not be aerated; sufficient gaseous exchange occurs between the air-water interface within the tube. Aeration could provide cultures with a source of contamination. Stock culture tubes can be conveniently kept in wire racks on shelves directly in front of fluorescent bulbs. To insure aseptic conditions, all stock culture tubes should be autoclaved, with medium, prior to inoculation. Stock cultures should be maintained for about one month and then transferred to create a new culture line. Aseptic techniques should be used for transfer and inoculation of algae whenever possible; this requires use of a sterile transfer hood.

According to Table 6, one 2.0 mℓ aliquot of stock culture is used to inoculate the subsequent culture phase, the small flask. These culture vessels are 125 mℓ Pyrex™ brand Erlenmeyer flasks, also having a screwtop cap. A rotating small flask culture is maintained, as previously mentioned, to restrict handling of stock cultures to a minimum. Figure 2 depicts transfer of algae within the culture system. As shown, a new small flask culture line is started from a suitable stock culture. A light intensity of 1500 lux is suitable for small flask

Table 6
ENRICHMENT LEVELS

Culture		Volume	Days cultured	Enrichment level
Stock culture		15 mℓ	30	F/8
Small flask	2.0 mℓ	50 mℓ	4	F/2
Carboy	50 mℓ	12 ℓ	4	F/2
Polytank	12 ℓ	300 ℓ	4	F/2

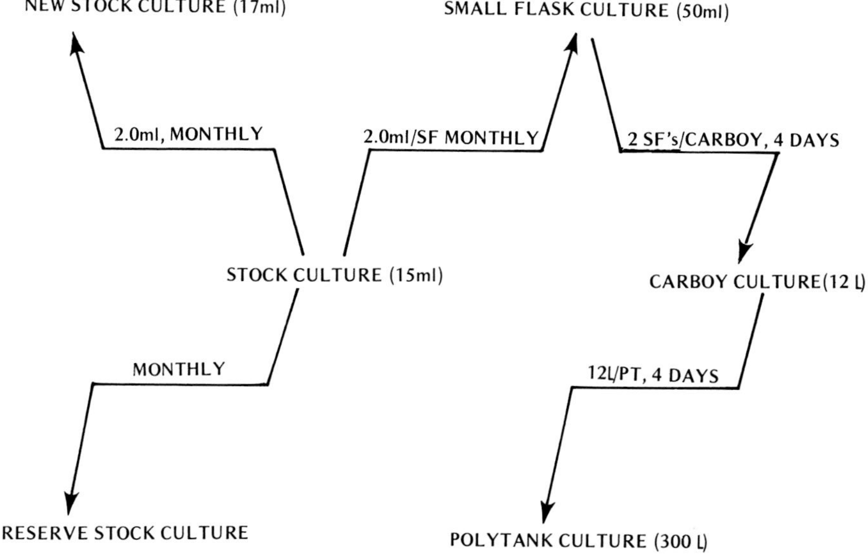

FIGURE 2. Algae transfer.

growth and can be accomplished by growing small flasks on a shelf directly in front of fluorescent lighting. Aeration is, again, not necessary; however, flasks should be periodically shaken - especially when nonmotile species are being cultured. Shaking enhances growth by reducing shading, which results from settling. Species with a higher settling rate should, obviously, be agitated more frequently. For preparation of small flasks, always add media prior to autoclaving. This reduces the need for aseptic addition of nutrients after autoclaving. An interesting note concerns silicate enrichment through leaching as a result of the autoclaving process. If boro-silicate culture tubes and small flasks are being used and are autoclaved, the addition of silicate is not required. This pertains only to the culture of diatoms. Transfer of inoculum for small flasks, whether originating from another small flask or from a stock culture, should occur under an aseptic transfer hood.

Once small flasks are approximately 4 days old, they are used as inoculum for the subsequent culture phase - the carboy. Carboys are autoclaveable 12- or 20-ℓ glass bottles (Pyrex™) specially modified for the culture of algae. The modifications are made according to Guillard's specifications shown in Figure 3.[6] Such an apparatus serves several purposes. The inoculation port utilizes a pressure differential allowing virtually aseptic inoculation of culture. The aeration standpipe reduces settling of cells on the bottom of the carboy. Constant aeration, at this culture level, assures homogeneity of nutrients, allows cells to frequently encounter a high light intensity, and also provides small amounts of CO_2 for growth. Since aeration is used in this phase, it is advisable to filter air prior to use. This can be accomplished

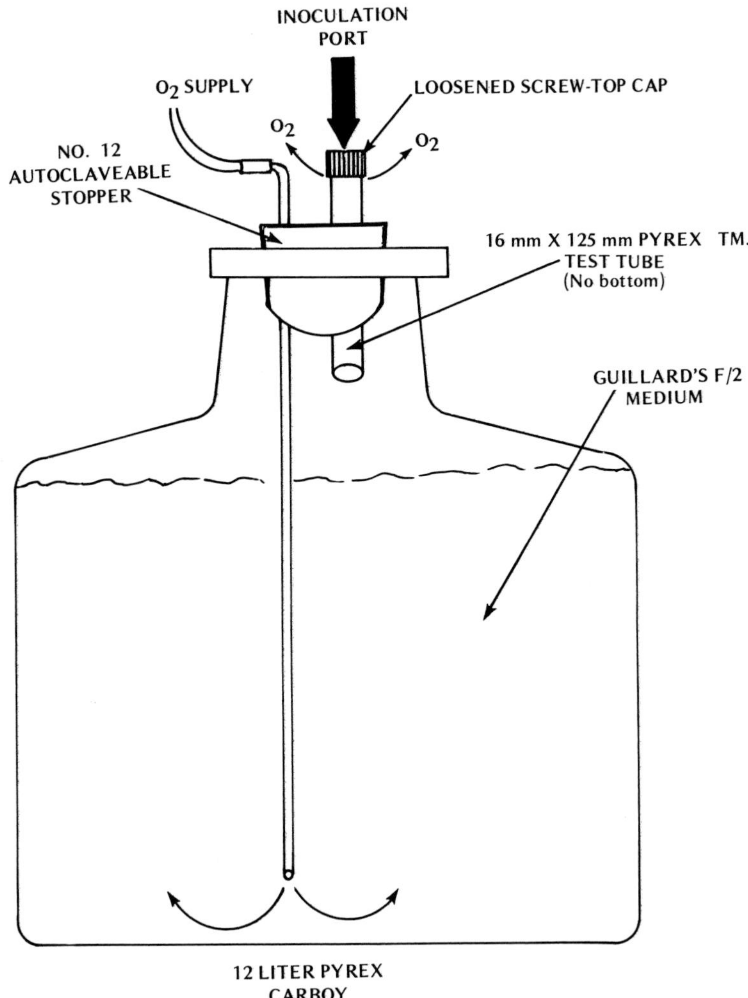

FIGURE 3. Carboy culture apparatus.

by use of an air filter such as the one shown in Figure 4. It is also advisable to strategically situate condensation traps throughout the system to vent excess water that might dilute the culture medium. Carboy cultures can be grown on a large shelf apparatus as shown in Figure 5. This shelf, or rack, can hold up to ten, 12-ℓ carboys plus provide adequate illumination. Illumination used consists of four, 72 in., 40 W, cool white, fluorescent bulbs situated directly behind the carboys. The rack, itself, is painted white for reflective purposes. The enrichment level for carboys, as shown in Table 6, is F/2 (Guillard's F medium diluted to half strength). This level is conducive to high growth rates required during this phase. Carboys, along with medium and aeration apparatus are also autoclaved. Addition of silicate is also necessary for such a large culture volume.

After another 4 days' growth, carboys are ready for use as inoculum for the final culture phase, the polytank or large volume culture tank. These tanks can be cylindrical, conical, or rectangular in shape. A conical shape reduces settling and, therefore, fouling by metabolites but is an inefficient use of space and may be important only for the culture of nonmotile species. Cylindrical tanks are used in some laboratories, but are often difficult to illuminate. Rectangular tanks are the most efficient use of space and, if a sufficient amount of aeration

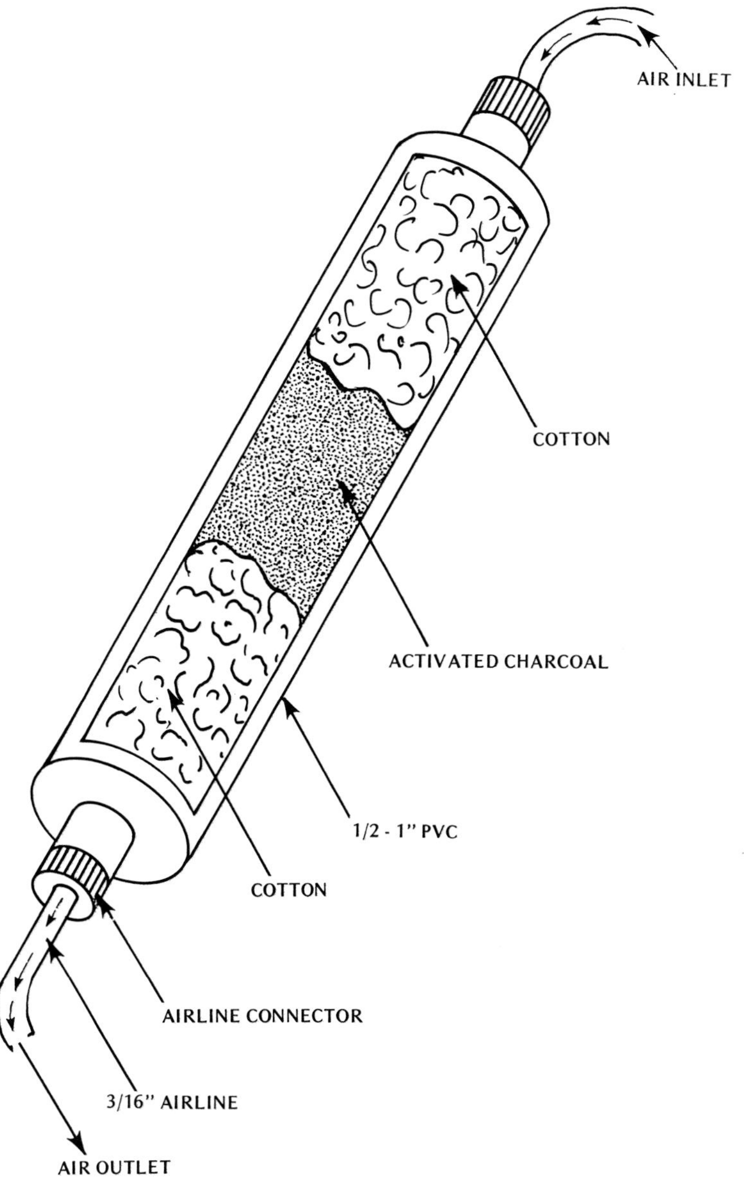

FIGURE 4. Aeration filter.

is available, prove suitable for most culture needs. Whatever tank is chosen, it should efficiently meet feeding requirements for the primary species being cultured. The Galveston Laboratory uses 300-ℓ tanks, although much larger tanks are probably just as suitable. These tanks can be made of a polymer-type plastic or constructed of plywood and coated with fiberglass. For reflective purposes, all intensive culture tanks should be painted white. Clear plastic cylindrical and rectangular tanks are sometimes used but are usually quite expensive. Plastic bags have also been used for the mass culture of algae; however, they must be suspended and are therefore subject to leaking due to constant stress. Also, with such bags, direct illumination of the water surface is not practical, requiring increased illumination. Illumination of polytanks is usually constant and provided by a light bank suspended directly above the tank (Figure 6). Constant illumination is most conducive to rapid growth in most

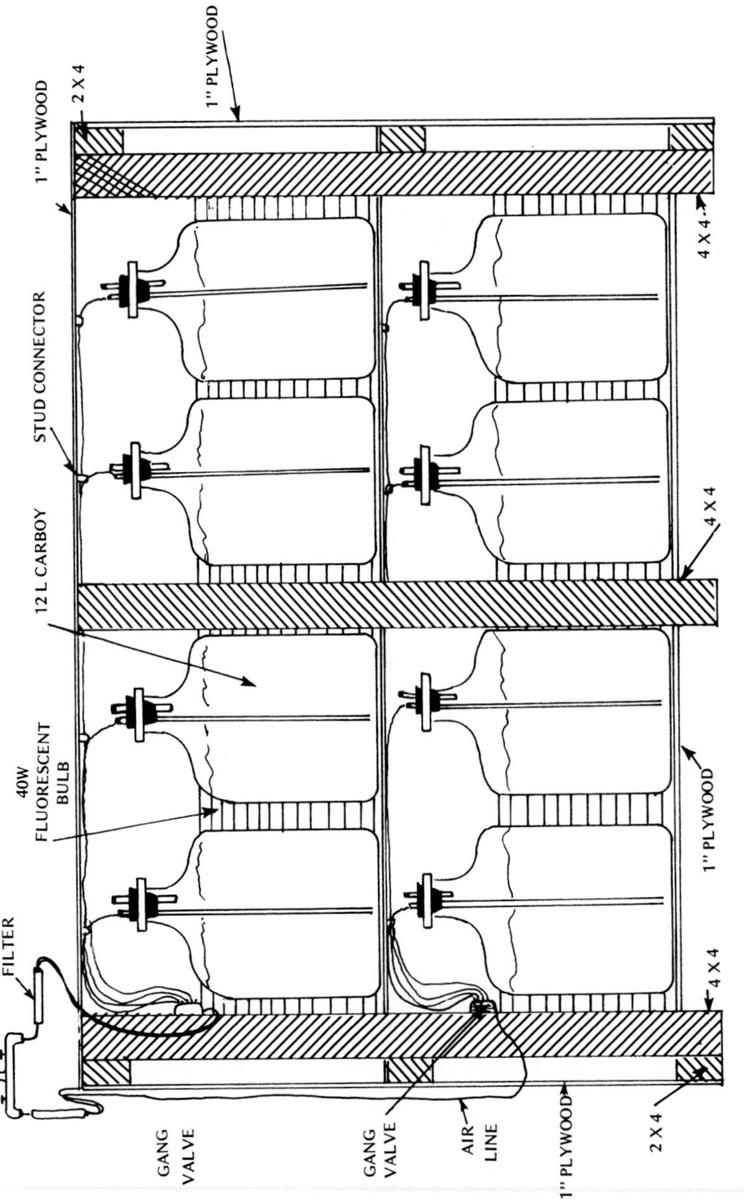

FIGURE 5. Carboy culture shelf.

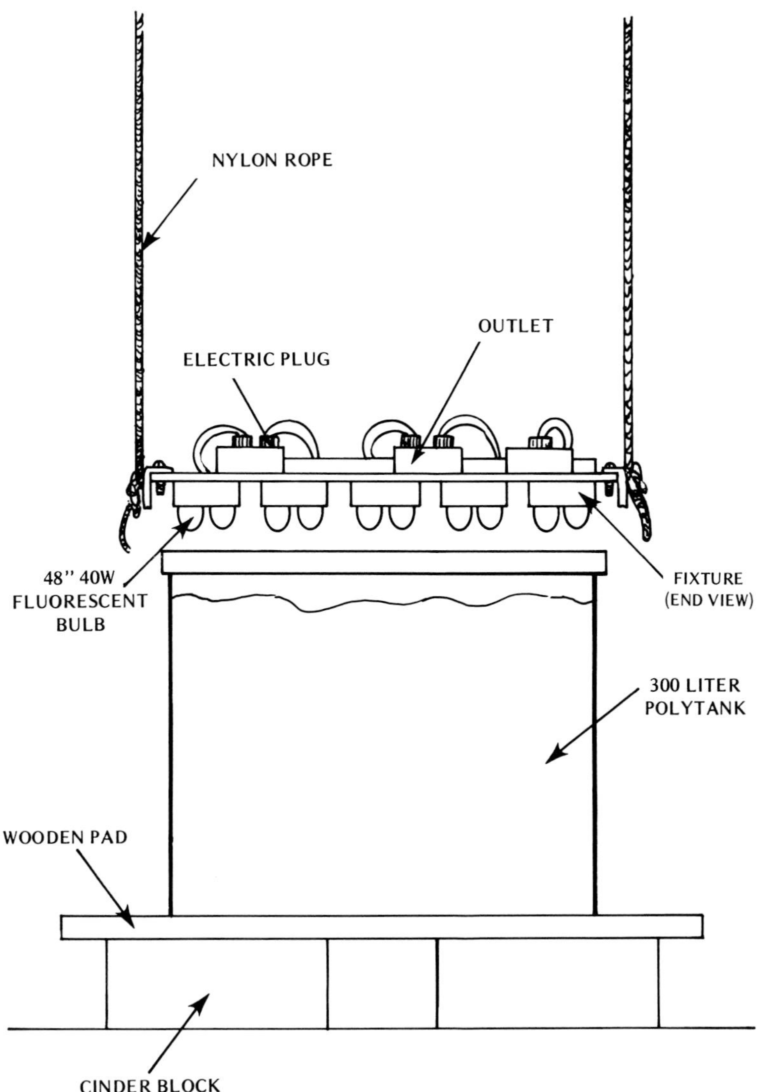

FIGURE 6. Polytank illumination.

cultures. Aeration for such tanks must provide adequate circulation for dispersal of nutrients and cells. At the Galveston Laboratory, a laminar-flow device is used for this purpose (Figure 7). Three weighted air stones are placed in one end of the tank directly beneath the laminar flow baffle. As the air from the airstones rises, it also creates an upwelling of water. As this air/water current strikes the baffle, laminar flow is created across the surface of the tank to the opposite end. Because water is pulled up off the bottom (upwelling), a downflow effect occurs at the opposite end of the tank. This results in substantial circulation within the tank. The nutrient level used for polytank culture at the Galveston Laboratory is, again, F/2 (Guillard's medium). A lower level may be used depending on use of the culture. If culture is to be rapidly depleted, there is no point in maintaining such a high enrichment level. Once the desired cell density is achieved, the culture is used as food for the subject animal.

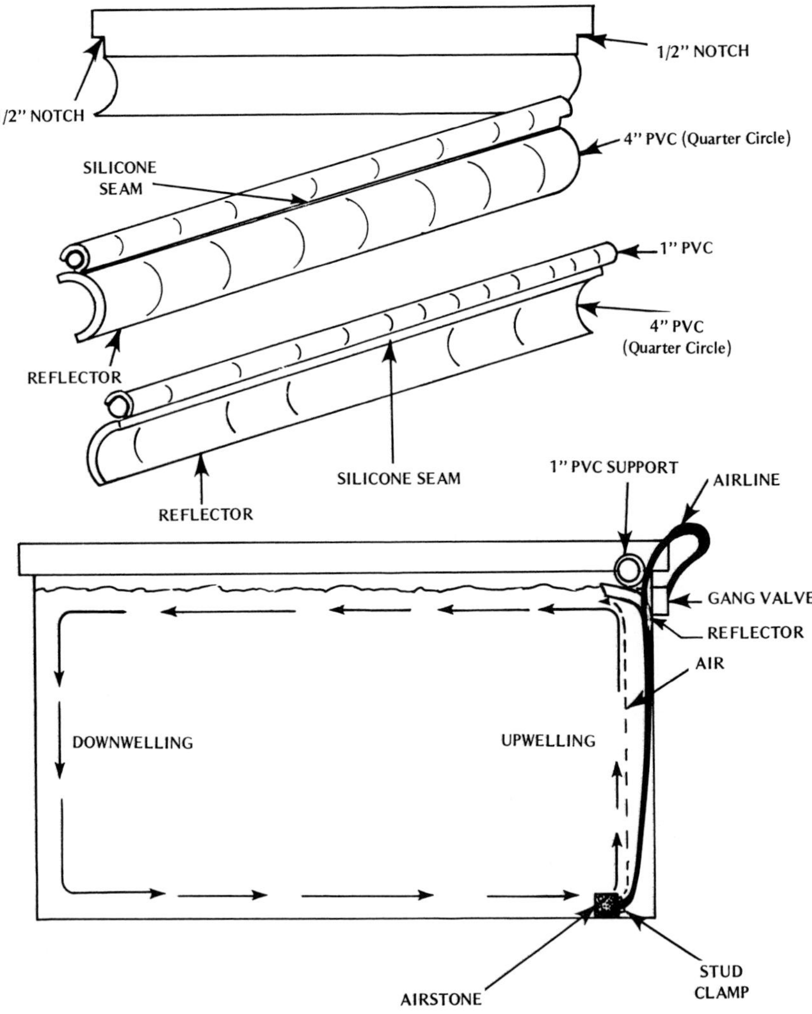

FIGURE 7. Laminar flow device.

SEAWATER FILTRATION

Adequate filtration of incoming seawater is necessary to insure a constant supply of clean water for the culture of algae. The level of filtration required is largely dependent on characteristics of the incoming seawater. The degree of filtration is, therefore, largely site-specific. In some areas, extensive filtration is required year-round, while in other areas, minimal precautions are taken. Areas where low levels of primary production exist typically utilize little filtration. If any doubt exists as to the presence of contaminants in incoming seawater, even if particulate matter is not present, steps should be taken to insure their removal prior to use. Once biological contamination is present in unialgal cultures, it is usually difficult to remove without reisolating the cultured species.

Figure 8 illustrates a typical filtration system. This system allows for separation of phytoplankton, zooplankton, and particulate matter as small as 1.0 μm from incoming seawater. This is accomplished by a series of filters of gradually decreasing porosity. The first filter, as shown in Figure 8, is usually a coarse-particle filter containing sand, gravel, or diatomaceous earth. These filters are pressurized and their filtration (separation) capacity is

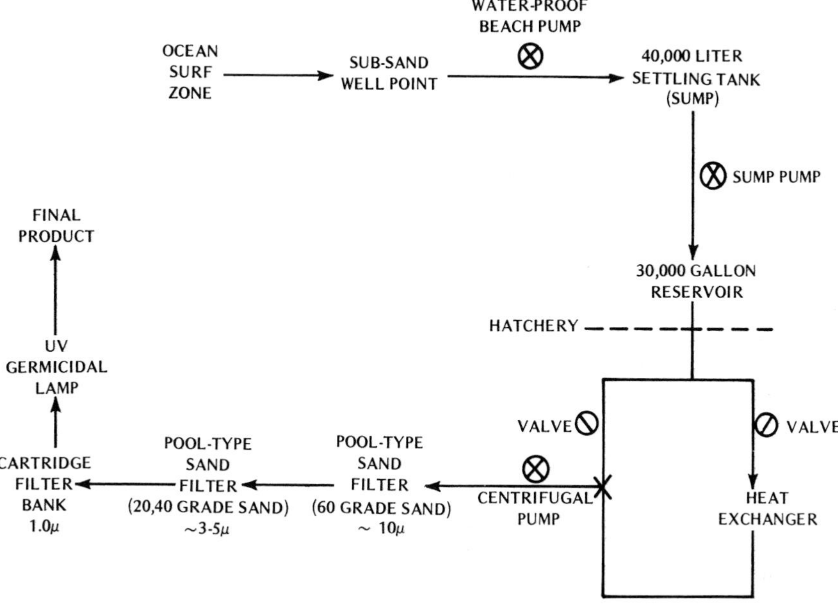

FIGURE 8. Typical seawater filtration system.

dependent on the grade or size of the particle used. Unfiltered water is passed through such filters, as previously mentioned, by pressure created by a pump situated prior to the filter. The most common pump type used for this purpose is a centrifugal pump with a neoprene impeller. These pumps can withstand a great deal of back pressure and, because of the composition of the impeller, are not considered a potential source of heavy metal or divalent cation toxicity. Also, the pumps are easily repaired; the most common cause of breakdown is deterioration of the impeller. The use of smaller porosity cartridge filters is almost a necessity to achieve filtration to the 1.0 μm level. These cartridges can be easily backwashed, which reduces operational costs by not having to purchase filters. A final step in treatment of seawater can involve UV irradiation. UV sterilization has been shown to reduce and sometimes eradicate certain contaminant species. Further work on this concept as well as that of ozonation is necessary, in the author's opinion, to warrant such measures. It is also suggested that an accredited engineer be used in designing the filter system. Then one can be assured all components in the system will work in harmony to produce the level of filtration and flow rate desired.

STERILIZATION

When conditions warrant, it is sometimes necessary to sterilize seawater used in growth media. In culture systems employing gradually increasing culture volumes, the maintenance of axenic base or stock cultures is required. If contamination should enter a production facility it can easily be transferred to subsequent larger volumes. Once contamination has taken place it is usually quite difficult to remove. If stock cultures become contaminated, they are often discarded and/or the cultured species is reisolated.

It is not practical to sterilize at all production levels. Usually, only smaller volumes of seawater are treated. The most common means of sterilization is autoclaving. Devices such as small pressure cookers up to large hospital autoclaves are used for this purpose. It must be emphasized, however, that technicians be thoroughly familiar with the operation of such devices as they are potentially dangerous. Standard sterilization conditions are described as

120°C and 20 psi, with sterilization time relative to the volume of water being treated. Sterilization of liquids less than 1 ℓ volume usually takes 15 min. Larger volumes such as 20 ℓ carboys require 45 min of treatment to achieve sterile conditions at the core of the vessel. The autoclaving of liquids also requires special preparation with regard to the vessel and media being treated. Containers should be made of high-grade glass such as Pyrex™ or Kimax™. If containers are also to be used as culture vessels, they should be transparent. Containers should never be filled to the top prior to autoclaving and should also be fitted with foam or gauze stoppers covered with aluminum foil. Such a stopper allows for equalization of pressure during the sterilization process. Autoclaveable screw caps are highly recommended as they can be sealed immediately after autoclaving and are generally easier to work with during transfers. After autoclaving, flasks should be allowed to cool to ambient temperature. Larger volume containers are usually cooled by aeration.

Another means of sterilization commonly used in algal culture is disinfection by chlorination. Chlorox™ is a suitable agent for this purpose and is generally considered a good, all-purpose disinfectant. Use of this oxidizing agent often applies to situations where autoclaving is not practical, as in large volume cultures. Chlorine is added to achieve a concentration of 1.0 to 2.0 ppm. After addition, the water is stirred to evenly distribute the disinfectant. Aeration is not used to avoid dissipation of the chlorine. The tank then sits for about 24 hr, followed by gentle aeration to remove residual chlorine. After 2 to 3 hr aeration, the tank is ready for inoculation. Chlorine can be neutralized, should aeration prove ineffective by the addition of sodium thiosulfate. Preliminary work with chlorine sterilization has proven successful for the culture of *Tetraselmis chuii* and *Isochrysis* sp. Should a high level of chlorine still exist in the tank after inoculation it will be evidenced by the deposition of dead cells on the side of the tank.

Another method of sterilization, previously mentioned, involves UV irradiation. Substantiation for its use is not available; however, it is generally felt that contaminants small enough to pass through the filtration system (bacteria, viruses, spores, etc.) could be eradicated by its use. Precautions should be taken to not exceed the flow-rate capacity of the UV light.

ASEPTIC TECHNIQUE

During the transfer of cultures (i.e., subculturing, inoculation), it is necessary to practice aseptic technique. The same rigid standards used in microbiological laboratories apply for the axenic culture of algae. This involves use of sterile graduated pipettes, forceps, an alcohol or Bunsen burner, UV light, disinfectants, and a transfer hood. Pipettes should be made of glass (autoclaveable) and should be graduated to insure equal aliquots of inoculum. They should be packed at the upper end with cotton to minimize oral contamination. One alternative to this technique is to use an automatic dispenser-type pipette and simply replace tips between transfers. Sterilized pipettes should be kept in airtight containers.

The following is a procedure used for the aseptic transfer of stock cultures at the Galveston Laboratory.

A. Materials needed:
   1. Aseptic transfer hood
   2. Spray disinfectant/solution
   3. Sterile pipettes plus canister
   4. Forceps
   5. Bunsen burner
   6. Log book

B. Procedure for aseptic transfer of algal stock cultures:
   1. Shut off UV light in transfer hood.

2. Open hood, spray down internal surfaces with disinfectant (Chlorox™, Amphyl™, zephryn chloride, etc.). Close hood and allow 5 min for effect.
3. Obtain fresh "new" stock culture tubes with media. Label accordingly and set in hood. A wire test rack is good for holding tubes.
4. Make appropriate notes in stock culture log. Outline transfers (species transferred, volume transferred, tubes inoculated, etc.).
5. Obtain stock cultures to be transferred. A determination as to which tubes are to be used should have been made prior to step (1). Stock cultures should be visually inspected for quality at least once per month, under similar aseptic conditions.
6. Arrange both "old" and "new" tubes in a strategic arrangement. This will minimize mistakes.
7. Light bunsen burner, adjust flame.
8. Remove first set of "old" tubes from racks and place them, alone, in a separate empty rack immediately in front of the working area.
9. Remove the lid from the pipette canister and flame the open end of the canister.
10. Simultaneously flame a pair of forceps and remove a sterile 1.0 mℓ graduated pipette. These pipettes are made of glass and are stuffed with a cotton end filter. Reflame open end of canister and replace lid.
11. Set the forceps down on a paper towel. While holding the pipette in the right hand, pick up the "old" stock culture tube in the left hand. Agitate by creating a vortex in the tube. Remove the cap with the little finger of the right hand while simultaneously holding the pipette.
12. Flame the open end of the "old" stock culture tube, tilt the tube slightly, and, with the right hand, insert the pipette into the tube.
13. Draw off 0.5 mℓ of culture from the middle of the tube and replace the cap (still held by the little finger of the right hand).
14. Pick up the "new" stock culture tube, remove cap in the same manner, and flame the open end of the tube. Tilt the tube and insert the partially full pipette.
15. Allow inoculum to drip into the fresh medium. Try not to touch the tip of the pipette to the inside of the tube. Do *not* blow into the pipette to remove residual inoculum.
16. Remove the pipette. Reflame the open end of the tube and replace the cap. Place the used pipette in a "used" pipette canister. This canister should be kept outside the hood.
17. Agitate the "new" stock culture tube and replace it in its rack.
18. Remove a new sterile pipette for the next transfer.
19. Upon finishing transfers, close hood and turn on UV light.

Transfer of small flasks is similar to the procedure used for the aseptic transfer of stock cultures, except a 2.0 mℓ volume of inoculum is used and, of course, the culture vessel is larger.

The following procedure describes the preparation and inoculation of carboys. The purpose in outlining such a procedure is to depict a guideline for the preparation of a culture vessel that undergoes autoclaving, yet is too large to be inoculated under an aseptic transfer hood.

C. Procedure for aseptic inoculation of carboys

1. Start autoclave, allow 45 min to warm up
2. Obtain clean, 12 ℓ glass carboys
3. Fill these carboys to the 12 ℓ mark with 1.0 μm filtered seawater. Close end by inserting a size 12 rubber stopper.
4. Obtain clean, carboy aeration apparatus.

5. If autoclave is of sufficient size to allow aeration apparatus to sit inside carboy, keep unit intact while autoclaving. Remove cap to inoculation port. Place some autoclaveable tubing over end of aeration standpipe and close tubing with pinch clamp.
6. If autoclave is too small for step 5, wrap aeration apparatus in autoclaveable paper, seal with autoclaveable tape.
7. Add 12 mℓ 1 *N* HCl to each carboy. This will disrupt the carbonate buffering system of the seawater and aid in the sterilization process.
8. Autoclave carboys and aeration apparatus on a liquid cycle using a slow exhaust. Allow 45 min sterilization once autoclave reaches a temperature of 121°C and a chamber pressure of 20 psi.
9. After autoclaving, immediately replace caps on carboy inoculation ports, screw caps down hand-tight, and secure aeration apparatus.
10. If step 9 does not apply, immediately after autoclaving, unwrap aeration apparatus and place in carboys as they are removed from the autoclave. Replace inoculation port caps. Add a small piece of tubing over end of aeration standpipe and close with pinch clamp.
11. Transfer carboys to stand, connect air lines, loosen caps, and initiate low-level aeration. Do not aerate excessively.
12. Increase aeration to a normal level once carboys have cooled. Allow 24 hr for cooling. Aeration will bring the pH back to normal (7.8 - 8.2).
13. Nutrients should be prepared and apportioned in small test tubes for use in enriching carboys. The following scheme can be used:

Nutrients	Tube volume	Nutrient concentration
Nitrates and phosphates	12 mℓ	500F (1000 × F/2)
Silicates	12 mℓ	500F (1000 × F/2)
Fe, trace metal mix	12 mℓ	500F (1000 × F/2)
Vitamins	12 mℓ	500F (1000 × F/2)

Store vitamins and trace metal mix in test tubes in racks in the refrigerator. Nitrate/phosphate and silicate tubes can be kept on a shelf near the carboy rack.

14. After carboys have cooled and are ready for inoculation, add one tube of each of the appropriate nutrients.
15. Remove two small flasks of four-day-old culture from the rotating small flask culture system, shake, and pour contents into a carboy through the inoculation port. The carboy is now inoculated.

NONASEPTIC TECHNIQUE

Polytank inoculations involve a minimal amount of aseptic technique. The inoculation itself cannot be done under stringent aseptic conditions and is further complicated by the use of unautoclaved seawater. To make matters worse, the polytank is not an enclosed culture vessel and is subject to airborne contamination. One can at least insure that the polytanks are clean prior to use. As a further note, it is advisable to disinfect the polytank culture area at least once every year. Disinfection can be accomplished by spraying the room with liquid Chlorox™. An insecticide sprayer using air or water pressure is good for this purpose. Allow the room to sit undisturbed for three to four days after spraying. Tanks should be removed and electrical fixtures should be taped closed prior to spraying.

Procedure for inoculation of carboys into 300-ℓ polytanks

1. Drain and remove dirty tank from culture area. Drainage of tanks can be by siphon/drainplug or pump arrangement. Rinse out the hose when through. Clean salt accumulation off light bulbs. Warm freshwater may be used for this purpose.
2. In the cleaning area, spray the tank with hot freshwater to remove as much sediment as possible. Rinse and drain tank.
3. Pour liquid Chlorox™ around and down the inside of the tank. Allow the liquid to accumulate in the bottom.
4. Use a stiff-bristled nylon brush to scrub the side(s) and bottom of the tank. It might be advisable to wear old clothes and gloves while performing this task. Dip the brush frequently into the Chlorox™ and thoroughly scrub the tank until all scum is removed. Drain the tank. Rinse repeatedly with hot water until chlorine odor is no longer present.
5. Allow the tank to dry by storing it upside down. Make sure the tank is elevated on either wooden two by fours or concrete cinder blocks.
6. When ready for use, return the tank to the culture area. Drop clean aeration lines, airstone, weights, and laminar flow device into the tank.
7. Treat the tank with UV light for 5 min. Figure 9 shows the UV irradiation device used by the Galveston Laboratory for this purpose. Remove apparatus when finished.
8. Install aeration and laminar flow apparatuses. Fill the tank to about one inch below the laminar flow level. Aerate.
9. Add nutrients by simply pouring them into the tank. Allow two minutes for dispersal.
10. Remove a suitable carboy from the rack. Remove the aeration apparatus and inoculate by simply pouring the carboy into the tank. The polytank has now been inoculated. Set aeration/turnover rate.

COUNTING/ANALYSIS OF ALGAL CULTURES

Analysis of algal cultures for feeding purposes usually involves visual inspection of the culture. The only counting/analysis procedure described in detail in this paper is for the simple hemocytometric technique. Other means such as Coulter® counters, turbidometers, and spectrophotometers (protein, carbohydrate, and chlorophyll a determinations) and cell volume have been used. Their drawbacks are that they usually require a great deal of technological expertise, expensive instruments are necessary, and they do not afford visual microscopic inspection of the culture which is helpful in detecting other problems.

The apparatus used for counting cells at the Galveston Laboratory is the American Optical™ hemocytometer. This device is routinely used in hospitals for blood serum analysis, but is quite suitable for counting algal cells. The specific device is shown in Figure 10.

A. Preparation of samples and hemocytometer for algal counts
   1. Obtain clean sample bottles (50 mℓ volumes are suitable).
   2. Label bottles appropriately.
   3. Place bottles on sample tray.
   4. Wash hands thoroughly and dry.
   5. Dip sample bottle into tank at an angle avoiding contact with culture by fingers.
   6. Repeat until all samples have been taken.
   7. Place sample tray in close proximity to compound scope.
   8. Prepare notebook for entry of data (a typical data sheet is shown in Figure 11).
   9. Prepare hemocytometers for loading by cleaning quartz slips with alcohol and lens paper.
   10. Clean hemocytometer using the same technique.
   11. Place cover slips on hemocytometer.

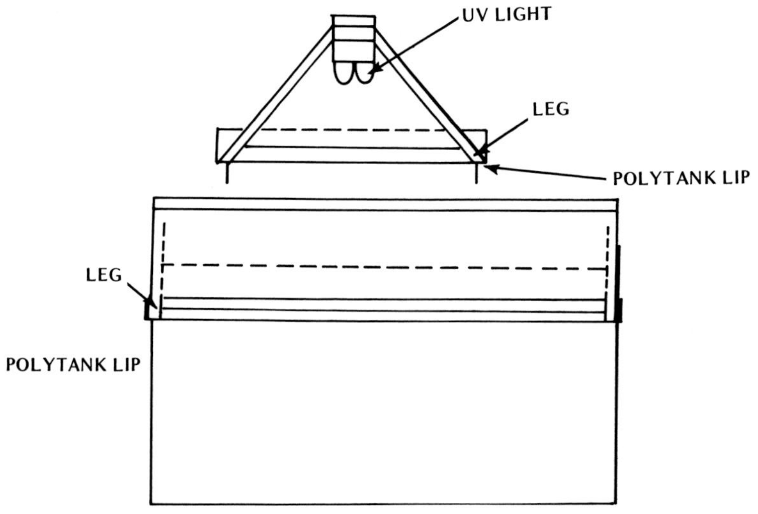

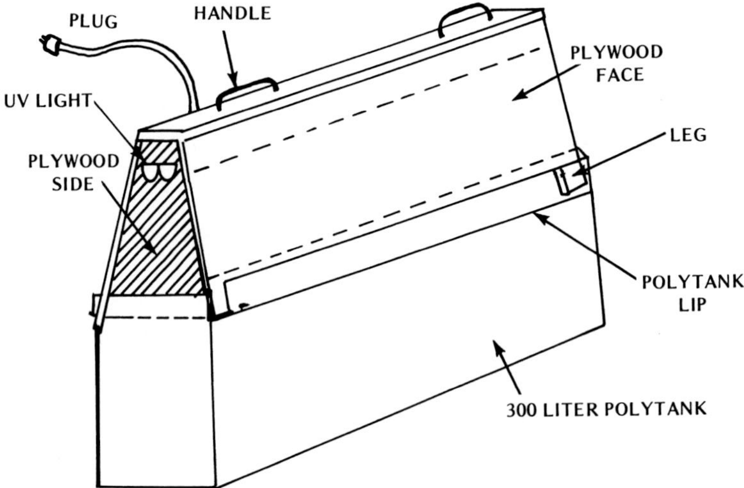

FIGURE 9. Polytank UV device.

12. Remove the first sample bottle from the tray. If motile cells are being counted, add 3-5 drops of 50% formalin to the bottle to preserve and demobilize cells. Shake bottle thoroughly.
13. Immediately remove sample from the bottle. A clean bulb-type pasteur pipette or inoculating loop may be used for this purpose.
14. Load hemocytometer by applying a drop of the sample to the loading port. Capillary action will disperse the sample over the grid surface. Load both chambers of the hemocytometer with sample.
15. Obtain a clean, bulbed pipette or flame sterilize the inoculating loop for the next sample.
16. Obtain a new, clean hemocytometer for the next sample.
17. Allow loaded hemocytometers to sit at least one minute for settling of cells.
18. The samples are now ready for counting.

B. Procedure for counting cells on the hemocytometer:
   1. Place loaded hemocytometer on stage of compound microscope.

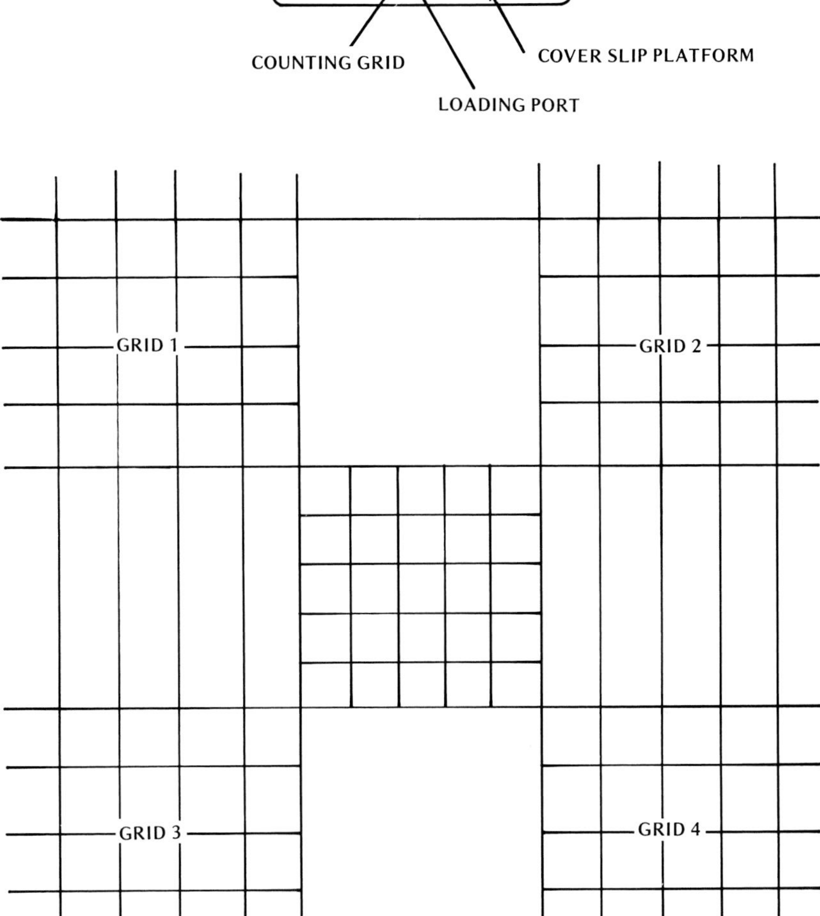

FIGURE 10. American optical™ hemocytometer.

2. Using a 10 × ocular adjust microscope to the 20 × objective. Raise objective up with the coarse adjustment knob.
3. Place slide under objective and focus onto the gridwork of the counting chamber (see Figure 10).
4. As shown in Figure 10, there are 5 counting grids to each chamber and two chambers to every hemocytometer. Depending on the number of cells per sample, either all or part of the chamber will be utilized. A good rule of thumb involves counting all four of the outer grids if less than 100 cells are counted per grid.

If more than 100 cells per grid are present, utilize the outer three rows of each grid to comprise approximately 400 total cells. Several variations to this technique can be used as long as symmetry is maintained. Count the same portion of grids to the right and left of

FIGURE 11

Sample algae data sheet.
POLYTANK DATA

Day: Monday   Time: 08:00
Date: 1/4/82   Lt: JMF

Tank #	Mode E/P	Run #	Specie used	Cell density (c/mℓ)	Visible contamination	
					Description	C/mℓ
1	Prod.	2	T-ISO	4.80×10^6	protozoan flagellate	$\sim 2 \times 10^4$
2	Prod.	1	TET	5.70×10^5	none	—
3	Exp.	1	SKEL	1.24×10^6	bacteria	moderate
4	Exp.	1	SKEL	1.38×10^6	none	—
5	Prod.	2	T-ISO	4.57×10^6	protozoan flagellate	$\sim 3 \times 10^4$
6	Prod.	1	T-ISO	8.74×10^5	none	—
7	Prod.	1	TET	5.62×10^5	none	—
8	Prod.	1	T-ISO	8.38×10^5	none	—
9	Prod.	2	TET	9.80×10^4	rod-shaped bacteria	light
10	Prod.	2	TET	9.45×10^4	none	—
11	Exp.	1	THAL.	3.72×10^5	bacteria	heavy
12	Exp.	1	THAL.	1.97×10^6	none	—

Comments: Tank 11 appears to be crashing.

center as above and below. The following guidelines describe the translation of the count to a cells/mℓ value:

Grids counted	Percent grid counted	Total cells	Division factor	Multiplication factor	Cells per mℓ
4	100	n	4	10^4	$-(10^4)$
4	75	n	3	10^4	$-(10^4)$
4	50	n	2	10^4	$-(10^4)$
4	25	n	1	10^4	$n\ (10^4)$

The equations in the cells per mℓ column can be used to determine cells per mℓ in the sample. A generalized equation for this purpose is

$$\left(\frac{n}{x}\right)\frac{(10{,}000)}{1} = \text{cells per m}\ell$$

where n = total cells in count
x = division factor based on percent of each of all four grids counted

For example, if 480 cells (n) were counted using 75% of all 4 grids, the cells per mℓ value would be determined as

$$\left(\frac{480}{3}\right)\left(\frac{10{,}000}{1}\right) = 1{,}600{,}000 \text{ cells per m}\ell$$

5. The usual procedure calls for loading two separate subsamples from the sample bottle per hemocytometer and expressing cell density as an average of the two counts.
6. In counting cells, only count cells which are more than halfway inside the determined counting area.

7. Record the count and make note of any contaminants observed. For this purpose it is advisable to observe a live sample of the culture. Some cells, especially protozoans, are difficult to distinguish in a nonliving state.
8. Remove the hemocytometers from the microscope stage. Carefully remove the cover slip. Clean both cover slip and hemocytometer carefully with lens paper.
9. Store in box until next sampling period.

ALGAL STORAGE

Further development of crustacean aquaculture will depend, somewhat, on the ability of the farmer to consistently provide high-quality larval foods, particularly algae, for extended periods of time. The advantages of various storage techniques are quite obvious. Long-term stockpiling of algae could greatly reduce the dependence of hatcheries on the production of live algal cultures.[13] Frozen or freeze-dried algae could serve, even at minimal level, as a constantly available reserve. Furthermore, the feeding process would be greatly simplified. Careful maintenance of high density algal cultures would no longer be a concern. This would also result in a marked decrease in operational costs.

Algal cultures may be stored for extended period in two forms - as a frozen concentrate or as a freeze-dried powder. The processing of algae for storage in a frozen form by the addition of cryoprotective agents is known as cryopreservation. Cryopreservation of algae involves the sequence seen in Figure 12.

High density algal cultures can be concentrated by either chemical flocculation or by centrifugation (separation). Most flocculants, upon addition, cause cells to precipitate to the bottom of the culture vessel. The addition of certain other flocculants in combination with an electrical charge can, however, result in buoyancy of cells.[14] Harvesting is accomplished by siphoning off the supernatant (former technique) or by skimming cells off the surface (latter technique).

Harvesting by centrifugation involves the use of an in-line cream separator for concentration of cells.[13] High density algal culture is pumped into the bowl of a standard dairy cream separator. The flow rate from the bowl into the centrifuge head (separator) is adjusted by a spigot-type valve. The flow rate is specifically determined by the algal species being harvested and the centrifugation rate of the separator. Cells are deposited by centrifugal force on the wall of the centrifuge head. The head is then removed and the thick algal "paste" is removed, resuspended in a known volume of water and mechanically stirred, and the cell density of the slurry is determined by dilution and hemocytometric inspection. Results are recorded as total cell biomass harvested and concentrate cell density. At this point cryoprotective agents such as glucose/glycerol and DMSO (dimethylsulfoxide) are added to the concentrate. These agents are used to maintain cellular integrity during the freezing process. The increase in volume is noted and its effect on slurry cell density is determined.

A tentative procedure for the addition of glucose and glycerol for cryoprotective purposes is as follows.[15]

Glucose:
1. Concentration 2% V/V
2. Add 0.25% every 10 min until 1% glucose (V/V) has been added
3. Add remaining 1% (V/V) glucose at T_{40} minutes

Glycerol:
1. Concentration 2% (V/V)
2. All glycerol added simultaneously with final addition of glucose at T_{40} minutes

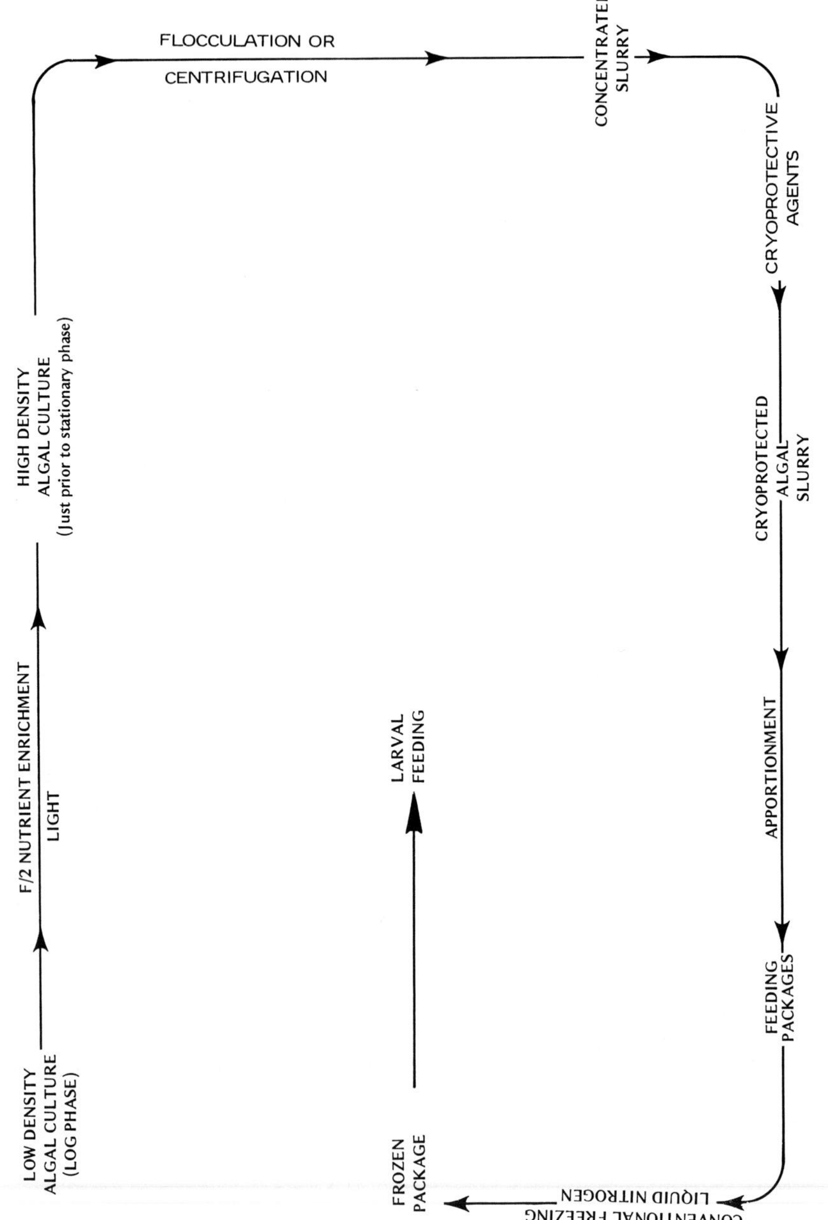

FIGURE 12. Schematic for cryopreservation of algae.

The prepared algal slurry is then apportioned into plastic tubes. The amount added to each tube depends upon the effect desired within the larval rearing tank. *Tetraselmis chuii* is usually stored in batches yielding 10,000 cells per mℓ per 2000 ℓ larval rearing tank. Freezing such a small amount of cells will reduce the possibility of overfeeding a tank. It might even be advisable, therefore, to freeze several levels of algae (for example, with *T. chuii,* prepare 5000; 10,000; and 20,000 cells per mℓ packages). Feeding studies using cryopreserved algae have been performed at the Galveston Laboratory with good results.[16]

It must also be noted that some success with feeding nonpreserved frozen algae has occurred.[17] Preliminary investigation has shown that *Tetraselmis chuii* maintains cellular integrity throughout the freezing process (and upon resuspension) without the use of cryoprotective agents.

Procedure for feeding frozen algae

1. Count algal cell density in larval rearing tanks. If the density is lower than that prescribed in the feeding guideline, consider supplementation.
2. Remove a sufficient quantity of frozen algae from the freezer and allow it to thaw in a graduated pitcher or cylinder, making note of volume and total number of cells (biomass).
3. Stir the concentrate and remove an appropriate volume for feeding. This volume can be determined by the following equation:

$$\frac{(Vc)\,(Cf)}{(Cc)} = V_f$$

where:

V_f = Unknown volume of concentrate required for feeding
Vc = Total volume of concentrate
Cf = Total cells needed for feeding
Cc = Total cells in concentrate

It must be emphasized, however, that considerable work is still needed in the area of algal cryopreservation. Cryogenic regimes appear to be as specific as the species involved. Some cryoprotective agents, as aqueous solutions, enhance bacterial growth. Also, some species of algae do not maintain cellular integrity as well as other species during centrifugation.

The freeze-drying of algal slurries (resulting from flocculation or centrifugation) is another means of preparing algae for storage. This process involves the use of freeze-drying equipment which is quite expensive but yields a dried, storable product. This product usually requires micronization to a suitable size ($\gtrless$ 10μ) prior to use. Furthermore, it is suggested that the final micronized product be suspended in water and stirred for one-half hour prior to feeding. This step reduces inherent clumping. For feeding purposes, the dry weight of the species being used must be determined. A known biomass of cells must be freeze-dried and its weight expressed as cells per milligram. (In many ways the feeding procedure for freeze-dried algae resembles that of the technique employed for the feeding of yeast.)

ALGAL CULTURE ECONOMICS

The cost of growing algae intensively is far greater than that by extensive means. Extensive culture utilizes endemic species, natural sunlight as a light source, and involves little maintenance. Intensive culture, in its true sense, is quite involved. Climate controlled culture rooms, continuous fluorescent illumination, the use of stringent aseptic techniques and constant maintenance and cleaning will provide fairly pure mass cultures of algae. These

Table 7
NUTRIENT COST PER POLYTANK

Nutrients	Cost per milligram ($U.S.)	Mg per 300-ℓ tank	Cost per 300-ℓ tank ($U.S.)
Na NO_3	1.00×10^{-5}	22,500	$0.225000
$NaH_2PO_4 \cdot 2H_2O$	2.48×10^{-5}	1,500	0.037000
$Na_2SiO_3 \cdot 9H_2O$	3.80×10^{-5}	9,000	0.342000
EDTA	2.55×10^{-5}	1,308	0.033000
$FeCl_3 \cdot 6H_2O$	7.55×10^{-5}	945	0.071000
$CuSO_4 \cdot 5H_20$	3.17×10^{-5}	3	0.000095
$ZnSO_4 \cdot 7H_2O$	3.10×10^{-5}	7.5	0.000233
$CoCl_2 \cdot 6H_2O$	1.57×10^{-4}	3	0.000470
$MnCl_2 \cdot 4H_2O$	5.00×10^{-5}	54	0.002700
$Na_2MoO_4 \cdot 2H_2O$	1.64×10^{-4}	1.8	0.000295
Thiamin HCl (B_1)	2.61×10^{-4}	30	0.007815
Cyanocobalamin (B_{12})	4.45×10^{-2}	0.15	0.006675
			$0.726283 ~73¢

Note: Average cell density (T-ISO) = 5.0×10^6 c/mℓ
Total cells per 300-ℓ tank = 1.5×10^{12} cells
Approximate cells produced per $ = 2.1×10^{12} cells
Nutrients

factors, however, contribute to substantial operational costs. The cost of nutrient enrichment of intensive cultures as compared to traditional extensive fertilization is quite high, although not apparent. Table 8 outlines the costs of nutrient enrichment for a 300 ℓ polytank. Also outlined is an approximation of the number of cells produced per nutrient dollar using Guillard's F/2 nutrient regime.[6]

ACKNOWLEDGMENTS

The author would like to thank Diane Hensley for scientific illustrations and the administration and staff of NMFS-Galveston and TAMU.

REFERENCES

1. **Mock, C. R., Revera, D. B., and Fontaine, C. T.,** The larval culture of *Penaeus stylirostris* using modification of the Galveston technique, *WMS*, 11, 102, 1980.
2. **Cook, H. L.,** A method for rearing penaeid shrimp larvae for experimental studies, *FAO Fish. Rep.*, (57), 3, 709, 1969.
3. **Hudinaga, M.,** The reproduction and development and rearing of larval *Penaeus japonicus* (Bate), *Jpn. J. Zoo.*, 305, 1942.
4. **Shigeno, K.,** Problems of prawn culture in Japan, Overseas Tech. Cooperation Agency, Tokyo, Japan, 37, 1972.
5. **Simon, C. M.,** The culture of the diatom *Chaetoceros gracilis* and its use as a food for penaeid protozoeal larvae, *Aquaculture*, 14, 105, 1978.
6. **Guillard, R. R. L.,** Culture of phytoplankton for feeding marine invertebrates, *Culture of Marine Invertebrate Animals*, Plenum Press, New York, 1975, 29.
7. **McLachlan, J.,** Growth media-marine, *Handbook of Phycological Methods, Culture Methods and Growth Measurement*, Stein, J. R., Ed., Cambridge University Press, London, 1973, 25.

8. **Gates, J. A. and Wilson, W. B.,** The toxicity of *Gonyaulax monilata* to *Musil cephalus, Limnol. Oceanogr.*, 5(2), 171, 1960.
9. **Stein, J. R.,** *Handbook of Phycological Methods, Culture Methods and Growth Measurement,* Cambridge University Press, London, 1973, 448.
10. **Fogg, G. E.,** *Algal Cultures and Phytoplankton Ecology,* University of Wisconsin Press, Madison, 1965, 11.
11. **Roels, O. A., Lawrence, A. L., Farmer, M., and Van Hemelryck, L.,** Organic production potential of artificial upwelling marine culture, *Process. Biochemistry,* Feb. Vol. 12(2), 18, 1978.
12. **Hostetter, H. and Hoshaw, R. W.,** Asexual development patterns of the diatom, *Stauroneis anceps, Culture J. Phycol.*, 8, 289, 1972.
13. **Mock, C. R.,** Larval culture of penaeid shrimp at the Galveston Biological Laboratory, Contribution No. 344 from the National Marine Fisheries Service, Galveston Laboratory, Galveston, Texas.
14. **De Pauwe, N.,** Personal communication, Laboratory for Biological Research of Aquatic Pollution, Laboratory for Mariculture, State University of Ghent, Belgium, 1981.
15. **Lawrence, A. L.,** Personal communication, Texas A&M Shrimp Mariculture Project, P. O. Drawer Q, Port Aransas, Texas, 78373, 1980.
16. **Brown, A.,** Experimental techniques for preserving diatoms used as food for larval *Penaeus aztecus. Proceedings of the National Shellfisheries Association,* Vol. 62, June 1972, 21.
17. **Mock, C. R.,** Personal communication, National Marine Fisheries Service, 4700 Ave. U., Galveston, Texas, 77550, 1980.

LARVAL FOODS FOR PENAEID PRAWNS

I-Chiu Liao, Huei-Meei Su, and Jaw-Hwa Lin

INTRODUCTION

During early July 1933, Hudinaga succeeded in the artificial spawning and hatching of the Kuruma prawn *Penaeus japonicus.*[1,2] However, the majority of the larvae died and only a few of them survived after the mysis stages. In 1940, after several years of continuous experiments to improve the survival rate of the prawn, Hudinaga eventually succeeded in growing the prawn to a marketable size. However, the number of prawns which reached a stockable size was very small and the artificial culture of the Kuruma prawn was not successful enough for commercial enterprise.

In 1941, methods for the isolation and pure culture of *Skeletonema costatum* were developed by Matsue.[3] Feeding *S. costatum* to the Kuruma prawn in the zoea stage resulted in a survival rate of 30% to the mysis stages, compared to a previous survival rate of only 1%. The fact that *S. costatum* was found to be a suitable food for prawn zoea stages was of critical importance. In 1956, Hudinaga used nauplii of the brine shrimp *Artemia salina* to feed the mysis larvae and early postlarvae of prawns with good result. By providing *S. costatum* for the zoea stages, and *Artemia salina* nauplii for the mysis stages and early postlarvae, 10 kg of prawns of marketable size were produced for the first time in 1958.

Skeletonema costatum is difficult to culture during the summer season.[4] A number of tanks are required to culture *S. costatum* and even more effort is needed to collect and feed this diatom to the prawn larvae. As a result, Hudinaga and Kittaka developed the same tank culture method in 1964. They added nutrient salts to the larval rearing tank to propagate the plankton-like diatom, thus lowering the cost and making it possible to produce a large number of young prawns in a single operation by a relatively simple method.

In recent years several successful cases of propagation of penaeid prawns were reported in the world by using almost similar culture procedures and methods except that different species of phytoplankton, rotifers, and artificial foods were used to feed the prawn larvae. Information on the methods developed by different research workers to rear the larval stages of various penaeid prawn species is listed in Table 1.

ALGAE

Algae used to feed the prawn larvae are cultured by two different methods. One is the "same tank" method, also called the large scale production method[17] or, outdoor method,[5] while the other is called the separate tank method which is also referred to as the small scale production or indoor method.[6] In the same tank method, the algae are cultured as food in the same water as that of larvae by using sunlight and fertilization. In addition, natural food such as copepods, polychaetes, etc., are propagated in the same tank and are available to the prawn larvae. In the separate tank method, one algal species (Table 2, methods C, D) or a mixture of algae (Table 2, method E) is cultured in a separate tank. The cultured algae are then either provided fresh or from refrigerated material to the prawn larvae. Some of the best algae used today as food for prawn larvae is the green flagellate *Tetraselmis* sp., and the diatoms *S. costatum, Chaetoceros* sp., and *Thalassiosira weissflogii.*

The separate tank method ensures greater larval densities, requires less space and water, and offers greater reliability, while the same tank method is simple, less labor intensive, requires no separate algal culture, and is also much less dependent on use of *Artemia,* the dried cysts of which are expensive and will be difficult to obtain. The question of which of these two methods is economically superior, clearly depends on local needs and limitations.

Table 1
SUMMARY OF LARVAL REARING OF PENAEID PRAWN WITH EMPHASIS ON LARVAL FEED

Species name	Volume of larval tank (ℓ)	Food item	Food quantity	Stages	Survival rate (%)	Algae production method[a]	Ref.
Penaeus aztecus	15	*Skeletonema costatum*	600—1,000 × 10^3 cells/mℓ	Z_1—M_1	50	C	8
		Artemia salina nauplii		Z_3—P_1			
P. aztecus	946	*S. costatum*	200—400 × 10^3 cells/mℓ	N_6—M_1	75	C	9
		A. salina nauplii		M_1—P_5			
P. aztecus	946	Freeze-dried *S. costatum*	100—1,000 × 10^3 cells/mℓ	Z_1—M_1	50	D	10
		Frozen *S. costatum*	1,000 × 10^3 cells/mℓ	Z_1—M_1	60—80	D	10
		Frozen *Thalassiosira* sp.	100 × 10^3 cells/mℓ	Z_1—M_1	70—85	D	10
P. duorarum	15	*S. costatum*	600—1,000 × 10^3 cells/ℓ	Z_1—M_1		C	8
		A. salina nauplii		Z_3—P_1			
P. indicus	50, 140	*Thalassiosira weissflogii*	4 × 10^3 cells/mℓ	N_6—Z_1	95.6	C	11, 12
	8,000		7 × 10^3 cells/mℓ	Z_1—Z_3			
			2 × 10^3 cells/mℓ	Z_3—M_3			
		Brachionus plicatilis	5 individuals/mℓ	Z_3—P_1			
P. indicus	16,000	Diatoms	5—10 × 10^3 cells/mℓ	Z_1—P_{20}	21.4	E	13
		Bread yeast	2 g/ton/feeding × 2 feedings/day	Z_1—P_5			
		B. plicatilis	10—25 individuals/mℓ	Z_1—P_4			
		A. salina nauplii	5 g/10,000 larvae/day	M_2—P_5			
P. japonicus	50,000	*S. costatum*		Z_1—M_1	10—50	B	7
		Rotifer		M_1—P_2			
		A. salina nauplii		M_1—P_2			
		Copepoda		M_1—P_2			
P. japonicus	57,000	Diatoms		Z_1—P_{20}	38.7	A	14
	200,000	*B. plicatilis*		M_1—M_3			
		A. salina nauplii		M_3—P_3			
		Copepoda		P_3—P_{20}			
P. japonicus	60,000	*Chaetoceros* sp.		Z_1—P_1	57.6	B	2
P. japonicus	1,000	Marine yeast	20—40 g/ton/N_6—P_{20}	30—90			15
	10,000	*B. plicatilis*	Z_3—P_{20}				

P. japonicus	500	*Chaetoceros rigidus*	50—300 $\times$ 10^3 cells/mℓ	Z_1—Z_3	35.8	C	16
P. japonicus	500	Soy-cake particle	0.16 mg/zoea/day	Z_1—Z_3	85		16
P. japonicus	500	Soy-cake particles + *C. rigidus*		Z_1—Z_3	72.5		16
P. japonicus	100,000	Diatoms		Z_1—M_3		A	18a[b]
		A. salina nauplii		M_1—P_{10}			
P. japonicus	200,000	Diatoms	5—8 $\times$ 10^3 cells/mℓ	Z_1—P_{20}	20—30	A	1, 14, 17
		A. salina nauplii					18b[b]
P. japonicus	100—250 $\times$ 10^3	Diatoms		Z_1—M_3		A	18c[b]
		B. plicatilis		Z_1—M_3			
		A. salina nauplii		Z_3—P_5			
P. japonicus	110,000	Diatom (*Chaetoceros*)		Z_1—M_3		A(summer)	18d[b]
		B. plicatilis		Z_3—P_5		B(spring)	
		A. salina nauplii		M_1—P_5			
P. japonicus	2,400,000	*Chaetoceros* sp.		Z_1—M_3		B	18e[b]
		B. plicatilis	350 individuals/larva	Z_2—Z_3			
		A. salina nauplii	150 individuals/larva	M_1—P_3			
P. marginatus	2.8	*Chlorella* sp.	250—300 $\times$ 10^3 cells/mℓ	Z_1—P_8	60	C	19
		A. salina nauplii	1—3 individuals/mℓ	Z_3—P_{10}			
P. marginatus	2.8	Wild phytoplankton		Z_1—P_8	70	C	19
		A. salina nauplii	1—3 individuals/mℓ	Z_3—P_{10}			
P. merguiensis	30—150	*Tetraselmis suecica*	75 $\times$ 10^3 cells/mℓ	Z_1—M_1	57.8	C	20
		A. salina nauplii	2—5 individuals/mℓ	M_1—P_1			
P. merguiensis	1,500	*Chaetoceros* sp.		Z_1—M_2		B	18i[b]
		B. plicatilis		M_1—P_4			
		A. salina nauplii		P_1—P_4			
P. monodon	500—1,000	*S. costatum*	5 $\times$ 10^3 cells/mℓ	Z_1—M_3	30—40	C	6, 7, 21, 22, 23
		A. salina nauplii		M_1—P_5			
		Rotifer		M_1—P_1			
P. monodon	16,000	Mixed diatom	5—10 $\times$ 10^3 cells/mℓ	Z_1—P_{20}	20	E	23
	60,000	Bread yeast	2 g/ton/daily $\times$ 2	Z_1—P_5			
		B. plicatilis	10—25 individuals/mℓ	Z_1—P_4			
		A. salina nauplii	5 g/10,000 individuals/ larvae/day	M_3—P_4			
P. monodon	2	Mixed diatom	10—30 $\times$ 10^3 cells/mℓ	Z_1—Z_3	90	E	24
	2	Bread yeast		Z_1—Z_3	88		

Table 1 (continued)
SUMMARY OF LARVAL REARING OF PENAEID PRAWN WITH EMPHASIS ON LARVAL FEED

Species name	Volume of larval tank (ℓ)	Food item	Food quantity	Stages	Survival rate (%)	Algae production method[a]	Ref.
P. monodon	16,000	Bread yeast	16 g/ton/day	Z_1—M_1	20		23
	60,000		12 g/ton/day	M_1—P_1			
			8 g/ton/day	P_1—P_4			
			4 g/ton/day	P_4—P_{10}			
		B. plicatilis		M_1—P_1			
		A salina nauplii	5 g/100,000 larvae/day	P_1—P_4			
P. monodon	1,000	Mixed diatom	20 × 10^3 cells/mℓ	Z_1—M_1		C, E	25, 26, 27
		B. plicatilis	8—30 individuals/mℓ	M_1—P_2			
		Bread yeast	2 g/day	M_1—P_2			
P. monodon	1,000	Mixed diatom	1—10 × 10^3 cells/mℓ	Z_1—M_2	50	E	27
	500	Mixed diatom +	1—3 × 10^3 cells/mℓ	Z_1—M_2	45		
		bread yeast	1 g/ton	Z_1—M_2			
		Mixed diatom +	1—5 × 10^3 cells/mℓ	Z_1—M_2	45		
		B. plicatilis	2—4 individuals/larva	Z_1—P_{10}			
P. monodon	200,000	*Chaetoceros calcitrans*	10—100 × 10^3 cells/mℓ	Z_1—P_{10}	50 (wet	C	28, 29
		Bread yeast	3—5 g/ton/feeding ×	Z_1—Z_3	season)		
			2 feedings/day		18 (dry		
		B. plicatilis	2—5 individuals/mℓ	Z_1—P_5	season)		
		A. salina nauplii	1 g/10,000 larvae/day	M_1—P_5			
P. monodon	500	*Cylindrotheca* sp.	80—150 × 10^3 cells/ℓ	Z_1—Z_3	60	C	30
		Tetraselmis sp.	10—40 × 10^3 cells/mℓ	Z_3—P_2			
		Rotifer	10 individuals/mℓ	M_1—P_2			
		A. salina nauplii	5 individuals/mℓ	P_1—P_4			
P. monodon	2.5—150	*T. suecica*	10—50 × 10^3 cells/mℓ	Z_1—Z_3			31
		Isochrysis galbana	20—250 × 10^3 cells/mℓ	Z_1—Z_3			
		A. salina nauplii	5 individuals/mℓ	M_1—M_3			
P. monodon	50—200	*C. calcitrans*	10—100 × 10^3 cells/mℓ	Z_1—P_5		C(large tank)	18f[b]
	× 10^3,	*B. plicatilis*	5 individuals/mℓ	M_1—P_5		D(2,000 ℓ)	
	2,000	*A. salina* nauplii		P_1—P_3			

P. monodon	16—60 × 10^3	Diatoms	5—10 × 10^3 cells/mℓ	Z_1—P_7		E	18g[b]
		B. plicatilis		Z_3—P_3			
P. monodon	40—100 × 10^3	*Chaetoceros* sp.		Z_1—P_5		B	18h[b]
		B. plicatilis		M_1—M_3			
		A. salina nauplii		M_3—P_5			
P. semisulcatus	500	*S. costatum*	5 × 10^3 cells/mℓ	Z_1—M_1	4—50	C	6, 7
		A. salina nauplii	80 individuals/larva/day	M_1—P_5			
		Rotifer		M_1—P_1			
P. semisulcatus	20	*T. suecica*		Z_1—P_1			32
		Phaeodactylum tricornutum		Z_1—P_1			
		A. salina nauplii	4—20 individuals/larva/day	M_1—P_1			
		Panagrellus (nematoda)	18—80 individuals/larva/day	M_1—P_1			
P. setiferus	15	*S. costatum*	600—1,000 × 10^3 cells/mℓ	Z_1—M_1	50	C	8
		A. salina nauplii		Z_3—P_1			
P. setiferus	30,000	Bread yeast	400—500 cells/mℓ	N_6—Z_1			33
		A. salina nauplii		Z_1—P_1			
P. stylirostris	40,000	*C. gracilis*	30—100 × 10^3 cells/mℓ	Z_1—M_1	84.8	B	34
P. stylirostris	2,000	Frozen *S. costatum*	5-100 × 10^3 cells/mℓ	N_6—Z_3	81	D	35
		Frozen *Thalassiosira* sp.	5,000 cells/mℓ	Z_2—Z_3			
		Bread yeast		Z_2—M_3			
		Frozen *Tetraselmis* sp.		Z_3—M_1			
		Frozen *A. salina* nauplii	2 individuals/mℓ	M_1—P_2			
		A. salina cysts and nauplii	2 individuals/mℓ	M_2—P_2			
P. vannamei	40,000	*C. gracilis*	30—100 × 10^3 cells/mℓ	Z_1—M_1	79.3	B	34

[a] Algae production methods:

A. Mixed algal population induced by larval tank fertilization
B. Culture of algae added to the fertilized larval tank
C. Culture of algae added to unfertilized larval tank
D. Concentrated *Chaetoceros* sp. cells added to unfertilized larval tank
E. A mixed algae culture added to unfertilized larval tank

[b] Reference number 18a-i

a-i indicate the different hatcheries reported under Reference 18.

Table 2
A COMPARISON OF DIFFERENT ALGAE CULTURE METHODS

Algae production method	Volume of larval tank (ℓ)	Fertilization of larval tank	Culture condition	Algae concentration in larval tank (cells/mℓ)	Survival rate (%) of prawn larvae		Algae species reported
					Z_1—M_1	Z_1—$P_{5\text{-}10}$	
A	16—250 × 10^3	With fertilization	No inoculation of algae into larval tank	5—70 × 10^3		20—40	*Melosira, Netzschia, Skeletonema, Thalassiosira, Rhizosolenia, Chaetoceros*
B	1—110 × 10^3	With fertilization	Inoculation of algae culture into larval tank	30—100 × 10^3	84.3	10—60	*Skeletonema costatum, Chaetoceros* sp., *Chaetoceros gracilis*
C	2—200,000	Without fertilization	Separate culture of unialgae	5—1,400 × 10^3	35—95.6	18—60	*Skeletonema costatum, Chaetoceros rigidus, Chaetoceros calcitrans, Tetraselmis suecica, Cyclotella* sp., *Phaeodactylum tricornutum*
D	1—2 × 10^3	Without fertilization	Separate culture of unialgae, concentrated by chemicals or frozenpreserved, before added into larval tank	10—200 × 10^3	50—85		*Chaetoceros calcitrans, Skeletonema costatum, Thalassiosira* sp., *Tetraselmis chuii*
E	2—60,000	Without fertilization	Separate culture of mixed algae	1—30 × 10^3	50—90	20	*Chaetoceros, Skeletonema, Thalassiosira, Rhizosolenia, Nitzschia, Navicula, Pleurosigma, Amphiprora, Coscinodiscus, Bacillaria, Thalassiothrix*

Comparable data for the two kinds of systems, as currently practiced, are largely lacking, mainly because of the different species of both algae and prawns used, but also due to factors such as location and the methodology followed. A comparison of larval tank, survival rate, fertilization, algal concentration, and species of algae involved is made in Table 2.

The Same Tank Method

After Hudinaga and Kittaka developed the same tank method to mass produce the prawn larvae of *P. japonicus,* this technique was widely followed and gradually replaced the separate tank method as the method of choice in the production of prawn larvae of *P. japonicus* in Japan. It is now also used in Taiwan for *P. japonicus,* in Hawaii for *P. stylirostris* and *P. vannamei,* and in other places for other penaeid prawn species.

Types of Tanks Used and Their Preparation

The culture of algae is generally carried out in rectangular concrete tanks of 15 to 250 ton capacity (1 ton equal 1,000 ℓ). Circular fiberglass tanks and vinyl-lined steel swimming pools were also used. One Florida company doubled their hatchery capacity by putting 30 ton circular fiberglass tanks inside their 200 ton concrete tanks. The prawn larvae are then reared in the fiberglass tanks and, simultaneously, in the surrounding concrete ones. Vigorous aeration is maintained in all kinds of tanks, usually by means of at least one airstone per 1 to 6 m^2 of bottom area.

The most common type of tank used (10 × 10 × 2 m) corresponds to that used by the pioneer Hudinaga. The floors of the tanks have a slope of 3/100 toward the drains. Aeration pipes are provided at a density of one airstone per 5 m^2 of floor space.[17] Tanks are used without roofs or they are covered with a translucent or transparent roof.

Clean tanks are fully or partially filled with seawater to a depth of 0.6-2 m before or on the day the gravid female prawns are introduced.[1,14,17,18b] To exclude predators and excessive amounts of debris, which may interfere with the hatching of the eggs, inflowing water is filtered through sand or through an 80 to 100 mesh synthetic fiber net. The water in the tank is kept at the same level without changing until several days after metamorphosis of larvae into the postlarval stage, and then only one fifth of the total volume of water is exchanged daily with fresh seawater.[17] Alternatively, quantities of seawater of 10 cm depth or 5 m^3 is added on specific days until the tank is fully filled, after which the seawater is partially changed every day (Table 3).[18b,34]

Fertilization

The propagation of diatoms and the continuation of propagation once introduced into the larval tank are very important. Diatom propagation in the tanks can be controlled to some extent by adjusting the quantity and quality of the fertilizer or by changing the illumination intensity.

The commonly used fertilizers are potassium nitrate and dibasic potassium phosphate. Generally 200 g of potassium nitrate and 20 g of dibasic potassium phosphate are applied daily per 200-ton tank. Fertilization commences on the day when nauplii appear or, the following day. This is continued until the changing of tank water commences at the postlarval stage. Sometimes the quality and quantity of fertilizers and the sequences of fertilization are slightly changed because of the maintenance of water quality and the abundance of the diatoms in the larval tanks (Table 3).

Algal Species Grown

Usually the algal blooms in the larval tanks are made up of several species partly because of the water quality but also due to the season in which the cultures are made. As many as 30 diatom species have been identified from a single culture tank.[36] Among them, *Skele-*

Table 3
FERTILIZATION OF LARVAL TANK

Amount of fertilizer	Sequence of fertilization	Larval tank and water treatment	Remarks
KNO_3 200 gm K_2HPO_4 20 gm	Fertilization was applied on the day after spawning, when nauplii appear, or sometimes on the following day, and was continued daily at least until the larvae were given *Artemia* or prepared food.[1,14]	10 × 10 × 2 m^3 rectangular concrete tank. The water supply was filtered through 80-100 mesh screen net. From the time of the placement of the parent prawn in the culture tank to the early part of the postlarvae stage, the water was not changed, but after this period, 1/5 of total amount of water was changed each day.	Half amount of fertilizer added before egg hatching did not effect the normal hatching. Trace of $FeCl_3 \cdot 6H_2O$ was added in some cases.
NS 100 gm = 100 g KNO_3 + 10 gm Na_2HPO_4 + 5 mℓ K_2SiO_3 Solution (19% SiO_2, 9% K_2O)	Fertilization commenced at the second day of zoea or later and there were only three additions of the fertilizer for the whole three day period. The total fertilizer addition is only NS 700 gm.[18b]	10 × 10 × 2 m^3 rectangular concrete tank. The tank contained seawater of 60 cm depth at beginning, seawater of 10 cm depth/day was added until 180 cm depth, after which seawater was changed every day.	The management is atypical of conventional method (above listed) due to the high nutrient levels in its water supply. In case diatom bloom failed, a relatively pure mass culture of *Chaetoceros* in separate tank was added to the larval tank.[18d]
D[a] (grams) / R[b] (grams) $NaNO_3$[c] 21 / 34 NH_4NO_3[c] 9.9 / 39.6 NaH_2PO_4[c] 2.5 / 10 Na_2SiO_3 3.75 / 15 $FeCl_3$[c] 0.325 / 1.3 EDTA[c] 200 / 200	Fertilization commenced on the morning of the day when nauplii were added into the larval tank, then it was applied daily during dry season, every two days during rainy season, from the 3rd day until the day when the larvae attained the mysis stage.	40 m^3 epoxy-painted reinforced concrete tank with an average depth of 1.7 m. Tanks were filled with seawater to 25 m^3 at beginning, then 5 m^3 of seawater each was added at day 2, 3 and 9 to reach 40 m^3 seawater in total.	500-600 ℓ of *Chaetoceros gracilis* at 0.7-1.3 × 10^6 cells/mℓ were inoculated after the tanks were filled with 25 m^3 seawater.[34]

[a] Dry season.
[b] Rainy season.
[c] The amount of fertilizer was calculated from the data of Reference 34.

tonema, Melosira, Thalassiosira, Nitzschia, Rhizosolenia are the numerically dominant diatoms.[2] In case of green algal blooms, the larvae are subjected to high mortalities.[17] On occasion a pure culture of *S. costatum, C. gracilis* or *Chaetoceros* sp. is inoculated to grow in the larval tanks to promote algal blooms of these species.

Algal Concentration

Algal concentration in the larval tanks can not easily be controlled in the same tank method. A minimum of 5 $\times$ 10^3 cells/mℓ is generally sufficient for the larvae but levels exceeding 70 $\times$ 10^3 cells/mℓ has been recorded.

For the culture of *P. stylirostris* and *P. vannamei,* concentrations of 30 $\times$ 10^3 — 100 $\times$ 10^3 cells per milliliter of *Chaetoceros gracilis* assures good growth and survival of the zoea stages. The risk of mass larval mortalities following algal collapse from levels above 100 $\times$ 10^3 cells per milliliter is, however, great.

Deterioration of algal cultures at levels less than 100 $\times$ 10^3 cells per milliliter does not appear to harm healthy larvae.[34] However, collapse of algal blooms above 100 $\times$ 10^3 cells per milliliter can cause significant mortalities in shrimp larvae in the afternoon. The following conditions were observed in the afternoon in a bloom of *C. gracilis* which was about to collapse during the same evening: (1) progressive decline in cell density in successive samples; (2) absence of cells in the process of division in the culture; (3) appearance in the sample of plasmolyzed cells; (4) onset of aggregation of diatoms. The addition of EDTA, nutrient solutions, fresh diatom cultures, and new water into the larval tank, and increased aeration did not prevent the collapse of the bloom, but at least prevented larval mortalities.[34]

In some Japanese hatcheries, soy-cake powder or bread yeast have been used in cases where blooms failed to develop. Algal bloom failures may be due to the lack of chelated iron, so ferric chloride is sometimes added routinely, but the use of synthetic chelators such as EDTA usually are not successful in stimulating blooms.[5]

Conclusions

The same tank method provides the maximum use of natural conditions but local and seasonal influences upon the cultivation of prawns can not be avoided. In addition, there is a marked difference in both the survival and growth rates during the larval stages from nauplius to postlarvae among the various penaeid prawn species. In general, the nongrooved prawns such as *P. vannamei* and *P. setiferus* are more suitable for the same tank method compared to the grooved prawns such as *P. japonicus.*[17]

The Separate Tank Method

The first success in the artificial propagation of prawn larvae was obtained by employing the separate tank method. Algae used as food in this case was *Skeletonema costatum,* cultured in a separate tank. Since then, several species of algae were experimented with and were found suitable as food for larval penaeids.[2,11,18,20,28,31,34,35] These algal species and the densities in which they were used in the larval tanks, as well as the penaeid species involved in each case, are listed in Table 4. From this information it is clear that *S. costatum* is at present the most commonly used species. In addition, the species *Chaetoceros calcitrans, Chaetoceros gracilis, Chaetoceros* sp. and *Tetraselmis suecica* also appear to be among the most suitable as food for the propagation of penaeid prawn larvae.

Comparison of Qualities of S. costatum, C. calcitrans, *and* T. chuii *for Prawn Larval Food*

S. costatum is widely used in research and in commercial hatcheries to grow larvae of *P. monodon* in Taiwan.[6,7,21,22] *C. calcitrans* is used in the Philippines (SEAFDEC) for the mass culture of *P. monodon* larvae. In Galveston, *T. chuii* is the prime species used on an experimental basis to grow penaeid larvae.

Table 4
ALGAE CONCENTRATION IN LARVAL TANK OF PENAEID PRAWN (× 10^3 CELLS/mℓ)

Penaeid species	Algae species							
	Skeletonema costatum	*Tetraselmis suecica*	*Chaetoceros calcitrans*	*Chaetoceros rigidus*	*Chaetoceros* sp.	*Thalassiosira weissflogii*	*Cylindrotheca* sp.	Mixed diatom or phytoplankton
Penaeus aztecus	100—1,000	—	—	—	—	—	—	—
P. indicus	—	—	—	—	—	2—4	—	5—10
P. japonicus	—	—	—	50—300	100	—	—	—
P. monodon	5	10—50	10—100	—	50	—	80—150	5—30
P. marginatus	—	—	—	—	—	—	—	250—300
P. merguiensis	—	75	—	—	+	—	—	—
P. semisulcatus	5	+	—	—	—	—	—	—
P. setiferus	600—1,000	—	—	—	—	—	—	—
P. stylirostris	50—200	—	—	—	—	—	—	—
P. teraoi	5	—	—	—	—	—	—	—

The respective exponential growth of *S. costatum,* a chain diatom, *C. calcitrans,* a unicellular diatom, and *T. chuii,* a unicellular green flagellate, are 3.16 to 4.44 (*S. costatum*), 4.29 to 4.65 (*C. calcitrans*), 0.95 to 1.32 (*T. chuii*).[37] Being unicellular, the cultures of *C. calcitrans* and *T. chuii* can both be easily contaminated by other phytoplankton species or by zooplankton. *S. costatum* is easy to collect with a plankton net. *C. calcitrans* and *T. chuii,* on the other hand must be concentrated either by centrifugation, or flocculation with chemicals. Even though *S. costatum* has some advantages over the other two species, it has three major disadvantages. The first is the difficulty to maintain a *Skeletonema* culture during summer, when water temperatures may rise above 30°C. The second disadvantage is the maintenance of its stock culture. The maximum period it can survive in a stock culture maintenance medium kept in darkness at 20°C is only 2 weeks, compared to 24 weeks for *T. maculata* and 8 weeks for *C. gracilis.*[38] *S. costatum* can be maintained for only two months frozen with protectant and stored at −20 to −22°C. In contrast *T. chuii* can be frozen for 4 months and *C. calcitrans* for 18 months.[39] From this it is clear that more attention should be paid to the maintenance of stock cultures of *S. costatum.* A third disadvantage is the relatively short period of harvest for *S. costatum.* When *S. costatum* is harvested in or after the stationary phase, it is not suitable for and may even be harmful to prawn larvae. The period of harvest is 2 to 3 days after inoculation and lasts for only 1 to 2 days. Normally, *S. costatum* is in its prime phase in the morning but less so in the afternoon, when its condition deteriorates rapidly. In the case when natural seawater is replaced by synthetic seawater, *Skeletonema* cultures can be maintained for 10 days. Unfortunately, the cost of synthetic seawater and the extra labor makes this method very expensive.

Algal Characteristics

Skeletonema costatum

S. costatum is a chain diatom, with cell size ranging from 4 to 15 μm. The cell size decreases gradually during consecutive cell divisions. *Skeletonema* normally reproduces asexually by cell division. When the cell size is less than 7 μm, sexual reproduction occurs, by the formation of auxospores. Auxospores produce a cell of maximum size after which the formation of the coarse chain is repeated. According to hatchery workers in Taiwan, coarser and browner cells are the best and thus preferred as feed for prawn larvae. According to their experience, the onset of decay of a *Skeletonema* culture usually occurs later in coarser chain lines. In practice the coarser chains of *Skeletonema* used as inocula for culture are selected by using a 150-mesh screen net.

More auxospores were found to be formed at a water temperature of 20°C than at either 15°C or 25°C. Auxospores occur in the salinity range of 20 to 35 ppt. During experiments at light intensities ranging from 100 to 5,000 lux, auxospores seldom formed below 500 lux, but were found more frequently with light intensities exceeding 1,000 lux. The highest incidence of auxospore formation occurs between 4,000 and 5,000 lux.[41]

The *Skeletonema* cells of decaying cultures are neither suitable as food for prawn larvae nor as inocula for subsequent cultures. The decaying culture is characterized by

1. The coagulation of cells, which adhere to the wall of culture bottle or by sedimentation to the bottom of culture tank
2. The presence of empty cells
3. Dark brown colored cells
4. Close arrangement of cells against each other
5. Circular cells
6. Shorter chains

In Taiwan, *S. costatum* blooms naturally in the coastal harbors, especially in Kaoshiung Harbor, where *S. costatum* is almost 99% dominant during certain seasons and at some localities. As a result, many hatcheries are located in the vicinity of the Kaohsiung Harbor because of the abundance of naturally occurring food and abundance of *Skeletonema* for stock cultures. The availability of *Skeletonema* in the Kaohsiung area was an important factor in the successful culture of *P. monodon* in that area.

Skeletonema is eurythermal. It grows at water temperatures ranging from 3 to 34°C, with an optimum temperature of 25 to 27°C. It grows well in salinities ranging from 15 to 34 ppt, with optimal growth being attained at 25 to 29 ppt. Light intensity has a more pronounced effect on its growth than temperature. The optimal light period varies between 10 hr of dark per 14 hr of light and 12 hr of dark per 12 hr of light. The growth rate of *Skeletonema* increases with light intensity progressing through 500 to 10,000 lux, and declines at intensities exceeding 10,000 lux.

Two types of seawater are used for *Skeletonema* cultures; natural seawater and synthetic seawater. The available synthetic seawater formulae used are the Suto medium,[42] NH medium,[43] and a commercial product known as "Instant Ocean" which is enriched with nutrients.[9] The formulae of the Suto medium is listed in Table 5.

The media available for *Skeletonema* cultures are listed in Table 6. Medium I is used by Matsue who isolated and cultured *Skeletonema* for the first time. Medium II was used by Liao during his success in the propagation of *P. monodon*.[7] While Medium III is a modification of II but KNO_3 is replaced by $CO(NH)_2$, as $CO(NH_2)_2$ which are good sources of nitrogen required for *Skeletonema* growth. Formula IV was provided by Lii, the owner of a prawn hatchery, where several million young prawns are produced annually. Formula V is known as the F medium [44] which is also used by the present author to maintain or culture *Skeletonema* on a small scale. Formula VI is a modification of the F medium, which was developed as a result of laboratory experiments using different amounts and sources of nitrogen, phosphorus, silicate, and iron. The fact that different combinations of these elements showed various effects on the growth of *Skeletonema* illustrates the need for further tests on the optimum culture medium for the mass culture of *Skeletonema*. The last formula (VII) is used by the Galveston Laboratory. In both quantity and quality the additives in I to VII differ only slightly. However, VII is different from the others because of the use of synthetic seawater.

When a *Skeletonema* culture, enriched with the F medium, is maintained between 25 and 30°C, and illuminated at an intensity of 5,000 to 10,000 lux, a growth of 1.0×10^6 cells per milliliter can be obtained in 2 to 3 days after inoculation. Alternatively, when a culture enriched with "Instant Ocean" synthetic seawater medium (Table 6, VII) is maintained at a temperature between 25 and 30°C, and illuminated at 8,074 to 10,765 lux, it can be kept in good condition for 10 days. Under these conditions a concentration of 8.0 to 10.0×10^6 cells per milliliter can be obtained.[9]

Chaetoceros sp.

There are many species of *Chaetoceros* available as food for larval penaeids. *C. calcitrans* is a nonchain-forming marine centric diatom, with a median cell volume of 50 μm^3 but a cell volume of 30 μm^3 may occur depending on the culture conditions.[45] The *C. calcitrans* isolated from the Philippines is 4 to 5 μm in diameter and is adaptable to varying environmental conditions. Apart from being a suitable food species for penaeid larvae, it is also an invaluable food source for bivalve larvae. *C. gracilis* is a solitary marine centric diatom. It is rectangular in shape, measuring 8-12 × 7-10 μm exclusive of the setae. It is cultured at AQUACOP in Tahiti, the Oceanic Institute in Hawaii, and at the Tungkang Marine Laboratory in Taiwan as food for penaeid larvae, bivalve larvae, rotifers, and copepods. An as yet unidentified species of a unicellular *Chaetoceros* also proved to be suitable for larvae of *P. japonicus*.

Table 5
SUTO MEDIUM FOR *SKELETONEMA COSTATUM*

NaCl	24 gm	Modified P-solution	
$MgSO_4 \cdot 7H_2O$	8 gm	Na_2EDTA	3 gm
KCl	0.7 gm	$FeCl_3 \cdot 6H_2O$	0.39 gm
$CaCl_2 \cdot 2H_2O$	0.368 gm	$ZnCl_2$	0.03 gm
$NaNO_3$	0.1 gm	$MnCl_2 \cdot 4H_2O$	0.43 gm
$Na_2HPO_4 \cdot 12H_2O$	0.01 gm	$CoCl_2 \cdot 6H_2O$	12.1 mg
$NaHCO_3$	0.168 gm	$CuSO_4 \cdot 5H_2O$	4.7 mg
Na_2SiO_3	0.004 gm	H_3BO_3	3.43 gm
Vitamin B_{12}	0.01—0.03 μg	$Na_2MoO_4 \cdot 2H_2O$	0.126 gm
Modified P-solution	1 mℓ	D.W.	1,000 mℓ
D.W.	1,000 mℓ		

Table 6
MEDIA USED TO CULTURE *SKELETONEMA COSTATUM*

	Practices						
Additives	**I**	**II**	**III**	**IV**	**V**	**VI**	**VII**
KNO_3	202 mg	100 mg					100 mg
$CO(NH_2)_2$[a]			60 mg	26 mg		30 mg	
$NaNO_3$					150 mg		
$FeNH_4(SO_4)_2$							10 mg
KH_2PO_4	7.7 mg						
$Na_2HPO_4 \cdot 12H_2O$		10 mg	10 mg				
$NaH_2PO_4 \cdot 2H_2O$					10 mg		
$Ca(H_2PO_4)_2$[a]				23 mg			
Na-glycerophosphate						10 mg	
Na_2SiO_3		10 mg	10 mg		10 mg	20 mg	11 mg
K_2SiO_3	13 mg						
KCl[a]				44 mg			
$FeCl_3 \cdot 6H_2O$	5 mg	5 mg					
Thiamin							100 μg
Vitamin B_{12}							1 ng
Micronutrient[b]					1 mℓ	1 mℓ	
Tris buffer[c]							400 mg
Na_2EDTA							10 mg
Seawater	N[d]	N[d]	N[d]	N[d]	N[d]	N[d]	28 ppt synthetic seawater[e] 1,000 mℓ
Reference number	3	6, 7, 22			44		9

[a] Fertilizer reagents.
[b] Micro nutrient is 7.3 gm Na_2EDTA, 6 gm $FeCl_3 \cdot 6H_2O$, 12 mg $Na_2MoO_4 \cdot 2H_2O$, 44 mg $ZnSO_4 \cdot 7H_2O$, 360 mg $MnCl_2 \cdot 4H_2O$, 20 mg $CoCl_2 \cdot 6H_2O$, 2 mg $CuSO_4 \cdot 5H_2O$ in 1,000 mℓ distilled water.
[c] Tris buffer is 2-amino-2 (hydroxymethyl)-1, 2-propanediol.
[d] Natural seawater 1,000 mℓ.
[e] Synthetic seawater is prepared by mixing commercially available salt ''Instant Ocean'' with tap water.

All these species of *Chaetoceros* are characterized by tolerance of high water temperatures. When a *C. gracilis* culture was kept at a water temperature of 40°C, there was no color development. When this same culture was then maintained at 20°C or 30°C, normal growth occurred. The maximum growth temperature for *C. gracilis* is 37°C, with optimum growth at temperatures ranging between 25 and 30°C. The mass culture of this species is generally performed within this range. The minimal salinity for its growth is 6 ppt, but it can grow

Table 7
THE COMPOSITION AND PREPARATION OF ERDSCHREIBER MEDIUM

Constituents

1. Seawater: autoclave 2 ℓ in a 3 ℓ borosilicate glass flat-bottomed-boiling flask with cotton wool plug at 1.06 kg/cm^2 for 20 min, stand for 2 days.
2. Soil extract: prepared as follows:
   a. 1 kg soil from a woodland or pasture area untreated with artificial fertilizers, insecticides etc. with 1 ℓ freshwater
   b. Autoclave at 1.06 kg/cm^2 for 60 min
   c. Decant
   d. Filter through Whatman No. 1 paper and then through glass fiber (GF/c) paper;
   e. Autoclave in 1 ℓ aliquots in polypropylene bottles at 1.06 kg/cm^2 for 20 min
   f. Store in deep freeze until required;
   g. Autoclave 100 mℓ in 500 mℓ borosilicate glass flat-bottomed boiling flask with cotton wool plug at 1.06 kg/cm^2 for 20 min
3. Nitrate/phosphate stock solution: dissolve 40 g $NaNO_3$ and 4 gm Na_2HPO_4 in 200 mℓ distilled water, autoclave in 500 mℓ flask at 1.06 kg/cm^2 for 20 min.
4. Silicate stock solution: dissolve 8 g $Na_2SiO_3{\cdot}5H_2O$ in 200 mℓ distilled water. Autoclave in 500 mℓ flask at 1.06 kg/cm^2 for 20 min.

Procedure

Add 100 mℓ soil extract (2) to 2 ℓ seawater (1). With sterile pipette add 2 mℓ nitrate/phosphate stock (3) and 2 mℓ silicate stock (4). Decant 250 mℓ into 8 empty autoclaved 500 mℓ flasks with cotton wool plugs. Use bunsen burner to flame necks of flasks immediately before and after decanting/pipetting.

well at salinities up to 50 ppt with optimum growth occurring between 17 and 25 ppt. The growth rate of *Chaetoceros gracilis* increases under illuminations of 500 to 10,000 lux.

Generally, natural seawater is used for the mass culture of *Chaetoceros*. In order to prevent contamination, 16 g/ton calcium hypochlorite[18f] or 150 ppm sodium hypochlorite solution (10% chlorine)[2] may be added to the culture tank to sterilize the seawater. This is followed after 12 hr by the addition of 40 to 45 g/ton sodium thiosulfate to neutralize the chlorine residue, after which the pure diatom culture is inoculated. At Conwy, seawater used for 2 to 200 ℓ cultures of *Chaetoceros* is filtered through a 0.45 μm filter.

The media used for *Skeletonema* cultures are also suitable for *Chaetoceros*. In addition, the Erdschreiber medium (Table 7) used for stock culture, the Walne medium (Table 8) for 2 to 200 ℓ cultures at the Conwy Fisheries Station, the modified F Medium (Table 9) used by the Oceanic Institute, as well as the SEAFDEC medium (Table 10) are all suitable media for *Chaetoceros* culture.

A 200 ℓ culture of *C. calcitrans* maintained at 21°C, 15,000 lux, aerated with 5 ℓ/min of 1% CO_2, and enriched with the medium, with an initial density of 1 $\times$ 10^6 cells per milliliter was found to reach a density of 28 to 30 $\times$ 10^6 cells per milliliter in 3 to 4 days at the Conwy Fisheries Station. The 200 ℓ culture vessel was specially constructed, and is a tall, narrow cylinder fitted with an internal fluorescent light source. However, the maximum densities obtained at the Oceanic Institute, SEAFDEC, and at the Tungkang Marine Laboratory are only 4 to 6 $\times$ 10^6 cells per milliliter. This appears to be a consequence of the simpler culture tanks, culture preparations, and culture conditions, which differ from those applied at Conwy.

Dense cultures of *C. calcitrans* were harvested at SEAFDEC in the Philippines by using 150 ppm[18f] aluminum sulfate as the flocculating agent. The diatom slurry (25 $\times$ 10^6 cells per milliliter) was placed in plastic bags and stored in a refrigerator at 0°C. These frozen diatoms were acceptable to penaeid larvae. The diatoms also survived and reproduced in the rearing medium. Also, there was no significant differences between larvae fed with

Table 8
FORMULA OF WALNE MEDIUM FOR ALGAE CULTURE

Stock A

$FeCl_3 \cdot 6H_2O$	1.30 g
$MnCl_2 \cdot 4H_2O$	0.36 g
H_3BO_3	33.60 g
EDTA (Na salt)	45.00 g
$NaH_2PO_4 \cdot 2H_2O$	20.00 g
$NaNO_3$	100.00 g
Trace metal solution[a]	1.0 mℓ
Distilled water to	1,000 mℓ

Add 2 mℓ Stock A per ℓ of seawater for *Chaetoceros calcitrans*, 1 mℓ Stock A per ℓ of seawater for *Tetraselmis suecica*.

Stock B

Vitamin B_{12} (Cyanocobalamin)	10 mg
Vitamin B_1 (Thiamin)	200 mg
Distilled water to	100 mℓ

This solution should be acidified to pH 4.5 before autoclaving. Add 0.1 mℓ Stock B per ℓ of seawater.

Stock C

$Na_2SiO_3 \cdot 5H_2O$	4.0 g
Distilled water to	100.0 mℓ

Add 2 mℓ Stock C per ℓ of seawater for diatom culture only.

[a] Trace metal solution

$ZnCl_2$	2.1 g
$CoCl_2 \cdot 6H_2O$	2.0 g
$(NH_4)_6Mo_7O_{24} \cdot 4H_2O$	0.9 g
$CuSO_4 \cdot 5H_2O$	2.0 g
Distilled water to	100.0 mℓ

Acidify with sufficient concentrated HCl to obtain a clear solution.

frozen and fresh *C. calcitrans*.[28] By freezing the harvested *Chaetoceros* with cryoprotectants at −20 to −22°C, it was possible to preserve viability as long as 18 months.[39] Sun-dried *Chaetoceros* also showed satisfactory results as penaeid larval food and thus may be used as a substitute or supplement for fresh algae during periods of scarcity.[29]

Tetraselmis chuii

T. chuii is a green flagellate which was isolated from water samples collected from the East Lagoon in Galveston in 1965. *Tetraselmis* is also known as *Platymonas*. It is cultured at Galveston and SEAFDEC. Another species, *T. suecica*, is cultured at AQUACOP and Conwy for prawn and bivalve larvae.

The constituents of the medium used for *Tetraselmis* cultures varies. At Galveston NH medium is used for stock culture. For large scale culture, enriched artificial seawater is employed. Both media are the same as that for *Skeletonema* except for the absence of Na_2SiO_3, Walne medium (Table 8) and SEAFDEC medium (Table 10) are both suitable for *Tetraselmis* cultures.

Table 9
FORMULA OF THE MODIFIED "F" MEDIUM FOR *CHAETOCEROS GRACILIS*

Nutrient	Concentration (mg/ℓ)
NH_4NO_3	79.2
NaH_2PO_4	10.0
Na_2SiO_3	15.0
$FeCl_3$	1.3
EDTA	10.0
Trace metals of "F" medium	Table 6, V

Table 10
NUTRIENTS USED FOR COMMON ALGAE CULTURE IN SEAFDEC

Nutrient	Concentration (mg/ℓ)
Urea[a]	100
K_2HPO_4 or N:P (16:20)[a]	10 or 20
$FeCl_3$	2
$NaSiO_3$	2
Vitamin B_1	0.01
Vitamin B_{12}	0.01
Agrimin	1

[a] These two salts only are used for large tank culture of *Chaetoceros.*

The suitable culture salinity for *Tetraselmis* in natural seawater ranges between 15 to 36 ppt and for artificial seawater between 22 to 36 ppt. Outdoor mass cultures grow at temperatures between 15 to 33°C under natural light conditions. Typical densities produced in the mass culture tanks range between 275 to 450 $\times$ 10^3 cells per milliliter, with a record of 1,500 $\times$ 10^3 cells per milliliter.[47]

Tetraselmis cultures are harvested by means of centrifugation with a milk separator. The concentrated algal cells are then either used immediately as food or may be preserved in a frozen condition for future use. The frozen cells are thawed out, diluted, and added continuously into the larval tanks with small peristaltic pumps from a refrigerated and constantly stirred reservoir. Frozen *Tetraselmis* is not as good a food as the live algae, but is more convenient to use and reliable due to its availability.[47]

Concepts of Algal Culture

Obtaining the Proper Algae

The best algae cultured for prawn larvae are isolated from local strains. Good examples are *S. costatum,* isolated in Taiwan, and *C. calcitrans* in the Philippines. Otherwise suitable algae characterized by their adaptability to local conditions are alternatively obtained from algal culture collections around the world. For more information on the isolation and purification of algae and culture collection centers, the *Handbook of Phycological Methods Culture Methods and Growth Measurement* is a useful reference.[44]

Maintenance of the Stock Culture

It is important to maintain the stock culture under favorable conditions and to subculture at intervals to retain vigorous growth so that it can be used as inocula for mass culture. It is very important to sterilize the glassware, culture media, materials, and equipment used and to operate under sterile conditions.

Knowledge of Culture Conditions

A knowledge of existing culture conditions like temperature, salinity, light intensity, light period, culture medium, and aeration and the procedure to expand from stock cultures to large scale cultures, must either be obtained from publications or through experimentation.

Prediction of the Development of the Culture

Cell densities and the times to harvest of cultures must be predictable. Only then can sufficient food be prepared on a timely basis for prawn larvae.

Culture Methods Practiced for S. costatum at the Tungkang Marine Laboratory

Stock Culture

S. costatum is isolated and purified by using the micropipette method. The purified stock culture is then maintained in a 250 mℓ glass flask containing 100 mℓ of the F medium (Table 6, IV) but with no aeration. The F medium is prepared in individual bottles by dissolving the proper amount of $NaNO_3$ and $NaH_2PO_4{\cdot}2H_2O$, Na_2SiO_3 and micronutrients separately in distilled water after which they are autoclaved. One mℓ of each of these solutions is needed for 1 ℓ of culture water. One hundred milliliters of autoclaved seawater and 0.1 mℓ of each of the three stock solutions are added to one 250 mℓ flask. One mℓ of *Skeletonema* culture, which is not in a decaying phase, is then inoculated. The time of the onset of decay and the time of the transfer depend clearly on environmental factors. At 20°C in a 12 dark/12 light period growth chamber, the time of transfer mainly depends on the light intensity. The culture can be maintained at 1,000 lux for 20 to 30 days before transfer, 4 to 5 days after inoculation at 5,000 lux and the third day at 10,000 lux.

As mentioned previously, natural blooms of *Skeletonema* in Kaohsiung Harbor also serve as stock culture, maintained by nature.

Mass Culture

One glass carboy containing 1 ℓ of autoclaved seawater enriched with F medium is inoculated with one stock culture of 100 mℓ. At 25 to 30°C, and 5,000 to 9,000 lux, maximum growth occurs after 3 to 4 days. One glass beaker containing 10 ℓ of boiled seawater, enriched with F medium is inoculated with two to three stock cultures of 100 mℓ each, or, with 200 to 300 mℓ of 1 ℓ culture. At 25 to 30°C and 2,600 to 6,000 lux, maximum growth occurs after 3 to 4 days, after which transfer is made to the large culture tanks. A fiberglass circular tank of one ton is filled with sand filtered seawater to the 800 ℓ mark. Two 10-ℓ quantities of *Skeletonema* culture, or 5 to 10 ℓ of the concentrated algae collected from the Kaohsiung Harbor with a 150 mesh plankton net is then inoculated. The nutrient solutions are prepared by dissolving the proper respective amounts (Table 6, II, III) of KNO_3 or $CO(NH_2)_2$ and $Na_2HPO_4{\cdot}12H_2O$ in 1 ℓ of freshwater and Na_2SiO_3 in another 1 ℓ of freshwater (if Na_2SiO_3 is dissolved in seawater, a white precipitate will form). These substances are then all added the same morning to the culture tank before or after the inocula. After 2 to 3 days and before the decay of culture begins, 10 to 20 ℓ of *Skeletonema* culture or 5 to 10 ℓ of the concentrated *Skeletonema* collected with a 150 mesh net, is transferred to another one-ton tank and the same procedure repeated as reported above. The one-ton cultures of *Skeletonema* are used to inoculate 10- to 100-ton mass cultures by using a .01 to .02 ratio of inoculation. The seawater of the large tank culture is prepared to contain a 25 to 30 ppt salinity by diluting with fresh water. Mild aeration by air stones is provided to the medium in order to avoid the breaking up of the diatom chains. The one ton tanks are covered under a plastic roof, while the larger concrete tanks are outdoors. Shading or applying artificial light to obtain optimum light intensities is practiced.

Harvest

We do not feed *Skeletonema* obtained in Kaohsiung Harbor to prawn larvae until 2 to 3 transfers have been made. This step is necessary because of oil and other water pollutants present in the original sample. The *Skeletonema* culture must then be harvested before the decay of the culture commences. It is harvested completely with a 150 mesh plankton net or a common nylon cloth bag, at the end of the exponential phase. The concentrated algae are then resuspended into the larval tank as feed for the prawn larvae. Ideally the concentrated algae must be utilized immediately, otherwise, it is not suitable and may become harmful to the larvae.

Conclusions

For culture methods of the other algae, refer to the reference papers.[34,35,40,45,46,47] There is no single algal species which is the best in every respect for culture and use as feed for larval penaeids in the world. Some algae are more easily cultured and maintained, while others are more tolerable to high temperatures. Some species are difficult to harvest, while others are plagued with decay. The question of which species is most suitable, largely depends on certain algal characteristics, the culture conditions, and local environmental factors as well as the food requirements of the species of penaeids propagated.

ROTIFERS

There are many species belonging to class Rotifera, but the one most suitable for mass culture appears to be *Brachionus plicatilis*. It is a very important and indispensable food for marine fish larvae and has also been used on a large scale during the last decade as food for the mysis stages of penaeid prawns. *B. plicatilis* reproduces asexually (parthenogenetically) under favorable conditions by laying 1 to 2 large eggs (80 to 100 × 110 to 130 μm) which hatch into amictic females. Factors such as high densities, kinds of food, water temperature, salinity, light penetration, and water quality together with the genetic characters of rotifers are all factors which affect sexual reproduction. At the onset of the sexual reproduction phase, mictic females, bearing 1 to 6 small eggs (50 to 70 × 80-100 μm) which hatch into males, are formed from the large eggs of an amictic female. Then the mictic female is fertilized by a male to produce one or two dormant eggs. A dormant egg must undergo a period of dormancy before it can be hatched into an amictic female. From this amictic female asexual reproduction continues. This is a generalized description of a life cycle of *B. plicatilis*.[48,49,50] Accordingly, to increase the population density of rotifers, it is necessary to keep them in the asexual reproduction phase by excluding the factors which induce sexual reproduction of rotifers in mass culture.

The size of *B. plicatilis* ranges between 100 to 400 μm. Rotifers can be separated into two strains, the L-type (230 to 320 μm) and the S-type (140 to 220 μm). These two strains are not the same and their respective sizes can not change.[51] However, the size of each type (L or S type) is affected by temperature, both being shorter in summer than in winter.[52]

The maximum growth rate (μ max) of *B. plicatilis* is finite and depends on the size of rotifer. The larger the size the slower the growth rate. The μ max of one strain of *B. plicatilis* of 150 to 200 μm is about 0.69 day^{-1} while that of another strain of 300 to 350 μm is about 0.40 day^{-1}.[53]

B. plicatilis is eurythermal. When the water temperature declines below 10°C, dormant eggs are formed. It still grows at 15°C but does not reproduce. Between 15 and 35°C the growth rate of *B. plicatilis* increases with temperature. However its life span is shorter at high water temperatures. The optimum water temperature for its growth is 22 to 30°C.[54,55]

B. plicatilis occurs naturally in eel ponds with salinities above 3.7 ppt. Females with eggs can survive salinities as high as 98 ppt. However, the optimum range of salinity for this species is 10 to 35 ppt (Cl 6 to 18 ppt).[54,56]

Beyond the range of pH 5.0 to 10.0, *B. plicatilis* can not survive. There is no difference in the filter rate of *B. plicatilis* between pH 6.0 to 9.0.[56] Although the optimum range of pH is known to be 7.5 to 8.0, the pH value in the mass culture of *B. plicatilis* is not specifically controlled. The same applies to light intensities.

Food used as feed for *B. plicatilis* is varied and includes microalgae, bacteria, yeast and microorganic particles. The best foods are the marine *Chlorella* and bread yeast. *B. plicatilis,* fed solely on bread yeast is as good a diet for *P. japonicus* as rotifers fed with *Chlorella* as demonstrated in feeding experiments.[18d] However, the former shows some nutrient deficiency when used as feed for fish larvae.[53]

Bread yeast is commercially available in Taiwan but is more expensive (about NT$ 30 for 500 g) than *Chlorella.* The marine algae *Chlorella* can also be isolated from brackish water ponds. *Chlorella* isolated from the fish ponds of the Tungkang Marine Laboratory grows at salinities between 0 to 35 ppt, with an optimum range of 10 to 20 ppt. The maximum temperature for *Chlorella* growth is 37°C. Although it can survive at 40°C, there is no color development. Its optimum temperature is 25 to 35°C. However, along with the increase in temperature, Protozoa became increasingly dominant in *Chlorella* culture water, especially during the summer season. More attention should therefore be paid to the prevention of contamination of *Chlorella* cultures by Protozoa under high temperature conditions.

Chlorella stock cultures are maintained in 250-mℓ glass flasks containing 100 mℓ of the F medium (Table 6, V) or Walne's medium (Table 8) at 20°C, with a 12 hr dark/12 hr light photo period, and at 2,000 lux of light intensity. Subcultures made once every fortnight are necessary to maintain algal vigor.

Mass cultures of *Chlorella* are made from stock cultures as follows. One 100-mℓ volume of the stock culture is inoculated into 1-ℓ glass carboys containing the same medium as that of the stock culture. Cultures are kept at 25-30°C, and 6,000-9,000 lux of light intensity, with a photoperiod of 12 hr dark/12 hr light, and with vigorous aeration. Maximum growth occurs in 4 to 5 days. Then a 1-ℓ culture with 2×10^7 cells per milliliter is inoculated into one large glass beaker containing 10-ℓ boiled seawater enriched with crude agricultural rtilizers, i.e., ammonium sulfate 150 mg/ℓ, urea 7.5 mg/ℓ and calcium superphosphate 25 mg/ℓ. The 10-ℓ culture is placed near a window and aerated with an air stone or kept under the same conditions as that of 1-ℓ culture. After 4 to 5 days, one 10-ℓ culture with 1 to 1.5×10^7 cells per milliliter is inoculated into one 500 ℓ transparent plastic outdoor tank containing 200-ℓ sand filtered seawater enriched with the same crude agricultural fertilizers. In 4 to 5 clear weather days another 250 ℓ sand filtered seawater is added while the tank water is enriched with the already mentioned nutrient salts. The mass culture of *Chlorella* in large tanks is then expanded further from the 450 ℓ cultures with inoculation at a ratio of 1/5 to 1/10. In addition to the fertilizers used at the Tungkang Marine Laboratory, there are also other nutrients used for the mass culture of marine *Chlorella* (Table 11).

Two methods exist for the mass culture of *B. plicatilis,* according to the size of the tank and the process of harvesting. One is the changing tank method and the other is the partial harvest method.[58] Tanks of 0.5- to 3-ton capacities are used in the changing tank method. At the onset, one tank (tank A) is inoculated with *Chlorella.* After the *Chlorella* densities reach 1×10^7 cells per milliliter, *B. plicatilis* is inoculated at a density of 10 to 20 individuals per milliliter. When green water is used and becomes clear, bread yeast is given twice daily at a ratio of one gram of yeast to 10^6 rotifers. When the density of *B. plicatilis* exceeds 100 individuals per milliliter (about 5 to 7 days after inoculation), much of the culture is harvested to feed the larvae. A small volume is retained to serve as inocula for another *Chlorella* tank (tank B). Thus the process is transferred from tank A to tank B, B to C, C to D, and so forth. As described, the repetition of the procedure, commencing with the culture of *Chlorella,* followed by inoculation with rotifers, and then the harvest, is called the method of changing tanks. The individual capacity and the number of tanks used depend on the quantities

Table 11
MEDIA USED FOR MASS CULTURE OF MARINE *CHLORELLA*

Fertilizer[a]	Concentration (mg/ℓ)			
Ammonium sulfate	150	100	300	100
Urea	7.5	5	—	10—15
Calcium superphosphate	25	15	50	—
Clewat 32[b]	—	5	—	—
N:P 16:20 fertilizer	—	—	—	10—15
User	T.M.L.	Japan (at start)	Japan (at present)	SEAFDEC

[a] All fertilizers are locally made agricultural fertilizer.
[b] Clewat 32 is a commercial product composed of micronutrients.

of rotifers needed. This method is simple and reliable. It usually does not require much space but is more labor intensive. In the partial harvest method, a number of separate large tanks for the mass culture of *Chlorella* and rotifers are needed e.g., two 200-ton *Chlorella* culture tanks along with several 40-ton rotifer culture tanks. Initially, the rotifer culture tank is inoculated with *Chlorella* and brought to a density of 1 to 2 $\times$ 10^7 cells per milliliter. The rotifers are then inoculated at a density of 10 to 20 individuals per milliliter. The same procedure as that described for the changing tank method is then followed except for the use of larger culture tanks. As soon as the rotifer density is beyond 100 individuals per milliliter, 1/5 to 1/3 of the culture is harvested depending on the reproduction rate of the rotifers. An equal quantity of the *Chlorella* culture (1 to 2 $\times$ 10^7 cells per milliliter) is then added. Thus by harvesting part of the culture and then introducing an equal quantity of cultured *Chlorella,* large numbers of rotifers can be harvested daily. However, due to changing water quality, the rotifer culture tank can only be maintained at peak productivity for 15 to 30 days. It must then be completely harvested and the tank cleaned, after which the same procedure as reported above is repeated. The fact that mass cultures of *Chlorella* may at times be contaminated with rotifers, and then fail, is a serious problem in the mass culture of rotifers.

Rotifers can alternatively be cultured solely by being fed on bread yeast. A 3-ton concrete tank containing 2 tons of seawater is sprayed with 200 g bread yeast suspension and the rotifers are inoculated at a level of 10 to 20 individuals per milliliter in the culture tank. The bread yeast, dissolved in freshwater, is added twice daily according to the ratio 1 g of yeast to 10^6 rotifers. After 7 to 10 days, the rotifer density usually exceeds 100 individuals per milliliter. At this stage they can be completely or partially harvested as feed for larvae.

Contamination by copepods or ciliates in the rotifer culture tank is a serious problem. Both compete for food with rotifers and inhibit, in many cases, the growth of the latter. In addition, some copepod species are predators of rotifers. Accordingly, attention must be paid to prevent contamination by copepods and ciliates.

Rotifers are harvested with a 200-mesh (75 μm) plankton net and rinsed with seawater. They are then released into larval tanks. If the salinity of the larval tank is the same or slightly different from that of the rotifer culture tank, most of the rotifers in the larval tank may survive for one day. If the salinity of the larval tank is under or above 15 ppt of that of the rotifer culture tank, only 50% of the rotifers may survive for one day. The density of rotifers applied to the larval tank may range between 5 to 30 individuals per milliliter (Table 1). With more advanced developments in the utilization of rotifers in the future, it is expected that a combination of *B. plicatilis* and an artificial diet would be able to replace the brine shrimp in the feeding program of penaeid prawn larvae.

Table 12
LIST OF *ARTEMIA* CYST SOURCES STUDIED

Abbreviation used	Geographical origin	Commercial origin
SFB	San Francisco Bay, California, USA	San Francisco Bay Brand Company
SPB	San Pablo Bay, California, USA	San Francisco Bay Brand Company
BRAZIL	Macau, Brazil	1. Companhia Industrial Do Rio Grande Do Norte 2. Aquafauna, Inc.
PHIL.	Barotac Nuevo, Panay, Philippines	Ceramar Agro-marine Industries
GSL	Great Salt Lake, Utah, U.S.	1. Sander's Brine Shrimp Company 2. Bio-marine Research
AUSTR.	Shark Bay, Australia	Artemia, Inc.
CAN	Chaplin Lake, Canada	Jungle Laboratories Corporation
FRANCE	Lavalduc, France	Compagnie Des Salins Du Midi et Des Salines de L'est
CHINA	Tien-Tsin, People's Republic of China	China National Foodstuffs Export Corporation
ITALY	Margherita Di Savoia, Italy	
R.A.C.[a]	Reference *Artemia* Cysts	International Study on *Artemia*, *Artemia* Reference Center

[a] See Reference 76.

BRINE SHRIMP

The nauplii of the brine shrimp, *Artemia,* constitute an ideal food for shrimp larvae which can handle a 0.5 mm size prey. *Artemia* has the unique property of being commercially available in dry powder form. This means that, except for a 24 hr hatching incubation of the dry cysts, one is entirely independent of live stock culture maintenance. After Seal (1933) and Rollefsen (1939) reported the value of freshly hatched nauplii of *Artemia salina* as food for fish fry, the exploitation of *Artemia* cysts has gradually increased.[60] Until about 5 years ago, commercial supplies of cysts were from two natural sources in the U.S. and Canada. The increasing demands for cysts by aquarium hobbyists and aquaculture hatcheries soon exceeded commercial supplies. As a result, a number of studies on the effective utilization of *Artemia* cysts in aquaculture have been conducted in recent years.

Today, there are many different geographical strains of *Artemia salina* L. More than 50 brine shrimp strains from various continents and countries have so far been registered, namely from Algeria, Kenya, Tunisia, Argentina, Brazil, Canada, Mexico, Peru, Puerto Rico, U.S., Venezuela, India, Iran, Iraq, Israel, Japan, China, Australia, Bulgaria, France, Italy, Spain, U.S.S.R., and others. At present, about ten commercial *Artemia* harvester-distributors exist, selling brands of different qualities (Table 12).[61,62] To evaluate the potency of *Artemia* cysts, standard indexes for judging the quality of the cysts of a certain brain or batch of cysts are employed, using seawater in standard conditions (35 ppt, 25°C). The following aspects are considered.[63]

1. Hatching efficiency (HE): number of hatched nauplii obtained from 1 g of *Artemia* cysts.
2. Hatching percentage (H %): number of hatched nauplii against the total number of cysts used.
3. T_0: incubation time until appearance of first nauplii.
4. T_{90}: incubation time until 90% of nauplii have hatched.
5. Individual dry weight of instar I nauplii and energy content.
6. Hatching output: total naupliar biomass and energy produced from 1 g of *Artemia* cysts (Table 13).[62]

Table 13
DATA ON HATCHING EFFICIENCY (HE), HATCHING PERCENTAGE (H%), HATCHING RATE (T_0, T_{90}), HATCHING OUTPUT, INDIVIDUAL DRY WEIGHT AND ENERGY CONTENT OF *ARTEMIA* NAUPLII HATCHED IN STANDARD CONDITIONS (35 ppt, 25°C) FROM DIFFERENT *ARTEMIA* CYST SOURCES

Abbreviation used	Batch no. or year of harvest	Hatching efficiency (nauplii/g)	Hatching percentage (H%)	Hatching rate		Individual dry weight (in µg)	Individual energy content (10^{-3} joule)	Hatching naupliar biomass (mg/g cysts)	Output naupliar energy (joule/g cysts)
				T_0 (hr)	T^{90} (hr)				
SFB	288—2596	267,200	71.4	15.0	20.5	1.63	366	435.5	9,780
	288—2606	259,200	—	16.4	23.2	—	—	—	—
	236—2016	249,600	—	25.8	37.6	—	—	—	—
SPB	1628	259,200	84.3	13.9	20.1	1.92	429	497.7	11,120
BRAZIL	871172	304,000	82.0	15.7	23.7	1.74	392	529.0	11,917
	87500	182,400	—	16.0	29.1	—	—	—	—
	May '78	297,600	—	16.4	21.9	—	—	—	—
PHIL	1978	214,000	78.0	14.7	22.0	1.68	382	359.5	8,175
GSL	1977	106,000	43.9	14.1	21.7	2.42	541	256.6	5,735
AUSTR	114	217,600	87.5	20.3	28.1	2.47	576	537.5	12,534
CAN	1978	65,600	19.5	14.3	33.0	2.04	448	133.8	2,937
ARG	1977	193,600	62.8	16.1	22.6	1.72	379	333.0	7,337
FRANCE	1979	182,400	75.8	19.5	30.5	3.08	670	561.8	12,221
CHINA	1978	129,600	73.5	16.0	37.2	3.09	681	400.5	8,826
ITALY	1977	137,600	77.2	18.7	25.3	3.33	725	458.2	9,976
R.A.C.[76]		211,000	45.7	18.0	32.2	1.78	403	375.6	8,503

From the point of view of the aquaculturist, *Artemia* cysts should be selected with high HE values and H %. However the shorter the T_0 and T_{90} value, the better the cysts. A shorter T_0 and T_{90} period guarantees higher reliability of obtaining supplies of nauplii on a timely basis and a reduction in the cost of the hatchery. On the other hand, *Artemia* cysts which yield heavier individual nauplii and higher hatching output should be chosen when the nutritional value as feed for fish and crustacea juveniles is considered.

The production of *Artemia* nauplii by incubation of the cysts in seawater is a simple procedure. The following conditions for optimum hatching are suggested.

First, cylindroconical containers should be used to keep all cysts suspended in the water with aeration. Second, the water used for hatching should be preferably maintained at a temperature of 30°C and, where possible, at a salinity level of that of natural seawater.[64] However, a salinity of 5 ppt has recently been demonstrated to increase not only the hatching efficiency, but the energy content of the hatched nauplii.[62,64] Third, the pH value of water should be 8 to 9. Saturated dissolved oxygen conditions must exist and a cyst density of less than 10 g/ℓ is recommended.[64] A continuous illumination of about 1,000 lux over the containers ensures a maximal hatching output.[62,64] By early harvesting, one can obtain larval food of the best quality and with the highest caloric content. After the second molting, which normally occurs within 24 hr after hatching at 25°C, the individual dry weight and caloric value of *Artemia* nauplii decreases by 20% and 27%, respectively.[65] This is one of the main reasons why the nauplii have to be fed to the fish and prawn larvae as soon as possible upon completion of the hatching process. After hatching, the harvesting of the nauplii is carried out with a 150 μm screen net. The larvae are then thoroughly rinsed to remove all dissolved organic materials, mainly glycerol, before they are released into prawn or fish larval tanks.[66]

If *Artemia* cysts are decapsulated before hatching the hatching rate will improve, although the individual naupliar dry weight does not show much variation.[62] Decapsulation of cysts is recommended as this technique has at least three other major advantages, namely disinfection of the *Artemia*, facilitation of separating cyst shells from the hatched nauplii, and last but not least, the potential use of decapsulated cysts as a direct food source for fish and crustacean larvae.[66,67,68] The shell surfaces of *Artemia* cysts are usually covered with bacteria and other contaminants,[69] and can easily introduce detrimental organisms such as Anthozoa and Ciliata to prawn larval cultures.[70] The digestive tract of the prawn larvae feeding on *Artemia* nauplii may also be clogged by the empty shells and unhatched eggs because of their indigestibility.[71]

Decapsulation of *Artemia* cysts was first described and improved upon by Sorgeloos et al.,[66] and consists of the following steps.

1. Preparation of the decapsulation solution — bleaching powder ($Ca(OCl)_2$ as active ingredient) is dissolved in water, and aerated for 10 min. Then technical CaO or Na_2CO_3 is added to stabilize the pH of the decapsulation solution and the solution is aerated for another 10 min. The solution mixture is then stored overnight for precipitation and cooling. The supernatant is siphoned off the next morning and used for decapsulation.
2. The ratio of cysts to bleaching powder is 5 g bleaching powder per 10 g cysts, and the ratio of cysts to sodium carbonate or calcium oxide is 7 g technical Na_2CO_3 or 3 g technical CaO per 10 g cysts.
3. During the entire decapsulation treatment period, the cysts are kept in a decapsulation container, which is a cylindroconical tank, competely made of stainless steel mesh (150 μm mesh size), equipped with an aeration system to optimize the circulation of the cysts within the container. The only manual work required during the decapsulation process consists of the consecutive transfers of the decapsulation container to sequential baths of seawater, hypochlorite, tap water, chloric acid, and finally tap water. These steps are itemized below.

a. At first, hydrate the *Artemia* cysts in a sea bath for 1 hr and transfer cysts to the decapsulation bath, where the cysts are kept for 5 to 10 minutes to allow a complete reaction to take place. During this step, the hypochlorite is kept at a temperature below 35°C by continuous circulation through a cooling element which consists of a copper coil submerged in a bath of salt and ice.
b. Transfer the decapsulated cysts to the washing bath, and wash thoroughly with tap water.
c. Resuspend the decapsulated cysts in the deactivation bath (0.1 *N* of HCl or HAc solution) for a few minutes for the deactivation of the chlorine residues which remain adsorbed on the decapsulated cysts even after thorough washing with tap water.
d. Finally, wash thoroughly with tap water. The decapsulated cysts are now ready for incubation under optimal hatching conditions.[66,67,71,72,73]

Newly hatched larvae measure about 0.45 mm in length, 0.17 mm in width and 0.01 mg in weight. Fine nets of 150 μm mesh size are usually used to collect *Artemia* nauplii. They are ideal food organisms for penaeid prawns such as *Penaeus japonicus, P. monodon* and *P. indicus,* especially for the mysis to postlarval stages. The number of *Artemia* nauplii is kept at a density of 1 to 5 nauplii per milliliter culture medium.

Preadult *Artemia* can be used as food in aquaculture as a substitute for artificial feed. Live *Artemia* are better food for postlarvae than most artificial feeds and their use leads to better water quality and improved survival of prawn larvae.

Artemia feed on protozoa, microalgae, yeast, or bacteria, and are omnivorous. They are also a typical nonselective particle filter-feeder but sometimes eat large particles of food by tearing it apart. They can therefore be fed directly with the following foods: chicken feed, rice bran, minced fish, and chopped chicken dung. In addition, when these foods become rotten, they turn into fertilizers which can be reutilized to produce natural food for *Artemia* such as bacteria, yeast, algae, and different kinds of microorganisms.[74] Therefore, the cost to culture *Artemia* can be low.

It is a well-known fact that preadult *Artemia* have a higher nutritive value than freshly hatched nauplii. *Artemia* nutritive value changes considerably during growth: the fat content decreases from ±20% to less than 10% of its dry weight while the protein content increases from ±42% to over 60%. Whereas nauplii are deficient in histidine, methionine, phenylalanine, and threonine, adults are rich in all essential amino acids.[64] The brine shrimp is marvelously euryhaline, eurythermal, and is a typical euroxybiont. They may survive for one day in completely freshwater, grow in salinities from 5 to 150 ppt,[74] and are adaptable to water temperatures of 6 to 35°C.[75] When dissolved oxygen is less than 1 ppm, brine shrimp can still survive.[75] As a result of all these characteristics, preadult *Artemia* is now widely used as food for both freshwater or marine prawn larvae during live transportation from the Republic of China to other countries. In the future, therefore, the use of *Artemia* as a food for prawn or fish larvae in aquaculture can be expected to increase.

REFERENCES

1. **Fujinaga, M.,** Kuruma shrimp (*Penaeus japonicus*) cultivation in Japan, in *Proceedings of the World Scientific Conference on the Biology and Culture of Shrimps and Prawns,* FAO Fisheries Report No. 57, Vol. 3, Mistakidis, M. N., Ed., Mexico, 1967, 811.
2. **Shigueno, K.,** Advance in the prawn culture (*Penaeus japonicus* Bate), in *FAO Suisan Zoshoku Kokusai Kaigi Ronbun Shu,* Suisan Cho, Tokyo, 1976, 8. (In Japanese.)
3. **Matsue, Y.,** Culture of marine diatom *"Skeletonema costatum* (Grev.) Cleve," in *Suisangaku no Gai Kan,* Japan Society of Fisheries, Tokyo, 1954, 1. (In Japanese.)
4. **Kittaka, J.,** Food and growth of penaeid shrimp, in *Proc. First Int. Conf. Aquaculture Nutr.,* 1975, 249.
5. **Heinen, J. M.,** An introduction to culture methods for larval and postlarval penaeid shrimp, *Proc. World Maricul. Soc.,* 7, 333, 1976.
6. **Liao, I. C.,** On the artificial propagation of five species of prawns, *China Fish. Mon.,* 205, 3, 1970.
7. **Liao, I. C. and Huang, T. L.,** Experiments on propagation and culture of prawns in Taiwan, in *Coastal Aquaculture in the Indo-Pacific Region,* Pillay, T. V. R., Ed., Fishing News Books, Farnham, Surrey, England, 1973, 328.
8. **Cook, H. L.,** A method of rearing penaeid shrimp larvae for experimental studies, in *Proceedings the World Scientific Conference on the Biology and Cutlure of Shrimps and Prawns,* FAO Fisheries Report No. 57, Vol. 3, Mistakidis, M. N., Ed., Mexico, 1967, 709.
9. **Mock, C. R. and Murphy, M. A.,** Techniques for raising penaeid shrimp from the egg to postlarvae, *Proc. World Maricul. Soc.,* 1, 143, 1970.
10. **Brown, A., Jr.,** Experimental techniques for preserving diatoms used as food for larval *Penaeus aztecus,* in *Proc. of the National Shellfisheries Association,* Vol. 62, Walpole, Me., 1972, 21.
11. **Emmerson, W. D.,** Ingestion, growth and development of *Penaeus indicus* larvae as a function of *Thalassiosira weissflogii* cell concentration, *Mar. Biol.,* 58, 65, 1980.
12. **Emmerson, W. D. and Andrews, B.,** The effect of stocking density on the growth, development and survival of *Penaeus indicus* Milne Edwards larvae, *Aquaculture,* 23, 45, 1981.
13. **Anonymous,** Preliminary studies on the mass production of the fry of white prawn, *Penaeus indicus* Milne Edwards and its life history, in the *Mindanao State University Institute of Fisheries Research and Development (MSU-IFRD) Philippines, Annual Report,* 1975, 19.
14. **Shigueno, K.,** Problems on prawn culture (*Penaeus japonicus* Bate) in *Suisan Zo Yoshoku Sosho* Vol. 19, Nikon Suisan Sigen Hogo Kyokai, Tokyo, 1969, 6. (In Japanese.)
15. **Furukawa, I.,** (Larval culture of the penaeid shrimp, *Penaeus japonicus,* fed marine yeast), *Fish Cult.,* 105, 38, 1972. (In Japanese.)
16. **Hirata, H., Mori, Y., and Watanabe, W.,** Rearing of prawn larvae, *Penaeus japonicus,* fed soy-cake particles and diatoms, *Marine Biol.,* 29, 9, 1975.
17. **Hudinaga, M. and Kittaka, J.,** Local and seasonal influences on the large scale production method for penaeid shrimp larvae, *Bull. Jpn. Soc. Sci. Fish.,* 41, 843, 1975.
18. **Maguire, G. B.,** A report on the prawn farming industries of Japan, the Philippines and Thailand, New South Wales State Fisheries, 1979, 110; (a) Nagasaki Prefecture Research Station, Japan; (b) Fujinaga Penaeid Shrimp Institute, Aio, Japan; (c) Yamaguchi Prefecture, Naikai Fisheries Farming Center, Aio, Japan; (d) Tarumizu Culture Center, Kagoshima, Japan; (e) SISFFA Hatchery, Shibushi, Japan; (f) SEAFDEC, Tigbuan, the Philippines; (g) Mindanao State University Institute of Fisheries Research Center, Naawan, the Philippines; (g) Phuket Fisheries Research Station, Phuket Island, Thailand; (i) Songkhla Marine Fisheries Station, Songkhla, Thailand.
19. **Gopalakrishnan, K.,** Larval rearing of red shrimp *Penaeus marginatus, Aquaculture,* 9, 145, 1976.
20. **Beard, T. W., Wickins, J. F., and Arnstein, O. R.,** The breeding and growth of *Penaeus merguiensis* de Man in laboratory recirculation system, *Aquaculture,* 10, 275, 1977.
21. **Liao, I. C., Huang, T. L., and Katsutani, K.,** Summary of a preliminary report on artificial propagation of *Penaeus monodon* Fabricius, *JCRR Fisheries Series,* 8, 67, 1969.
22. **Liao, I. C. and Chin, L. P.,** Manual on propagation and cultivation of grass prawn, *Penaeus monodon,* Tungkang Marine Laboratory, Taiwan, 1980.
23. **Anonymous,** Mass production of the fry of Sugpo, *Penaeus monodon* Fabricius and some notes on its biology, in *The Mindanao State University Institute of Fisheries Research and Development (MSU-IFRD), Philippines, Annual Report,* 1975, 1.
24. **Anonymous,** The effect of food types on the survival of the zoea stages of *Penaeus monodon* Fabricius, in *The Mindanao State University Institute of Fisheries Research and Development, Philippines, Annual Report,* 1975, 79.
25. **Anonymous,** Preliminary studies on seed production of *Penaeus monodon,* in *Aquaculture Department, Southeast Asian Fisheries Development Center (SEAFDEC), Philippines, Annual Report,* 1974, 27.
26. **Anonymous,** Domestication of *Penaeus monodon,* in *Southeast Asian Fisheries Development Center, Annual Report,* 1975, 5.

27. **Anonymous,** Studies on the prawn, in *Southeast Asian Fisheries Development Center, Annual Report,* 1975, 13.
28. **Anonymous,** Prawn program, in *Southeast Asian Fisheries Development Center, Annual Report,* 1976, 13.
29. **Anonymous,** Prawn program, in *Southeast Asian Fisheries Development Center, Annual Report,* 1976, 11.
30. **Aquacop,** Reproduction in captivity and growth of *Penaeus monodon* Fabricius in Polynesia, *Proc. World Maricul. Soc.,* 7, 927, 1977.
31. **Beard, T. W. and Wickins, J. F.,** Breeding of *Penaeus monodon* Fabricius in laboratory recirculation systems, *Aquaculture,* 20, 79, 1980.
32. **Samocha, T. and Lewninsohn, C.,** A preliminary report on rearing peraeid shrimps in Israel, *Aquaculture,* 10, 291, 1977.
33. **Ward, D. G., Middleditch, B. S., Missler, S. R., and Lawrence, A. L.,** Fatty acid changes during larval development of *Penaeus setiferus, Proc. World Maricul. Soc.,* 10, 464, 1979.
34. **Simon, C. M.,** The culture of the diatom *Chaetoceros gracilis* and its use as a food for penaeid protozoea larvae, *Aquaculture,* 14, 105, 1978.
35. **Mock, C. R., Revera, D. B., and Fontaine, C. T.,** The larval culture of *Penaeus stylirostris* using modification of the Galveston Laboratory Technique, *Proc. World Maricul. Soc.,* 11, 102, 1980.
36. **Bardach, J. E., Ryther, J. H., and Mclarney, W. O.,** *Aquaculture, the Farming and Husbandry of Freshwater and Marine Organisms,* Wiley-Interscience, New York, 1972, 587.
37. **Aujiro, E. J.,** Growth phases of cultured algae used as larval food, in *Southeast Asian Fisheries Development Center Q. Res. Rep.,* 4, 1, 5, 1980.
38. **Antia, N. J. and Cheng, J. Y.,** The survival of axenic cultures of marine planktonic algae from prolonged exposure to darkness at 20°C, *Phycologia,* 9, 179, 1970.
39. **Aujero, F. and Millamena, O.,** Viability of frozen algae used as food for larval penaeids, *Southeast Asian Fisheries Development Center Q. Res. Rep.,* 3, 4, 11, 1979.
40. **Anonymous,** A Culture Method of *Skeletonema,* National Marine Fisheries Service Biological Laboratory, Galveston, Texas.
41. **Migita, S.,** Sexual reproduction of centric diatom, *Skeletonema costatum, Bull. Jpn. Soc. Sci. Fish.,* 33, 5, 392, 1967. (In Japanese.)
42. **Suto, S.,** An artificial medium for the diatom, *Skeletonema costatum, Aquiculture,* 7, 17, 1959. (In Japanese.)
43. **Gates, J. A. and Wilson, W. B.,** The toxicity of *Gonyaulax monilata* Howell to *Mugil cephalus, Limnology and Oceanography,* 5, 171, 1960.
44. **McLachlan, J.,** Growth media-marine, in *Handbook of Phycological Methods Culture Methods and Growth Measurements,* Stein, J. R., Ed., Cambridge University Press, London, 1973, chap. 2.
45. **Liang, L.,** Recommended procedures for the culture of *Chaetoceros calcitrans, Fish. Res. Tech. Rep.,* No. 53, 8, 1979.
46. **Helm, M. M., Liang, L., and Jones, E.,** The development of a 200 ℓ algal culture vessel at Conwy, *Fish. Res. Tech. Rep.,* No. 53, 1, 1979.
47. **Griffith, G. W., Murphy Kenslow, M. A., and Ross, L. A.,** A mass culture method for *Tetraselmis* sp. A promising food for larval crustaceans, *Proc. World Maricul. Soc.,* 4, 289, 1973.
48. **Hino, A. and Hirano, R.,** Ecological studies on the mechanism of bisexual reproduction in the rotifer *Brachionus plicatilis.* I. General aspects of bisexual reproduction inducing factors, *Bull. Jpn. Soc. Sci. Fish.,* 42, 1093, 1976.
49. **Hino, A. and Hirano, R.,** Ecological studies on the mechanism of bisexual reproduction in the rotifer *Brachionus plicatilis.* II. Effects of cumulative parthenogenetic generation on the frequency of bisexual reproduction, *Bull. Jpn. Soc. Sci. Fish.,* 43, 1147, 1977.
50. **Lubzens, E., Fisher, R., and Berdugo-white, V.,** Induction of sexual reproduction and resting egg production in *Brachionus plicatilis* reared in sea water, *Hydrobiologia,* 73, 55, 1980.
51. **Hino, A. and Hirano, R.,** On the change of body size of *Brachionus plicatilis,* in *Abstract of Jpn. Soc. Sci. Fish. Spring Meeting,* 1973. (In Japanese.)
52. **Fukusho, K. and Iwamoto, H.,** Cyclomorphosis in size of the cultured rotifer, *Brachionus plicatilis, Bull. Natl. Res. Inst. Aquaculture,* 1, 29, 1980. (In Japanese.)
53. **Endo, K.,** The culture of food organism for fish larvae, *Fish Culture,* 14, 86, 1977. (In Japanese.)
54. **Ito, T.,** On the culture of mixohaline rotifer *Brachionus plicatilis* O. F. Muller in the sea water, *Report of Faculty of Fisheries Prefectural Univ. of Mie,* 3, 708, 1960. (In Japanese.)
55. **Hirayama, K. and Kusano, T.,** Fundamental studies on physiology of rotifer for its mass culture. II. Influence of water temperature on population growth of rotifer, *Bull. Jpn. Soc. Sci. Fish.,* 38, 1357, 1972.
56. **Hirayama, K. and Ogawa, S.,** Fundamental studies on physiology of rotifer for its mass culture. I. Filter feeding of rotifer, *Bull. Jpn. Soc. Sci. Fish.,* 38, 1207, 1972.
57. **Watanabe, T.,** The nutrient value of food organisms used for fish larvae, *Symposium of Marine Science,* 10, 46, 1978. (In Japanese.)
58. **Hirata, H.,** The culture method of *Brachionus plicatilis, Fish Culture,* 17, 3, 35, 1980. (In Japanese.)

59. **Fukusho, K., Hara, O., and Yoshio, J.,** Mass production of the rotifer, *Brachionus plicatilis,* by feeding *Chlorella* sp. and yeast using large-scale outdoor tanks, *Aquaculture,* 24, 3, 96, 1976. (In Japanese.)
60. **Sorgeloos, P., Baeza-Mesa, M., Claus, C., Vandeputte, G., Benijts, F., Bossuyt, E., Bruggeman, E., Persoone, G., and Versichele, D.,** *Artemia salina* as live food in aquaculture, in *Fundamental and Applied Research on the Brine Shrimp Artemia salina (L.), in Belgium,* Jaspers, E., Ed., European Mariculture Society Spec. Publ. No. 2, Bredene, Belgium, 1977, 37.
61. **Vos, J.,** *Brine Shrimp (Artemia salina) Inoculation in Tropical Salt Ponds: A Preliminary Guide for Use in Thailand,* FAO Report THA: 75/008/79/WP/4, Publications of the National Freshwater Prawn Research and Training Centre, Freshwater Fisheries Division, Department of Fisheries, Ministry of Agriculture and Cooperatives, Bangpagong, Chacheongsao, Thailand, 1979, 2.
62. **Vanhaecke, P. and Sorgeloos, P.,** Hatching data on 10 commercial sources of brine shrimp cysts and re-evaluation of the "hatching efficiency" concept, presented at Proc. 12th Annual Meeting World Mariculture Society, Seattle, March 8-10, 1981.
63. **Sorgeloos, P., Persoone, G., Baeza-Mesa, M., Bossuyt, E., and Bruggeman, E.,** The use of *Artemia* cysts in aquaculture: the concept of "Hatching Efficiency" and description of a new method for cysts processing, in *Proc. 9th Annual Meeting World Mariculture Society,* Avault, J. W., Jr., Ed., Louisiana State University, Baton Rouge, La., 1978, 715.
64. **Sorgeloos, P.,** The use of the brine shrimp *Artemia* in aquaculture, in *The Brine Shrimp Artemia,* Vol. 3, Persoone, G., Sorgeloos, P., Roels, O., and Jaspers, E., Eds., Universa Press, Wetteren, Belgium, 1980, 25.
65. **Benijts, F., Vanvoorden, E., and Sorgeloos, P.,** Changes in the biochemical composition of the early larval stages of the brine shrimp *Artemia salina* L., in *Proc. 10th European Symposium on Marine Biology,* Vol. 1, Persoone, G. and Jaspers, E., Eds., Universa Press, Wetteren, Belgium, 1976, 1.
66. **Sorgeloos, P. et al.,** Decapsulation method for *Artemia* cysts, in *The Culture and Use of Brine Shrimp, Artemia salina, as Food for Hatchery-raised Larval Prawns, Shrimps, and Fish in Southeast Asia,* FAO Report THA: 75/008/78/WP/3, Publication of the National Freshwater Prawn Research and Training Centre, Freshwater Fisheries Division, Department of Fisheries, Ministry of Agriculture and Cooperatives, Bangpagong, Chacheongsao, Thailand, 1978, Annex D.
67. **Bruggeman, E., Sorgeloos, P., and Vanhaecke, P.,** Improvements in the decapsulation technique of *Artemia* cysts, in *The Brine Shrimp Artemia,* Vol. 3, Persoone, G., Sorgeloos, P., Roels, O., and Jaspers, E., Eds., Universa Press, Wetteren, Belgium, 1980, 261.
68. **Claus, C., Benijts, F., and Sorgeloos, P.,** Comparative study of different geographical strains of the brine shrimp *Artemia salina,* in *Fundamental and Applied Research on the Brine Shrimp Artemia salina (L.),* in Belgium, Jaspers, E., Ed., European Mariculture Society Spec. Publ. No. 2, Bredene, Belgium, 1977, 91.
69. **Wheeler, R., Yudin, A. I., and Clark, W. H., Jr.,** Hatching events in the cysts of *Artemia salina, Aquaculture,* 18, 59, 1979.
70. **Sorgeloos, P.,** *The Culture and Use of Brine Shrimp, Artemia salina, as Food for Hatchery-raised Larval Prawns, Shrimp, and Fish in Southeast Asia,* FAO Report THA: 75/008/78/WP/3, Publication of the National Freshwater Prawn Research and Training Centre, Freshwater Fisheries Division, Department of Fisheries, Ministry of Agriculture and Cooperatives, Bangpagong, Chacheongsao, Thailand, 1978, 18.
71. **Sorgeloos, P., Bassuyt, E., Laviña, E., Baeza-Mesa, M., and Persoone, G.,** Decapsulation of *Artemia* cysts: A simple technique for the improvement of the use of brine shrimp in aquaculture, *Aquaculture,* 12, 311, 1977.
72. **Bruggeman, E., Baeza-Mesa, E., Bossuyt, E., and Sorgeloos, P.,** Improvements in the decapsulation of *Artemia* cyst, in *Cultivation of Fish Fry and its Live Food,* Styczynska-Jurewicz, E., Backiel, T., Jaspers, E., and Persoone, G., Eds., European Mariculture Society Spec. Publ. No. 4, Bredene, Belgium, 1979, 309.
73. **Tunsutapanich, A.,** An improved technique for decapsulation and preservation of *Artemia* cysts (brine shrimp eggs) developed at the Chachoengsao Fisheries Station, FAO Report THA: 75/008/79/WP/6, *Thai Fish. Gaz.,* 32, 181, 1979.
74. **Tunsutapanich, A.,** Cyst production of *Artemia* in Thai salt ponds, in *Giant Prawn 1980,* Bangkok, Thailand, June, 1980. IFS Provisional Report No. 9, International Foundation for Science, Stockholm, Sweden, Research session I: 10, 133.
75. **Persoone, G. and Sorgeloos, P.,** General aspects of the ecology and biogeography of *Artemia,* in *The Brine Shrimp Artemia,* Vol. 3, Persoone, G., Sorgeloos, P., Roel, O., and Jaspers, E., Eds., Universa Press, Wetteren, Belgium, 1980, 3.
76. **Sorgeloos, P.,** Availability of Reference *Artemia* cysts, *Aquaculture,* 23, 381, 1981.

THE USE OF BRINE SHRIMP *ARTEMIA* IN CRUSTACEAN HATCHERIES AND NURSERIES

Patrick Sorgeloos, Etienne Bossuyt, Patrick Lavens, Philippe Léger, Paul Vanhaecke, and Danny Versichele

During the past 10 years, the *Artemia* Reference Center (ARC) at the State University of Ghent (Belgium) has studied the fundamental and applied biology of brine shrimp, *Artemia*, related to its use as a food source in aquaculture. The ARC works in close cooperation with aquaculture institutes in the third world and coordinates the activities of the "International Study on *Artemia*", an interdisciplinary group of laboratories in Europe and the U.S. that is studying *Artemia* strains for their potential use in aquaculture. Research at the ARC focuses on the following topics: (1) techniques for harvesting, processing, storage, hatching, and decapsulation of cysts; (2) characterization of geographical strains, i.e., biometrics and biochemistry of cysts and nauplii, nutritional value of nauplii for various predator species, larval growth and conversion rates, temperature-salinity tolerance ranges for cultured larvae, and reproductive characteristics; (3) intensive culturing of adult *Artemia* in batch and flow-through systems on cheap food sources; (4) controlled production of cysts and nauplii under laboratory conditions; (5) nutritional value analyses of different *Artemia* preparations in a standard predator-prey food-chain with the mysid *Mysidopis bahia;* (6) extensive production of *Artemia* biomass in manured ponds; and (7) transplantation and inoculation of *Artemia* in solar salt operations.

In crustacean hatcheries *Artemia* is used as a practical and suitable larval food source either in the form of 0.4-mm nauplii that can be produced easily from commercially available dry cysts,[1,2] or as ±1-cm adults that can be harvested from natural sources or be produced in very high densities on cheap food in controlled culture systems. Live, freshly hatched nauplii are the dominant form under which *Artemia* are offered as food source for the larvae of crab, prawn, and shrimp.[3-5] Recent findings, however, indicate that important savings and improved hatchery/nursery outputs can be expected from the use of frozen nauplii,[6] decapsulated cysts,[6-10] and even (pre-) adult *Artemia*.[2,6,11] Lobster larvae are mostly fed with very expensive live or frozen adult *Artemia* that are purchased from aquarium pet dealers.[12,13] Cheap *Artemia*,[2] cultured on the spot in intensive production systems with inexpensive feeds, could contribute to important savings in lobster farming.

USE OF FRESHLY HATCHED NAUPLII

Selection of a Suitable Source of *Artemia* Cysts

To date *Artemia* cysts are commercially available from various production sources in America, Asia, Australia, and Europe.[14] An updated list of the most important harvestors/distributors of brine shrimp cysts is given in Table 1. As long as there is no uniform pricing system nor a standard guarantee service for specific quality characteristics,[2] it is up to the customer to negotiate with the cyst dealers for a cyst product that is acceptable for his needs. Aside from various business aspects such as price, payment conditions, shipment and insurance costs, etc., the following criteria should be taken into consideration in selecting a specific cyst product: (1) cyst hatching quality, (2) naupliar size, (3) naupliar food value, and (4) cyst packaging conditions.

Hatching Quality

An acceptable cyst product should contain minimal quantities of impurities (e.g., sand, cracked shells, plumes, salt crystals, etc.).[15] Incubated at 25°C in 35-ppt seawater, the first

Table 1
UPDATED LIST OF *ARTEMIA* CYST DEALERS[a]

Aquafauna Bio-Marine Inc.
P.O. Box 5, Hawthorne, Calif., 90250; phone (213)973-5275; cable "Biomarine", Hawthorne, Calif.
(distributor of Great Salt Lake, Utah)

Aquarium Products
180L Penrod Court, Glen Burnie, Md., 21061; phone (301)761-2100;
(harvestor-distributor of sources from Argentina and Colombia)

Artemia Inc.
P.O. Box 2891, Castro Valley, Calif., 94546; phone (415)582-8866;
(harvestor-distributor of Shark Bay, Australia; distributor of Macau, Brazil)

China National Cereals, Oils & Foodstuffs Import & Export Corp. — Tientsin Branch
No. 134 Chih Feng Road, Tientsin, PR China; cable "Foodco", Tientsin; telex 22503 TJFDS CN
(harvestor-distributor of Tientsin-PR China)

Henrique Lage Marinocultura Ltda.
Rua Henrique Lage s/n, Imburanas, 59.500 Macau-RN, Brazil; phone (084)521-1150; telex 842187 HLSN BR
(harvestor-distributor of Macau, Brazil)

Jungle Laboratories Corporation
P.O. Box 66, Comfort, Tex., 78013; phone (512)995-2789
(harvestor-distributor of Chaplin Lake, Canada)

Salmac-Sosal
Av. Cunha da Mata, 126/128, Mossoró-RN, Brazil; phone (084)321-3122; telex 811811.
(harvestor-distributor of Salinas Guanabara and Salinas Sosal, Brazil)

Sander's Brine Shrimp Co.
1255 West 4600 South, Ogden, Utah, 84403; phone (801)393-5027
(harvestor-distributor of Great Salt Lake, Utah)

San Francisco Bay Brand Inc.
8239 Enterprise Drive, Newark, Calif., 94560; phone (415)792-7200; telex 171627 NANETLX-SUVL-792-7200
(harvestor-distributor of San Francisco Bay, Calif.)

[a] Some information compiled from Sorgeloos.[2]

nauplii should appear after 15 to 20 hr cyst incubation and the last nauplii should have hatched within less than 10 hr thereafter.[16,17] Two more hatching criteria that should be taken into account are the hatching efficiency and the hatching output.[15,16] The former refers to the number of nauplii that can be produced per gram cyst product and can easily be determined following a technique outlined by Sorgeloos et al.[15] The hatching output gives the total naupliar biomass (in milligrams dry weight) produced per gram cyst product and is to be computed from the hatching efficiency figure and the individual naupliar dry weight for the corresponding geographical source of *Artemia* (Table 2). An illustration of variation in hatching characteristics among commercial cyst sources and even among batches from the same brand is provided in Figures 1 and 2.

Naupliar Size

Data on naupliar length of commercial *Artemia* strains are given in Table 2. As long as nauplius size does not interfere with the ingestion mechanism of the predator, one may expect that the use of larger nauplii (with a high individual weight) will be beneficial;[18] the predator will, indeed, spend less energy in taking up a smaller number of larger nauplii to fulfill its food requirements.[19]

Table 2
SIZE AND INDIVIDUAL DRY WEIGHT OF *ARTEMIA* NAUPLII HATCHED IN STANDARD CONDITIONS (35 ppt, 25°C) FROM DIFFERENT CYST SOURCES[a]

Cyst source[b]	Nauplius length (μm)	Nauplius dry weight (μg)
San Francisco Bay, Calif. (SFB)	428	1.63
Macau, Brazil (BRAZIL)	447	1.74
Great Salt Lake, Utah (GSL)	486	2.42
Shark Bay, Australia (AUSTR)	458	2.47
Chaplin Lake, Canada (CAN)	475	2.04
Buenos Aires, Argentina (ARG)	431	1.72
Lavalduc, France (FRANCE)	509	3.08
Tientsin, PR China (CHINA)	515	3.09
Margherita di Savoia, Italy (ITALY)	517	3.33
Reference *Artemia* Cysts (RAC)	448	1.78

[a] Data compiled from Vanhaecke and Sorgeloos.[16,18]
[b] In parentheses abbreviations used in Figures 1 and 2.

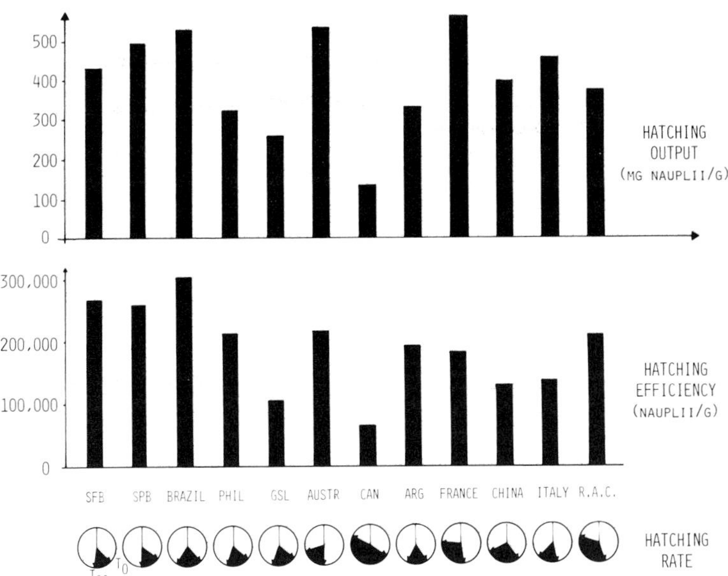

FIGURE 1. Hatching characteristics of *Artemia* cysts from different geographical origins when incubated in 35-ppt seawater at 25°C (legend to abbreviations in Table 2; T_0 and T_{90} values refer to the time lapse in hours from incubation of the cysts until appearance of the first nauplii, respectively, the moment at which 90% of the hatching efficiency is reached). (Modified from Vanhaecke, P. and Sorgeloos, P., *Aquaculture*, 30, 43, 1983. With permission.)

Naupliar Food Value

It has been shown on repeated occasions that *Artemia* nauplii from different geographical origins do not perform equally well as a food source for various predators. For detailed reviews on this subject see Sorgeloos[1,2] and Beck and Bengtson.[20] In general terms, good

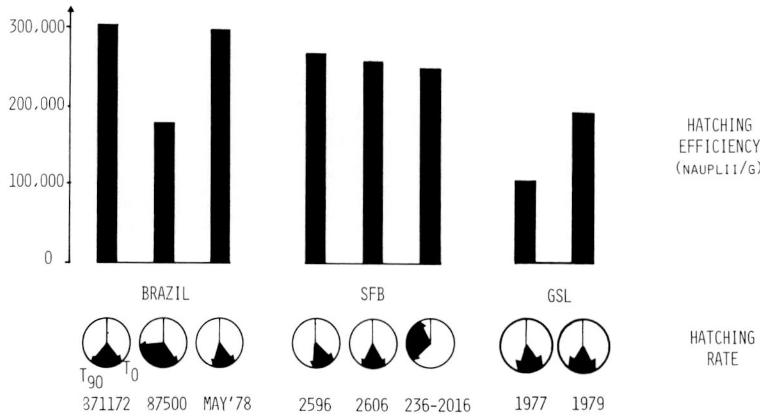

FIGURE 2. Variation in hatching rates and efficiencies among cyst samples from three commercial brands; legend to T_0 and T_{90} abbreviations as in Figure 1). (Modified from Vanhaecke, P. and Sorgeloos, P., *Aquaculture*, 30, 43, 1983. With permission.)

results were obtained with *Artemia* from Lavalduc (France), Macau (Brazil), Margherita di Savoia (Italy), San Francisco Bay (California), Shark Bay (Australia), and Tientsin (People's Republic of China). Intermediate results were obtained with nauplii from Chaplin Lake (Canada) origin. Specific samples from San Francisco Bay Brand (SFBB) origin cause high mortalities in all marine test species.[2,21] Figure 3 and Table 3 illustrate an interesting correlation that was found with the mysid *Mysidopsis bahia* between poor larval survival and low levels in the polyunsaturated fatty acid 20:5ω3 in the suspected SFBB batches.[21] Although Great Salt Lake (Utah) *Artemia* is a good food for mysid larvae,[22] it caused more than 80% mortality during the zoea-megalopa molting period in the crabs *Rhithropanopeus harrissii* and *Cancer irroratus*.[23] So far, it is not clear which parameter(s) determine(s) the nutritional suitability of a given *Artemia* source for a specific predator species; e.g., differences in contamination levels or polyunsaturated fatty acid patterns, synergistic effects of the latter, increased sensitivity for nutritional deficiencies in predator species that undergo metamorphosis, etc. For data on the biochemistry of various *Artemia* strains the reader should consult References 24 to 29. As long as specific literature data on the nutritional value of different *Artemia* strains for commercially important crustacean species are lacking, one should be very careful in switching from one *Artemia* source to another, i.e., a comparative bioassay should be run with an intercalibration product such as, e.g., Reference *Artemia* Cysts which are available from the *Artemia* Reference Center.[30]

Cyst Packaging Conditions

Artemia cysts should be stored dry (water level below 9%, preferably between 2 and 5%) and in oxygen-free conditions.[31] In this regard it is important to know the packaging conditions of commercial cyst products. When not stored in vacuum-sealed cans or under nitrogen atmosphere, cyst hatching rates and efficiencies start to drop after a few months storage in air-tight cans at room temperature.[17] Deep freezing cysts increases the period of time that cysts remain viable.[32] However, Kinne[5] and Hessinger and Hessinger[33] draw attention to the fact that following low temperature storage, the cysts should be exposed to room temperature for 1 to 2 weeks prior to incubation, otherwise poor nauplius yields will prevail.

Hatching Techniques

Best hatching results with high densities of cysts can be achieved with funnel-shaped containers that are aerated from the bottom. We mostly use heat-sealed plastic bags of up

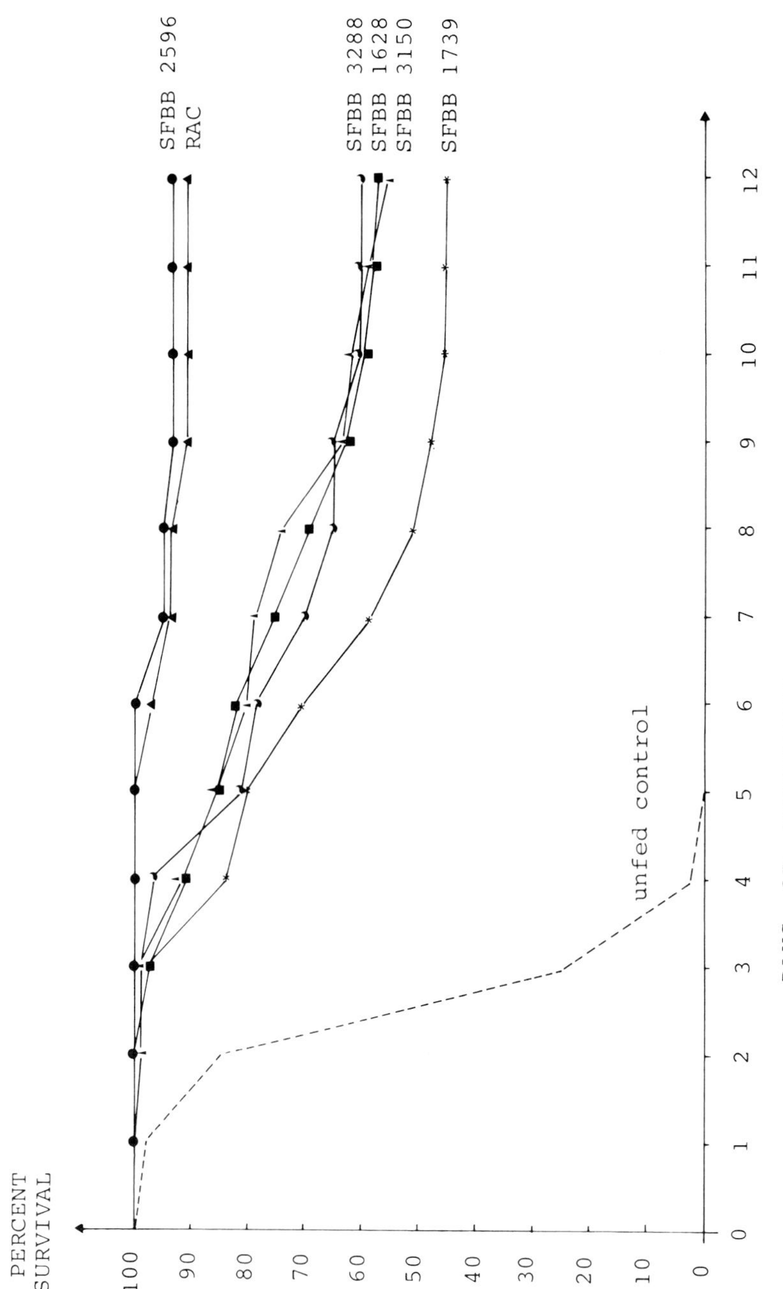

FIGURE 3. Survival of *Mysidopsis bahia* larvae fed nauplii from Reference *Artemia* Cysts and from various cyst batches of San Francisco Bay Brand origin. (Modified from Léger, Ph., Vanhaecke, P., Simpson, K. L., and Sorgeloos, P., manuscript submitted for publication.)

Table 3
FATTY ACID COMPOSITION OF FRESHLY HATCHED *ARTEMIA* NAUPLII FROM SAN FRANCISCO BAY BRAND ORIGIN AND FROM REFERENCE *ARTEMIA* CYSTS[a]

<table>
<tr><th>FAME[b]</th><th>SFB 1628</th><th>SFB 1739</th><th>SFB 3288</th><th>SFB 3150</th><th>SFB 2596</th><th>R.A.C.[c]</th></tr>
<tr><td>14:0</td><td>0.45</td><td>0.50</td><td>0.59</td><td>0.64</td><td>1.33</td><td>1.86</td></tr>
<tr><td>14:1</td><td>1.91</td><td>1.43</td><td>2.05</td><td>1.33</td><td>0.91</td><td>2.23</td></tr>
<tr><td>15:0</td><td>Trace</td><td>Trace</td><td>Trace</td><td>Trace</td><td>0.26</td><td>0.83</td></tr>
<tr><td>15:1</td><td>0.61</td><td>0.33</td><td>0.45</td><td>0.25</td><td>0.21</td><td>0.94</td></tr>
<tr><td>16:0</td><td>9.27</td><td>11.55</td><td>10.94</td><td>12.94</td><td>13.02</td><td>13.65</td></tr>
<tr><td>16:1ω7</td><td>4.82</td><td>4.45</td><td>5.43</td><td>4.68</td><td>21.94</td><td>16.39</td></tr>
<tr><td>16:2ω7—17:0</td><td>0.71</td><td>0.52</td><td>0.48</td><td>0.29</td><td>—</td><td>2.22</td></tr>
<tr><td>16:3ω4—17:1ω8</td><td>0.87</td><td>1.00</td><td>1.05</td><td>1.26</td><td>0.84</td><td>3.66</td></tr>
<tr><td>18:0</td><td>3.63</td><td>3.54</td><td>3.17</td><td>2.99</td><td>2.96</td><td>3.24</td></tr>
<tr><td>18:1ω7/ω9</td><td>26.97</td><td>28.99</td><td>28.74</td><td>28.81</td><td>34.13</td><td>31.19</td></tr>
<tr><td>18:2ω6</td><td>9.85</td><td>10.37</td><td>9.43</td><td>10.15</td><td>4.71</td><td>9.78</td></tr>
<tr><td>18:3ω3/ω6—20:0[d]</td><td>31.01</td><td>28.23</td><td>27.69</td><td>29.02</td><td>7.83</td><td>1.30</td></tr>
<tr><td>20:1ω7/ω9 }</td><td rowspan="2">6.98</td><td rowspan="2">5.77</td><td rowspan="2">5.43</td><td rowspan="2">4.56</td><td>—</td><td>0.94</td></tr>
<tr><td>18:4ω3 }</td><td>1.91 }</td><td>—</td></tr>
<tr><td>21:0</td><td>—</td><td>—</td><td>—</td><td>—</td><td>0.15 }</td><td rowspan="2">0.31</td></tr>
<tr><td>20:2w6/w9</td><td>0.33</td><td>0.41</td><td>0.34</td><td>0.25</td><td>—</td></tr>
<tr><td>20:3ω3</td><td>0.13</td><td>0.10</td><td>0.20</td><td>0.13</td><td>0.11</td><td>0.16</td></tr>
<tr><td>20:4ω3/ω6</td><td>0.48</td><td>1.25</td><td>1.61</td><td>1.85</td><td>1.90</td><td>4.24</td></tr>
<tr><td>22:1</td><td>1.65</td><td>1.09</td><td>1.07</td><td>0.86</td><td>0.34</td><td>0.06</td></tr>
<tr><td>20:5ω3</td><td>0.15</td><td>0.51</td><td>1.35</td><td>1.25</td><td>7.91</td><td>7.05</td></tr>
<tr><td>24:0</td><td>—</td><td>—</td><td>—</td><td>—</td><td>—</td><td>—</td></tr>
<tr><td>24:1</td><td>—</td><td>—</td><td>—</td><td>—</td><td>—</td><td>—</td></tr>
<tr><td>22:5ω3</td><td>—</td><td>—</td><td>—</td><td>—</td><td>—</td><td>—</td></tr>
<tr><td>22:6ω3</td><td>—</td><td>—</td><td>—</td><td>—</td><td>—</td><td>—</td></tr>
</table>

[a] Data compiled from Léger et al.[21]
[b] Fatty acid methyl ester.
[c] Reference *Artemia* Cysts.
[d] More than 99% 18:3ω3.

to 20-ℓ volume made from polyethylene sheets or 75-ℓ tanks made of transparent PVC. Schematic drawings and dimensions of hatching containers that are currently in use in our laboratory are found in Figure 4. The data provided here are valid for hatching cyst densities of up to 5 g/ℓ. For reasons of practical convenience natural seawater is mainly used to hatch cysts. However, at 5-ppt salinity the hatching rate increases and for some sources of cysts higher hatching efficiencies are obtained and the nauplii have a higher energy content.[1,16] We, therefore, advise the use of either natural seawater diluted with tap water to 5 ppt and enriched with 2 g $NaHCO_3$ per liter hatching medium, or artificial hatching solution as prepared following the formula in Table 4. Increased buffer capacities are essential in order to assure the pH levels do not drop below 8.0.[1] Temperature should be kept constant within the range of 25 to 30°C. Oxygen levels above 2 mg/ℓ can be maintained at aeration rates of approximately 7 ℓ air per minute in the 20-ℓ hatching container and approximately 20 ℓ/min in the 75-ℓ tank.

With some cyst products hatching solutions turn turbid shortly after hydration as a result of organic impurities associated with the cysts. Bacterial development and consequent increases in oxygen demand during hatching can be minimized by a 1-hr presoak of these cysts in tap water followed by a thorough washing of the hydrated cysts on a 125-μm screen prior to their transfer into the hatching solution.

20 l PLASTIC BAG
(heat-sealed with polyethylene sheet)

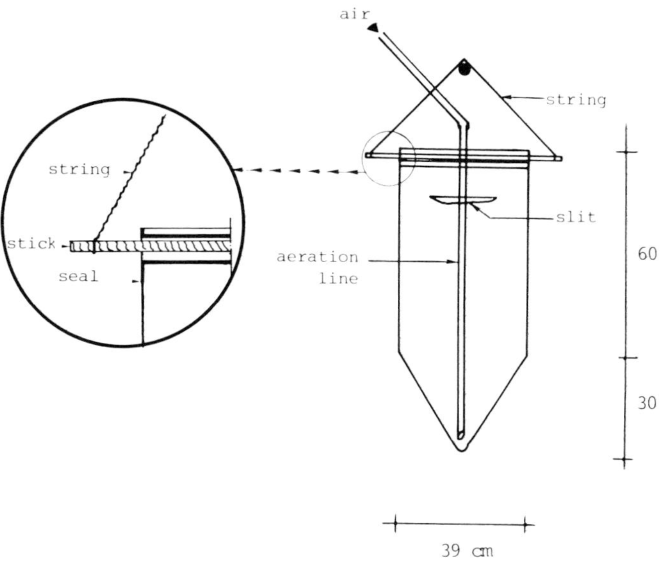

75 l TANK
(made of transparent PVC)

FIGURE 4. Schematic drawings and dimensions of hatching containers.

Illumination of the cysts, at least during the first hours following their complete hydration, is essential for maximum hatching results.[1,34] Considering the differences which are observed among brine shrimp strains,[35] a continuous illumination of about 1000 lux assures optimal results. This can be attained by illuminating the hatching suspension at a distance of about 20 cm with two (20-ℓ bag) or four (75-ℓ tank) 60-W fluorescent light tubes.

Since instar I nauplii depend completely on their energy reserves for nutrition, they should

Table 4
ARTIFICIAL SEAWATERS USED IN *ARTEMIA* HATCHING AND CULTURING

	Hatching medium[a]	Culturing medium[a]
Evaporated sea salt (NaCl)	5.0	31.08
$MgSO_4$[b]	1.3	7.74
$MgCl_2$	1.0	6.09
$CaCl_2$	0.3	1.53
KCl[b]	0.2	0.97
$NaHCO_3$[b]	2.0	2.00

[a] In gram salts (technical grade) per liter tap water.
[b] Should be dissolved separately in warm tap water before addition to the solution of other salts.

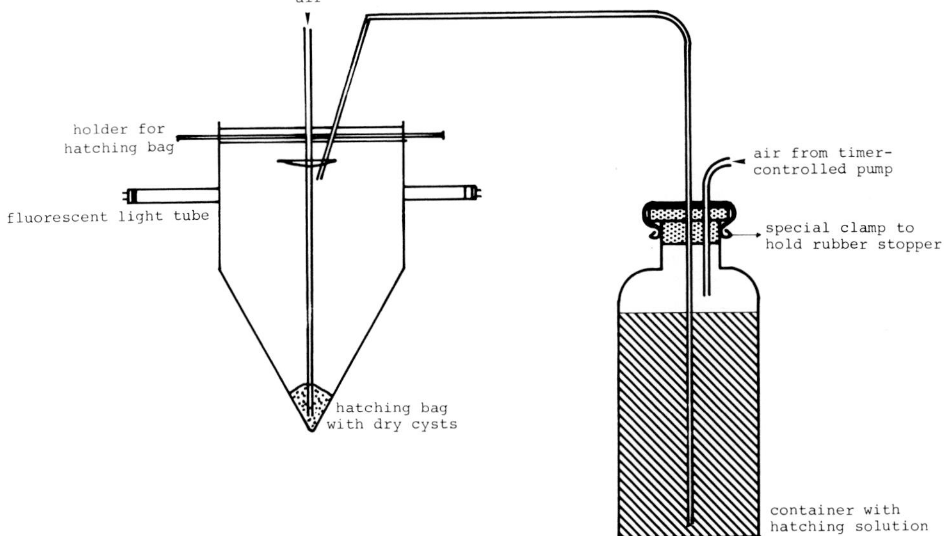

FIGURE 5. Set-up for automatic incubation of *Artemia* cysts. (Modified from Léger, Ph. and Sorgeloos, P., *Aquaculture Eng.*, 1, 42, 1982. With permission.)

be harvested and fed to the crustacean larvae in their highest energy form, i.e., as soon as possible after hatching.[36] The most economical use of *Artemia* cysts, thus, requires incubation of the cysts under constant conditions with regard to the parameters outlined above to allow harvesting of the maximum number of instar I nauplii after a well-defined incubation period. With those strains that have a poor hatching synchrony (e.g., Chaplin Lake, Canada; see Figure 1), a first harvesting of nauplii is done several hours before maximal hatching efficiency is reached in order to assure optimal use of produced nauplii.

A simple technique for standardization in cyst hatching is described by Léger and Sorgeloos;[37] i.e., at the preset time incubation of the cysts is done automatically by timer-controlled pumping of seawater into the hatching tank that already contains the dry cysts (see Figure 5).

Harvesting of *Artemia* nauplii, more or less free from empty shells and unhatched cysts, is done after about 5- to 10-min interruption of the aeration; empty cyst shells float at the surface while the nauplii concentrate in the lower part of the funnel. Siphoning should be started at the very bottom of the funnel in order to first drain off the heavy debris and unhatched cysts that have accumulated just underneath the nauplii. Since nauplii are positively phototactic, their concentration can eventually be obtained and increased by shading the upper part of the hatching container with a black plastic sheet and assuring that light reaches the lower part of the funnel only. Flotation of the cyst shells can also be improved by increasing the salinity up to 35 ppt shortly before harvesting through addition of saturated brine or crude salt. The sudden salinity change does not harm the *Artemia* nauplii.[38]

The circular separator boxes for *Artemia* nauplii that have been described by Persoone and Sorgeloos[39,40] are very useful in the laboratory but much too complex for commercial-scale application in crustacean hatcheries.

Some strains of *Artemia* (e.g., Chaplin Lake, Canada; Great Salt Lake, Utah; etc.) can hardly be separated following any of the above-mentioned techniques. With these cysts, in particular, and whenever contamination with empty cyst shells has to be avoided, the use of decapsulated cysts is advisable.

In order to prevent contamination of the culture tank with glycerol,[41] hatching metabolites, and bacteria,[42] harvested *Artemia* nauplii should be washed thoroughly on a 125-μm screen prior to their transfer to the larval culture tanks.

The frequency of hatching operations, with the consequent labor involved, can be greatly reduced by storage of instar I nauplii in the refrigerator (0 to 4°C) in aerated containers at densities of up to 15,000 nauplii per milliliter for up to 48 hr.[43] Except with Chaplin Lake (Canada) and Buenos Aires (Argentina) *Artemia,* naupliar viability remains over 90% (even 24 hr after transfer of the cooled nauplii into the culture tank at 25°C). Energetic losses are minimal (no significant differences after 24 hr, ±8% drop in individual dry weight after 48-hr storage) and decreases in nutritional value for mysid and carp larvae are insignificant respectively minimal (wet weight gain after 2 weeks culturing is 8% less in carps fed with 48-hr stored nauplii as compared to controls fed freshly hatched *Artemia*). This storage technique provides opportunities for automation in food distribution. A system that has been in continuous operation for more than 2 years for feeding mysids at 3-hr intervals during the daytime with stored *Artemia* nauplii is outlined in Figure 6. At the end of each food distribution, feed lines are automatically rinsed with seawater, i.e., the electromagnetic valve opens about 30 sec before the peristaltic pumps stop their activity and seawater is drained into the food distribution tube; as a result of the hydrostatic pressure of the seawater inflow, the one-way valve closes off the supply of nauplii and clean seawater is pumped through the feed lines.

Decapsulation of Cysts

As mentioned earlier complete separation of *Artemia* nauplii from their cyst shells is not always possible. The latter shells, when ingested by the crustacean predator, might cause deleterious effects,[44] particularly because of their high bacterial load.[45] The hard shell or chorion of *Artemia* cysts can be removed without affecting the viability of the embryos by short exposure of the hydrated cysts to a hypochlorite solution; this process is called cyst decapsulation.[44] The use of decapsulated cysts not only eliminates naupliar separation problems, but several other advantages recommend application of the decapsulation technique, i.e., disinfection of the cysts,[44] improved hatching results (see Table 5),[46] lower threshold for light stimulation at the onset of the hatching metabolism,[35] and, last but not least, the fact that decapsulated cysts appear to be as good a food source as freshly hatched nauplii for a variety of organisms: *Scylla serrata, Portunus pelagicus,*[7] *Penaeus monodon, P. indicus, Metapenaeus ensis, M. endevouri, Macrobrachium rosenbergii,*[8] *Metapenaeus monoceros,*[9] and for *P. kerathurus.*[10] A major handicap with regard to the latter application, however, is that as a result of the loss of the chorion, which ensures the buoyancy of the cyst,[44] decapsulated cysts settle out of suspension; extra circulation, e.g., with air-water lifts, is, thus, needed to keep this food source in the water column.[6]

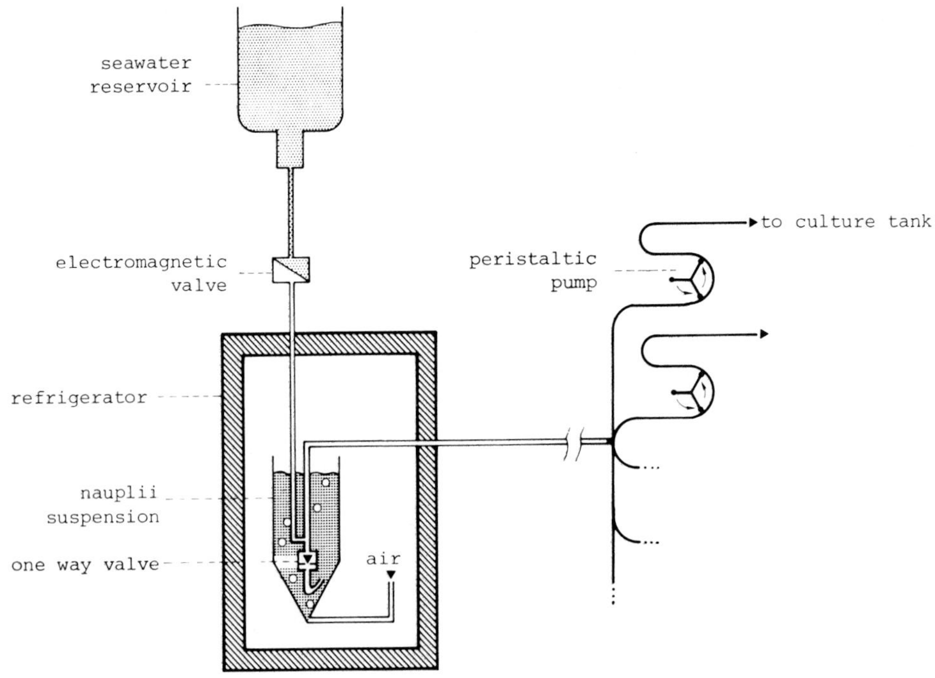

FIGURE 6. Schematic diagram of automatic distribution system for *Artemia* nauplii. (Modified from Léger, Ph. and Sorgeloos, P., *Aquaculture Eng.*, 1, 42, 1982. With permission.)

Table 5
IMPROVED HATCHING QUALITY OF *ARTEMIA* CYSTS AS A RESULT OF DECAPSULATION[a]

Cyst source	Hatchability[b]	Naupliar dry weight[b]	Hatching output[b]
San Francisco Bay, Calif.	+15	+7	+23
Macau, Brazil	+12	+2	+14
Great Salt Lake, Utah	+24	−2	+21
Shark Bay, Australia	+4	+6	+10
Chaplin Lake, Canada	+132	+5	+144
Buenos Aires, Argentina	+35	+10	+49
Lavalduc, France	+2	+0	+2
Tientsin, PR China	+4	−1	+2
Margherita di Savoia, Italy	+10	+8	+19
Galera Zamba, Colombia	+14	+0	+13
Barotac Nuevo, Philippines	+11	+6	+19
Reference *Artemia* Cysts	+59	+1	+29

[a] Data compiled from Vanhaecke and Sorgeloos[16] and Bruggeman et al.[46]
[b] Percent difference with untreated cysts.

The decapsulation procedure involves the following consecutive steps: (1) hydration of the cysts, (2) treatment in hypochlorite solution, (3) washing and deactivation of chlorine residues, followed by either (4) direct use or (5) dehydration for storage.

Cyst Hydration

Complete removal of the chorion can only be performed when the cysts are spherical; in most strains full hydration is reached after 2-hr exposure to freshwater or seawater (maximum

35 ppt) at 25°C (hydration time increases with decreasing temperatures). We recommend the use of the same funnel-shaped containers used for *Artemia* hatching in order to assure a homogeneous swelling rate of the cysts. As soon as full hydration is reached the cysts should be transferred to the hypochlorite solution, i.e., they are filtered over a 120-μm screen, washed to remove impurities, and drained to remove excess water. Prolonged exposure (hours) to optimal hatching conditions previous to the hypochlorite treatment appears to negatively affect the hatching rate and efficiency of these decapsulated cysts. Hydrated cysts that cannot be treated immediately may be stored for a few hours (eventually overnight) in the refrigerator (0 to 4°C).

Treatment in Decapsulation Solution

Two sources of hypochlorite can be used, either liquid bleach NaOCl or bleaching powder Ca $(OCl)_2$. The activity of the latter product is usually correctly mentioned on the label of commercial products (mostly 70% by weight activity). The activity of NaOCl solutions can be determined by various methods.[47] Provided the solution is fresh or has been stored properly, the most useful method is refractive index determination, e.g., with a refractometer. A conversion table for various units of measurements is given in Table 6.

The weight of active product per volume of decapsulation solution, per gram of dry cysts to be treated, is identical for both hypochlorite products, i.e., 0.5 g active product per gram of cysts and 14 mℓ decapsulation solution per gram of cysts. The chemical, however, that is added to increase the pH level of the decapsulation solution above 10 is different; with NaOCl 0.15 g technical grade NaOH (0.33 mℓ of a 40% solution) is added per gram of cysts and with $Ca(OCl)_2$ 0.67 g Na_2CO_3 or 0.4 g CaO per gram of cysts.

Decapsulation solutions are to be made up with 35-ppt seawater and should be cooled (eventually by addition of ice) as to have a working temperature of 15 to 20°C. Practical examples for the preparation of decapsulation solution using either liquid bleach or bleaching powder as hypochlorite source are given in Table 7. With the $Ca(OCl)_2$ decapsulation solution it is critical to first dissolve the bleaching powder and only then add the CaO or Na_2CO_3. After thorough mixing for about 10 min, e.g., through strong aeration, the suspension should be allowed to settle and only the supernatant be used.

The hydrated but drained cysts are transferred into the decapsulation solution and should be kept in continuous suspension, e.g., by manual stirring or by air bubbling from the bottom of a conical tank. Within a few minutes the exothermic oxidation reaction starts, foam develops, and as the chorions dissolve a gradual color change in the cysts is observed from dark brown into grey when using Ca-hypochlorite, or via grey into orange with Na-hypochlorite.

When treating quantities of 500-g cysts and more, the suspensions's temperature should be checked regularly and eventually ice added in order to prevent the lethal cyst temperature of 40°C. The decapsulation treatment only requires 10 to 15 min.

Washing and Deactivation Treatment

As soon as the chorions have been completely dissolved (stereoscopic microscope inspection or when no more change in color nor increase in temperature occurs) the decapsulated cysts are filtered off and extensively washed on a 120-μm screen with tap water (or seawater) until there is no more chlorine smell. Toxic residues that remain adsorbed to the decapsulated cysts are to be deactivated with thiosulfate,[47] or more efficiently in a bath of chloric or acetic acid;[46] i.e., the cysts are dipped a couple of times in a 0.1 *N* HCl or HOAc tap water (or seawater) solution. This deactivation should last less than half a minute after which treatment the cysts are again washed with tap water or seawater.

Upon transfer to tap water or seawater the decapsulated cysts sink. Cysts that have not been treated properly and still contain chorion fragments can be skimmed from the water surface; upon dehydration in brine (see Dehydration and Storage) the latter cysts can be added to a next batch of cysts that have to be decapsulated.

Table 6
THE ACTIVITY OF NaOCl SOLUTIONS OF DIFFERENT STRENGTHS DETERMINED BY DENSIMETER OR REFRACTOMETER[a]

Specific gravity at 15°C	Chlorometric[b] (°)	Activity[c] (%)	Chlorine per liter (g)	Refractive index[d]
—	10.0	3.21	32.2	1.3451(66)
—	16.3	5.25	52.5	1.3518(104)
1.099	20.4	6.56	65.6	1.3562(129)
1.116	24.3	7.81	78.1	1.3604(152)
1.133	28.6	9.19	91.9	1.3650(175)
1.151	32.7	10.51	105.1	1.3694(—)
1.170	37.7	12.12	121.2	—
1.189	41.0	13.18	131.8	—
1.209	46.8	15.04	150.4	—
1.220	50.5	16.23	162.3	—

[a] Data compiled from Bruggeman et al.[47]
[b] French chlorometric degrees = 3.111 × % activity.
[c] English chlorometric degrees = 0.3214 × ° chlorometric.
[d] Scale accuracy of 1.3740; higher concentrations to be analyzed upon dilution with tap water; in parenthesis salinity reading on refractometer; conversion from refractive index reading (x) into concentration of active product in g/ℓ (y) following equation y = 3000 x −4003.

Table 7
PRACTICAL EXAMPLES OF COMPOSITION OF DECAPSULATION SOLUTIONS MADE UP WITH EITHER NaOCl OR $Ca(OCl)_2$ AS ACTIVE AGENTS FOR THE TREATMENT OF 100-G DRY CYSTS

Example 1 — use of concentrated liquid bleach
  Refractive index of the diluted NaOCl solution (one part NaOCl per one part tap water): 1.3604
  Concentration of active product in NaOCl-solution:
    [(3000 × 1.3604) − 4003] × 2 = ±156 g/ℓ
  Amount of active product needed for decapsulation of 100-g cysts:
    100 × 0.5 = 50 g
  Volume of liquid bleach needed: 50 × 1000/156 = 320 mℓ
  Amount of NaOH needed: 0.15 × 100 = 15 g or 0.33 × 100 = 33 mℓ 40% solution
  Total volume of decapsulation solution: 14 × 100 = ±1400 mℓ
  Volume seawater needed: 1400 − 320 − 33 = ±1050 mℓ
  Decapsulation solution is thus made up of:
    320 mℓ liquid bleach
    33 mℓ 40% NaOH solution
    1050 mℓ seawater

Example 2 — use of bleaching powder
  Activity of product: 70% (by weight) active ingredients
  Amount of product needed for the decapsulation of 100-g cysts:
    (100 × 0.5) × 100/70 = 71 g bleaching powder
  Amount of Na_2CO_3 needed: 0.67 × 100 = 67 g (or 0.4 × 100 = 40 g CaO)
  Volume of decapsulation solution (= volume of seawater needed): 14 × 100 = 1400 mℓ
  Decapsulation solution is thus made up of:
    1400 mℓ seawater,
    71 g bleaching powder,
    67 g Na_2CO_3 (or 40 g CaO)

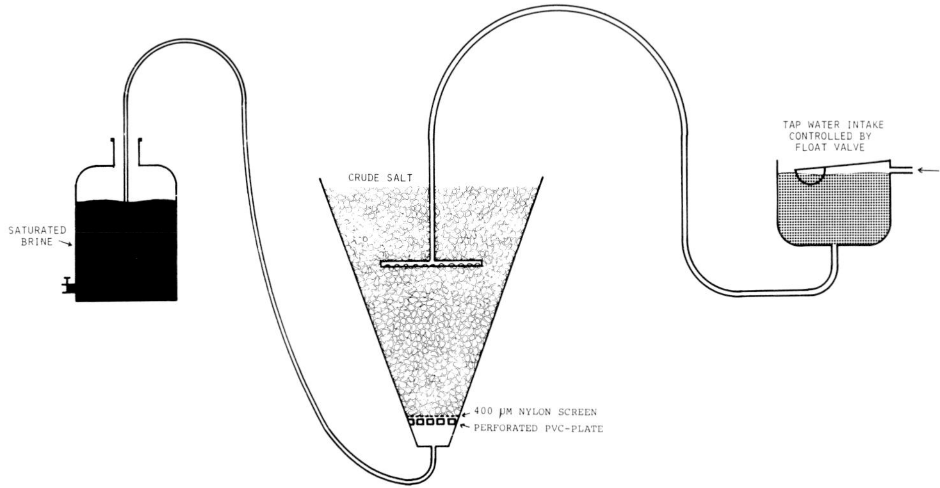

FIGURE 7. Schematic diagram of Sterling Brinomat. (Modified from Spotte, S., *Fish and Invertebrate Culture. Water Management in Closed Systems,* Wiley-Interscience, New York, 1970. With permission.)

Direct Use of Decapsulated Cysts

The hydrated decapsulated cysts can be offered directly as food source to the predator larvae (see Decapsulation of Cysts). If needed they can be stored for a few days in the refrigerator at 0 to 4°C. As mentioned earlier it is critical to assure sufficient circulation in the culture tank as to maintain these cysts in suspension. Since decapsulated cysts are considerably smaller than freshly hatched nauplii, it might be possible that for some predators these cysts could be offered to an earlier larval stage: e.g., in San Francisco Bay (CA-USA) *Artemia*, decapsulated cysts have an average diameter of only 210 μm whereas nauplii have a length of 428 μm or when refering to volume, nauplii are about 50% larger than decapsulated cysts.

The dehydrated decapsulated cysts can also be incubated in optimal hatching conditions for the production of nauplii (see Hatching Techniques). Upon hatching, separation of cyst shells has become superfluous; i.e., nauplii are filtered off, washed free from glycerol, hatching metabolites and bacteria, and are offered to the predator larvae.

Dehydration and Storage

Upon completion of the deactivation and washing procedures, decapsulated cysts that have to be stored need to be dehydrated, i.e., the cysts are drained on a 120-μm screen and transferred into a saturated brine solution at a rate of 1 g (dry) cysts per 10 mℓ.

The preparation of brine can be automated by use of a modified Sterling Brinomat[48] as schematically outlined in Figure 7. In this apparatus the saturated, and at the same time filtered, brine is made up continuously and automatically by gravity flow of tap water through packed layers of bulk salt.

During overnight dehydration (or a minimum of 3 hr at room temperature) cysts have released most of their cellular water and, upon interruption of the aeration, the now coffee-bean-shaped decapsulated cysts settle out. Cysts are drained off on a 120-μm screen, transferred to plastic containers, topped with fresh brine, and stored in the refrigerator or freezer. Since the water content of cysts in a NaCl-saturated brine ($\gtrsim$330 g/ℓ) approximates 20%, maximal cyst hatchability is guaranteed for storage periods of only a few months.[46] For long-term storage, water levels in the cysts should be below 10% which is assured when using $MgCl_2$ brine (1670 g $MgCl_2 \cdot 6\ H_2O$/ℓ tap water).[49] Prior to their use, either as a direct food source or for hatching, the dehydrated decapsulated cysts are filtered off on a 120-μm

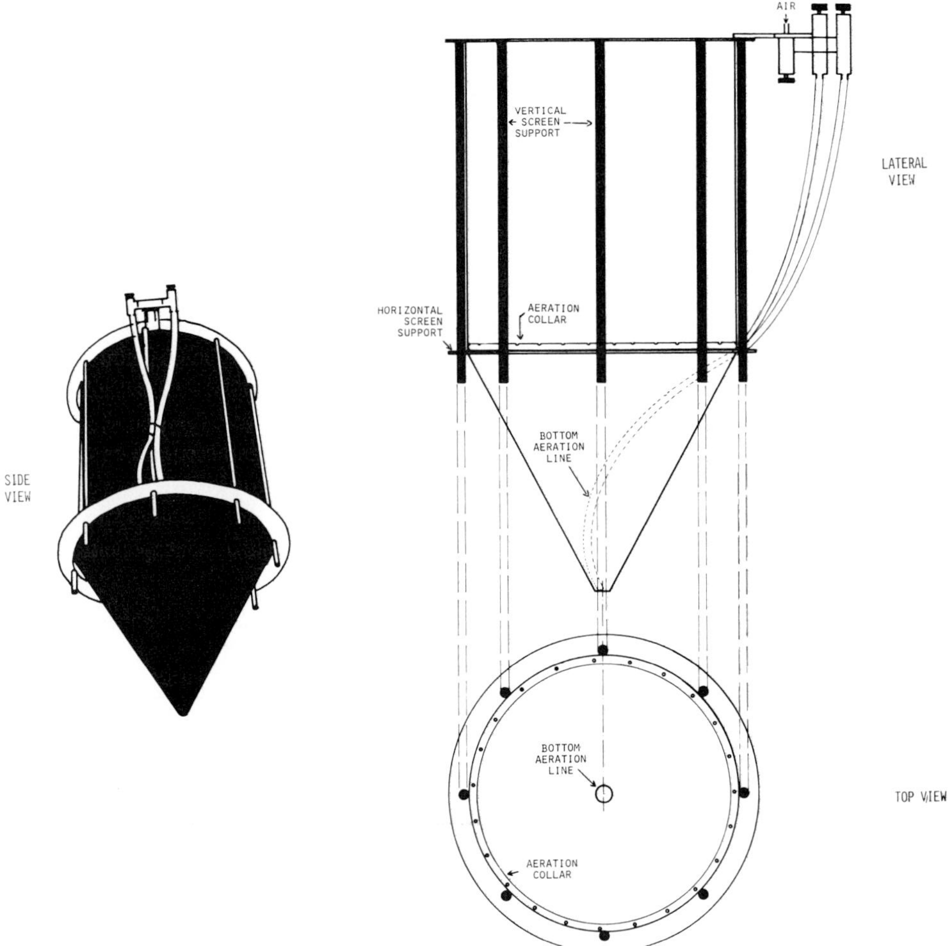

FIGURE 8. Schematic drawings of special decapsulation container. (From Bruggeman, E., Sorgeloos, P., and Vanhaecke, P., *The Brine Shrimp Artemia,* Vol. 3, Persoone, G., Sorgeloos, P., Roels, O., and Jaspers, E., Eds., Universa Press, Wetteren, Belgium, 1980, 261. With permission.)

screen and rinsed with tap water in order to remove all traces of brine. The other operations are under Direct Use of Decapsulated Cysts (see also Cyst Packaging Conditions).

Since decapsulated cysts lose their hatchability when exposed to UV light, it is advised to perform the decapsulation treatment and to keep the decapsulated cysts away from direct sunlight.

Routine decapsulations and especially treatments with large cyst quantities can be facilitated by use of a special decapsulation container (see Figure 8).[46] During the entire treatment cysts are kept in suspension in the cylindroconical stainless steel container which, as schematically outlined in Figure 9, is progressively transferred from one processing bath to another, i.e., freshwater hydration, hypochlorite decapsulation, freshwater washing, acid deactivation, freshwater washing, and finally brine dehydration. In the decapsulation step, the hypochlorite is kept at a temperature below 35°C by continuous circulation through a cooling element. In order to optimize both the circulation of the cysts within the container and the exchange of the internal solution with the external medium, the tank is equipped with two separate aeration systems. The first is a tube extending to the bottom of the funnel, and the second is a perforated tube installed in the lower part of the cylinder acting as an

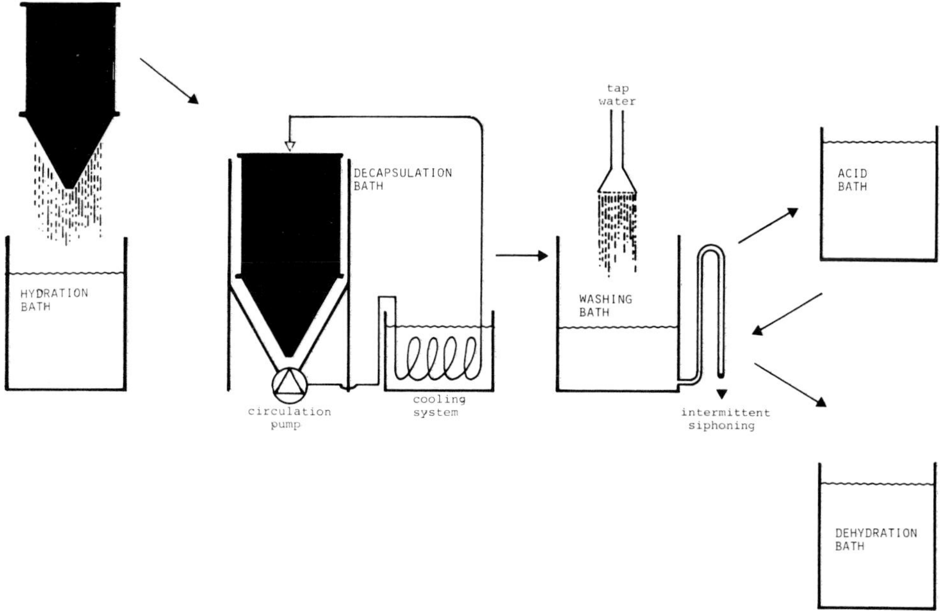

FIGURE 9. Schematic diagram of consecutive steps in the decapsulation treatment with cylindroconical decapsulation container. (From Bruggeman, E., Sorgeloos, P., and Vanhaecke, P., *The Brine Shrimp Artemia,* Vol. 3, Persoone, G., Sorgeloos, P., Roels, O., and Jaspers, E., Eds., Universa Press, Wetteren, Belgium, 1980, 261. With permission.)

aeration collar. The prototype container shown in Figure 8 is proportioned to treat 1 kg of cysts.

INTENSIVE CULTURING OF *ARTEMIA* FROM NAUPLIUS TO ADULT STAGE

Batch Production without Water Renewal

Of the various techniques which we have tested out for growing *Artemia* larvae in batch culture from nauplius to adult without any water renewal, the air-water lift (AWL) operated raceway, originally described by Mock et al. to culture postlarval *Penaeus,* proved to be the most suitable.[50-53] The construction and operation of the culturing tank is, indeed, simple and high production results can be obtained.[54,55]

An *Artemia* raceway essentially consists of a rectangular tank with a central partitioning to which the AWLs can be fixed. The corners of the tank may be curved, although this is not absolutely necessary. In order to assure an optimal water circulation the central partitioning should not be closer to the wall of the small side than 1 to 1/5 of the channel width. The most important parameter for the configuration of the tank is the height/width ratio, which should be kept close to 1. For optimal water circulation, using axial blowers, the water depth should not exceed 1 m. Three types of AWL raceways which we use routinely are schematically outlined in Figure 10: a 300-ℓ polyethylene container, a 2-m^3 concrete tank, and a 5-m^3 tank, the basic structure of which consists of aluminum plates supported at the outside by angle irons; the latter two tanks are insulated with styrofoam, 5 cm thick, and made watertight with 0.85-mm PVC liner.

AWLs are constructed with sanitary PVC pipes and elbows in such a way that the lower part of the tubes, cut at an angle of 45°, stands on the bottom of the raceway, and that the elbow's outflow is half submerged (see Figure 11). Aeration lines consist of 6 to 10-mm

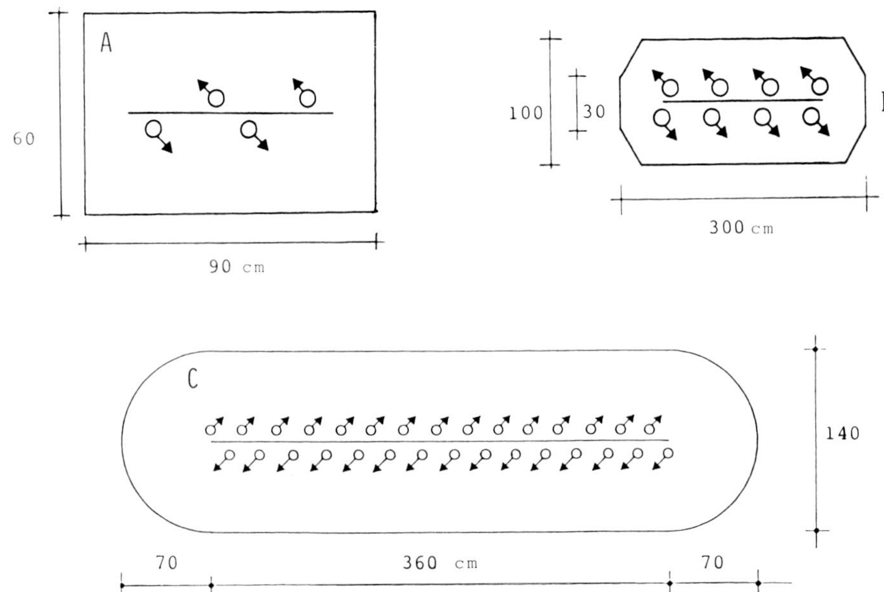

FIGURE 10. Top views and dimensions of three raceway systems from *Artemia* culturing. (A) 300-ℓ container: 60-cm water depth, diameter AWL 4 cm, ±7ℓ air/min/AWL; (B) 2-m^3 tank: 75-cm water depth, diameter AWL 5 cm, ±14 ℓ air/min/AWL; (C) 5-m^3 tank: 80-cm water depth, diameter AWL 5 cm, ±16 ℓ air/min/AWL. (From Bossuyt, E. and Sorgeloos, P., *The Brine Shrimp Artemia*, Vol. 3, Persoone, G., Sorgeloos, P., Roels, O., and Jaspers, E., Eds., Universa Press, Wetteren, Belgium, 1980, 133. With permission.)

polyethylene tubing that fits into a central air-distribution cylinder and extends into the AWL via an opening in its elbow. By making this opening slightly smaller than the outer diameter of the tubing, the aeration line can be raised or lowered at will and its tight fitting in the elbow prevents any undesired displacement. The aeration lines should extend as deep as possible in the AWLs to assure the best water lift effect.

The optimal culture temperature for brine shrimp is situated in the range of 25 to 30°C. Heaters and thermostats can be directly immersed in the culture water. In the 2-m^3 raceway a copper coil tubing, covered by an aluminum plate 1 mm thick, is installed underneath the PVC liner on the bottom of the raceway. A thermostat immersed in the culture medium controls pumping of freshwater from a 50°C boiler through the heat exchanger in a closed system. In the 5-m^3 raceway, we have combined the function of heat-exchanger and central partitioning by using plate radiators, commonly used in domestic central heating but epoxy coated. Water evaporation, which is an important cause of heat losses, can be greatly reduced by placing an insulated cover on top of the raceway; the nontransparent cover is, furthermore, beneficial for food conversion since brine shrimp grow faster in darkness than in light.[56]

Natural seawater enriched with bicarbonate (2 g $NaHCO_3$/ℓ) or artificial seawater prepared following the formula in Table 4 is used as the culture medium.

Although the intensive culture techniques outlined here are applicable for other strains of brine shrimp, we have found that best production results are obtained with Great Salt Lake (Utah) *Artemia*.[55,57] Raceway cultures are inoculated in the late afternoon with freshly hatched nauplii at a rate of 10,000/ℓ. Attention is paid to minimize contamination with empty shells which provide substrate for development of bacterial floccules.

A wide range of live and inert products can be successfully used to culture brine shrimp;[40,58] nonetheless, suitable feeds for intensive culturing in AWL raceways are limited to those products that are available under the form of a dry powder or at least a wet paste, have a

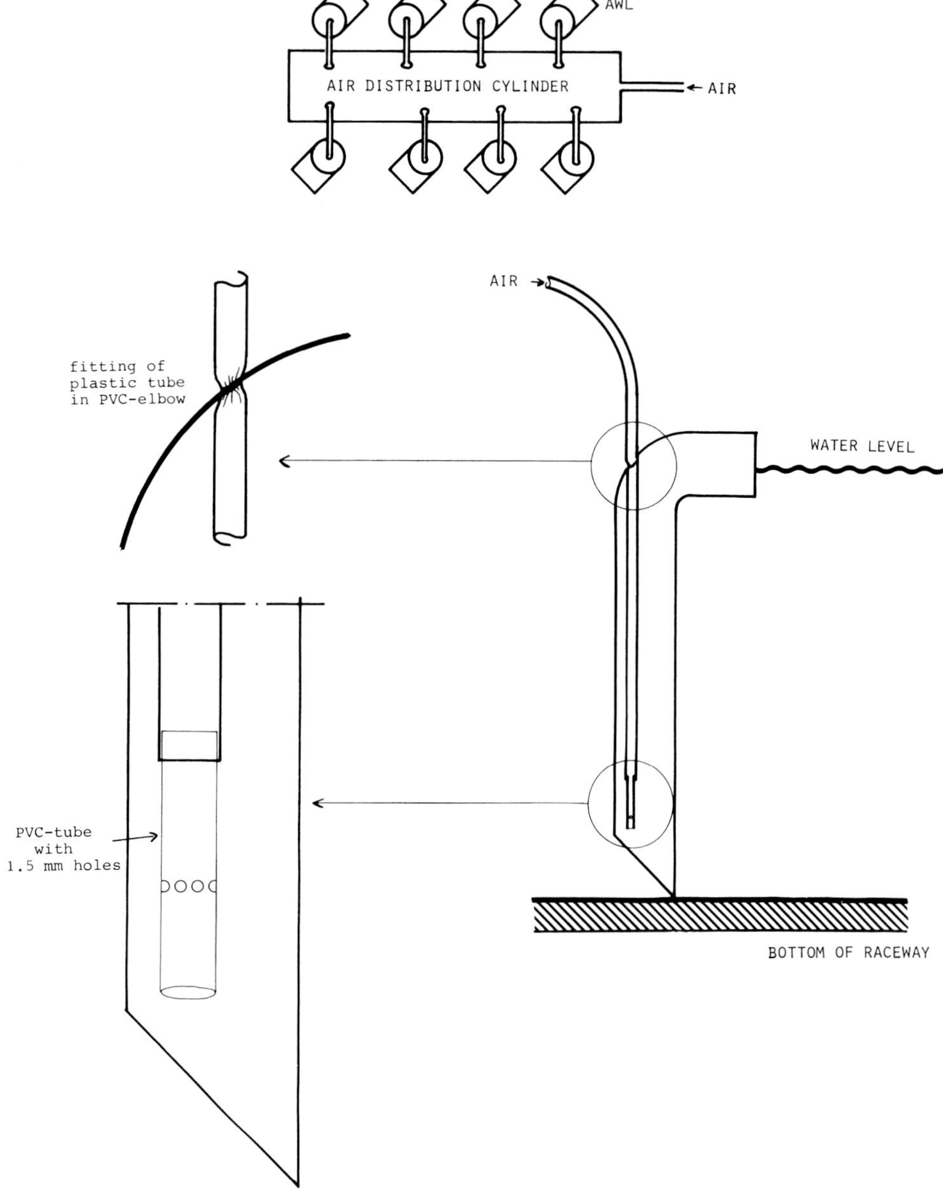

FIGURE 11. Details on the installation of air lines in AWL raceways. (Modified from Bossuyt, E. and Sorgeloos, P., *The Brine Shrimp Artemia,* Vol. 3, Persoone, G., Sorgeloos, P., Roels, O., and Jaspers, E., Eds., Universa Press, Wetteren, Belgium, 1980, 133. With permission.)

particle size smaller than 50 μm, and have but a minimal solubility in the culture medium,[54] e.g., micronized rice bran, spray-dried *Spirulina*, and other dried algae, yeasts, etc. The information given below is valid for micronized rice bran with which we have the most experience.[54,55,58,59]

Since *Artemia* is a continuous filter-feeder, the fastest growth and the most efficient food conversion are obtained at constant food densities. The transparency of the culture medium was found to be a very useful parameter for the determination of the optimal food dose.[31] With rice bran suspension transparency levels should be maintained within the range 15 to

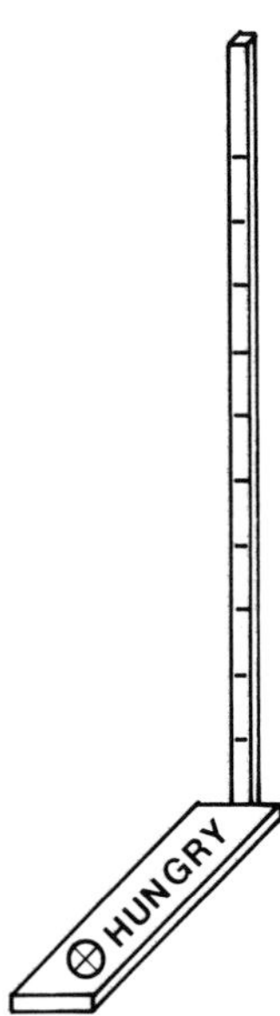

FIGURE 12. Transparency stick for use in *Artemia* culturing. (Modified from Bossuyt, E. and Sorgeloos, P., *The Brine Shrimp Artemia,* Vol. 3, Persoone, G., Sorgeloos, P., Roels, O., and Jaspers, E., Eds., Universa Press, Wetteren, Belgium, 1980, 133. With permission.)

20 cm. In practice this can be estimated with a transparency stick, in fact, a type of Secchi disc (Figure 12); i.e., the stick is submerged and lowered until the bottom plate is just visible; rice bran suspension is then added until the transparency measured with the stick is approximately 15 cm. Since food densities have to be kept constant throughout day and night, distributions of food should be frequent (at least once an hour) and, thus, automated. This can be achieved with timer-controlled peristaltic pumps, small aquarium pumps, etc. A cheap and reliable system that we have used for several years on various types of AWL raceways is represented schematically in Figure 13. Micronized rice bran is suspended in saturated NaCl brine at a rate of 50 to 100 g/ℓ in a conically or V-shaped container (e.g., same container as used for hatching cysts, see Figure 4). This suspension is gently aerated in order to prevent bacterial decomposition and particle sedimentation in the food stock. A siphon, curved at the end, delivers food suspension to a T-piece, one end of which is connected to an air inlet, the other one to the food distribution tube extending above the water level in the food reservoir. An electric clock activates the air pump at preset time intervals and for preset time periods. This triggers food distribution to the raceway. Full automatic food distribution can be achieved with an electronic transparency meter as described by Versichele et al.[60]

Since fecal pellets and exuviae physically hamper food uptake by the *Artemia* and affect the water quality, particulate wastes are continuously removed from the culture medium from day 4 onwards. This primary water treatment is performed with a modified version of the plate separator developed by Mock et al.[51] for solid waste removal from postlarval *Penaeus* cultures. Schematic drawings and characteristics of the plate separator that we use to treat a 2-m^3 AWL raceway are given in Figure 14 and Table 8, respectively. The separator consists of a rectangular tank subdivided into a small inflow and a large settling compartment

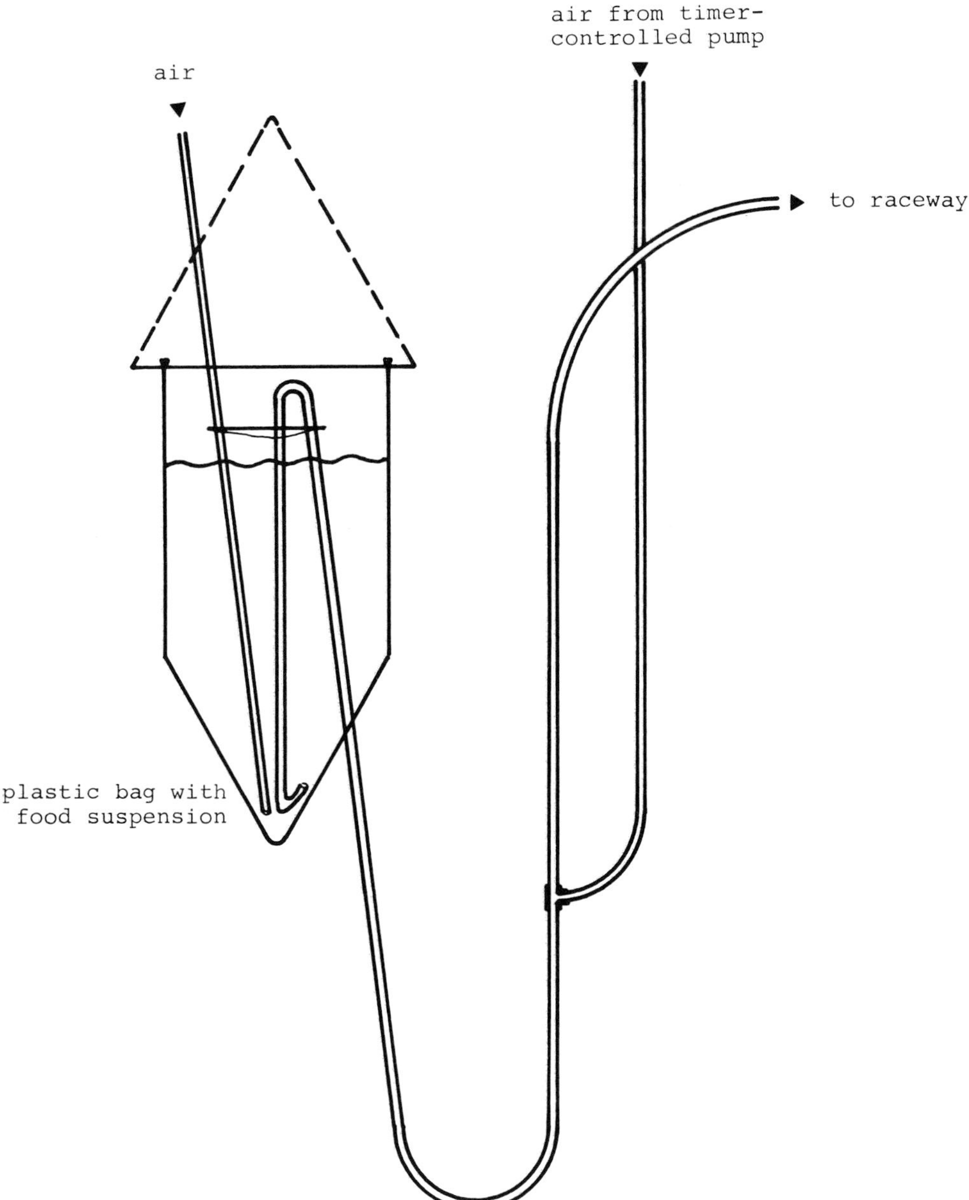

FIGURE 13. Schematic diagram of air-lift operated food distribution system. (Modified from Bossuyt, E. and Sorgeloos, P., *The Brine Shrimp Artemia*, Vol. 3, Persoone, G., Sorgeloos, P., Roels, O., and Jaspers, E., Eds., Universa Press, Wetteren, Belgium, 1980, 133. With permission.)

interconnected by an open space at the sloped bottom part of the tank. In the settling compartment several plates with a rough surface are mounted in an inclined position oriented in the direction of the waterflow through the separator tank. Sedimentation of the waste is optimized by adjusting the retention time of the suspension in the separator tank. Working at 20- to 30-min retention time, fecal pellets and exuviae settle out while small food particles remain in suspension and are drained back into the culture medium. Once every other day accumulated wastes are siphoned out. In order to assure that only suspended particles and no *Artemia* are pumped into the plate separator a filter-screen system is installed inside the

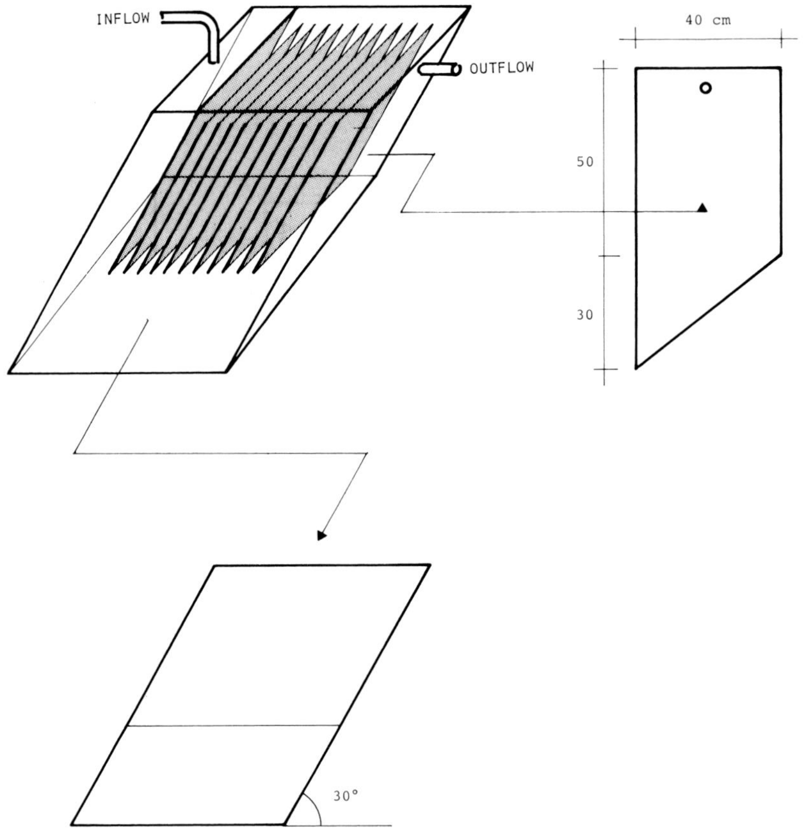

FIGURE 14. Different views of plate separator used for primary water treatment of 2-m³ *Artemia* raceway. (Modified from Bossuyt, E. and Sorgeloos, P., *The Brine Shrimp Artemia,* Vol. 3, Persoone, G., Sorgeloos, P., Roels, O., and Jaspers, E., Eds., Universa Press, Wetteren, Belgium, 1980, 133. With permission.)

Table 8
CHARACTERISTICS OF PLATE SEPARATOR USED FOR PRIMARY WATER TREATMENT OF 2-m³ *ARTEMIA* RACEWAY[a]

Dimensions	
Length	60 cm
Width	40 cm
Depth	50—80 cm
Number of plates	11
Inclination of plates	30°
Inclination of bottom part	30—40°
Inflow rate	6 ℓ/min
Retention time	25 min

[a] Data compiled from Bossuyt and Sorgeloos.[54]

culture tank. We use a PVC-frame construction, as schematically outlined in Figure 15, over which a filter-screen bag made of nylon can be glided. This filter-screen system fits into an aeration collar that is made of PVC tubing and is fixed to the bottom of the raceway.

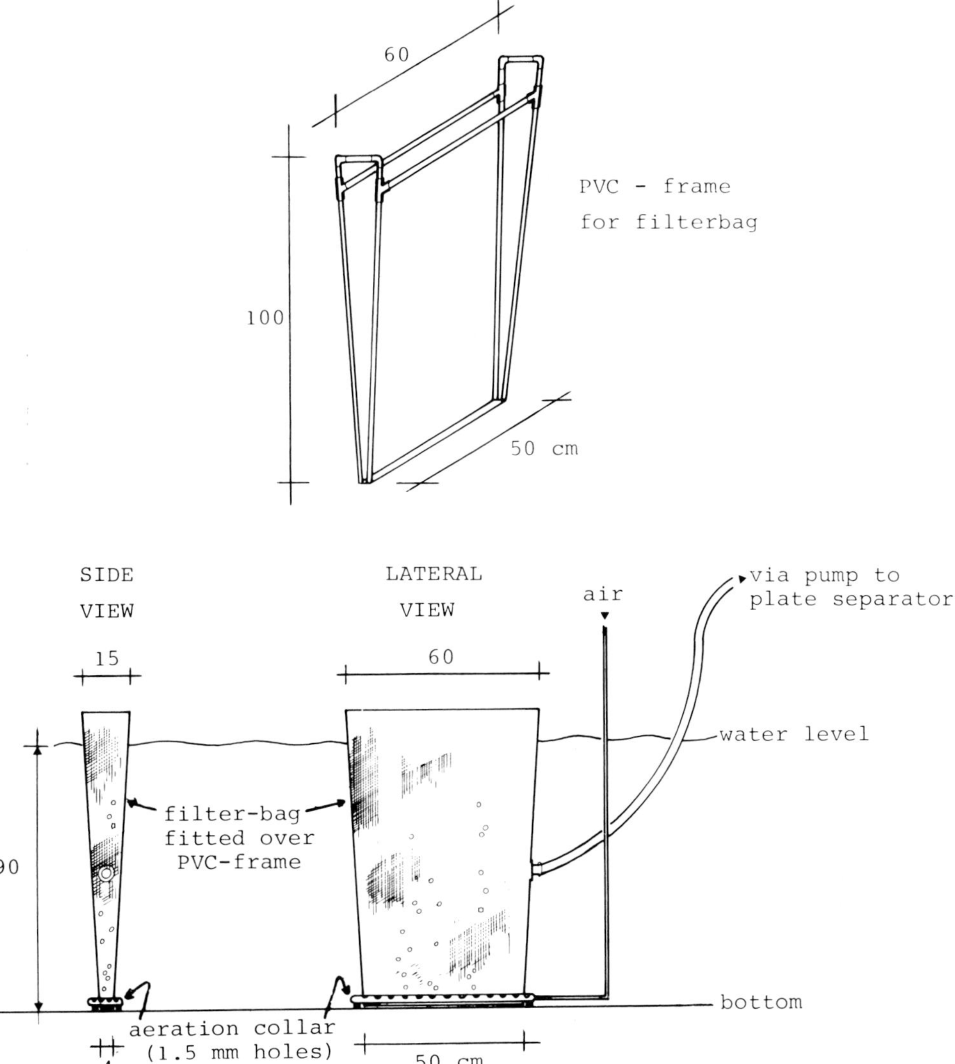

FIGURE 15. Different views of filter-screen system used for primary water treatment of 2-m^3 *Artemia* raceway. (Modified from Brisset, P., Versichele, D., Bossuyt, E., De Ruyck, L., and Sorgeloos, P., *Aquacultural Eng.*, 1, 115, 1982. With permission.)

The continuous rising of a curtain of air bubbles against the sides of the filter bag efficiently reduces clogging. As the animals grow we progressively switch to filter bags with a larger mesh width, i.e., 200, 250, 350, and 450 μm.

From day 6 onward, when the animal's thoracopods become functional and as a consequence their filter-feeding efficiency greatly increases, transparency levels, now measured inside the filter system, are allowed to fluctuate within the range of 20 to 25 cm. Dissolved oxygen and pH are monitored regularly; the aeration rates are increased or more $NaHCO_3$ is added when DO or pH drops below 2 mg/ℓ and 7.5, respectively.

After 2 weeks of culturing, larvae have reached an average length of ≳8 mm. Average production yields amount to 5 kg wet *Artemia* per cubic meter of culture media with gross food consumption of ≳4 kg rice bran (dry micronized product). Harvesting of the preadults can be facilitated by taking advantage of the special "surface respiration" behavior in

Artemia,[61] i.e., the raceway's aeration is turned off and the DO level is allowed to drop progressively in the undisturbed culture. After approximately 30 min, the oxygen concentration in the culture medium has become critical and the *Artemia* concentrate in very dense numbers at the water's surface from which they can be easily harvested with a scoop net.

Details on the biochemistry of cultured *Artemia* are given by Dobbeleir et al.[58] and Sorgeloos et al.[59] Amino acid levels are very similar to those found in natural brine shrimp. Fatty acid patterns, however, are very different but can be modified by diet manipulations at the end of the production period.[58] In culture tests with lobster larvae it was proven that *Artemia* fed with a monodiet of rice bran is as good a food as brine shrimp harvested from the wild.[62] Further tests, however, are needed to evaluate the nutritional value for different predator species of *Artemia* cultured with inert feeds.

Flow-Through Culturing

It is obvious that a continuous renewal of culture water, with consequent removal of all particulate and dissolved metabolites, will lead to high production capacities with brine shrimp. Application of the flow-through culture technique, however, is limited to those situations where large volumes of sufficiently warm seawater (or brine) are available at a reasonable price, e.g., desalination plants, thermal effluents from power plants and geothermal wells, artificial upwelling projects, tertiary treatment systems, etc.

The most important equipment in the flow-through culture tank is the special filter system that has to restrain the larvae in the culture tank. We have initially worked with cylindrical units but have recently switched to the inclined rectangular filter which is more efficient in operation and better suited for scaling up.[31,63-65] Its construction has been outlined earlier (see Batch Production without Water Renewal and Figure 15). This filter system is installed in the center of the culture tank. Aside from being the holder for the filter system, the aeration collar performs three functions: the rising air bubbles clean the filter screen, aerate the culture water, and keep food and larvae in homogeneous suspension.

Cultures are set up at naupliar densities of 20,000/ℓ. The retention time of culture water within the *Artemia* tank is set at a rate that assures efficient removal of fecal pellets, i.e., starting at 4-hr retention time, the water inflow rate is gradually increased over the 2-week culturing period to end at a constant rate of 1-hr retention time from day 10 onwards. Cultures are set up with a filter bag of 130 μm mesh width. As the animals grow it is critical to switch to progressively larger mesh sizes, i.e., filter bags are replaced on day 3, 6, and 9 and mesh widths used are 225, 300, and 400 μm, respectively.

The spectrum of potentially suitable feeds for flow-through culturing of *Artemia* is larger than for batch culturing without water renewal; e.g., feeds with a high solubility in water such as whey and soybean can be considered. In general, feeding strategies and techniques are similar to those described under Batch Production without Water Renewal. Concentrated food suspension is added at a minimum of 15-min intervals, amounts and rates being controlled by transparency readings made inside the filter system.

Although our experience with flow-through culturing is much more limited compared to batch production without water renewal, we already achieve harvests of 25 kg *Artemia* per cubic meter tank after 2 weeks culturing at 25°C; 18-kg micronized rice bran was used as a food source (conversion efficiency of 0.7 to 1) and about 150-m^3 seawater as culture medium.

Marine algae cultures can be used as a combined culture medium and food source in flow-through culturing of *Artemia*.[63] A list of suitable algal species was reported by D'Agostino.[66] As outlined in detail by Tobias et al.,[63] retention times of the algal culture in the *Artemia* tank have to be adjusted in proportion to the cell concentration in the culture's effluent, i.e., the most efficient use of the algal food is attained when the cell concentration in the effluent approximates the so-called "critical minimum cell concentration" which is specific to the algal species. Production yields are limited by the cell density in the algal culture; e.g., in the St. Croix Artificial Upwelling Project retention times at the end of the 2-week culturing period had to be lowered to a few minutes only to assure enough food for an *Artemia* density of 18,000/ℓ. Maximal harvest averaged 25 kg/m^3 for a total volume of $\gtrsim$4000-m^3 culture of *Chaetoceros curvisetus* (STX-167) at a cell density of 4.5 × 10^4/mℓ.[67]

REFERENCES

1. **Sorgeloos, P.,** The use of the brine shrimp *Artemia* in aquaculture, in *The Brine Shrimp Artemia,* Vol. 3, Persoone, G., Sorgeloos, P., Roels, O., and Jaspers, E., Eds., Universa Press, Wetteren, Belgium, 1980, 25.
2. **Sorgeloos, P.,** Live animal food for larval rearing in aquaculture: the brine shrimp *Artemia,* in *Realism in Aquaculture: Achievements, Constraints, Perspectives,* Bilio, M., Rosenthal, H., and Lindermann, C. J., Eds., in press.
3. **Glude, J. B.,** *The Freshwater Prawn Macrobrachium rosenbergii (de Man). A Literature Review and Analysis of the Use of Thermal Effluents in the Culture of the Freshwater Prawn,* Glude, J. B., Ed., Seattle, Wash., 1978.
4. **Glude, J. B.,** *The Marine Shrimp Penaeus spp. A Literature Review and Analysis of the Use of Thermal Effluents in the Culture of Penaeid Shrimp,* Glude, J. B., Ed., Seattle, Wash., 1978.
5. **Kinne, O.,** *Marine Ecology,* Vol. 3, Part 2, John Wiley & Sons, New York, 1977.
6. **Mock, C. R., Fontaine, C. T., and Revera, D. B.,** Improvements in rearing larval penaeid shrimp by the Galveston Laboratory method, in *The Brine Shrimp Artemia,* Vol. 3, Persoone, G., Sorgeloos, P., Roels, O., and Jaspers, E., Eds., Universa Press, Wetteren, Belgium, 1980, 331.
7. **Laviña, A.,** unpublished data, 1978.
8. **Laviña, E. and Figueroa, R.,** The use of decapsulated brine shrimp eggs as food for shrimp larvae, manuscript to be submitted for publication.
9. **Royan, J. P.,** Decapsulated brine shrimp cysts — an ideal feed for shrimps in aquaculture, *Indian J. Mar. Sci.,* 9, 125, 1980.
10. **Rodriguez Martin, A. and Rodriguez, R.,** personal communication, 1981.
11. **Palmegiano, G. B. and Trotta, P.,** *Artemia salina* as pabulum for growing penaeids under laboratory conditions, poster paper presented at the World Conference on Aquaculture, Venice, September 21 to 25, 1981.
12. **Van Olst, J. C., Carlberg, J. M., and Hughes, J. T.,** Aquaculture, in *The Biology and Management of Lobsters,* Vol. 2, Cobb, J. S. and Phillips, B. F., Eds., Academic Press, New York, 1980, 333.
13. **Glude, J. B.,** *The American Lobster Homarus americanus Milne-Edwards. A Literature Review and Analysis of the Use of Thermal Effluent in the Culture of the American Lobster,* Glude, J. B., Ed., Seattle, Wash., 1978.
14. **Sorgeloos, P.,** Improvements in availability and use of *Artemia* as food source for *Macrobrachium,* in *Giant Prawn 1980,* Provisional Report No. 9, International Foundation for Science, Stockholm, Sweden, 1980, 123.
15. **Sorgeloos, P., Persoone, G., Baeza-Mesa, M., Bossuyt, E., and Bruggeman, E.,** The use of *Artemia* cysts in aquaculture: the concept of "hatching efficiency" and description of a new method for cyst processing, in *Proc. 9th Annual Meeting World Mariculture Society,* Avault, J. W., Jr., Ed., Louisiana State University, Baton Rouge, 1978, 715.
16. **Vanhaecke, P. and Sorgeloos, P.,** International Study on *Artemia.* XIX. Hatching data for 10 commercial sources of brine shrimp cysts and re-evaluation of the "hatching efficiency" concept, *Aquaculture,* 30, 43, 1983.
17. **Vanhaecke, P. and Sorgeloos, P.,** International Study on *Artemia.* XVIII. The hatching rate of *Artemia* cysts — a comparative study, *Aquacultural Eng.,* 1, 263, 1982.
18. **Vanhaecke, P. and Sorgeloos, P.,** International Study on *Artemia.* IV. The biometrics of *Artemia* strains from different geographical origin, in *The Brine Shrimp Artemia,* Vol. 3, Persoone, G., Sorgeloos, P., Roels, O., and Jaspers, E., Eds., Universa Press, Wetteren, Belgium, 1980, 393.
19. **Vanhaecke, P. and Sorgeloos, P.,** International Study on *Artemia.* XXX. Bio-economical evaluation of the nutritional value of 9 *Artemia* strains for carp larvae *(Cyprinus carpio* L.), *Aquaculture,* in press.

20. **Beck, A. D. and Bengtson, D. A.,** International Study on *Artemia.* XXII. Nutrition in aquatic toxicology: diet quality of geographical strains of the brine shrimp *Artemia* ssp., in *Aquatic Toxicology and Hazard Evaluation: Fifth Conference,* Pearson, J. G., Forster, R. B., and Bishop, W. E., Eds., American Society for Testing and Materials, STP 766, Philadelphia, 1981, 139.
21. **Léger, Ph., Vanhaecke, P., Simpson, K. L., and Sorgeloos, P.,** International Study on *Artemia.* XXV. Nutritional value of *Artemia* nauplii from suspected San Francisco Bay brand sources for the mysid *Mysidopsis bahia* (Molenock), manuscript to be submitted for publication.
22. **Johns, D. M., Berry, W. J., and Walton, W.,** International Study on *Artemia.* XVI. Survival, growth and reproductive potential of the mysid, *Mysidopsis bahia* Molenock fed various geographical collections of the brine shrimp, *Artemia, J. Exp. Mar. Biol. Ecol.,* 53, 209, 1981.
23. **Johns, D. M., Peters, M. E., and Beck, A. D.,** International Study on *Artemia.* VI. Nutritional value of geographical and temporal strains of Artemia: Effects on survival and growth of two species of Brachyuran larvae, in *The Brine Shrimp Artemia,* Vol. 3, Persoone, G., Sorgeloos, P., Roels, O., and Jaspers, E., Eds., Universa Press, Wetteren, Belgium, 1980, 291.
24. **Fujita, S., Watanabe, T., and Kitajima, C.,** Nutritional quality of *Artemia* from different localities as a living feed for marine fish from the viewpoint of essential fatty acids, in *The Brine Shrimp Artemia,* Vol. 3, Persoone, G., Sorgeloos, P., Roels, O., and Jaspers, E., Eds., Universa Press, Wetteren, Belgium, 1980, 277.
25. **Olney, C. E., Schauer, P. S., McLean, S., You Lu, and Simpson, K. L.,** International Study on *Artemia.* VIII. Comparison of the chlorinated hydrocarbons and heavy metals in five different strains of newly hatched *Artemia* and a laboratory-reared marine fish, in *The Brine Shrimp Artemia,* Vol. 3, Persoone, G., Sorgeloos, P., Roels, O., and Jaspers, E., Eds., Universa Press, Wetteren, Belgium, 1980, 343.
26. **Schauer, P. S., Johns, D. M., Olney, C. E., and Simpson, K. L.,** International Study on *Artemia.* IX. Lipid level, energy content and fatty acid composition of the cysts and newly hatched nauplii from five geographical strains of *Artemia,* in *The Brine Shrimp Artemia,* Vol. 3, Persoone, G., Sorgeloos, P., Roels, O., and Jaspers, E., Eds., Universa Press, Wetteren, Belgium, 1980, 365.
27. **Seidel, D. R., Johns, D. M., Schauer, P. S., and Olney, C. E.,** International Study on *Artemia.* XXVI. The value of the nauplii from Reference *Artemia* Cysts and four geographical collections of *Artemia* as a food source for mud crab, *Rhithropanopeus harrissi* larvae, *Mar. Ecol. Prog. Ser.,* submitted for publication.
28. **Blust, R.,** Het voorkomen van Fe, Cu en Ca in *Artemia salina* L., thesis, University of Antwerp (UIA), Belgium, 1981.
29. **Jones, A. and Houde, E.,** Mass rearing of fish fry for aquaculture, in *Realism in Aquaculture: Achievements, Constraints, Perspectives,* Bilio, M., Rosenthal, H., and Lindermann, C. J., Eds., in press.
30. **Sorgeloos, P.,** Availability of Reference *Artemia* Cysts, *Aquaculture,* 23, 381, 1981.
31. **Sorgeloos, P.,** The Culture and Use of Brine Shrimp *Artemia salina* as Food for Hatchery-raised Larval Prawns, Shrimp and Fish in South East Asia, FAO Rep., THA/75/008/78/WP3, 1978.
32. **Bowen, S. T.,** Genetics of *Artemia salina.* III. Effects of X-irradiation and of freezing upon cysts, *Biol. Bull.,* 125, 431, 1963.
33. **Hessinger, D. A. and Hessinger, J. A.,** Methods for rearing sea anemones in the laboratory, in *Marine Invertebrates,* National Academy of Sciences, Washington, D.C., 1981, 153.
34. **Sorgeloos, P.,** First report on the triggering effect of light on the hatching mechanism of *Artemia salina* dry cysts, *Mar. Biol.,* 22, 75, 1973.
35. **Vanhaecke, P., Cooreman, A., and Sorgeloos, P.,** International Study on *Artemia.* XV. Effects of light intensity on hatching rate of *Artemia* cysts from different geographical origin, *Mar. Ecol. Prog. Ser.,* 5, 111, 1981.
36. **Benijts, F., Vandeputte, G., and Sorgeloos, P.,** Energetic aspects of the metabolism of hydrated *Artemia* cysts, in *Fundamental and Applied Research on the Brine Shrimp, Artemia salina* (L.) *in Belgium,* Jaspers, E. and Persoone, G., Eds., European Mariculture Society Spec. Publ. No. 2, Bredene, Belgium, 1977, 79.
37. **Léger, Ph. and Sorgeloos, P.,** Automation in stock-culture maintenance and juvenile separation of the mysid *Mysidopsis bahia* (Molenock), *Aquaculture Eng.,* 1, 42, 1982.
38. **De los Santos, C., Jr., Sorgeloos, P., Laviña, E., and Bernardino, A.,** Successful inoculation of *Artemia* and production of cysts in man-made salterns in the Philippines, in *The Brine Shrimp Artemia,* Vol. 3, Persoone, G., Sorgeloos, P., Roels, O., and Jaspers, E., Eds., Universa Press, Wetteren, Belgium, 1980, 159.
39. **Persoone, G. and Sorgeloos, P.,** An improved separator box for *Artemia* nauplii and other phototactic invertebrates, *Helgoländer Wiss. Meeresunters.,* 23, 243, 1972.
40. **Sorgeloos, P. and Persoone, G.,** Technological improvements for the cultivation of invertebrates as food for fishes and crustaceans. II. Hatching and culturing of the brine shrimp, *Artemia salina* L., *Aquaculture,* 6, 303, 1975.
41. **Clegg, J. S.,** The control of emergence and metabolism by external osmotic pressure and the role of free glycerol in developing cysts of *Artemia salina, J. Exp. Biol.,* 41, 879, 1964.

42. **Gilmour, A., McCallum, M. F., and Allan, M. C.,** Antibiotic sensitivity of bacteria isolated from the canned eggs of the Californian brine shrimp *(Artemia salina), Aquaculture,* 6, 221, 1975.
43. **Léger, Ph., Vanhaecke, P., and Sorgeloos, P.,** International Study on *Artemia.* XXIV. Cold storage of live *Artemia* nauplii from various geographical sources: Potentials and limits in aquaculture, *Aquacultural Eng.,* in press.
44. **Sorgeloos, P., Bossuyt, E., Laviña, E., Baeza-Mesa, M., and Persoone, G.,** Decapsulation of *Artemia* cysts: a simple technique for the improvement of the use of brine shrimp in aquaculture, *Aquaculture,* 12, 311, 1977.
45. **Wheeler, R., Yudin, A. I., and Clark, W. H., Jr.,** Hatching events in the cysts of *Artemia salina, Aquaculture,* 18, 59, 1979.
46. **Bruggeman, E., Sorgeloos, P., and Vanhaecke, P.,** Improvements in the decapsulation technique of *Artemia* cysts, in *The Brine Shrimp Artemia,* Vol. 3, Persoone, G., Sorgeloos, P., Roels, O., and Jaspers, E., Eds., Universa Press, Wetteren, Belgium, 1980, 261.
47. **Bruggeman, E., Baeza-Mesa, M., Bossuyt, E., and Sorgeloos, P.,** Improvements in the decapsulation of *Artemia* cysts, in *Cultivation of Fish Fry and Its Live Food,* Styczynska-Jurewicz, E., Backiel, T., Jaspers, E., and Persoone, G., Eds., European Mariculture Society Spec. Publ. No. 4, Bredene, Belgium, 1980, 309.
48. **Spotte, S.,** *Fish and Invertebrate Culture. Water Management in Closed Systems,* Wiley-Interscience, New York, 1970
49. **Clegg, J. S. and Cavagnaro, J.,** Interrelationships between water and cellular metabolism in *Artemia* cysts. IV. ATP and cyst hydration, *J. Biophys. Biochem. Cytol.,* 88, 159, 1976.
50. **Mock, C. R., Neal, R. A., and Salser, B. R.,** A closed raceway for the culture of shrimp, in *Proc. 4th Annual Meeting World Mariculture Society,* Avault, J. W., Jr., Ed., Louisiana State University, Baton Rouge, 1973, 247.
51. **Mock, C. R., Ross, L. R., and Salser, B. R.,** Design and preliminary evaluation of a closed system for shrimp culture, in *Proc. 8th Annual Meeting World Mariculture Society,* Avault, J. W., Jr., Ed., Louisiana State University, Baton Rouge, 1977, 335.
52. **Sorgeloos, P., Baeza-Mesa, M., Bossuyt, E., Bruggeman, E., Persoone, G., Uyttersprot, G., and Versichele, D.,** Automatized High Density Culturing of the Brine Shrimp, *Artemia salina L.,* presented at the 8th Annual Workshop of the World Mariculture Society, San Jose, January 9 to 13, 1977, unpublished manuscript.
53. **Sorgeloos, P., Baeza-Mesa, M., Benijts, F., Bossuyt, E., Bruggeman, E., Claus, C., Persoone, G., Vandeputte, G., and Versichele, D.,** The use of the brine shrimp, *Artemia salina* in aquaculture, *Actes Colloq. C.N.E.X.O.,* 4, 21, 1977.
54. **Bossuyt, E. and Sorgeloos, P.,** Technological aspects of the batch culturing of *Artemia* in high densities, in *The Brine Shrimp Artemia,* Vol. 3, Persoone, G., Sorgeloos, P., Roels, O., and Jaspers, E., Eds., Universa Press, Wetteren, Belgium, 1980, 133.
55. **Bossuyt, E. and Sorgeloos, P.,** Batch Production of Adult *Artemia* in 2 m^3 and 5 m^3 Air-Water-Lift Operated Raceways, poster paper presented at the World Conference on Aquaculture, Venice, September 21 to 25, 1981.
56. **Sorgeloos, P.,** The influence of light on the growth rate of larvae of the brine shrimp, *Artemia salina L., Biol. Jaarb.,* 40, 317, 1972.
57. **Vanhaecke, P. and Sorgeloos, P.,** International Study on *Artemia.* XIV. Growth and survival of *Artemia* larvae of different geographical origin in a standard culture test, *Mar. Ecol. Prog. Ser.,* 3, 303, 1980.
58. **Dobbeleir, J., Adam, N., Bossuyt, E., Bruggeman, E., and Sorgeloos, P.,** New aspects of the use of inert diets for high density culturing of brine shrimp, in *The Brine Shrimp Artemia,* Vol. 3, Persoone, G., Sorgeloos, P., Roels, O., and Jaspers, E., Eds., Universa Press, Wetteren, Belgium, 1980, 165.
59. **Sorgeloos, P., Baeza-Mesa, M., Bossuyt, E., Bruggeman, E., Dobbeleir, J., Versichele, D., Laviña, E., and Bernardino, A.,** The culture of *Artemia* on ricebran: the conversion of a wasteproduct into highly nutritive animal protein, *Aquaculture,* 21, 393, 1980.
60. **Versichele, D., Bossuyt, E., and Sorgeloos, P.,** An inexpensive turbidimeter for the automatic culturing of filter-feeders, in *Cultivation of Fish Fry and Its Live Food,* Styczynska-Jurewicz, E., Backiel, T., Jaspers E., and Persoone, G., Eds., European Mariculture Society Spec. Publ. No. 4, Bredene, Belgium, 1980, 415.
61. **Sorgeloos, P.,** Het Gebruik van het Pekelkreeftje *Artemia* Spec. in de Aquakultuur, thesis, State University of Ghent, Belgium, 1979.
62. **Dobbeleir, J.,** Unpublished data, 1979.
63. **Tobias, W. J., Sorgeloos, P., Bossuyt, E., and Roels, O. A.,** The technical feasibility of mass-culturing *Artemia salina* in the St. Croix "artificial upwelling" mariculture system, in *Proc. 10th Annual Meeting World Mariculture Society,* Avault, J. W., Jr., Ed., Louisiana State University, Baton Rouge, 1979, 203.

64. **Tobias, W. J., Sorgeloos, P., Roels, O. A., and Sharfstein, B. A.,** International Study on *Artemia*. XIII. A comparison of production data of 17 geographical strains of *Artemia* in the St. Croix artificial upwelling-mariculture system, in *The Brine Shrimp Artemia,* Vol. 3, Persoone, G., Sorgeloos, P., Roels, O., and Jaspers, E., Eds., Universa Press, Wetteren, Belgium, 1980, 383.
65. **Brisset, P., Versichele, D., Bossuyt, E., De Ruyck, L., and Sorgeloos, P.,** High density flow through culturing of brine shrimp *Artemia* on inert feeds — preliminary results with new culture system, *Aquacultural Eng.,* 1, 115, 1982.
66. **D'Agostino, A. S.,** The vital requirements of *Artemia,* physiology and nutrition, in *The Brine Shrimp Artemia,* Vol. 2, Persoone, G., Sorgeloos, P., Roels, O., and Jaspers, E., Eds., Universa Press, Wetteren, Belgium, 1980, 55.
67. **Sharfstein, B. A., Tobias, W. J., and Roels, O. A.,** The technical feasibility of mass culturing of the brine shrimp, *Artemia* sp., in the St. Croix artificial upwelling system. II. Maximum stocking density and yields, and growth efficiency in flow-through raceway culture, manuscript to be submitted for publication.

THE CULTURE OF LARVAL FOODS AT THE OCEANIC INSTITUTE, WAIMANALO, HAWAII

Donald W. Anderson and Noel Smith

INTRODUCTION

The Oceanic Institute is located on the southeast point of the island of Oahu. It is a private institute that has been involved in developing aquaculture techniques for the production of crustaceans and finfish since the early 1970s. The Institute has developed hatcheries for marine shrimp and finfish that can be considered prototypes for future commercial hatcheries. It is one of the few facilities that has both Japanese-style as well as Galveston-style penaeid shrimp hatchery systems.

In order to support the shrimp and fish hatchery systems, several types of algae and rotifer culture systems have been tested. The following sections describe the systems presently in use for the production of algae and rotifers. These systems can be scaled up or down to meet the specific needs of any facility.

PHYTOPLANKTON CULTURE

The primary function of the Oceanic Institute Phytoplankton Program is the production of monocultures of phytoplankton for use as larval feeds in our marine shrimp, baitfish and mahimahi programs. The principal algae genera we work with are *Chaetoceros, Chlorella,* Tahitian *Isochrysis, Phaeodactylum, Tetraselmis* and *Thalassiosira*. We also culture *Spirulina* which we hope to use as an ingredient in processed aquaculture feeds.

The foundation of the program is our Algae Culture Room. Its function is to maintain stocks and to provide inoculum for outdoor tanks. This facility was designed so that several species could be raised simultaneously without fear of contamination. The room consists of six compartments off a main passageway inside our finfish hatchery building. Each compartment can be used for a different algae species and consists of a bank of algae cylinders on either side. Each bank consists of four 35-ℓ culture cylinders mounted directly above four 175-ℓ cylinders. All algal cylinders are 18 in. in diameter. Cylinders are made of fiberglass cones with a PVC valve on the bottom. The sides are made of alcinite. Small cylinders have sides which are 25.4 cm high, and large cylinders have sides which are 122 cm high. Removable cylinder lids are made of fiberglass. Each cylinder is aerated through an air line held down by a ceramic weight.

Each compartment is supplied with salt water, fresh water and an air conditioning duct. All plumbing is PVC. The seawater is filtered through an ultraviolet sterilizing filter. The room temperature is maintained at 18 to 22° C, and its total capacity is 10,000 ℓ.

Six, 2-cm-long high-intensity daylight fluorescent bulbs are mounted horizontally between each adjacent bank of cylinders. Each set of lights therefore illuminates two banks of cylinders. A sheet of glass is mounted on both sides of each set of lights. The glass acts as a wall between adjacent cylinder banks, preventing nearby cylinders from contaminating each other. The glass further acts as a protective barrier, protecting the lights from splashing salt water. This was a major problem in our old algae facility in which fixtures became corroded and electrical shortages occurred.

A pipe running the length of the room has a valve at each compartment. Cultures from the cylinder flow by gravity through this pipe to a continuous-flow centrifuge, and the shrimp and mahimahi hatcheries.

Each cylinder normally runs on a 3- to 5-day cycle. When a 35-ℓ cylinder is approaching maximum density, 5 ℓ are used to inoculate another 35-ℓ cylinder. The remaining 30 ℓ are dropped to the 175-ℓ cylinder below.

The larger algae species such as *Tetraselmis* are cultured to about 1.0×10^6 cells per milliliter. The smaller algae, such as *Chlorella* and *Chaetoceros*, normally have a peak density of 1.5 to 2.5×10^6 cells per milliliter.

A modified Guillards F/2 nutrient formula is used for all species except *Spirulina*. Silicates and vitamins are added to the diatom cultures. Most algae is raised in full strength seawater except for the *Chlorella* which is cultured at 20‰ salinity.

Pure stock cultures of all species are maintained in an enclosed cabinet with its own light source. The cultures are kept in 125-mℓ Erlenmeyer flasks. Transfers are made once a week using sterile techniques. Two-liter Erlenmeyer flasks are used for periodically starting small cylinders with a fresh culture. Then flasks are aerated with a pipette inserted through a rubber stopper in the mouth of the flask.

Two 5000-ℓ round fiberglass tanks are located in the baitfish hatchery, adjacent to the culture room. Each tank has two pairs of 8-ft lights overhead and is covered with a clear plexiglass lid. Aeration is through a ring of PVC pipe on the bottom of the tank.

These tanks are inoculated with approximately 500 ℓ of phytoplankton from the culture room. Seawater is continually added at a rate of about 1000 ℓ per day. The strategy is to maintain a constant cell density while increasing the volume. An F/4 nutrient media is used.

A 20 ft diameter fiberglass tank is located outside. Eventually this tank may be covered. It is inoculated using cultures from one or two 5000 ℓ tanks. The volume is brought up to 30,000 ℓ within 36 to 48 hrs. Seawater is then continually added at a rate of about 25,000 ℓ per day, depending upon the weather conditions. The culture should maintain a constant density for 7 to 10 days. The overflow leaves through a standpipe where it is then piped to the continuous-flow centrifuge. An F/8 media formula is used.

In the near future, most emphasis at the Oceanic Institute will be placed on the use of centrifuged algae. Most phytoplankton produced in our larger tanks will be centrifuged and stored either in a liquid form or as a frozen slurry. The programs will be able to use the product at their own pace.

ROTIFER CULTURE

The Oceanic Institute uses rotifers extensively in hatchery operations. Hatchery operations for marine shrimp and finfish create a great need for larval feeds. Rotifer production runs through the year and plays an important part in the success of these different programs. To supply rotifers on a large scale, we have used several methods and found the following to be most dependable in supplying our needs.

Culture Method

Tank Description

Rotifer production tanks are rectangular and are constructed of 3/4-in. plywood. Wood surfaces on all tanks are laminated with resin and painted. Tanks are raised 0.5 m off of a concrete slab by 4-in. beams. The total capacity of the tanks is 1000 ℓ. Six of these tanks were built and placed side by side. All are plumbed and have one central discharge for drainage. Each tank has a 2-in. drain that may be opened or closed by a ball valve.

Aeration is supplied by an electric blower built by Fuji Electric of Japan. Air is supplied by running it through an air bar which is made of 1-in. PVC pipe. The bar is perforated with one row of 0.5-mm air holes every 10 cm and runs the length of the tank. It should be weighted down so it remains stationary on the bottom. This type of air system provides good circulation, a constant level of oxygen, and disperses rotifers throughout the water column.

Location and Environmental Conditions

Tanks are located outdoors and are shaded by translucent, corrugated roofing (see Figure

FIGURE 1. Rotifer production tanks with overhead shading at Oceanic Institute. The tank dimensions are length — 2.45 m, width — 1.20 m, and depth — .60 m.

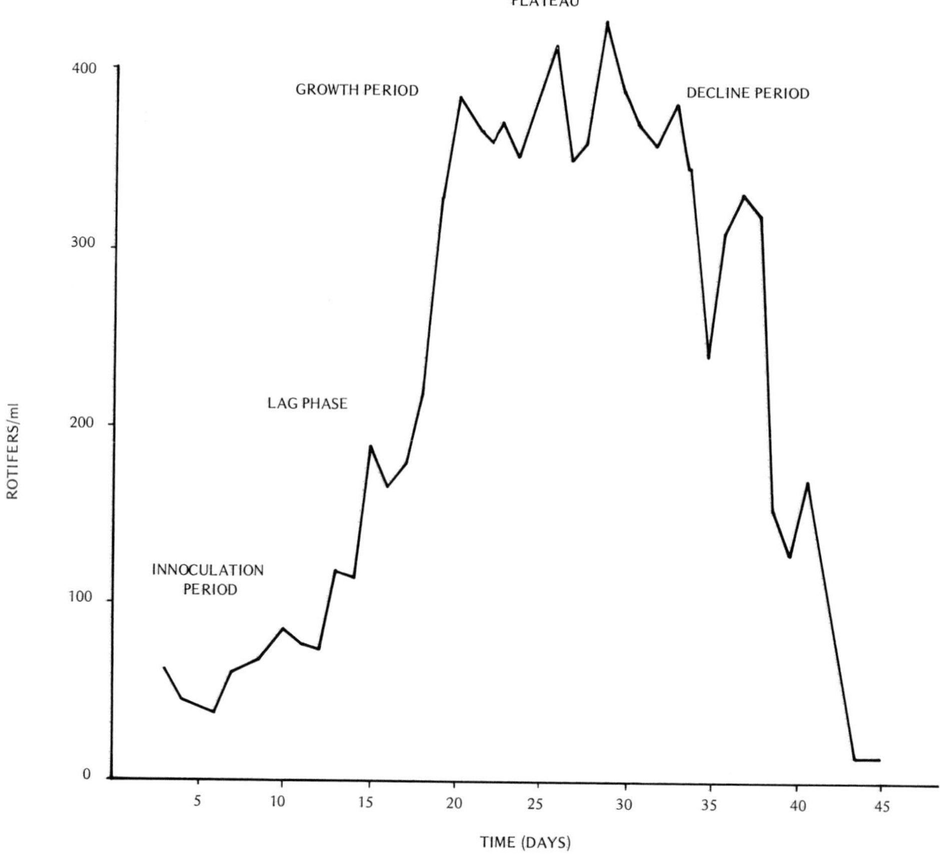

FIGURE 2. Typical rotifer culture growth curve at Oceanic Institute.

1). Besides providing shade, the roof helps in maintaining salinity by diverting any rain water. Climatic changes are variable and sunlight and air temperatures differ during the year. Low temperatures of 18°C during the winter months and highs of 27°C in the summertime are common. Light intensities range from 150 to 300 fc and change often during the day. The source of seawater at the Oceanic Institute is tapped in from a well adjacent to the shoreline. Salinity of the incoming seawater remains constant at 31 to 32 ppt. Water temperatures range from 22 to 25°C and pH ranges from 7.4 to 7.6.

Procedure for Initiating Cultures

A starting volume of 400 ℓ is used and volume adjustments are made as rotifer density increases. By increasing water volume 100 ℓ at a time and watching densities closely, good production at a volume of 900 ℓ may be reached in 15 to 20 days (Figure 2).

Both organic and inorganic nutrients are added to cultures. Nutrients help to boost the algal bloom and enrich rotifer health and condition. Manure taken from cattle feedlots is used along with sodium nitrate and phosphate. To a starting volume of 400 ℓ of seawater, 100 g manure, 20 g sodium nitrate, and 2 g phosphate is used. This same amount is added three times a week for the first 2 weeks. After fertilizing, one may notice a drop in rotifer densities; but after each drop a sharp increase usually follows.

Stocking Cultures

When stocking cultures, densities and overall condition of stock are important. Rates of increase are affected and fast reproduction may decrease due to poor health conditions. Condition of the population may be checked by viewing rotifers under a microscope at 40 × power. Size of rotifers should be between 100 and 175 μm. A good majority of rotifers should consist of amictic females which are gravid and large in size. These females may be carrying from three to five egg sacs. They should be actively swimming and of good color. The number of rotifers stocked initially into culture should be between 50 and 75/mℓ.

Counts and Observations

Upon stocking a culture, an initial count of rotifers per milliliter should be taken. Records and observations should also be initiated on a twice-daily basis. Counting is done by taking a 50-mℓ sample and viewing it under a dissecting scope. A 1-mℓ sample is counted in a 1-mℓ pipette. Rotifers contained in this pipette are counted individually, thus giving the count per milliliter. Observing other organisms present should be done daily to assure good conditions. Ciliates and dinoflagellates may hamper good production because of competition in feeding.

Feed Rates

Feed used in this method is brewer's yeast that is purchased from Anheuser-Busch. This yeast comes in 1-lb blocks at a relatively low cost. Feeding is done twice a day at the rate of 1 g per million rotifers. We varied feed rates and closely observed resultant production. Rates of 50, 100, and 150% were used and we found that overfeeding was much more of a problem than starvation. By counting rotifers per milliliter and calculating volume and total population, feed amounts can be determined. After weighing the amount to be fed, the yeast is dissolved in freshwater and added to the tank. Tanks should be stirred shortly after feeding.

Harvesting

Once good densities are reached and populations are stable, harvesting may be started. A harvest of 25% volume is possible and by rotating harvests good production will be steady and stable. Siphon hoses are used along with different-sized mesh to screen rotifers from the culture. Water is passed through three different-sized screens to remove sediment and debris from the harvested rotifers. The smallest-sized mesh, 60 μm, is used to catch rotifers that are being harvested, while 250- and 175-μm screens are used to catch all debris. After harvesting each culture, water should be brought back to its original volume.

Water Quality

Measurements that are taken consist of temperature, salinity, and pH. Over a 3-month period temperatures ranged from 18 to 23°C. Good production was maintained through this period. Salinities recorded went from 26 to 31 ppt. Temperature and salinity appear to affect productivity the least. The pH was determined daily as it affected cultures the most. Cultures start at a pH of between 8.4 to 8.1 and drop slowly for the first 2 weeks. When pH drops to 7.6, rotifer production should be stable. If pH drops below pH 7, bad conditions may result. When conditions are bad, tanks should be drained and refilled with clean seawater. Feed rates should be cut to stop further problems in culture water quality. As the pH changes back to 7.6, higher feed rates can be used once again. By watching pH closely, good stable cultures will be maintained.

Section II
Maturation, Spawning, Hatchery, and Grow-Out Techniques for Crustaceans

CONSTITUTION OF BROODSTOCK, MATURATION, SPAWNING, AND HATCHING SYSTEMS FOR PENAEID SHRIMPS IN THE CENTRE OCEANOLOGIQUE DU PACIFIQUE

Aquacop
Centre Océanologique du Pacifique

INTRODUCTION

The Centre National pour l'Exploitation des Océans (CNEXO) has developed a tropical aquaculture program at the Centre Océanologique du Pacifique (COP), located in Tahiti (French Polynesia), and the Station d'Aquaculture de St-Vincent, located in New Caledonia.

Tropical areas seem to be highly appropriate for the development of penaeid shrimp farming, due to high temperatures which allow full production throughout the year. Since there is no indigenous species of penaeid shrimp for commercial culture in the waters of Tahiti, it has been particularly necessary for the COP to control the complete cycle of reproduction in captivity. The COP is located in the Vairao lagoon where the water is largely renewed by wave action above the barrier reef; this provides a true oceanic water which gives a stable environment.

About 40 scientists and technicians work in the Center to develop rearing techniques for marine and freshwater species in the tropical environment. The Center includes: an algal laboratory with two algal culture rooms, nutrition laboratory to determine and produce the right feed for the different species, a water chemistry laboratory to study and control water quality in the rearing systems, a pathological laboratory to prevent and treat diseases, and a technological unit for the design, construction, and maintenance of water pumping systems, ponds, and tanks.

For the penaeid culture program the COP has two hatcheries with 20 m^3 of total rearing volume (tanks of 0.5 m^3, 0.8 m^3, and 2 m^3) allowing a theoretical production of 4×10^6 postlarvae per month. The program also maintains 11 700-m^2 grow-out ponds (earthen and concrete ponds) and a matuiation area with 12 12-m^3 tanks. Separate rooms are maintained for spawning adult shrimp and hatching of eggs produced.

The St-Vincent Station facilities consist of one hatchery with 13 m^3 of total rearing volume; 22,200 m^2 of grow-out ponds (from 1200 to 10,000 m^2); one 75,000-m^2 pond; maturation room with nine 10-m^3 tanks; and spawning and hatching rooms.

Since 1973 nine shrimp species have been tested at the COP and St-Vincent Stations. Five of these nine species are now in production: *Penaeus merguiensis, P. indicus, P. monodon, P. stylirostris,* and *P. vannamei*. The last three species are the main candidates for intensive culture in the tropical environment.

CONSTITUTION OF BROODSTOCK: POND GROWING FROM POSTLARVAE TO REPRODUCTIVE SIZE

History and Main Results (Table 1)

P. monodon

This large species from the Indo-Pacific was first introduced to COP in 1975 from Fiji and New Caledonia. From 1975 to 1979 enough breeders were obtained for experimental purposes by choosing the best 20- to 30-g males and females at the end of 6-months commercial grow-out. These shrimps were then grown to reproductive size (females >60 g) in other ponds at lower density. In 1978 and 1979 a disease we called "blue disease" affected an increasing number of these future breeders causing large mortalities. This disease

Table 1
SPECIES AND STRAINS REARED IN COP NUMBER OF GENERATIONS

Species	Date of introduction to COP	Origin (company)	Number of generations reared	Observations
P. monodon	1975	Fiji, New Caledonia	8	
	October 1978	Philippines (SEAFDEC)	1	Lost in 1979
P. merguiensis	1973	New Caledonia	15	
P. indicus	September 1980	Philippines (SEAFDEC)	2	
P. vannamei	July 1975	U.S. Ralston Purina (Crystal River)	6	
	January 1981	Ecuador (Semacua)	1	
P. stylirostris	November 1976 January 1977	Panama	4	Lost in 1980
	April 1978	Mexico	2	Lost in 1979
	April 1980	Mexico	3	
	January 1981	Ecuador (Semacua)	2	

was mainly attributed to a food deficiency, and a new plan for producing breeders from postlarvae was developed in 1980. It is based on three principles: low density, high quality food, and frequent changes of shrimp to new ponds. The plan was put into practice in 1980 to 1981 on four batches of *P. monodon* postlarvae separated by 3 to 4 months and proved to be better than the previous method. Other species were more or less integrated into this plan. We describe it in Proposed Diagram of Production (page 113).

P. merguiensis, P. indicus

These two species are found in the Indo-Pacific region along with *P. monodon,* but reach a smaller size when reared. *P. merguiensis* was introduced to COP from New Caledonia in 1973. Numerous generations have been easily produced at the COP and St-Vincent Stations. *P. indicus,* introduced to both stations from the Philippines (SEAFDEC) in 1980, seems to show the same ability to reproduce in captivity. With these two species, thousands of juveniles were obtained from spawnings in grow-out ponds. The females can spawn at a weight of 6 to 8 g.

P. vannamei, P. stylirostris

These species live along the west coast of America, from Mexico to Peru. In ponds they grow to an intermediate size between *P. merguiensis* and *P. monodon.* The spawning size of females is above 30 g for *P. vannamei* and 40 to 50 g for *P. stylirostris.* The COP received some *P. vannamei* postlarvae from Crystal River Hatchery (Florida) in July 1975. Spawnings from this stock made it possible to supply the St-Vincent Station with postlarvae in April 1976. A new shipment occurred in February 1977 from Agromarina Company (Panama). Since then, it has been possible to produce each year a sufficient number of postlarvae for some growing experiments and stock renewal at both the COP and St-Vincent Stations. *P. vannamei* is a hardy and long-lived species; but it does not reproduce well in captivity because of the failure of males to mate under captive conditions (cf. Maturation). In January of 1981 the COP received some *P. vannamei* postlarvae from Ecuador; they have been reared apart from those captive for several generations and are now spawning.

The COP first received *P. stylirostris* postlarvae in November 1976 and January 1977 from Panama. In August 1977, second generation postlarvae were sent to New Caledonia where the species proved to keep on growing at the lower temperatures (20 to 25°C) of the cold season. *P. stylirostris* was also difficult to breed in captivity due to lack of mating by males. The pure Panama strain was lost at COP in 1980 and *P. stylirostris* was perpetuated only as hybrids from crossing the Panama strain with the Mexican strain. Two generations were obtained from this hybrid cross. In April 1980 new *P. stylirostris* postlarvae arrived from Mexico; three generations of them have been reared in captivity until now. In January 1981, the COP received *P. stylirostris* postlarvae from Ecuador. Thanks to artificial insemination, it is no longer a problem to complete the reproductive cycle for *P. vannamei* and *P. stylirostris* (cf. Maturation).

Equipment, Methods, and Conditions of Rearing

Ponds and Water Networks

The ponds used for growing the future broodstock are rectangular, 400 to 10,000 m^2 in surface area, and of two types: the first type is 1.20 m deep and has compressed clay bottom and earthen dikes; the second, 1.80 m deep, has earthen dikes covered with a layer of concrete and polyester sheet and a bottom made of compacted coral. Some of these coral bottoms are partly covered with a layer of gravel and a layer of sand separated by a permeable synthetic cloth. Moderate aeration is provided through a longitudinal perforated PVC pipe. This type of pond is half covered with a shading screen hung 1 m above surface. The first type of pond, the only type at the St-Vincent Station, might be sufficient to rear the breeders.

Most ponds are equipped with seawater and freshwater inlet pipes joining into a single pipe for mixing the waters. In the second type "double bottom" ponds, the water flows into drain pipes embedded in the gravel, under the sand. Bottom and surface waters are evacuated at the other end of the pond through 2 PVC pipes provided with filter screens.

Control, Maintenance, Harvest Material, and Methods

Control, maintenance, harvest material, and methods for future breeders are almost the same as those used for commercial rearing. The only difference is special care provided to culture medium and animals in broodstock ponds.

Routine outdoor pond measurements are limited to water temperature and disappearance depth of a Secchi disc, morning and evening three times a week, and underwater visual observations of pond water, bottom, shrimp, and feed by scuba divers three times a week. Special measurements, using standard procedures, are on pH, dissolved oxygen, nutrients (nitrates, ammonia, phosphates, silicates), and organic matter concentration, as well as bacteriological and physiological studies of the culture medium and shrimp on an occasional basis. Microscopic observation and counting of algae cells and zooplankton are also made periodically.

The maintenance of full ponds consists of water inflow and salinity regulation, filter screens cleaning, and removal of bottom or surface algae, feces, and various wastes; this removal can be made by chaining, scraping, siphoning of bottom, and pulling floating nets on surface.

Maintenance of empty ponds consists of scraping, collecting, and evacuating various wastes and fouling and then raking the bottom by hand or mechanically so as to make it soft and dry.

Capturing future breeders is always done by nets to make observation counts and sexing easier. Nets include landing nets often used under water, cast nets, and several types of fixed nets such as traps, single-wing, single-trap eel nets, single-wing pluri-trap nets, and V-shaped wings, single-trap nets. Draining of ponds is done through a fishing box or tubular fishing net.

Rearing Conditions

These are summarized in Table 2 and Table 3.

Results

P. monodon

Between 5-day-old postlarvae (P_5) and 2 g mean weight, shrimp are grown at a density of 10 to 20/m^2 for about 2 months and survival is above 70% in most cases.

The results between 2 and 25 g are very diverse, so we will emphasize the four 1980 to 1981 test rearings: numbers 100, 108, 123, and 130. Survivals were, respectively, 79, 90, 74, and 27%. Mean weights after a 5-month growing period from 2 g were 21, 25, 26, and 24 g. Figure 1 shows that up to 20 to 25 g mean weight, the rate of growth at a density of 4/m^2 (n° OP1-2) and even 10/m^2 (n° 14) can be optimum.

At COP shrimp are not measured after they reach 25 g to avoid disturbing potential spawners. St-Vincent Station's samplings of *P. monodon* give the results shown in Table 4.

At St-Vincent Station, some *P. monodon* have reached the age of 30 months.

The survival of *P. monodon* larger than 25 g varies greatly from one rearing to another. The "blue disease" often caused mortalities in our rearings. Our best production of *P. monodon* spawners, n-126, was obtained in a concrete, double-bottom pond, at a density of three *P. monodon* per square meter, and provided 76% survival after a 6-month rearing period.

P. merguiensis, P. indicus

P. merguiensis is a slow growing species, under our rearing conditions, if culture density is greater than 1 shrimp per square meter. At the St-Vincent Station this species reaches about 2 g in 2 months from postlarvae, and 10 g in 4 months (males — 8 g, females — 12 g) when the density is under 1/m^2. In subsequent months the growth slows down. The biggest individuals recorded were 40-g females and 18-g males after 1 year. The survivals are variable, better in concrete aerated ponds than in earth ponds. Lifetime in earth ponds does not exceed 1 year. On the whole, 4 to 5 months of growing may be enough to obtain shrimps of reproductive size with this species.

P. indicus was the subject of only a few trials. At low density (under 2/m^2), it would reach 2 g in 40 days after metamorphosis. Like *P. merguiensis* it can grow from 2 to 10 g in 2 months, but growth remains better than *P. merguiensis* with time. Until now, our biggest females and males of this species weighed 34 and 19 g, respectively, at age 14 months. *P. indicus* are resistant and have a high percent of survival. Breeders can be obtained in less than a year.

P. stylirostris, P. vannamei

P. stylirostris is an extremely fast growing species at a low density. At 10/m^2 it can reach 2 g in less than 40 days after metamorphosis. After 2 g, at a density lower than 1/m^2, it reaches 25 g in 2 to 4 months. In 5 months its mean weight increases to 30 to 50 g (Figure 2). The Mexican strain seems smaller than the Ecuadorian and Panamanian ones, and Panama strain females reached 30 g in 6 months, 60 g in 12 months, and 80 g in 18 months at our New Caledonia Station.

At low density, survival is generally high (above 70%). Nevertheless, diseases such as "white pleura disease" can break out, especially with temperatures above 30°C, leading to mortalities. The lifetime of *P. stylirostris* in our conditions is 18 to 24 months. To reach the reproductive size, 6 to 8 months of growing can be sufficient.

P. vannamei is particularly adapted to intensive rearing. It grows from postlarvae to 2 g in an average of $^1/_2$ months at a density of 10 to 20/m^2. After 2 g, good growth was obtained

Table 2
REARING CONDITIONS OF FUTURE BREEDERS

Size of shrimp	Density (nb/m²)	Food	Water exchange (%/day)	Physicochemical conditions	Fauna and flora of the ponds
				T°(°C) 20—30	Flora
P_5—P_{20}	10—50	Alginated micropellets	Inf. 10	Sal. (%) 20—35	Chaetoceros
					Skeletonema
				Secchi (cm) sup. 50	Navicula
					Nitzchia
P_{20}—2 g	10—50	1 mm MOSOJ J1	Inf. 10	pH 8.4	Chlorella
					Cyanophycae
				DO saturation	Enteromorpha
					Fauna
				$N.NH_4$ 0.000 mg/ℓ	
					Rotifers
2—30 g	1—3	MOSOJ + Japanese pellet	10—30	$N.NO_2$ 0.000 mg/ℓ	Copepods
				$N.NO_3$ 0.005 mg/ℓ	Mysids
					Sergestids
					Carid shrimps
Sup. 30 g	1—3	Japanese pellet + (fresh food)	Sup. 30	$P.PO_4$ 0.040 mg/ℓ	Barnacles
					Aplysia
				Si—SiO_3 5 mg/ℓ	Mollies

Table 3
COP PELLETS COMPOSITION

Ingredients	MOSOJ 3 mm	MOSOJ J 1 mm
Fish meal	10	7
Concentrate dissolved fish proteins	5	5
Shrimp meal	15	12
Meat and bone meal	7	7
Blood meal	2	3
Alkane yeast		10
Brewer's yeast	10	
Copra cake		5
Soybean cake	20	24
Cod liver oil		6
Fish oil	6	
Wheat gluten	7	7
Cereals (wheat, corn, rice)	10	
Spirulina		2
Peptonal		5
Trocus (or Achatina)	2	2
Vitamins and salts	6	5
% Proteins	49	52.2
% Lipids	10	9.5

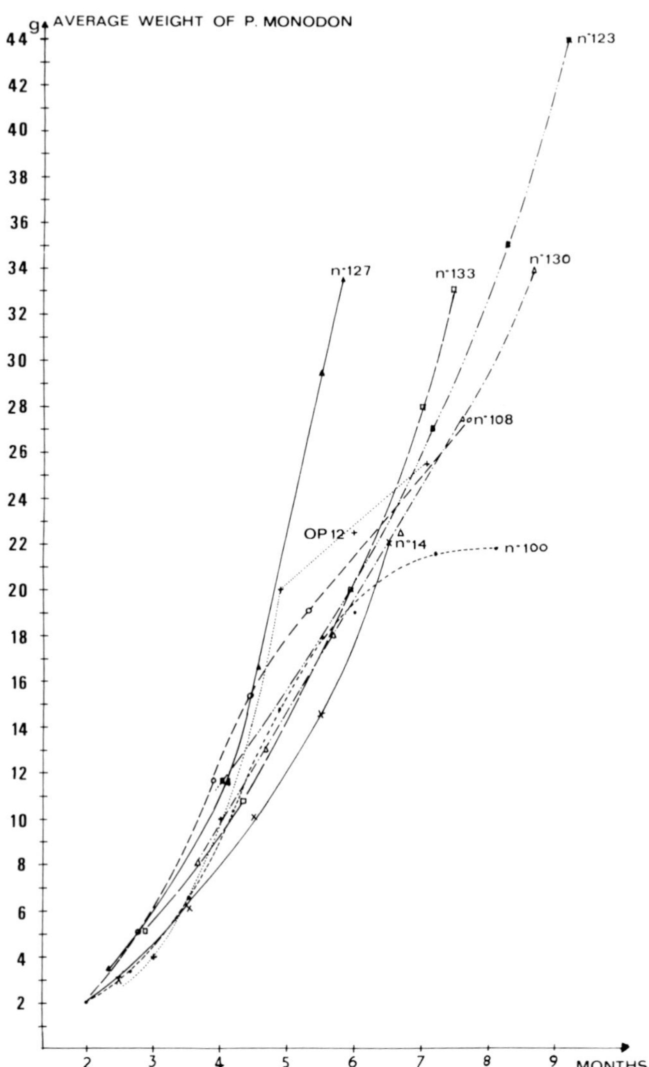

FIGURE 1. Low density *P. monodon* growth curves (after 2 g).

		Trial n°	*P. monodon* density
×	————————	14	10/m^2
●	- - - - - - - - - - - -	100	2/m^2
○	— — — — — — — —	108	1/m^2
■	— •• — •• — •• — ••	123	0.15/^{2}
△	— • — • — • — •	130	1/m^2
▲	————————	127	0.3/m^2
□	——— ——— ———	133	0.4/m^2
+	• • • • • • • • • • • •	OP 1—2	4/m^2

in small capacity tanks in high density conditions (up to 200/m^3). In concrete or earth ponds, the growth is slower, the mean weight increasing from 2 to 15 g in 4 to 5 months, whatever the density. In most rearings, survival is more than 70%. In three 1980 to 1981 special rearings for breeders, groups number 100, 123, and 130, the growth was faster (Figure 3). In groups number 100 and 130 the growth was exceptional, up to 20 g (2 to 20 g in less

Table 4
RATE OF GROWTH WITH TIME FOR *P. MONODON* AT COP AND ST-VINCENT STATIONS

	COP results			St-Vincent results		
Age (months)	0	2	7	12	18	24
Mean weight ♂ (g)	—	2	20	40	65	80
Mean weight ♀ (g)	—	2	30	70	110	150

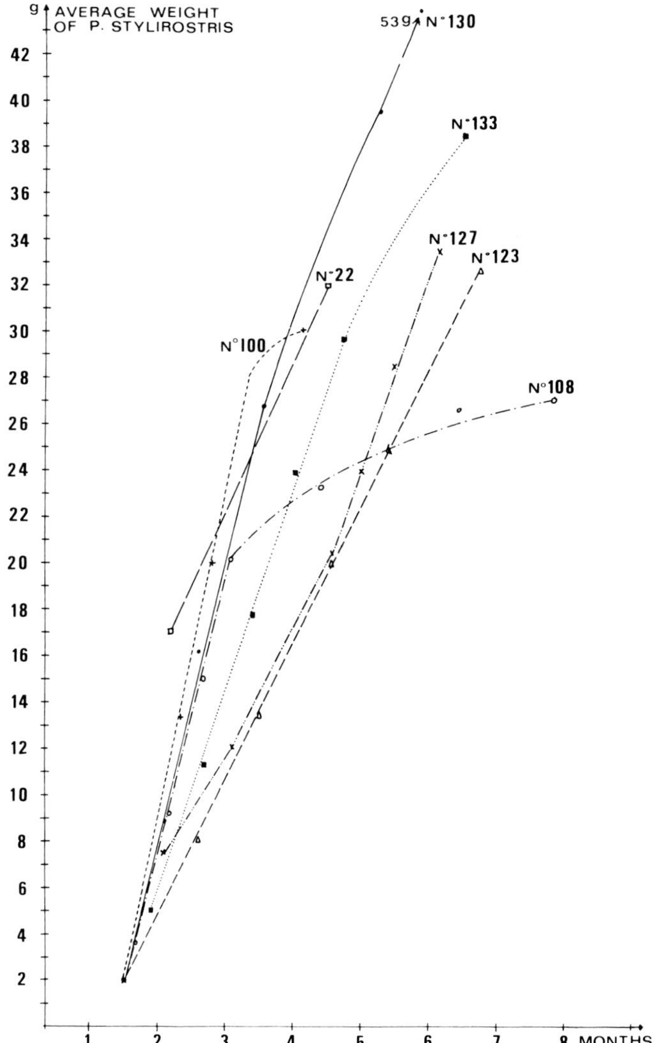

FIGURE 2. Low density *P. stylirostris* growth curves (after 2 g).

	Trial n°	*P. stylirostris* density
+	100	0.08/m²
○	108	0.8/m²
△	123	0.8/m²
●	130	0.6/m²
■	133	0.14/m²
×	127	2.7—1/m²
□	22	Inf. 0.05/m²

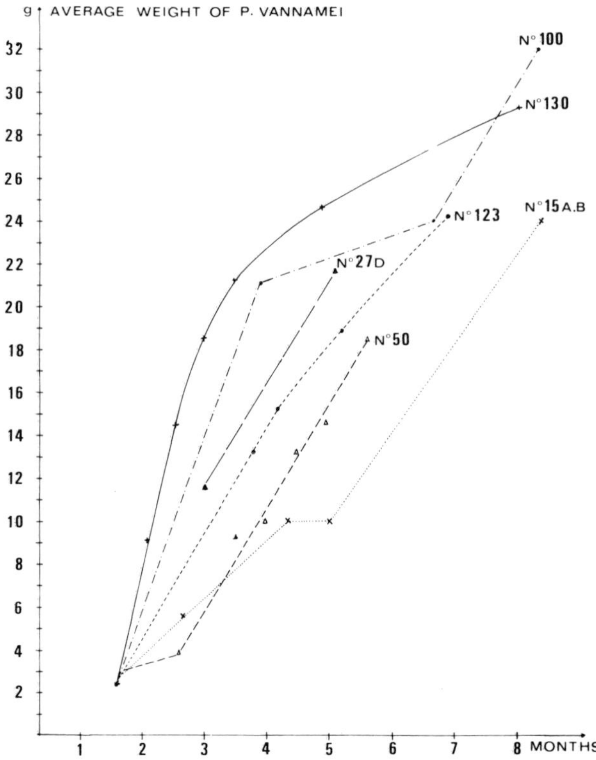

FIGURE 3. Low density *P. vannamei* growth curves (after 2 g).

		Trial n°	*P. vannamei* density
×	· · · · · · · · · · · ·	15 A, B	100—44/m^2
▲	—— —— ——	27 D	110/m^2
△	— — — — — — — — —	50	30/m^2
●	— · — · — · — ·	100	0.1/m^2
○	- - - - - - - - - - - - - - - -	123	0.7/m^2
+	————————	130	0.6/m^2

than 2 months) and thereafter slowed down; in group number 123, growth began slower but continued well after 20 g. The mean weight increased from 2 to 25 g in 5 months as an average in these three groups. Survivals were, respectively, 61, 72, and 82%. After 25 g, the growth is not precisely studied at COP; but for three batches of *P. vannamei* reared at St-Vincent Station, we obtained 14 to 17 g at 6 months, 29 to 37 g at 12 months, 42 to 44 g at 18 months, and 50 g at 24 months.

Some *P. vannamei* lived more than 30 months. As with *P. stylirostris, P. vannamei* require 6 to 8 months of growth in ponds after postlarvae to get breeders.

Proposed Diagram of Production

P. monodon

The trials since 1975 and principally those in 1980 to 1981 for *P. monodon* lead to a three-stage scheme of production for breeders as seen in Table 5.

In Figure 4 is a theoretical diagram providing for production of 400 breeders (200 males and 200 females) every third month. For such production, 3500 m^2 of pond are necessary.

Table 5
THREE-STAGE GROWOUT SCHEME FOR *P. MONODON* BROODSTOCK

	Stage 1	Stage 2	Stage 3
Time in stage (months)	2	5	5
Change in size	P_5—2 or 3 g	2—25 g	25—60 g
Initial density/m^2	10—20	2—3	1—2
Expected survival (%)	70	80	>60
Depth of pond (m)	1.20	1.20	1.50[a]
Feed type per conversion ratio	Alginated micropellet MOSOJ J_1/ℓ	MOSOJ + Japanese pellet per 2	MOSOJ Japanese pellet and fresh food
Percent water exchange/day	<10	10—30	>30

[a] Pond should be shaded.

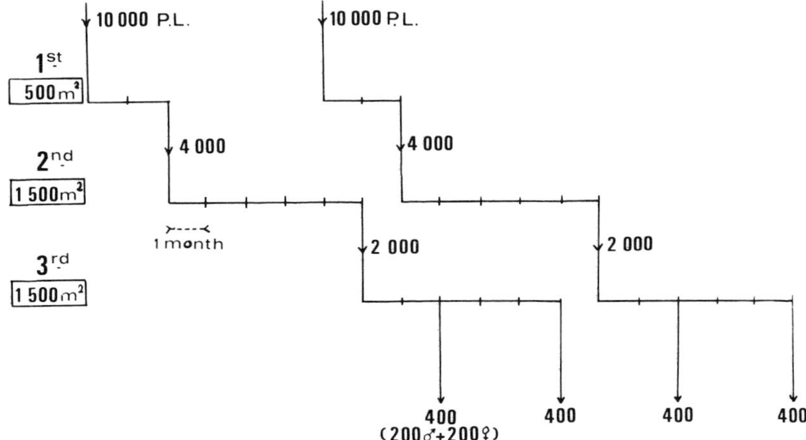

FIGURE 4. Proposed diagram of production for 400 breeders each third month.

This number of breeders should be able to support a hatchery production of 60 million postlarvae per year.

Other Species

Obtaining mature shrimps from *P. merguiensis* or *P. indicus* postlarvae is a fast and easy process so that it is not necessary, with these two species, to elaborate a precise rearing scheme. A low density might be a sufficient condition.

P. stylirostris and *P. vannamei* differ from *P. monodon* which requires 9 to 12 months to obtain reproduction size. Consequently, at the end of the second stage of the diagram proposed for *P. monodon,* shrimps can be passed directly into maturation tanks. The third stage pond can be used for holding, especially for *P. vannamei* which can live longer.

MATURATION

Equipment and Material

The maturation area (200 m^2) is covered with a greenhouse and holds 12 circular tanks. Tanks have a 4-m diameter and a 1.25-m height for a 1-m water depth (Figure 5). Holding

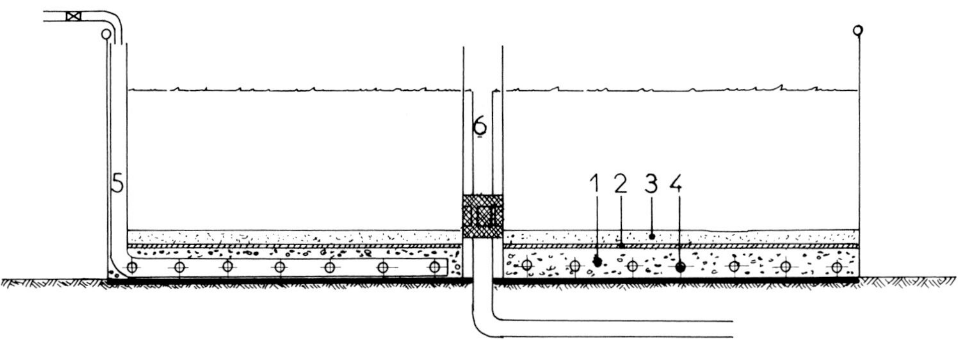

FIGURE 5. Maturation tank. (1) Gravel layer, (2) permeable synthetic cloth, (3) coral and sand layer, (4) agricultural drains, (5) water inlet pipe, (6) water outlet pipe.

capacity is about 12 m^3. Side and bottom are made of a black fiberglass sheet of 1-mm thickness. The substrate consists of a 10-cm layer of gravel separated from a 5- to 10-cm-thick second layer of coral sand by a permeable synthetic cloth to prevent gravel from mixing with sand.

Concentric, semirigid, plastic, perforated pipes (50-cm agricultural drains) fitted in a PVC tube (100-mm diameter) are embedded in the gravel. The water flows into these pipes and then passes through the sand. Thus, shrimp feces and unfed pellets are not trapped in the sand as in the classic circulation system using double bottom and airlift. The sediment remains soft without any reduction signs after several months of continuous flow. Water discharge is achieved by means of two concentric tubes allowing the bottom water to evacuate first. With a rate of complete exchange two or three times a day, the water remains perfectly clear.[1]

One half of the tank surface is covered with a black sheet of synthetic material (polyane), the second half is covered with a shading screen transmitting 10% of the incident light. These covers seem necessary to prevent a heavy development of benthic algae and also to subdue the solar light for nonburrowing species.

Four tanks are in a completely dark room illuminated by fluorescent light giving an intensity of about 100 lux at the water surface. A ''switch-clock'' provides an altered photoperiod with the artificial nighttime being during the natural daytime to facilitate personnel work times.

The food is distributed twice a day. In the morning, fresh food is used: frozen squid, live mussel *(Perna viridis),* and trocus *(Trocus niloticus);* during the afternoon 60% protein compound pellets are distributed. The daily feeding rate is between 3 and 5% wet weight of the shrimp biomass.

Temperature range throughout the year is 25.5 to 30°C, salinity fluctuates around 34 ppt, and pH is 8.2. There are 10 hr of light in July and 14 hr in December. Due to the water exchange rate in the tanks diurnal variations are negligible.

Animals

In the 400-m^2 concrete tanks, the largest and healthiest animals are selected by divers, caught with a landing net, and stocked in maturation tanks. Unilateral eyestalk ablation is practiced on the females by simple pinching of the eyestalk.[2] If the eyestalk ablation is done on healthy animals, no mortality occurs. Mortalities may occur for recently molted fales. Each female is then doubled tagged. A ring of a colored elastic silicone bearing a label is inserted around the remaining eyestalk and another label with a number is glued on the cephalothorax. After molting, the first tag stays on the animal while the other one is found on the discarded carapace (Figure 6). This technique allows us to follow each female

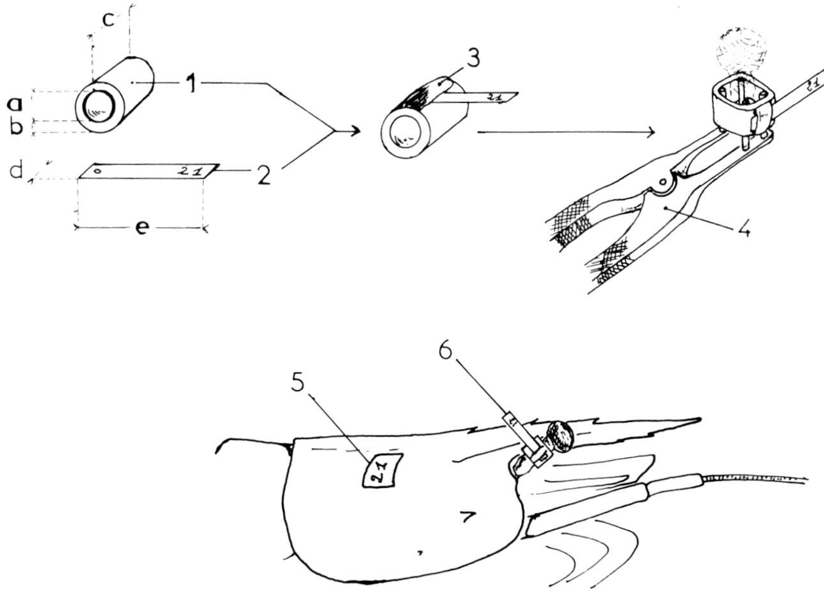

FIGURE 6. Female tags. (1) Silicone mastic cylinder — a = 3.2 mm for shrimp <40 g and 4 mm for shrimp >40 g, b = 1 mm, c = 4.5 mm; (2) plastic tag — d = 2.5 mm and e = 12 mm; (3) gluing the tag on the cylinder with silicone mastic; (4) grip to insert the cylinder on eyestalk; (5) cephalothorax tag; (6) eyestalk tag.

individually in a tank (Figure 7). Females are examined every day at dawn for ovarian development.

Results

P. monodon

The weight of the selected females used in maturation is over 60 g corresponding to 9- to 12-month-old animals. The number of shrimp per maturation tank is around 60 individuals with a sex ratio of 1:1.

In the maturation tanks adult *P. monodon* lie on the sand substrate and rarely burrow; their swimming activity is low day and night. The intermolt period for epedunculated females is about 3 weeks. An extension of this period is an indication of animals that are too old or weak. Above 25°C and for epedunculated females, maturation and spawning occur throughout the year although the ovarian development is minimal during the coldest period from July to September.

The observed courtship and mating behavior takes place just after the molting of the females when the shell is soft.[5-7] The success of the impregnation can be seen the next morning by the remains of the spermatophore which appears as a whitish jelly hanging from the thelycum splits.

After ablation some females develop full ovaries in 3 days; some others molt and development occurs 2 or 3 weeks later. The duration of the ovarian development is variable: 3 or 4 days for some animals, 2 weeks for the others. Usually in the second case, a regression of the gonad appears and the female molts.

Each female can give several spawnings in a short period of time between two molts; after each spawning the gonad is completely emptied but can start developing again the next day. The same stock of sperm is used for the different spawnings between two molts.

For *P. monodon,* which has a dark carapace, the ovarian development is examined with a waterproof handlight with permits viewing the ovaries without handling, thus, avoiding

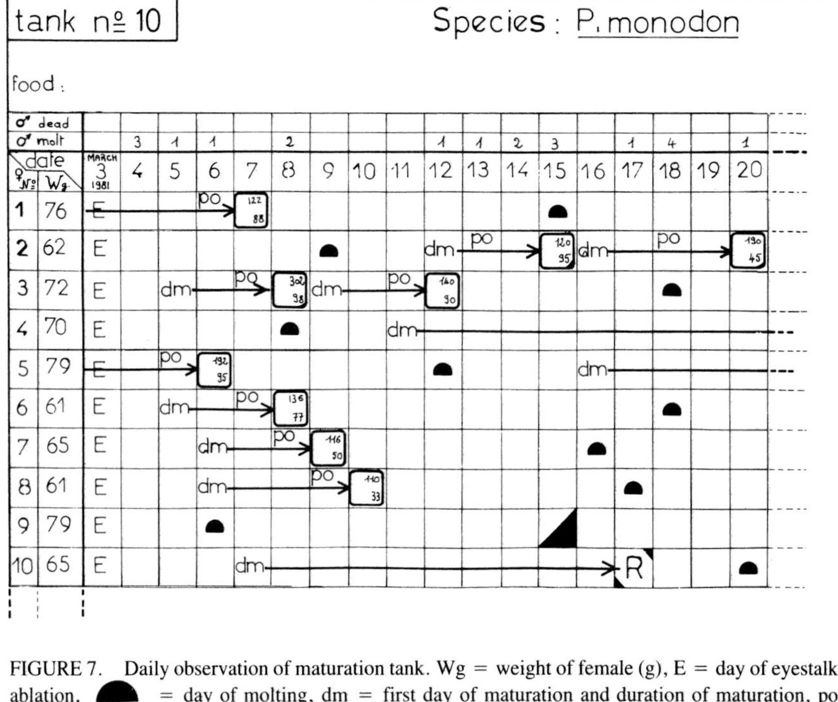

FIGURE 7. Daily observation of maturation tank. Wg = weight of female (g), E = day of eyestalk ablation, [molt symbol] = day of molting, dm = first day of maturation and duration of maturation, po = female placed in spawning tank, [box] = day of spawning: 120,000 eggs — 90% fertilized, [box] = day of regression, [death symbol] = day of death.

stress. When the ovarian color, shape, and texture indicate readiness to spawn, the animal is removed from the maturation tank and placed in a spawning tank.

The color of the developing gonad is first whitish, then it turns greenish and is dark green on spawning day. For some females, the ovary shows a large swelling in the first abdominal segment on the spawning day. For others spawning can occur while the ovaries are barely developed without swelling. More significant than the ovary swelling is its texture which must be granular; but this observation requires handling the animals with subsequent stress. The mean index of spawning per female and per month, for *P. monodon,* is near 1.

P. merguiensis, P. indicus

The females of *P. merguiensis* or *P. indicus* can spawn at a very small size (6 g) with or without eyestalk ablation. The number of shrimp in a maturation tank can be very high, due to the small size of the animal; 200 to 300 shrimp with a mean weight of 10 g are stocked in each tank. The sex ratio is generally 1:1 but it appears that only 25% males is sufficient to obtain regular mating.

For *P. indicus,* the intermolt period for unepedunculated females is about 13 days. Usually, females molting during the night are impregnated. In the morning, the white and expanded parts of the paired spermatophores are clearly visible outside the thelica.

For both species it seems unnecessary to perform eyestalk ablation. For 25-g unepedunculated *P. indicus,* the duration of the ovarian development can be 1 to 9 days, although, in most cases, it only takes 2 to 3 days.

The first sign of ovarian development is a black line visible through the abdominal exoskeleton due to the color of scattered melanophores. In the early ripe stage, a pale green mass is quite visible in the posterior part of the head and the anterior ovarian lobes are evident. One or two days later, the size of the ovaries has still increased; their color is darker green and sometimes expanded parts of the posterior lobes appear in the first and second abdominal segments. These females spawn during the night.

One tank containing 22 females of *P. indicus* produced 129 spawnings during a 66-day period, averaging 2.66 spawnings per female per month.

P. vannamei, P. stylirostris

The size of the selected females should be between 30 to 45 g for *P. vannamei,* 50 to 70 g for *P. stylirostris* (Panama or Ecuador strain), and 40 g for *P. stylirostris* (Mexican strain). There should be between 40 to 100 animals in each tank.

For *P. stylirostris* and *P. vannamei,* the males deposit the spermatophores only on the hard-shell females which will spawn a few hours later. In some periods the chasing behavior is very intense and in others very weak; this phenomenon seems independent of the presence of ripe females. Sometimes the spermatophores are attached in the wrong location possibly due to obstruction by another male.[5] The percentage of sperm deposited can be increased by separation of males and females and transferring only the ripe females into the tank with the males.[5] If no spermatophore is glued on a ripe female 1 or 2 hr before the suspected time of the spawning, artificial insemination can be effected. To accomplish this, spermatophores are removed from one male and the sperm jelly masses are separated from the corneous part of the spermatophore. Then, the two sperm masses are glued on the female between the third and fourth pair of thoracic legs.

The maturation dark room with altered photoperiod induces *P. stylirostris* and *P. vannamei* to spawn between 10 and 12 a.m. and, so, permits us to observe the complete process from artificial insemination to development of embryo.

Selection of ready-to-spawn females is simple for *P. stylirostris* and *P. vannamei:* their carapaces are translucent and the color of the ovaries can be seen without hand-netting the animals. The gonad, which is first whitish, then yellow, turns golden brown or greenish brown on the day of spawning and appears as a constriction in the top of the first abdominal segment.[5]

During the intermolt period each female can spawn several times; shrimps have spawned twice at 3-day intervals and some *P. stylirostris* have spawned four times between molts.

In one tank, 19 *P. stylirostris* (Mexican strain) females gave 57 spawnings in a 24-day period, for an average of 3.73 spawnings per female per month. *P. vannamei* does not spawn as readily and the index is around 1.

SPAWNING

Equipment and Material

The spawning room is equipped with ten spawning tanks. These 150-ℓ capacity tanks have a cylindroconical shape (Figure 8). A perforated plastic plate prevents females from entering the conical bottom and eating the eggs.[4] The tank is supplied with a constant filtered water flow. The overflow passes through a concentrator provided with a 100-μm mesh that retains the eggs.

The ripe females are transferred to spawning tanks daily after selection in the maturation tanks. Three to five shrimps *(P. merguiensis* or *P. indicus)* can spawn in the same tank, but for the larger species *(P. monodon* or *P. stylirostris)* it is better to put one female per tank.

After spawning, the tank is drained and the eggs are collected in the concentrator (Figure 9). The eggs are passed through a 335-μ plankton mesh which retains particles such as feces and then they are collected on a 160-μ plankton mesh.

The eggs are poured into a 10-ℓ bucket and five 1-mℓ samples are taken for counting.[4] About 100 eggs are examined under a microscope to determine the percentage of normal, abnormal, and unfecundated eggs.

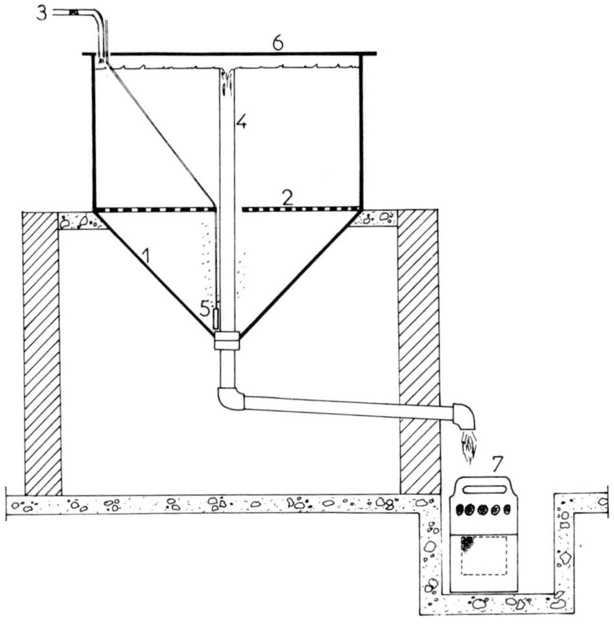

FIGURE 8. Spawning tank. (1) Conical bottom, (2) perforated plastic plate, (3) water inlet pipe, (4) water outlet pipe, (5) air-stone, (6) black cover, (7) egg concentrator.

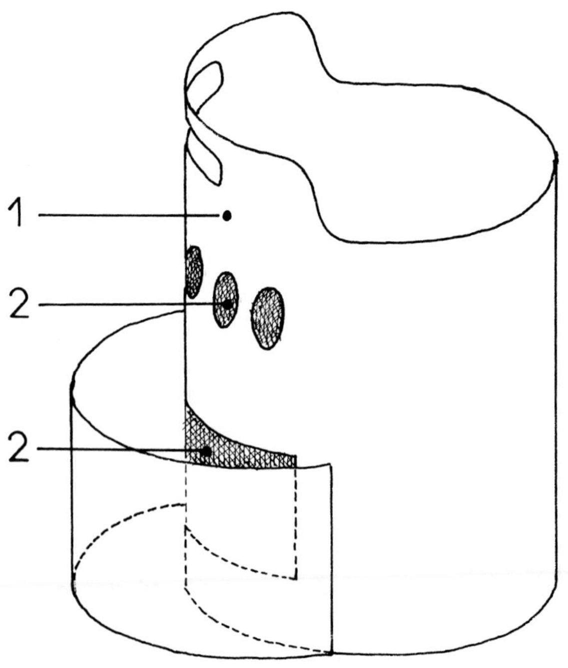

FIGURE 9. Egg concentrator. (1) 250-mm PVC pipe, (2) 160-μm plankton mesh.

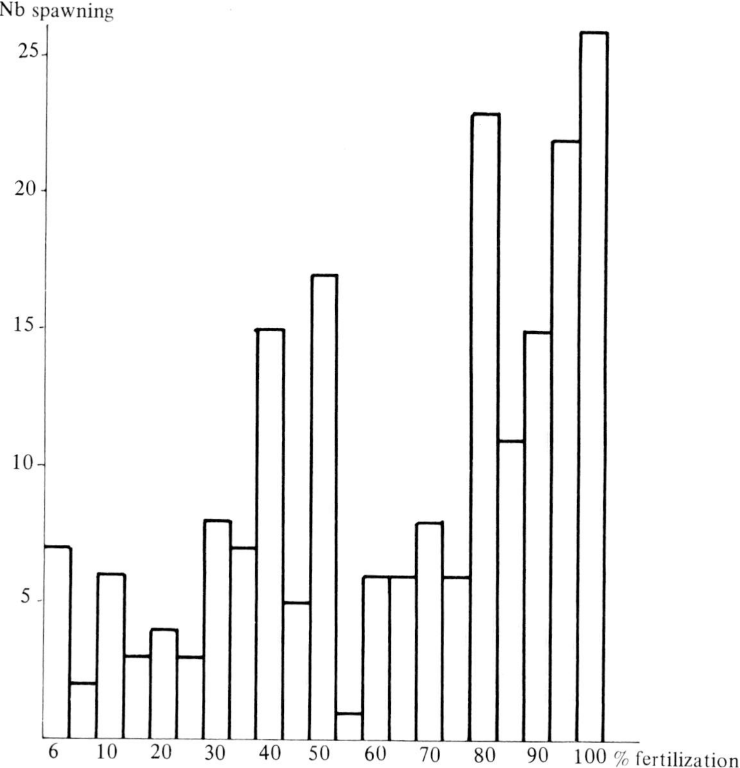

FIGURE 10. Percentage of fertilized eggs in relation to the spawning number observed from 01.01.81 to 06.30.81 on *P. monodon*.

Results

P. monodon

The number of eggs depends on the weight of the female; it varies from 100,000 to 600,000 for 60- to 130-g animals, averaging 175,000. This number has not decreased through successive generations.[6]

Under microscopic examination the eggs can appear as three forms:[3-6] unfertilized eggs with two or three big cells and many small ones; normally fertilized eggs with a fecundation membrane and the presence of well-developed nauplii with apparent setae; fertilized eggs in which development has stopped at different stages.

The percentage of normal eggs varies with the spawning; it can reach 100%, the average being around 60% (Figure 10). Generally, nonfertilization is due to a lack of sperm in the thelycum, but in some cases it occurs with well-impregnated females. This happened noticeably in 1977 when the females were fed only pellets and did not receive fresh food during the last days of the maturation process.[6]

P. merguiensis, P. indicus

For an 18-g *P. merguiensis,* the number of eggs per spawning is around 20,000. For 25-g *P. indicus,* the number of eggs is between 75,000 and 80,000. The percentage of fertilized eggs is generally very high for these species: more than 70% for *P. merguiensis* and more than 80% for *P. indicus.* The totally unfertilized spawnings are due to nonimpregnated females. Abnormal development of eggs has never been observed.

P. stylirostris, P. vannamei

The number of spawned eggs is from 60,000 to 200,000 for 30- to 45-g *P. vannamei,*

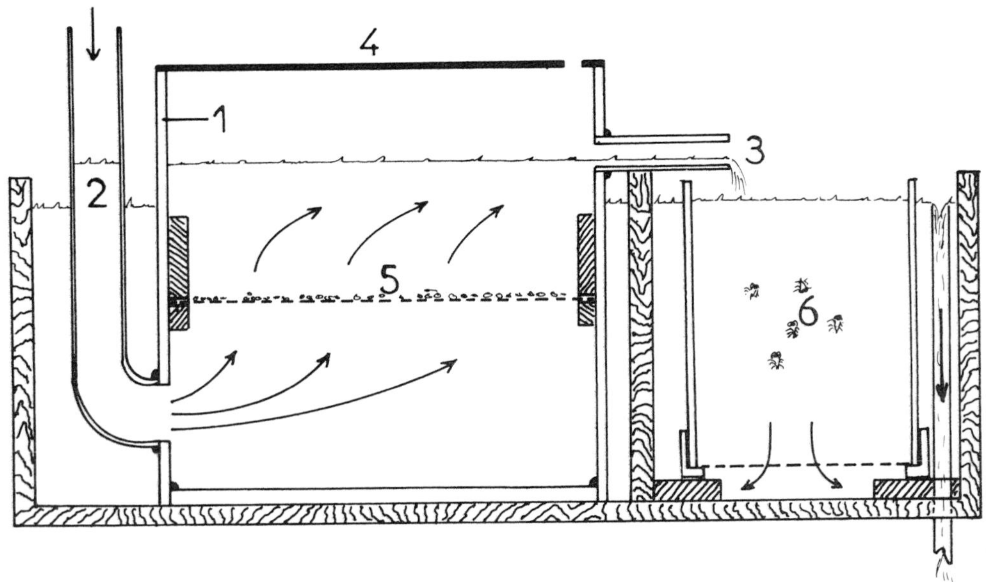

FIGURE 11. Hatching tank. (1) 315-mm PVC tube, (2) 40-mm inlet PVC pipe, (3) 20-mm outlet PVC pipe, (4) black fiberglass cover with a 10-mm hole, (5) 100-μm mesh screen supporting fertilized eggs, (6) 48-μm sieve concentrating nauplii, → = direction of water flow.

100,000 to 350,000 for 60- to 80-g *P. stylirostris* (Panama or Ecuador strain), and 70,000 to 100,000 for 30- to 40-g *P. stylirostris* (Mexican strain). The percentage of viable eggs is very low and most of the impregnated females give abnormal or unfertilized eggs.[5]

With the technique of artificial insemination, the percentage of fertilized eggs can be improved for *P. stylirostris* around 50%; but for *P. vannamei*, this percentage remains very low, near 10%. This low percentage seems to be due to sperm of poor quality. The males often have very small spermatophores or brown spermatophores with necrosis. The sperm mass removed from the spermatophore is very small and does not adhere very well on the female.

HATCHING

Equipment and Material

The hatching room is equipped with ten hatching tanks (Figure 11). These tanks are made from a 315-mm diameter PVC tube closed with a PVC disc bottom.

Halfway up, a 100-μ mesh screen is positioned. The water flows down through a 40-mm PVC pipe under the screen and flows out by a surface 20-mm PVC outlet into a 48-μ sieve.

The fertilized eggs are placed on the 100-μ mesh screen. When the nauplii hatch, they swim to the surface and are borne away by the overflow into the sieve.

Results

P. monodon

The nauplii hatch during the day after the spawning, between 2 and 3 p.m. at a temperature of 28°C. About 60% of the eggs are fertilized, and 70% of them produce nauplii.

For one *P. monodon* maturation tank, the results per month can be summarized in Table 6.

P. merguiensis, P. indicus

At 28°C, the nauplii hatch in the morning after the night of spawning. All fertilized eggs hatch.

Table 6
***P. MONODON* MEAN RESULT PER MONTH AND PER MATURATION TANK**

	Coefficient	Number
Female		30
Spawning	1/female	30
Eggs	175,000/female	5,250,000
Fertilized eggs	60%	3,150,000
Hatched nauplii	70%	2,200,000

P. stylirostris, P. vannamei

The duration of hatching time is the same as for *P. monodon.* For *P. stylirostris,* 100% of fertilized eggs hatch; but for *P. vannamei,* the number of hatched nauplii is lower than the number of normally fertilized eggs observed shortly after spawning. It appears that the embryonic development of some fertilized eggs stops after the third or fourth division. The reason for this is not known at present.

REFERENCES

1. Aquacop, Maturation and spawning in captivity of Penaeid shrimp: *Penaeus merguiensis* de Man, *Penaeus japonicus* de Bate, *Penaeus aztecus* de Ives, *Metapenaeus ensis* de Hann and *Penaeus semisulcatus* de Hann, in *Proc. World Maricul. Soc. 6,* Avault, J. W., Ed., Louisiana State University, Baton Rouge, 1975, 123.
2. Aquacop, Reproduction in captivity and growth of *Penaeus monodon* Fabricius in Polynesia, in *Proc. World Maricul. Soc. 8,* Avault, J. W., Ed., Louisiana State University, Baton Rouge, 1977, 293.
3. Aquacop, Observations sur la maturation et la reproduction en captivité des Crevettes Pénéides en milieu tropical, in *3rd Meeting I.C.E.S. Working Group on Mariculture, — Actes de Colloques du CNEXO 4,* Brest, France, May 10 to 13, 1977, 157.
4. Aquacop, Elevage larvaire de Pénéides en milieu tropical, in *3rd Meeting I.C.E.S. Working Group on Mariculture — Actes de Colloques du CNEXO 4,* Brest, France, May 10 to 13, 1977, 179.
5. Aquacop, Penaeid reared broodstock: closing the cycle on *P. monodon, P. stylirostris* and *P. vannamei,* in *Proc. World Maricul. Soc. 10,* Avault, J. W., Ed., Louisiana State University, Baton Rouge, La., 1979, 445.
6. Aquacop, Reared Broodstock of *Penaeus Monodon,* presented at Symp. on Coastal Aquaculture, Marine Biological Association of India, Cochin, January 12 to 18, 1980.
7. **Primavera, J. H.,** Notes on the courtship and mating behaviour in *Penaeus monodon* Fabricius, in *Contribution n° 29, Aquaculture Department, Southeast Asia Fisheries Development Center,* Iloilo City, Philippines.

PENAEID LARVAL REARING IN THE CENTRE OCEÁNOLOGIQUE DU PACIFIQUE

Aquacop
Centre Océanologique du Pacifique

The Centre Océanologique du Pacifique (COP), a branch of the Centre National pour l'Exploitation des Oceans (CNEXO), already described in other chapters of this handbook, has been working since 1973 on the culture of penaeid shrimp.

This chapter describes the larval rearing technique perfected at the COP and used for 7 years on eight species: *Penaeus aztecus, P. japonicus, Metapenaeus ensis, P. merguiensis, P. indicus, P. monodon, P. vannamei, and P. stylirostris.*

MATERIAL AND METHODS

The COP has two hatcheries. The first one is used mainly for experimental purposes and is equipped with eight 500-ℓ tanks and three 2-m^3 tanks. Two 500-ℓ tanks more can be operated on a closed system with biological filtration. The second hatchery is used for production and has four 800-ℓ tanks, three 2-m^3 tanks, and one 10-m^3 tank.

Larval Rearing Tanks

All larval rearing tanks are made of fiberglass and polyester resin.

500-ℓ Tanks (Figure 1)

In these cylindroconical tanks, the water renewal is made through a lateral overflow connected by a flexible hose to an immersed filter. Emptying the tank is done with a siphon. The aeration is ensured by a central aerator. These tanks are located within a concrete tank with running seawater as a temperature regulator.

800-ℓ and 2-m^3 Tanks (Figure 2)

These cylindroconical tanks have a 45° slope at the bottom. The drain is centrally located and operated by an exterior valve. A vertical PVC pipe is screwed on the central drain and has a quarter-turn opening system halfway up. For water changes, a filter fitted with a mesh adapted to the size of the larvae (205, 335, 500 μm) is slipped on this pipe and the opening system operated so as to empty half of the tank.

10-m^3 Tank

A 10-m^3 tank, elongated and U-shaped, has been tried and is presently in use in the production hatchery. This tank uses the same principle as the 2-m^3 tanks, the center filter being connected to the drain by a flexible hose and immersed horizontally in the tank.

Seawater and Air Networks

Seawater

The seawater is pumped from the lagoon in front of the COP at a depth of 5 m and has oceanic characteristics: temperature from 25 to 29°C, salinity of 35 ppt, and a pH of 8.2. This water is filtered through 5- and 0.5-μm cartridges and distributed to the larval rearing tanks without any other treatment. Heating is not necessary but could be useful in the cold months to shorten the larval period.

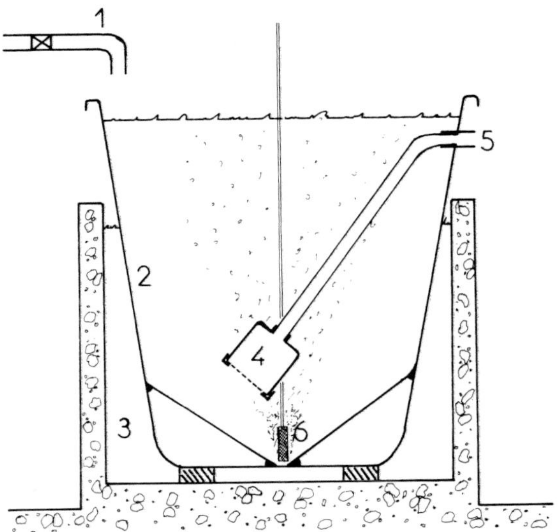

FIGURE 1. 500-ℓ Larval rearing tank. (1) Filtered seawater inlet, (2) larval rearing tank, (3) concrete tank, (4) immersed filter, (5) lateral overflow, (6) air-stone.

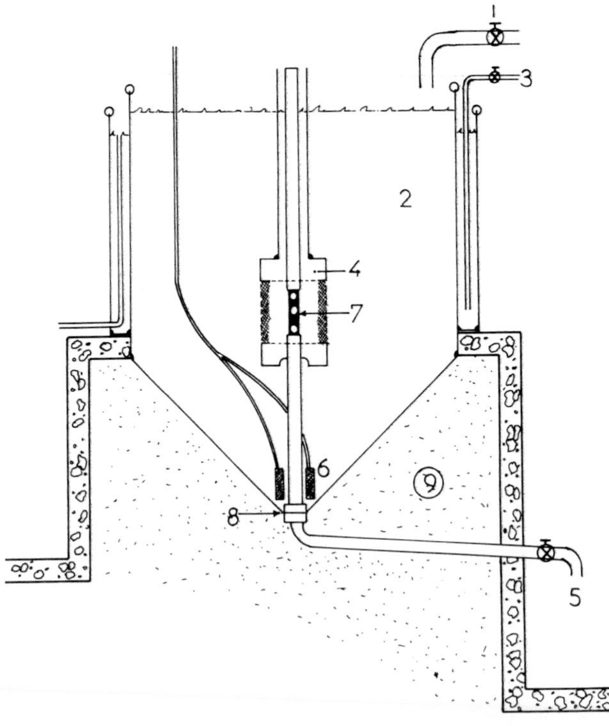

FIGURE 2. 2-m^3 Tank. (1) Filtered seawater inlet, (2) larval rearing tank, (3) constant temperature bath water inlet, (4) central filter, (5) water outlet, (6) air-stones, (7) quarter-turn opening system, (8) drain.

Air

The general air system of the COP is fed by a blower with a 600-m^3/hr capacity. The air is delivered to the larval rearing tanks through air stones at 0.2 bar pressure.

Algae

The algae culture methods used at the COP have been described in another chapter of this handbook. For penaeid larvae, three species of algae are used: *Isochrysis* sp., *Chaetoceros gracilis,* and *Platymonas* sp. These algae are distributed to the rearing tanks through a 25-μm mesh sieve to eliminate algal clumping that sometimes occurs in culture.

Artemia

A special room with five 150-ℓ tanks is used for *Artemia* nauplii production. The hatching tanks have a conical translucent bottom equipped with a valve. To recover the nauplii, aeration is stopped and the tank covered with a black top. The valve is opened and the nauplii, which are concentrated at the bottom, are washed through a 207-μm sieve that holds back the remaining unhatched cysts.

Treatments

Fungicide

Contamination of larvae by fungi (*Sirolpidium* sp. and *Lagenidium* sp.) is frequent and a continuous preventive treatment is necessary.[1] Treflan®, diluted to 5-ppm solution, is distributed constantly.

Antibiotics

Antibiotic treatments are used to control bacterial contamination, which result in necrosis of the larvae and often heavy mortalities.[1] *P. monodon* is particularly sensitive to bacterial necrosis. Many drugs have been tested but Chloramphenicol, the most effective, is used either preventively with doses of 2 to 6 ppm every 2 days according to the larval stages or curatively with doses of 2 to 10 ppm.

Daily Observations

Morning and afternoon, the following observations are made: counting of the larvae in a 1-ℓ sample; observation of color, behavior, and appearance of larvae; determination of larval stage; determination of presence of molts; determination of presence of weak and dead larvae; determination of necrosis and fungi; feeding counts of algae and *Artemia* nauplii.

LARVAL REARING

Technique

The technique presently in use is derived and adapted from the Galveston method.[2-5] Larvae are reared at high density (100 to 120 larvae per liter) until P_4 (4th day after the first postlarva appear).

The normal larval culture sequence, i.e., without any pathological problem or incident, does not vary significantly from one species to another. This sequence is illustrated in Table 1, which should be used as the primary guideline for treatment and care of larval shrimp.

Results

The survival between nauplius stage and the postlarvae P_4 is from 65 to 80% for *P. merguiensis, P. indicus, P. vannamei,* and *P. stylirostris.* For *P. monodon,* the survival is lower, around 45%, because of their susceptibility to bacterial attacks. The pathological mortalities are mainly from fungal attacks, bacterial necrosis, and the presence of bad-shaped

Table 1
LARVAL CULTURE SEQUENCE

Day	Stage	Feeding: *Isochrysis* (cells/mℓ)	Feeding: *Chaetoceros* (cells/mℓ)	Feeding: *Artemia* (nauplii/mℓ)	Water exchange (per day)	Treatments: Treflan® (mℓ/m³)	Treatments: Antibiotics (ppm)
D0	E.N	—	—	—	No	20	—
D1	N	—	—	—	No	2 × 20	—
D2	N—Z_1[a]	50,000	—	—	No	2 × 30	—
D3	Z_1	80,000	20,000	—	No	2 × 30	2
D4	Z_1—Z_2	80,000	20,000	—	No	2 × 30	—
D5	Z_2	50,000	50,000	—	No	2 × 30	2
D6	Z_3	50,000	50,000	—	No	2 × 40	—
D7	Z_3—M_1	—	80,000	—	Total	2 × 40	—
D8	M_1	—	50,000	0.2	1/2	2 × 40	—
D9	M_2	—	50,000	0.2	2/3	2 × 50	4
D10	M_3	—	30,000	0.5	2/3	2 × 50	—
D11	M_3—P_1[b]	—	30,000	1	2/3	2 × 50	6
D12	P_1	—	—	1	2/3	2 × 50	—
D13	P_2	—	—	2	2/3	2 × 50	—
D14	P_3	—	—	5	Total	2 × 50	10
D15	P_4		Harvesting				

[a] Z_1—Z_3 Represent protozoeal stages.
[b] P_1—P_4 Represent postlarval age in days.

nauplii.[1] Other diseases have been observed, especially those related to the nutrition of the larvae, i.e.,

1. Larvae that do not feed at stage Zoea 1
2. Larvae with a black stomach: a black plug is present in the stomach at the Z_1 or Z_2 stage and obstructs the digestive tract
3. Larvae with a grey stomach: the algae ingested look like they are not digested

For all these diseases, the therapy consists of antibiotic treatments and from the Zoea-3 stage onward complete draining of the tanks can be done. Larvae are retained on a sieve and put in a bowl where separation between live and dead larvae is done more easily.

The COP is also experimenting with the substitution of inert particles (microgranules and yeasts) for live food (algae and *Artemia*).

All the results described in this paper concern larvae coming from broodstock completely reared in captivity through successive generations. In hatcheries which are working with broodstocks constituted from adults caught in the wild, the survival is higher, around 80%. It seems the quality of the eggs obtained from captive broodstock is not as good as that of eggs obtained from wild animals, probably due to some deficiency in the feed.

REFERENCES

1. Aquacop, Observations on diseases of Crustacean cultures in Polynesia, in *Proc. World Maricul. Soc. 8,* Avault, J. W., Ed., Louisiana State University, Baton Rouge, 1977, 685.
2. **Mock, C. R. and Murphy, M. A.,** Technique for raising penaeid shrimp from eggs to post-larvae, in *Proc. World Maricul. Soc. 1,* Avault, J. W., Ed., Louisiana State University, Baton Rouge, 1971, 143.
3. **Mock, C. R. and Neal, R. A.,** Penaeid hatchery systems, in *F.A.O. Symp. on Aquaculture in Latin America,* Carpas 16/74/SE, 29, 1974, 9.
4. Aquacop, Elevage larvaire de Peneides en milieu tropical, in *3rd Meeting I.C.E.S. Working Group on Mariculture — Actes de Colloques du CNEXO 4,* Brest, France, May 10 to 13, 1977, 179.
5. Aquacop, Reproduction in captivity and growth on *Penaeus monodon Fabricius* in Polynesia, in *Proc. World Maricul. Soc. 8,* Avault, J. W., Ed., Louisiana State University, Baton Rouge, 1977, 927.

HATCHERY TECHNIQUES FOR PENAEID SHRIMP UTILIZED BY TEXAS A & M-NMFS GALVESTON LABORATORY PROGRAM

James P. McVey and Joe M. Fox

INTRODUCTION

The National Marine Fisheries Laboratory at Galveston has been the principal site for shrimp hatchery research in the U.S. Published works on shrimp hatchery techniques date back to 1969 and have continued until the present.[1-3] The laboratory Aquaculture Program has a staff of ten permanent employees and approximately 30 to 40% of the total research effort has been dedicated to development of shrimp hatchery techniques. The laboratory maintains cooperative agreements with the University of Houston and Texas A & M University. During recent years Texas A & M has frequently operated the shrimp hatchery in cooperation with NMFS staff. This has resulted in an excellent exchange of ideas and techniques for the operation of the shrimp hatchery.

This chapter describes the techniques presently in use in the hatchery in the combined NMFS-Texas A & M operation.

LABORATORY LOCATION AND CLIMATE

Galveston is located on a barrier island at the mouth of Galveston Bay (29° 30′ N and 94° W). Galveston Bay is one of the major shrimp-producing estuaries on the Gulf Coast and can be characterized as having large areas of relatively shallow *Spartina* marsh, silt-clay substrates, and substantial fluctuations in temperature and salinity. According to laboratory records summer water temperatures range from 27 to 32°C while winter temperatures may go as low as 5°C with an average closer to 15°C. Salinities range from 10 to 32 ppt depending on local rainfall. These conditions are sometimes limiting to shrimp hatchery operations and we, therefore, attempt to control these factors within the hatchery. As a result, the lab has adopted fairly small-scale intensive culture systems, which allow the maximum amount of environmental control during hatchery operations.

SEAWATER SYSTEM AND WATER PRETREATMENT

Seawater is obtained directly from the Gulf of Mexico through offshore, subsand wellpoints that effectively prefilter the water before it enters the system. A 3-hp submersible pump enclosed in a large stainless steel housing pushes water from the beach to the laboratory (approximately 300 m) where it drops into a 40,000-ℓ settling tank (sump). This relatively clear, clean seawater is then pumped to two 120,000-ℓ fiberglass-lined, redwood reservoirs where further settling occurs. These reservoirs are equipped with standpipes allowing gravity flow of the upper water column to the hatchery. The flow rate to the hatchery is increased with an in-line centrifugal pump, which allows the hatchery tanks to be filled more rapidly, and the use of pressurized filters. The first pressurized filter contains activated charcoal (1 to 4 mm), a large carbon particle, which filters out large particulate matter, large phytoplankton, and, because of its chemical properties, certain hydrocarbon toxins. The water is then passed through a 5.0-μm pressurized cartridge filter (Filter Spun, Amarillo, Tex.), and most phytoplankton and very fine particulate matter is removed at this point. The next step in water treatment involves the use of an in-line UV germicidal lamp immediately after the 5.0-μm cartridge filter. This consists of a UV lamp encased in waterproof quartz filament housing and has a capacity for treating 8.0 gpm. Water is then passed through another 5.0-

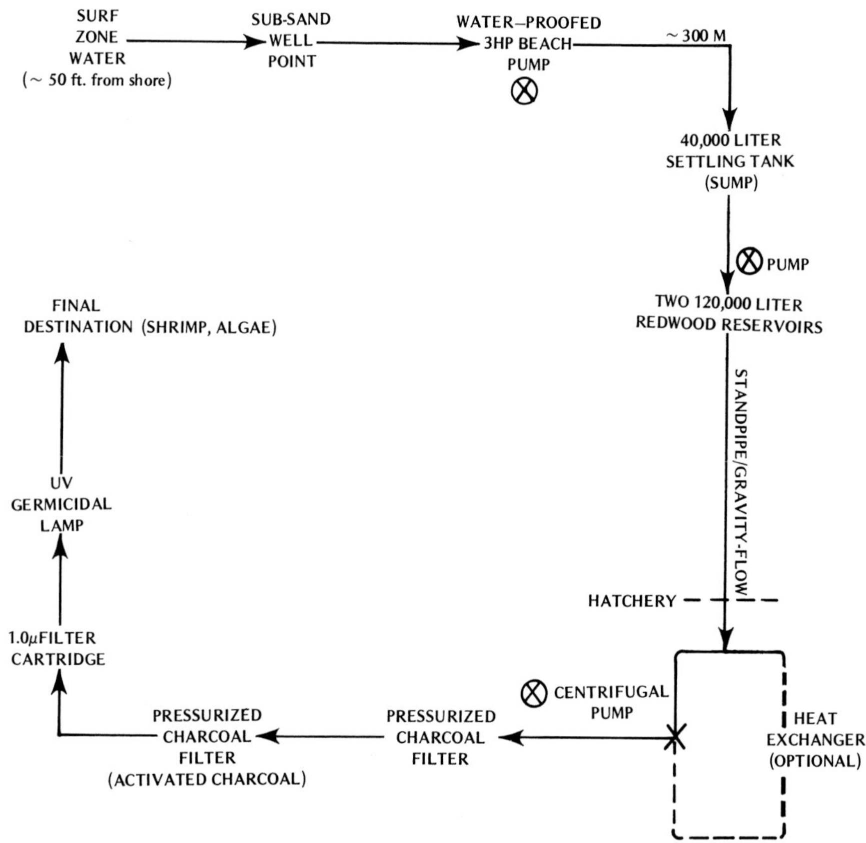

FIGURE 1. Schematic diagram of existing water treatment system for the NMFS-Galveston Laboratory.

μm cartridge filter prior to use in the hatchery tanks. See Figure 1 for a schematic of the seawater system.

SHRIMP HATCHERY SYSTEM

The shrimp hatchery system has already been described by Salser and Mock and other publications.[4-6] Figure 2 illustrates the hatchery system as it is used today. The present hatchery system consists of four 2000-ℓ (2 ton) conical tanks, each capable of producing 100,000 PL_5 shrimp (at a stocking density of 100/ℓ, 50% survival). The 2000-ℓ tank size has a capacity appropriate for a single spawn, a distinct advantage when the number of spawns in any one night is unpredictable.

Environmental Control

Temperature control — During colder months of the year it is necessary to heat the culture water to 28°C before it is placed in the culture tanks. The room temperature of the hatchery must also be maintained at 28°C. This temperature is used extensively throughout the world for hatchery operations.[7,8] Heating incoming seawater to 28°C is accomplished by an in-line heat exchanger which is situated inside the hatchery immediately before the pump used to pressurize the chemical filters (see Figure 1). The heat exchanger includes a thermostatically controlled boiler and can be by-passed if heat control is not needed. Heating inside the different hatchery rooms is accomplished with a central heating unit.

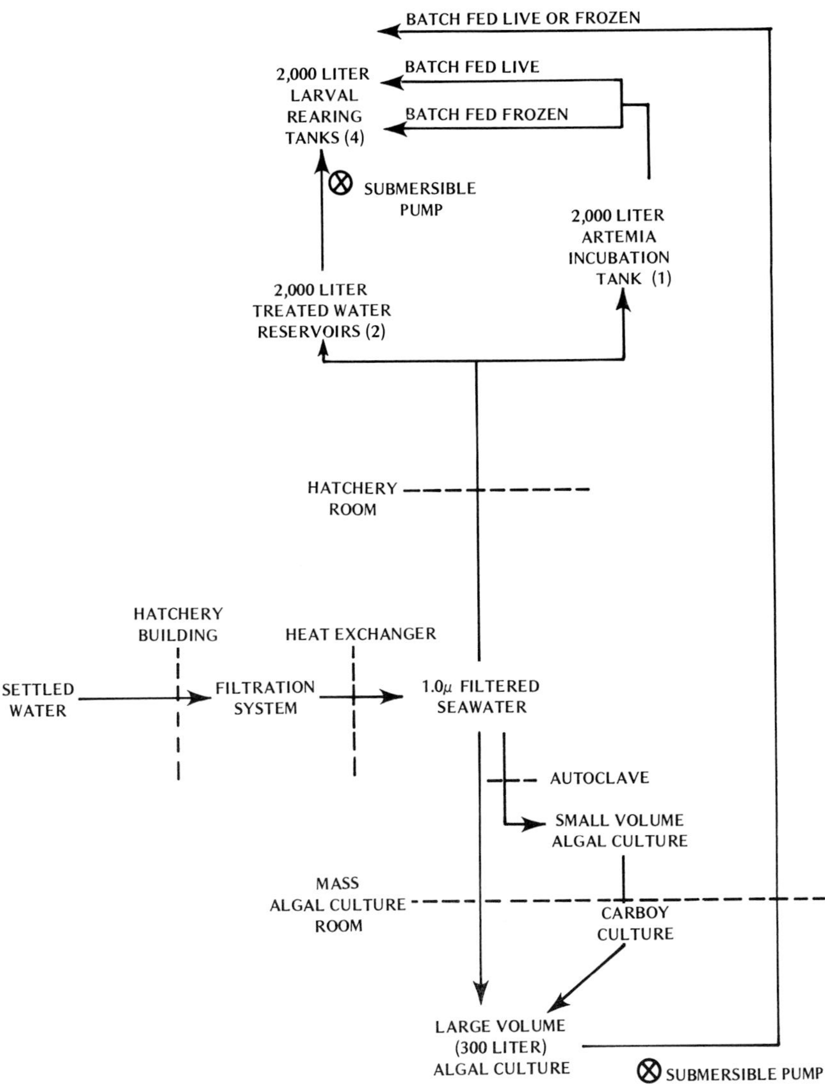

FIGURE 2. Schematic of NMFS-Galveston shrimp hatchery system, including algal culture system.

Salinity control — If incoming seawater is below 25 ppt, artificial sea salts are added to bring the salinity up to a suitable level (~ 30 ppt). Salinity does not appear to be critical as long as the range is between 25 and 35 ppt, and the incoming nauplii have been acclimated to the prevailing hatchery salinity.

Aeration

We have used a variety of aeration devices. We recommend a commercial, oil-free air blower with large volume and fairly low pressure for long-term dependability. It is important to provide several condensation traps throughout the system to bleed off any water. These blowers can be quite loud and should be situated accordingly.

SHRIMP HATCHERY OPERATION

Tank Preparation

Once a production run has been completed in a conical tank, the tank is immediately drained and sprayed down with hot water. This removes a large amount of sedimentation and accumulated organic material from the side of the tank as well as from the aeration equipment inside the tank (airstones, weights, airline, air-lift pumps). All equipment and materials inside the tank are then removed. Airstones and airlines are either replaced if no longer functional or, if functional, they are soaked in a dilute Chlorox® or HCL solution. If the airline is brittle, it is replaced. New airlines should be "cured" by submerging in fresh or salt water for 24 hr prior to use. This allows leaching of any toxins associated with the plastics used.

Stainless steel weights (or whatever weights are used) are soaked in a dilute Chlorox® solution, scrubbed, rinsed and reused. For experimental purposes, airline and airstones should be replaced after each run. Air-lift pumps should also be removed after each run and thoroughly scrubbed with Chlorox® before use. Submersible heaters are also removed and cleaned before replacing in the tank. The tank, once empty, should be ready for cleaning. The best method for cleaning large tanks is to stand outside the tank and use a long-handled, stiff-bristled brush. Once again, a dilute solution of Chlorox® can be used as a cleaning fluid. If cleaning the tank requires getting inside the tank, the worker should proceed cautiously. Shoes can introduce contamination and should be removed before entering the tank. Also, dilute Chlorox® will emit noxious fumes, therefore, one should plan to be in the tank for only a short period of time. The best practical approach is to avoid entering the tank. The entire internal surface area should be thoroughly scrubbed, rinsed in hot water, and allowed to dry before being equipped for the next run. Once the tank is dry, all aeration equipment can be replaced. As a precaution, all airlines should be bled prior to filling of the tank. Figure 3 illustrates the aeration system.

The tank is now ready for refilling. Filtered (~5.0-μm) seawater should be used. The tank is rinsed thoroughly with filtered seawater and allowed to drain. The drain is then closed and the tank is filled. Submersible heaters are replaced and set to keep the water at ~28°C. Aeration is initiated and balanced throughout the tank. A small amount of EDTA (10 mg/ℓ) is then added to chelate any toxic metals present in the water. (The chelate EDTA has been shown by Cook[1] to increase hatching rates of fertilized penaeid eggs.) The water level in the tank should be 2 to 3 in. below operating level to accommodate the additional volume of water which is added when algal foods are placed in the tank. Once the water temperature in the tank has reached ~28°C, it is ready for stocking.

When sufficient nauplii are available, they are acclimated, just prior to stocking, to conditions (salinity and temperature) existing in the tank. Algal foods are added according to a predetermined feeding guideline (Table 1). Cell density counts are made to confirm proper food density. Nauplii are then carefully transferred to the tank.

Acclimation and Stocking of Nauplii

Before larvae can be stocked in the hatchery they must first be acclimated to the physical conditions existing in the larval rearing tanks (i.e., temperature and salinity). If nauplii are in water that is within ± 0.5°C or ±1 ppt salinity no acclimation is necessary. If larger deviations are observed we use an acclimation chamber (see Figure 4). Nauplii are placed in the acclimation chamber with their original shipping water. Hatchery tank water is then delivered to the vessel very slowly by siphon. The design of the vessel allows the hatchery water to mix slowly with the shipping water while the nauplii remain concentrated in the central chamber. The rate of acclimation should not be greater than 1°C or 1 ppt/15 min. The rate is adjusted by controlling the flow of the siphon. Once temperature and salinity are equilibrated, the nauplii are ready to be transferred to the hatchery tank. This is done by lowering the water level in the central chamber of the acclimation vessel to the lower level of the mesh. Care must be taken to rinse the side of the container during the lowering process to keep nauplii from adhering to the sides. Once the nauplii are contained in the

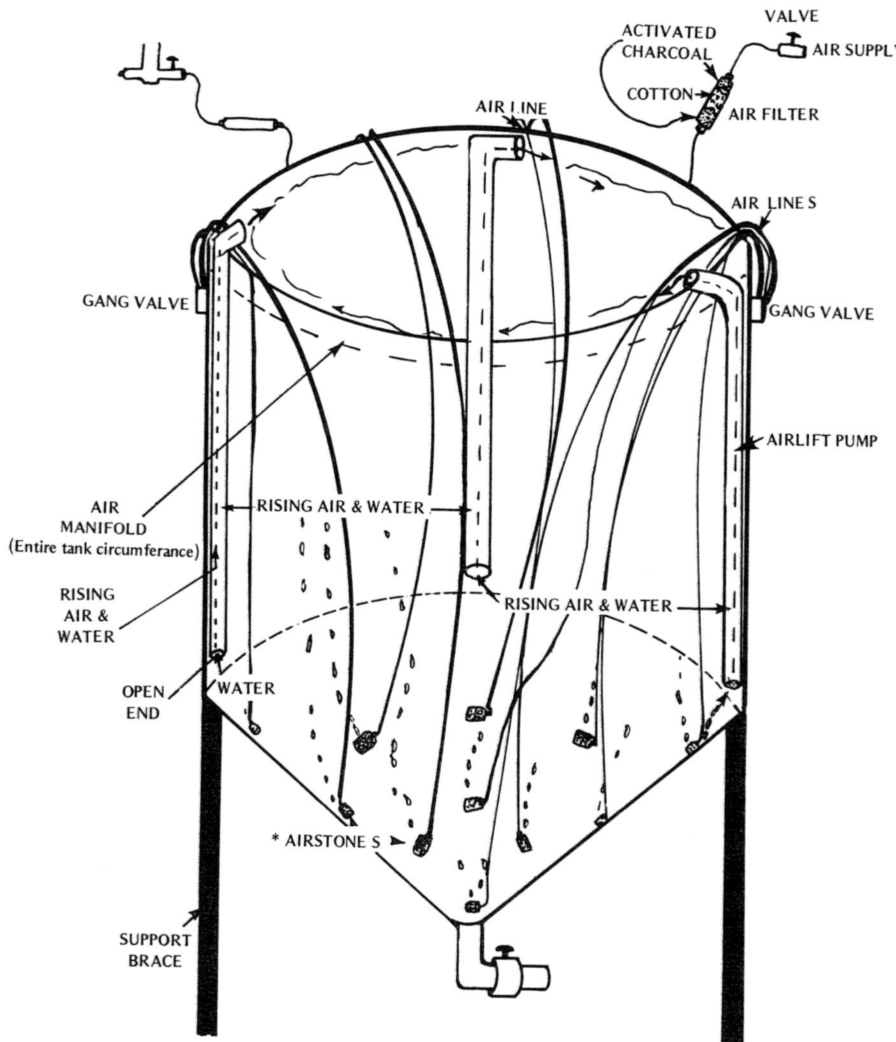

FIGURE 3. Diagram of typical 2000-ℓ conical bottom shrimp hatchery tank. Airstones provide aeration and mixing while air-lift pumps provide circular flow, aeration, and mixing. Drainage occurs through bottom fitting. An air manifold at air-water interface helps keep larvae from sticking to sides of tank above the waterline.

lower portion of the container, the container is lifted out and placed in the hatchery tank to release the nauplii by removing the bottom plug.

All nauplii should be inspected for quality and quantity before being placed in the acclimation chamber or hatchery tank. The nauplii should be examined for pigmentation, general activity, and physical deformities. It is also helpful to know the hatching rate and fertilization rate for a spawn. If there is any doubt as to the quality of nauplii, and if other spawns are available, poor quality nauplii should be discarded as hatchery operations are too expensive to utilize inferior nauplii. In addition, for experimental purposes, it is also advisable to use nauplii from a single spawn whenever possible, as stocking hatchery tanks with nauplii from several spawns inevitably leads to differential rates of development and possible cannibalism of advanced stages on less developed stages.

Table 1
FEEDING GUIDELINE AND EXPECTED SURVIVAL RATES FOR LARVAL PENAEID SHRIMP

Survival relative to stage (%)	Larval substage	Average time to obtain stage	Diet (cells/mℓ or *Artemia*/mℓ)
95	N_5	36 hr	75,000 *Isochrysis* sp. (80%) + yeast (20%)
80	P_1	48 hr	100,000 *Isochrysis* sp. (80%) + yeast (20%)
75	P_2	5 days	75,000 *Isochrysis* sp. (80%) + yeast (20%) 20,000 *Tetraselmis chuii*
70	P_3	7 days	50,000 *Isochrysis* sp. (80%) + yeast (20%) 40,000 *Tetraselmis chuii* 3.0 *Artemia* (frozen)
65	M_1	9 days	40,000 *Tetraselmis chuii* 3.0 *Artemia* (frozen)
60	M_2	10 days	20,000 *Tetraselmis chuii* 3.0 *Artemia* (frozen) 3.0 *Artemia* (live)
55	M_3	11 days	8.0 *Artemia* (live)
50	PL_1	12 days	10.0 *Artemia* (live)

Approximately 200,000 to 300,000 nauplii are, therefore, needed from a single spawn whenever possible (for the 2-ton tank). This can be achieved with the larger species such as *P. monodon, P. stylirostris*, and *P. vannamei*; but for smaller species, either smaller hatchery tanks or multiple spawns from the same night should be used.

For commercial hatcheries where square footage and operational costs are critical, it is more advantageous to use larger, multiple-spawn rearing tanks. If nauplii come from the same broodstock, rate of metamorphosis should be similar so that cannibalism will be minimal. (This statement assumes synchronous maturation of broodstock.)

Once the nauplii are in the hatchery tank a population count should be made and the nauplii should be fed according to the predetermined feeding guideline (see Table 1).

Feeding and Maintaining Larval Substages

Feeding Larval Substages

Table 1 provides feeding guidelines for all larval stages as well as the percent of expected survival and duration of each stage. Feed levels are determined each morning and evening and appropriate feeds added to the tank to bring the level of food to the prescribed amount. Algal levels are determined by taking small water samples with a beaker, stirring the sample for homogeneity, and pipetting enough sample for placement in a hemocytometer. The technique for determining the level of algae by hemocytometer count is explained in the chapter on algae culture at the Galveston lab (this volume). *Artemia* levels are determined by strip sampling (see next section) and concentrating a known volume of sample in a sieve and then counting total *Artemia*. If nauplii are consuming large amounts of feed, the feed levels may be determined three times a day and feed added in the late evening hours to assure an adequate level through the night.

Counting Larval Substages

It is very important to record larval survival during the course of the hatchery run in order to determine progress and the success of the run. Larval population counts are done each morning by either strip sampling or by Hensen-Stemple pipette. We prefer the strip sampling method.[2] Strip sampling involves submerging a 1-in. × 3-ft PVC pipe downward into the

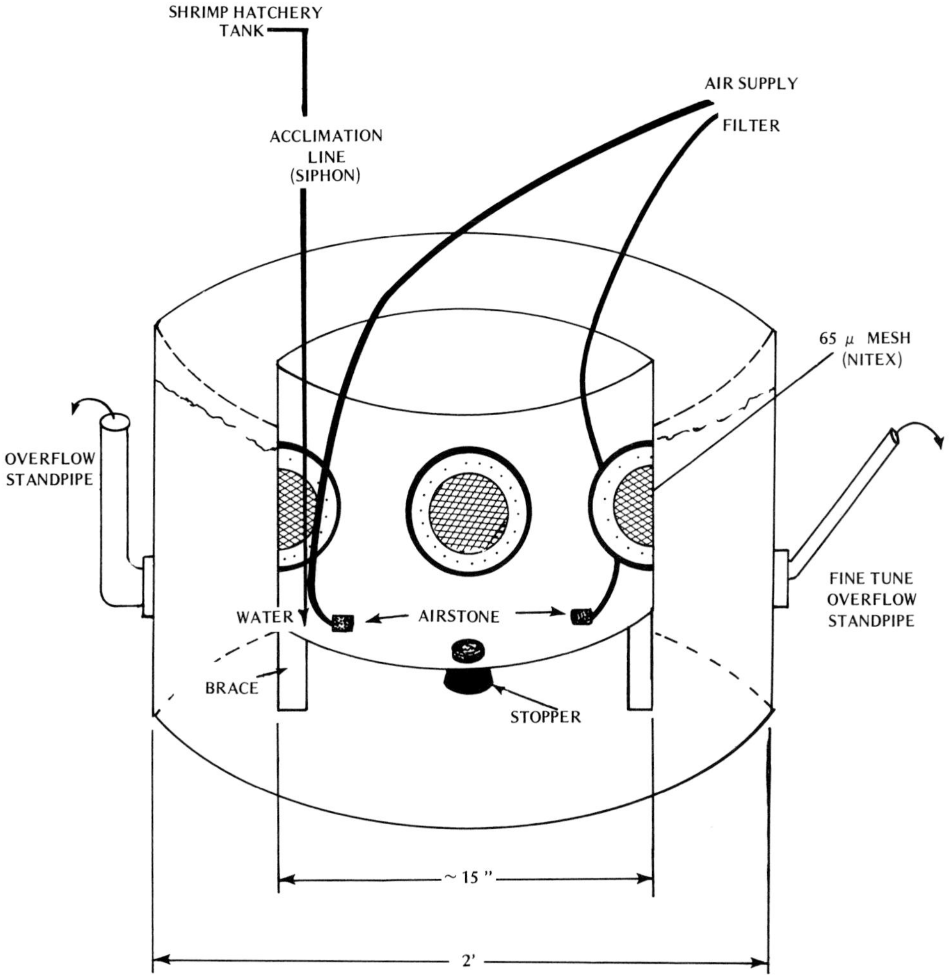

FIGURE 4. Acclimation and harvesting chamber. Shrimp are retained within inner chamber by nitex screening.

water column of the hatchery tank. Both ends of the pipe are kept open so that the pipe effectively samples the whole water column. Prior to removal, the upper end of the tube is capped with an appropriate size rubber stopper and the lower end is covered by hand. The tube is quickly removed and the contents slowly released into a graduated plastic pitcher (2-ℓ capacity). Two other areas of the water column are also sampled and combined with the first sample. The total volume sampled is usually between 1600 and 2000 mℓ. If the larvae appear unevenly distributed prior to sampling, the hatchery tank should be stirred with a large paddle to achieve homogeneity. However, care should be taken when stirring to minimize damage to the larvae.

Once a strip sample has been taken and the volume recorded, the contents of the pitcher are poured through a small 100-μm sieve. This sieve (Figure 5) is constructed of nitex screen and a 1-in. (level) section of 1 1/2-in. diameter PVC pipe. The screen is attached to the PVC pipe with PVC cement. Before pouring, the sieve is placed in a round or square gridded plastic petri dish. The sieve is held between the thumb and forefinger and slightly tilted so the water can flow easily out the bottom of the sieve. The petri dish and sieve are both tilted to allow a steady overflow of excess filtered water. All larvae from the strip sample are concentrated in the small sieve for counting. The sieve is then placed under a dissecting

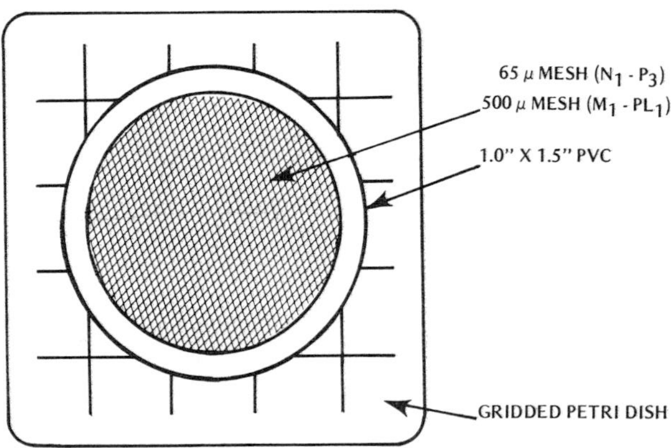

TOP VIEW

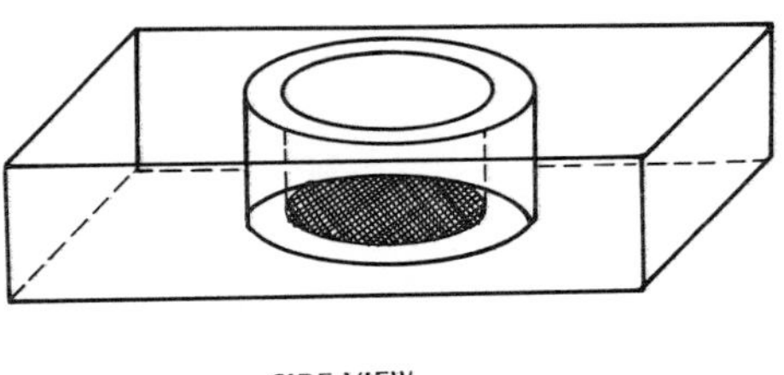

SIDE VIEW

FIGURE 5. Larval concentration sieve.

scope and the larvae evaluated for activity level. Once all "live" evaluations of the animals have been made, they are preserved by adding a few drops of 5% formalin solution to the sieve. The animals are then counted physically by removing individuals with a pipette or by visual grid examination (low concentrations). The number of larvae is counted and population is determined by the following equation:

$$\frac{\text{larvae per sample}}{\text{sample volume, m}\ell} = \frac{\text{larvae per tank (unknown)}}{\text{tank volume, m}\ell}$$

$$\text{larvae per tank} = \frac{\text{(larvae per sample) (tank volume in m}\ell)}{\text{m}\ell \text{ per sample}}$$

Example:

$$N_t = \frac{(100)\ (2{,}000{,}000)}{2000}$$

$$N_t = \frac{200{,}000{,}000}{2000}$$

$$N_t \cong 100{,}000 \text{ larvae per tank}$$

The Hensen-Stemple pipette method is only used for concentrated, homogeneous animal samples (larvae or *Artemia*) where the whole water column can be easily stirred. Counting highly concentrated larvae samples by Hensen-Stemple pipette usually requires dilution. For

FIGURE 6. Penaeid egg, 280 μm.

this purpose a 1.0- to 2.0-mℓ sampling chamber is used. The sample to be counted is stirred thoroughly with the pipette held in the "open" position. Once the water column has been homogenized, the sample is quickly taken. The closed pipette is rinsed to remove any larvae clinging to the outer surface and, if the sample is not too concentrated, it is placed in either a sieve or a gridded petri dish and counted, as with strip samples. If the sample is too concentrated as with harvested *Artemia*, it is diluted by placing the 1.0-mℓ sample in 999 mℓ of filtered seawater (1:1000 dilution) and stirred on a magnetic stirrer for approximately 1 min. A 1.0-mℓ sample is then taken from the dilution, evaluated, counted, and the result multiplied by 1000 to give the concentration in nauplii per milliliter of original sample. If the sample was taken from a 2000-ℓ hatchery tank, the number of nauplii per milliliter is multiplied by 2,000,000 to get the number per tank.

Identifying Larval Stages

Naupliar Stage

A gravid female shrimp, depending on size and species, can produce between 50,000 and 500,000 eggs per spawn. When maintained at an optimum temperature of approximately 28°C, fertilized eggs, which are about 280 μm (Figure 6), will usually hatch within 18 hr to produce the first larval stage, the nauplius. Freshly hatched nauplii are approximately 0.3 mm in length and are characterized as being totally planktonic, positively phototactic, and subsisting entirely on their own egg yolk. In most cases, the nauplius stage is divided into five substages N_1 to N_5. Early substages (N_1 to N_3) are differentiated by structural variations in the first antennae, the second antennae, and the caudal peduncle (Figures 7 and 8). Latter substages (N_4 and N_5) exhibit rudimentary feeding appendages with mandibles becoming

FIGURE 7. Nauplius (N_1), approximately 300 μm in size. Note posterior setae and setae at end of appendages only.

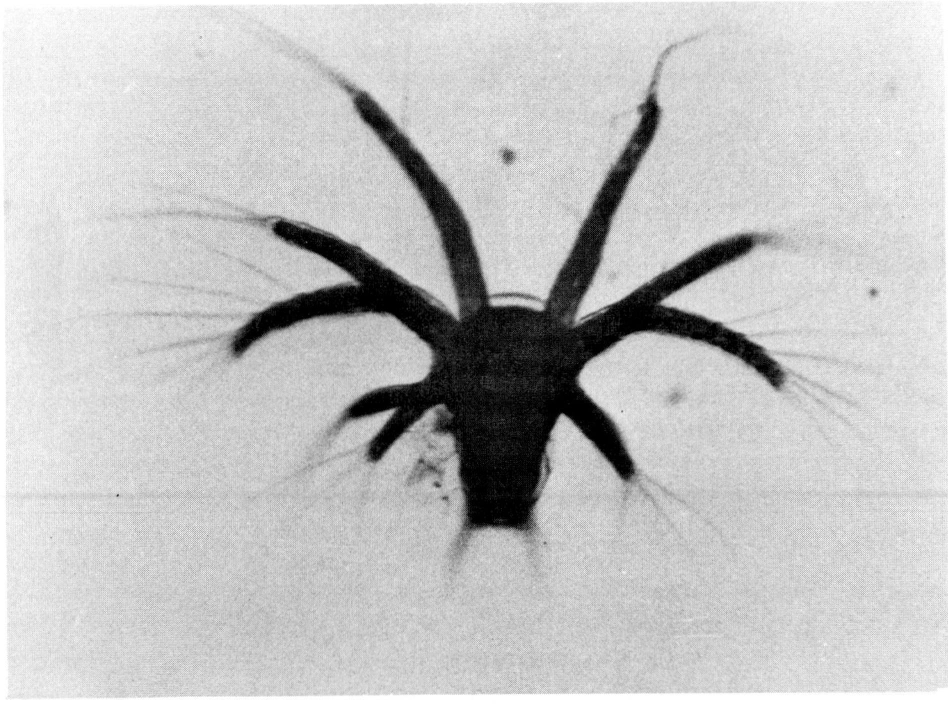

FIGURE 8. Nauplius (N_2). Note greater number of setae on appendages and posteriorly.

FIGURE 9. Nauplius (N_5). Notice elongation and bifurcation of future telson area. Appendages appear to be situated more to the anterior of the larva.

visible in the last substage, N_5 (Figure 9). At 28°C the nauplius stage normally lasts 36 to 48 hr.

Protozoeal Stage

The next stage, protozoea, denotes a radical change in the shape of the larvae. The body is more elongate, approximately 1.0 mm in length, and appears more suited for active swimming. The presence of feeding appendages suggests that the larvae are now capable of ingesting food, a major distinction from the nauplius stage. The major distinguishing characteristics of the protozoeal stage are shown in Figures 10 to 13. The protozoeal stage is classified into three substages.

1. P_1 — The first protozoea substage (P_1) larvae can be distinguished from N_5 larvae by several characteristics. A distinct, well-rounded carapace is present emphasizing the developing abdominal region. Feeding appendages are functional and antennae and caudal spines are even further developed. As with N_5 larvae, a single central eyespot is still present (Figure 10).
2. P_2 — Substage P_2 larvae are distinguished from P_1 larvae by the presence of two compound eyes, a rostrum with spines, and a further elongated abdomen. Antennae are no longer laterally situated (as in P_1), but are oriented toward the anterior (Figure 11).

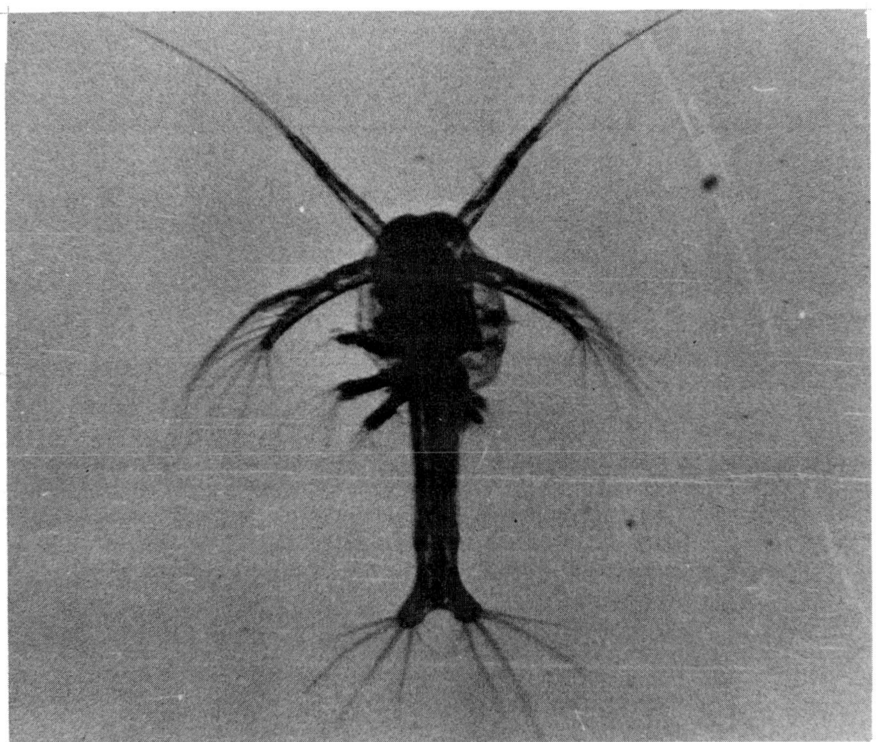

FIGURE 10. Protozoea (P_1), ventral view. Notice extension of the abdomen and development of the thoracic region.

3. P_3 — The abdominal segments of P_3 larvae contain dorsal and lateral spines which were absent in the second substage, P_2. Antennae are more setose (contain numerous hair-like bristles) while also being shorter and thicker (Figure 12). Vestigial pereipods or walking limbs posterior to the feeding appendages can also be seen. The most obvious characteristic of P_3 larvae is the presence of uropods, flat fan-like swimming appendages, anterior to the tail (Figure 13). Substage P_3 larvae are approximately 2.5 mm in length.

Since protozoea are planktonic filter feeders, they should not be fed an elusive diet; the diet must also be planktonic. A typical diet for protozoeal shrimp consists of marine phytoplankton, small unicellular plants approximately 3 to 30 μm in diameter. Protozoeal shrimp will, however, feed on any particle of suitable size for filtration. Because of this, yeast and soycake have also been used. Protozoea continue to subsist partially on yolk during P_1, but are primarily dependent on the environment for food during the latter substages. Under standard conditions the protozoeal stage lasts approximately 5 days.

Mysis Stage

The third larval stage is the *mysis* stage. In this stage the larvae have assumed an even more adult-like appearance and actively feed on both phytoplankton and zooplankton (microscopic marine animals). The major distinguishing characteristics of the three mysis substages are shown in Figures 14 to 19:

1. M_1 — The mysis larvae (M_1) are easily distinguished from the last protozoeal substage by an even further elongation of the body, development of functional pereipods,

FIGURE 11. Protozoea (P_2), ventral view. Notice first appearance of two compound eyes and further elongation of abdominal view.

and the appearance of another fan-like swimming appendage, the telson (Figure 14). No pleopods are present (Figure 15).

2. M_2 — Substage M_2 larvae are characterized as being longer than M_1 larvae, although the difference is subtle. Inspection of abdominal segments reveals the presence of vestigial unsegmented pleopods, another swimming appendage (Figures 16 and 17).
3. M_3 — Substage M_3 larvae (Figure 18) can be distinguished from M_2 larvae by appearance of the pleopods (Figure 19). These pleopods are composed of two segments and are nearly twice as long as those of M_2 larvae. The larvae are now approximately 4.0 mm in length.

Under normal conditions, the duration of the mysis stage is approximately 3 days. During the mysis substages there is a gradual transition from phytoplankton to zooplankton as the preferred food. Most hatcheries use brine shrimp, *A. salina*, as the standard zooplankton portion of the diet; however, other sources such as rotifers and nematodes have been tried.

Postlarval Stage

Once the shrimp have successfully developed past the mysis stage, they are considered

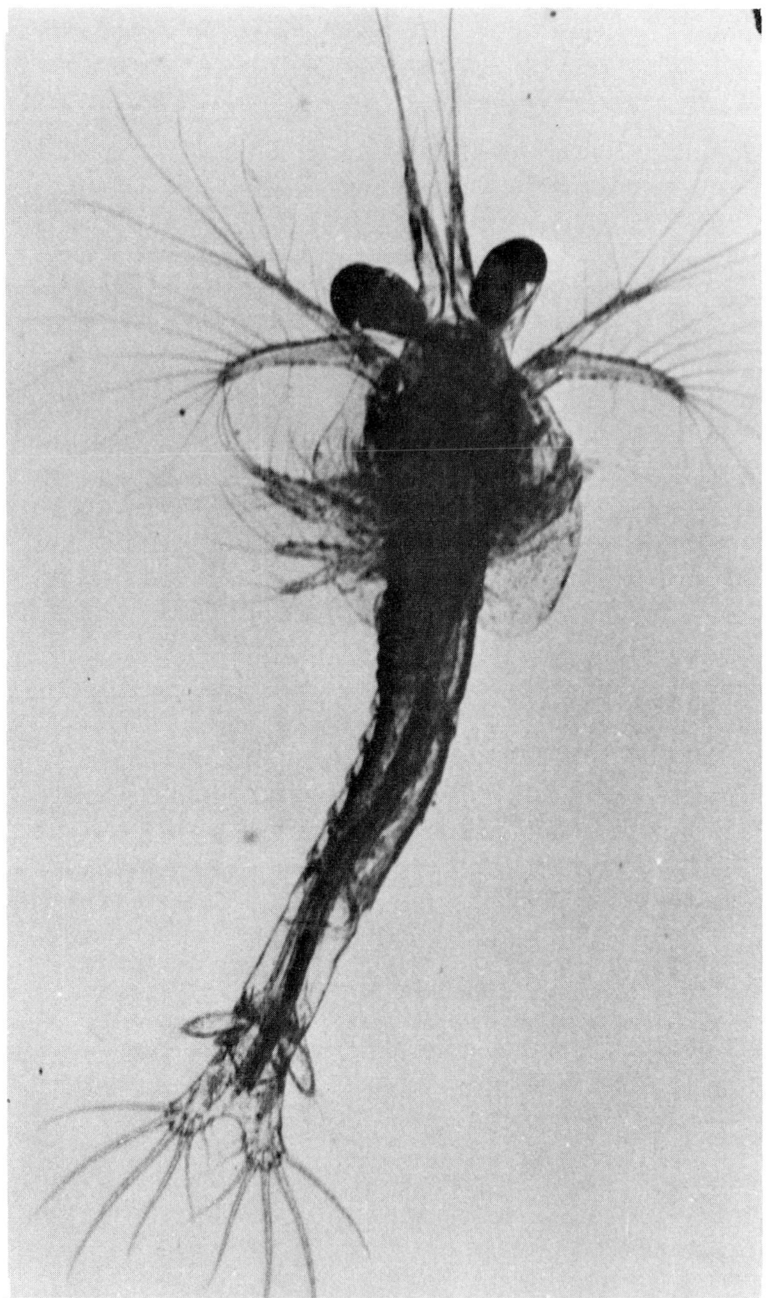

FIGURE 12. Protozoea (P_3), ventral view. Notice development of uropods on the posterior of the larva.

postlarvae. Postlarvae resemble adult shrimp and are approximately 4.5 mm in length (Figure 20). Postlarvae can be differentiated from M_3 larvae by the appearance of setae (hair-like bristles) on the abdominal pleopods (Figure 21). Early postlarvae are still partially pelagic, but mostly benthic in habits, while those 6 days old are predominantly benthic.

Water Exchange

Water quality is a critical factor in the rearing of penaeid larvae. If water is not exchanged

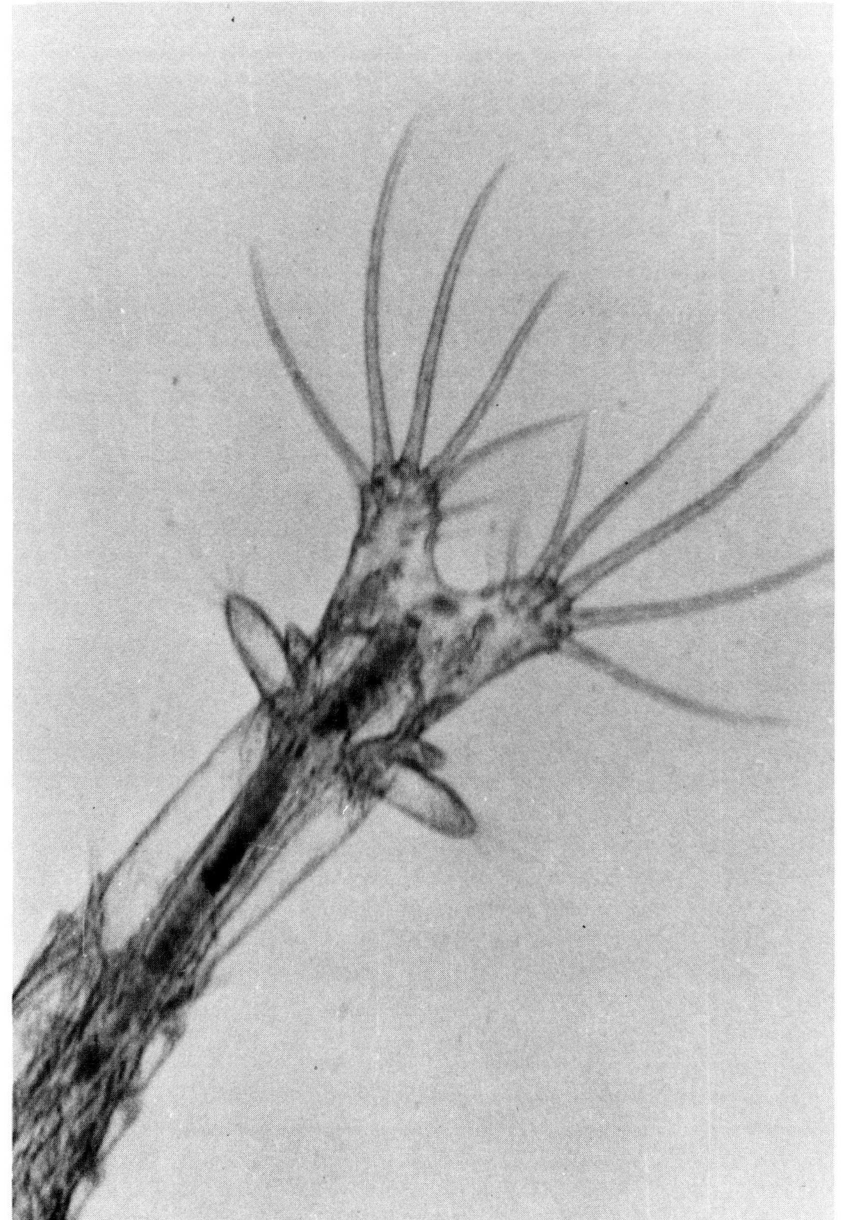

FIGURE 13. Protozoea (P_3), dorsal view of tail section, showing uropod development.

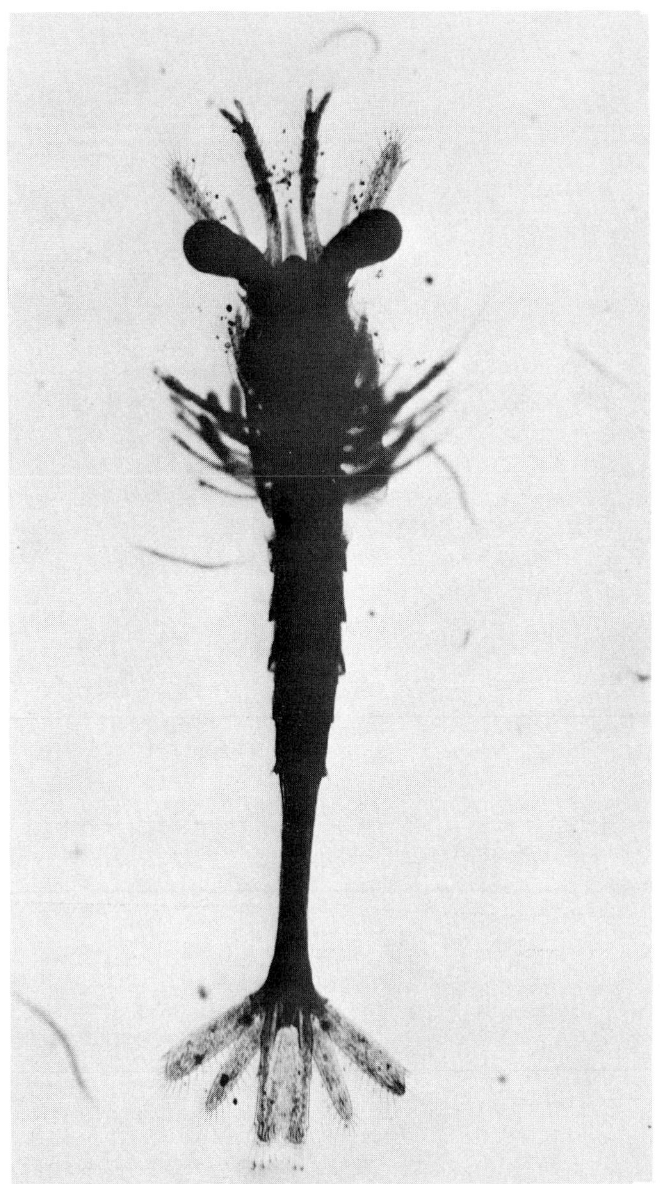

FIGURE 14. Mysis (M_1), dorsal view. Notice development of telson, further differentiation of uropods, lack of abdominal appendages (Pleopods).

on a daily basis, larval and algal metabolites can build up encouraging protozoan and bacterial contamination, thus, affecting survival. We initiate water exchange whenever we suspect water quality problems and every day after larvae reach the mysis stage. We suspect that metabolite build-up is minimal during nauplius and protozoea stages and water exchange during this time could be more detrimental than positive. We need to do more work in this area.

The exchange rate during the mysis and postlarval stages is 50% of the tank volume per day. If an unusually large amount of bacterial and/or protozoan contamination is present prior to exchange, larger volumes can be exchanged. Any water exchange should be weighed against possible stress to larvae and type of bacterial contamination. Some epizootic bacteria

FIGURE 15. Mysis (M_1), lateral view, showing lack of pleopod development.

FIGURE 16. Mysis (M_2), dorsal view. Notice development of antennae, elongation of rostrum, early formation of pleopods.

would not be reduced by the water exchange technique. All food and population counts should have already been completed before any water exchange takes place. Algal foods should not be added immediately prior to exchange to avoid loss of the new food during the exchange. Air lift pumps (Figure 3) should be turned off and a center screen (Figure 22) should be installed. When this screen is installed properly, the larvae will be retained as the "old" water is drained out of the tank. To verify the retention of larvae, the acclimation vessel (Figure 4) should be placed so that the drain line empties into it. The drain should then be opened one fourth of the way and water allowed to flow through the acclimation vessel and harvesting chamber for approximately 10 min. Effluent from the drain line should be checked for larvae that might be escaping or which were trapped within the screen. If no larvae are seen in three replicate effluent samples, the acclimation vessel can be removed, and its contents poured back into the hatchery tank. The drain line can then be placed in

FIGURE 17. Mysis (M_2), lateral view. Notice unsegmented pleopods.

FIGURE 18. Mysis (M_3), dorsal view. Notice development of chelipeds.

the floor drain and the valve opened up halfway. Check for deposition of larvae on center screen. If noticed, then the flow rate is too high. Exactly 50% of the total tank volume should be exchanged unless contamination is high. If the tank is not calibrated for volume, determine the volume of the cylindrical portion of the tank ($\pi r^2 \ell$) and determine volume per inch (liters per inch). A yardstick can then be used to measure water volume in the tank. During the entire water exchange, the culture screen should be kept rinsed off. Once the exchange has been completed, the drain is closed and the center screen removed and placed across the top of the tank. Larvae adhering to the screening are then washed back into the hatchery tank.

Refilling the tank is accomplished by adding treated (10 mg EDTA per liter), filtered (5.0 μm) seawater from a reservoir tank adjacent to the hatchery tank. We use a submersible

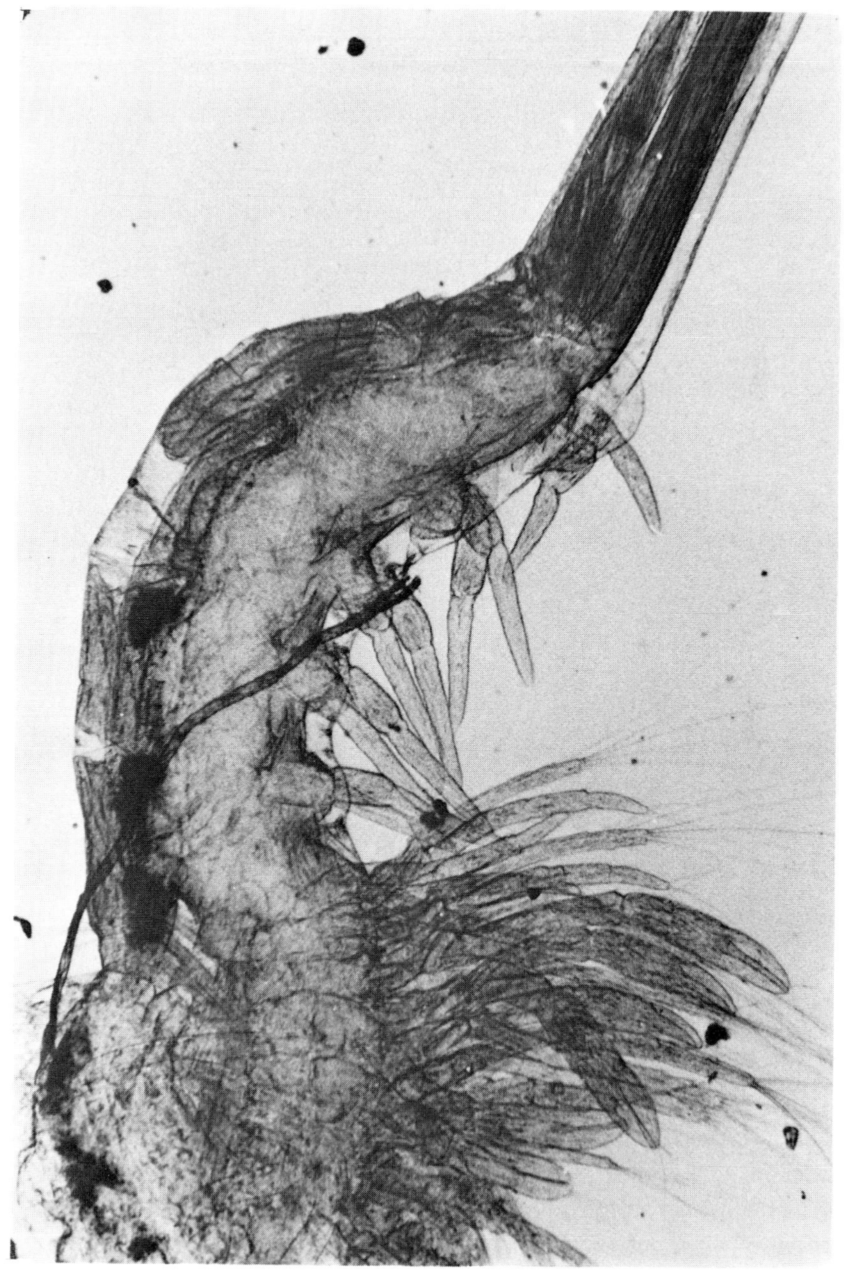

FIGURE 19. Mysis (M_3), lateral view. Notice segmented, nonsetose, pleopods.

FIGURE 20. Postlarva (PL_1), dorsal view. Notice well-developed pleopods, shrimp-like shape.

pump which is lowered into the reservoir tank for transferring the water. The hatchery tank is filled to a level approximately 2 to 3 in. below the effluent ports of the air-lift pumps. This is to allow for addition of algal food. Once a sufficient amount of food has been added, the tank can be topped up to operating level.

Water Treatments

Several chemical additives have been used to promote water quality or reduce contamination in the Galveston shrimp hatchery. The only additive presently in use is ethylenediaminetetraacetic acid (EDTA), a chelating agent. Other compounds such as the antibiotics Maracyn I®, Maracyn II®, and Terramycin have been used occasionally for control of bacterial contamination.

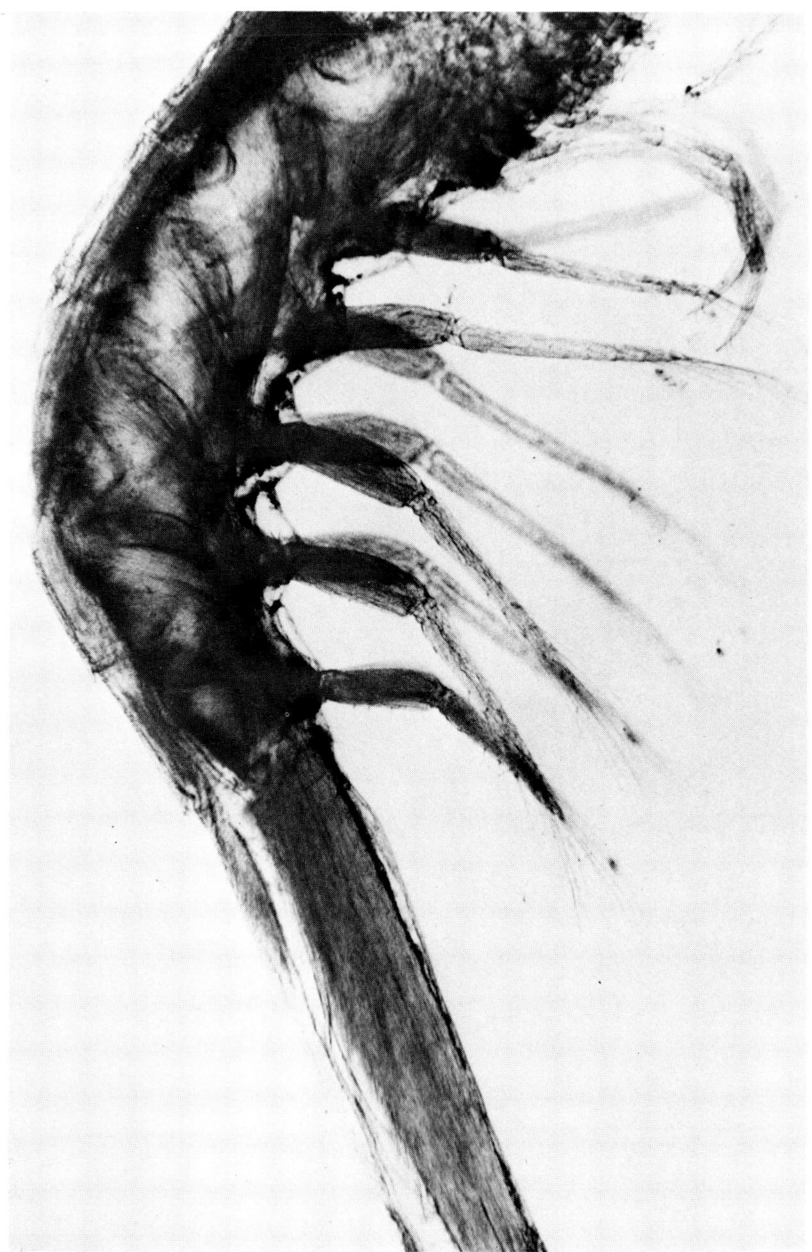

FIGURE 21. Postlarva (PL_1), lateral view. Notice development of setae on pleopods.

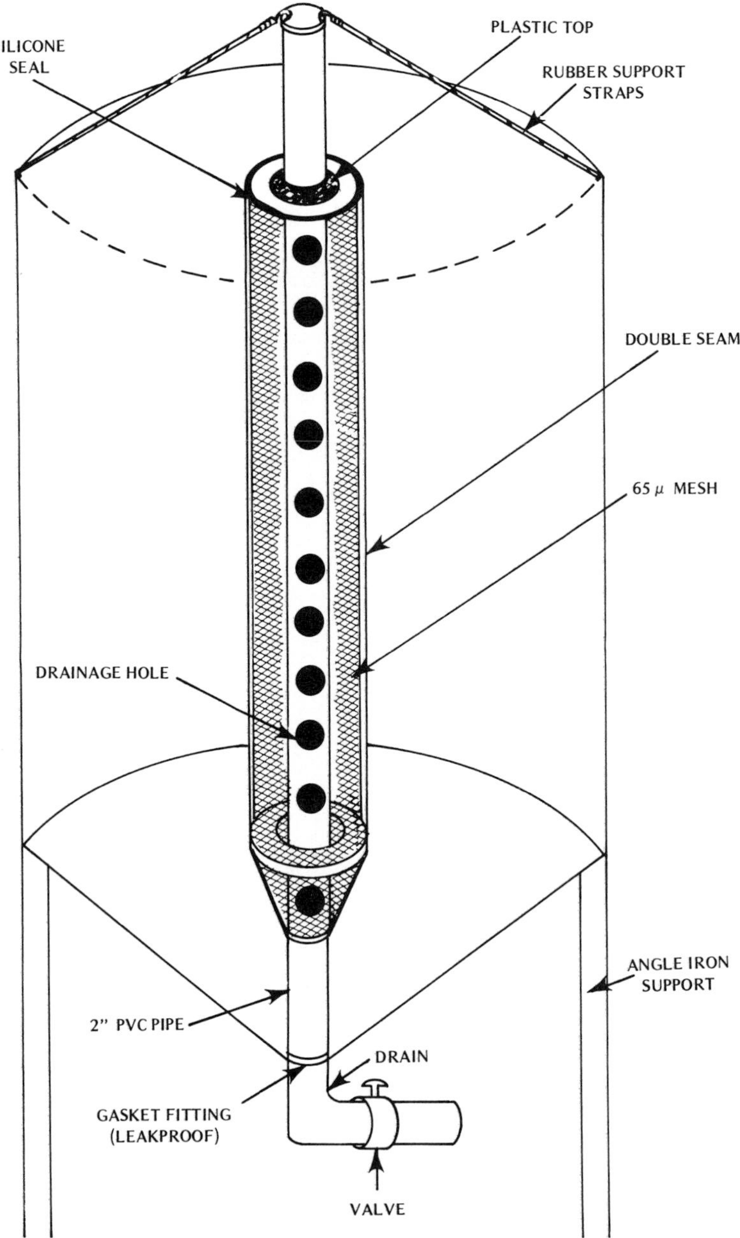

FIGURE 22. Center screen apparatus used during water exchange to retain shrimp larvae in tank.

The chelate, EDTA, has been shown by Cook[1] to increase hatching rates of fertilized penaeid eggs from wild females mated offshore (Gulf of Mexico) and brought into the laboratory to spawn in seawater containing EDTA. Though EDTA has been used since 1969 for the hatching of marine shrimp eggs, there has been little explanation of its beneficial effect on crustaceans.

EDTA and other chelates have received widespread use in the culture of marine algae.[9] Johnston reported that the presence of EDTA and the resulting chelated metals increased growth of marine phytoplankton cultures.[10] According to Johnston, EDTA increases the availability of trace metals essential to the growth of marine algae. Results from Sunda and Guillard,[11] Daney et al.,[12] and Ragan et al.[13] indicate that EDTA enhances algal growth by complexation, thus, making inaccessible toxic or growth-suppressing cations inherent in some seawater.[11]

Occasionally we have tried UV irradiation to control flagellates and other biotic contamination. Each tank can be equipped with its own UV lamp and the water recirculated through the UV device by drawing from the drain with the retention screen in place. This method has a drawback in that the retention screen becomes clogged and serves as an additional substrate for contaminants. A UV device would be more useful if the water to be treated could be drawn directly from the tank through a filter screen. We intend to define the use of UV light and its potential in future experiments.

Harvesting and Holding Postlarvae

Once shrimp larvae leave the mysis-III substage they metamorphose into postlarvae. The age of shrimp is then calculated from metamorphosis and a 1-day-old postlarvae is designated as a PL_1, a 2-day-old postlarvae as a PL_2, and so on.

The time required for the majority of larvae to metamorphose to postlarvae in a particular hatchery tank varies with the quality of larvae, culture conditions in the tank, and the feeding regime utilized. We usually harvest a tank 5 days after the first postlarvae appear. This usually results in an effective yield of 40 to 60%. Holding postlarvae longer in the tank would lead to mortalities due to cannibalism. Once the shrimp metamorphose to postlarvae they become more benthic in habits, and the conical bottom tank does not offer the surface area necessary for large numbers of postlarvae. Therefore, it is important to get postlarvae out of the hatchery as soon as possible.

In order to harvest, the entire contents of the hatchery tank are drained into the acclimation chamber (Figure 4). The acclimation tank is then transferred to a suitable holding tank and the larvae released. The holding tank should have adequate aeration and a larger bottom surface area to accommodate the postlarvae. While in the holding tank larvae should be fed on live *Artemia* and prepared flake foods until they can be converted to pelleted foods.

Postlarvae should be placed in nursery ponds or raceways as soon as possible after harvest. Prolonged holding in small containers leads to unnecessary mortalities. One alternative to the methods described would be to transfer M_1 larvae to hatchery raceways to provide them more room after metamorphosis. We hope to try this procedure in the future.

ACKNOWLEDGMENTS

The authors would like to thank UTMB-Galveston for the use of their photomicroscopic equipment, Dr. Charles Hooks for development and custom processing of photomicrographs, Diane Hensley for scientific illustrations, and the administration and staff of NMFS-Galveston and TAMU.

REFERENCES

1. **Cook, H. L. and Murphy, A. M.,** The culture of larval penaeid shrimp, *Trans. Am. Fish. Soc.*, 98(4), 751, 1969.
2. **Mock, C. R., Revera, D. B., and Fontaine, C. T.,** The larval culture of *Penaeus stylirostris* using modifications of the Galveston Technique, *W.M.S.*, 11, 102, 1980.
3. **Mock, C. R., Fontaine, C. T., and Revera, D. B.,** Improvements in rearing larval penaeid shrimp by the Galveston Laboratory method, *Brine Shrimp Artemia,* 3, 331, 1980.
4. **Salser, B. R. and Mock, C. R.,** Equipment used for the culture of larval penaeid shrimp at the National Marine Fisheries Service Galveston Laboratory, in Proc. 5th Congreso Nacional de Oceanographia, Guaymas, Sonora, Mexico, October 22 to 25, 1974.
5. **Mock, C. R.,** Larval culture of penaeid shrimp at the Galveston Biological Laboratory, NOAA Tech. Rep., *N.M.F.S. Circ.*, 388, 34, 1974.
6. **Neal, R. A.,** Penaeid shrimp culture research at the National Marine Fisheries Service Galveston Lab., in *Proc. FAO Tech. Conf. on Aquaculture, Kyoto, Japan,* 1975.
7. Aquacop, Elevage larvaire de Peneides en milieu tropical, in 3rd meeting of the I.C.E.S. Working group on mariculture, Brest, France, May 10 to 13, 1977, *Actes de Colloques du CNEXO,* 4, 1977, 179.
8. **Liao, I. C.,** On the artificial propagation of five species of prawns, *Chin. Fish. Mon.,* 205, 3, 1970.
9. **Stein, J. R.,** *Handbook of Phycological Methods, Culture Methods and Growth Measurements,* Cambridge University Press, New York, 1973.
10. **Johnston, E.,** Seawater, the natural medium of phytoplankton. II. Trace metals and chelation and general discussion, *J. Mar. Biol. Assoc. U.K.*, 44, 87, 1964.
11. **Sunda, W. and Guillard, R. R. L.,** The relationship between cupricion activity and the toxicity of copper to phytoplankton, *J. Mar. Res.*, 34, 511, 1976.
12. **Davey, E. W., Morgan, M. J., and Erickson, S. J.,** A biological measurement of the copper complexation capacity of seawater, *Limnol. Oceanogr.*, 18, 993, 1973.
13. **Ragan, M., Ragan, C., and Jensen, A.,** Natural chelators in sea water. Detoxification of Zn^{2+} by brown algal polyphenols, *J. Exp. Mar. Biol. Ecol.*, 44, 261, 1980.

MATURATION AND SPAWNING OF PENAEID PRAWNS IN TUNGKANG MARINE LABORATORY, TAIWAN

I-Chiu Liao and Yi-Peng Chen

The successful controlled maturation in captivity of some species of penaeid prawn has been accomplished at several laboratories in the world (see Table 1). Different kinds of treatment, such as control of physical factors of the environment, preparation of nutritive diets, and eyestalk ablation, are employed to promote ovarian maturation and spawning. At present, most of these experiments can induce spawning, and the rearing of postlarvae at various survival rates is now possible. In this section, a discussion is made and an outline given on the practical application of induced maturation and spawning of penaeid shrimps, especially those techniques followed for the grass prawn, *Penaeus monodon*, at the Tungkang Marine Laboratory (TML) in Taiwan, Republic of China.

Tungkang Marine Laboratory, initially named "Tungkang Shrimp Culture Center", was established in 1968. It was originally a section of the Tainan Station of the Taiwan Fisheries Research Institute (TFRI), but became an independent organization, directly under the TFRI in January 1971. This laboratory, with a land holding of 7 ha, is located on the southwest coast of Taiwan facing the Taiwan Strait, at a latitude of 21° 54′ N.

In its short span of 13 years, it has contributed substantially toward solving problems encountered in the large-scale culture of eight fish and prawn species of great commercial value to Taiwan.[1,2] This laboratory is also actively engaged in developing more advanced techniques which promise to promote further advancement of aquaculture in this country.

At the Tungkang Marine Laboratory there are adequate research facilities for the study of fish and prawn nutrition, physiology, and histology, as well as fisheries resources. Constant temperature rooms are available for the study of factors such as temperature tolerance limits of marine and freshwater fish and prawn species. In addition to a large green house equipped with indoor ponds and an algae cultivation division, the TML has more than 100 outside ponds of various capacities, ranging in size from 1 to 800 m^3. The TML also has a good library and provides extension services to prawn and fish farmers.

Even though techniques for the artificial propagation of some penaeid prawns in Taiwan were perfected in 1968,[3] the availability of spawners of *P. monodon* still depended mainly on the capture of wild gravid females from the sea.[4] Unfortunately, the number of mature females became fewer year by year.[5,6] To assure a stable supply of spawners and seed availability, it, therefore, became increasingly important to induce maturation and to spawn these prawns in captivity. Research on this topic began in 1973.[7] Fifteen-month-old laboratory-reared *P. penicillatus* Alcock were held in an indoor concrete tank of 7 × 2.5 × 1.5 m. One month later the gonads of two out of eight females reached the stage of ovarian maturation, with a Gonadal Somatic Index (GSI) of 8.99. At the same time, a mature pond-reared spawner of *P. monodon* was found, but spawning occurred without cleavage.[7] Also, the treatment of bilateral eyestalk ablation was carried out on *P. japonicus* but with a low survival rate.[7]

P. monodon is one of the largest prawns among Indo-Pacific penaeid species,[8-11] and the most suitable for mariculture. However, it seldom reaches sexual maturity in ponds. Therefore, trials to induce sexual maturity in *P. monodon* in captivity were carried out at the Tungkang Marine Laboratory during the last few years. Eyestalks of decapod crustaceans are known to have an X-organ secreting molting-inhibiting hormone (MIH) and gonad-inhibiting hormone (GIH).[12,13] So the application of eyestalk ablation to *P. monodon* was considered to be the best technique to reduce hormonal production of gonad-inhibiting hormone and to accelerate vitellogenesis.

Table 1
SUMMARY OF RESEARCH ON MATURATION AND SPAWNING OF PENAEID PRAWNS IN CAPTIVITY

Species	Country	Treatment of eyestalk ablation	Result			
			Maturation	Spawning	Hatching	Postlarvae
Penaeus aztecus	U.S.	+	+	?	?	?
	Tahiti	+	+	+	+	
P. californiensis	U.S.	−	+	+	+	+
P. duorarum	U.S.	+	+	?	?	?
P. japonicus	Tahiti	+	+	+	?	?
	Taiwan	+	+	?	?	?
P. kerathurus	Italy	+	+	+	+	+
	Spain	−	+	?	?	?
P. merguiensis	Tahiti	−	+	+	+	+
	Jepara	+	+	+	+	+
	Great Britain	−	+	+	+	+
	Fiji	−	+	+	+	+
P. monodon	Taiwan	−	+	+	−	−
	Great Britain	+	+	+	−	−
	Philippines	+	+	+	+	+
	Jepara	+	+	+	−	−
	Philippines	+	+	+	+	+
	Taiwan	+	+	+	+	+
	Philippines	+	+	+	+	+
	Tahiti	+	+	+	+	+
	Philippines	+	+	+	+	+
	Taiwan	+	+	+	+	+
	Tahiti	+	+	+	+	+
	Great Britain	+	+	+	+	+
P. orientalis	Great Britain	+	+	+	−	−
P. penicillatus	Taiwan	+	+	+	+	+
P. plebejus	Australia	+	+	+	−	−
P. semisulcatus	Tahiti	−	+	−	−	−
P. setiferus	U.S.	+	+	−	−	−
	U.S.	−	+	+	−	−
	U.S.	+	+	+	?	?
	U.S.	+	+	+	+	+
P. stylirostris	U.S.	+	+	+	−	−
	Tahiti	+	+	+	+	+
	U.S.	+	+	+	+	+
P. vannamei	Tahiti	+	+	+	+	+

Note: E.A. +: eyestalk ablation, E.A. −: no eyestalk ablation.

Females are ablated by several different methods:

1. Pinching eyestalk or by crushing the eyeball between the fingers[9,14,15]
2. Ligating the base of eyestalk[15]
3. Cutting off the eyestalk with surgical scissors[5,15,16]
4. Severing the eyestalk prior to cauterizing with a soldering iron[17]
5. Severing the eyeball with a razor[15]
6. Squeezing the eyestalk tissue[15]
7. Crushing the eyestalk after emptying the eyeball through an incision[15]
8. Removing the eyestalk with hot surgical clamps[18]
9. Penetrating the eyeball with a lancet[19]

P. monodon males mature spontaneously in captivity, therefore, eyestalk ablation is applied only to females, while the males remain untreated.

The eyestalk ablation technique currently employed at the TML is to unilaterally sever the eyestalk at its base with a heated surgical forceps. This seems to be the best procedure to keep the hemolymph from percolating from the wound. It has also been demonstrated that bilaterally ablated prawns show rapid ovarian development but such prawns usually suffer higher mortalities than those unilaterally ablated. Therefore, bilateral ablation is not considered practical.[10] Some ablations are conducted by holding the female prawns under water and some are not. In an attempt to reduce the mortality rate, the application of antibiotics to the wound[18] and the cooling of the water in which the prawns are held before and after ablation are reported.[17,24] Neither of these techniques are followed presently at the TML. After ablation, females are placed in indoor tanks or outdoor concrete ponds without any bottom substrate. It seems that nongrooved and nonburrowing penaeid prawns such as *P. monodon* do not require a soft bottom substrate.

The water used at the TML is pumped directly from a 7- to 10-m deep underground well near the seashore. During the holding period of prawns in winter the water is heated by automatically controlled electric heaters in the indoor tanks. The pH of the water fluctuates between 7.5 and 8.2 with a salinity which may vary from 17 to 31 ppt. This fluctuation is largely due to the effects of local rainfall. Airstones are provided to ponds and a dissolved oxygen concentration of more than 5 ppm is maintained. Illumination of the indoor ponds is 10% of natural daylight.

The maturation diet may consist of mashed squid mantles, chopped raw fish, trash whole shrimp, live marine worms, fresh oysters, different species of mussels or hard clams on half shell, and commercially prepared artificial feeds. These are administered at a daily rate of 10 to 15% of the total biomass of the prawn. Other kinds of foods include tuna fish,[10] toad,[8] alamang,[8] specially prepared feeds,[20,21] and commercial trout feeds.[17] Some additives were also used by other research workers to feed the spawners.[17] As for feeding frequency, the prawns are fed twice daily at 0800 and 1700 hours. Investigators elsewhere distribute their feed from one to four times daily at the rate of 3 to 30% of the prawn biomass per day.[8,10,11,16,19,22-28]

In general, the first maturation of prawn is observed approximately 2 to 3 weeks after the eyestalk or eyestalks are ablated. The average number of days between eyestalk ablation and the first spawning of *P. monodon* is 18.1.[19] Some researchers have reported that maturation or spawning could be obtained within 5 to 11 days after ablation, and that rematuration can occur within 3 to 5 days after the first spawning.[11]

Ovarian development in *P. monodon* is examined at night by scooping the shrimp out of the tank with a fine-meshed scoop net and observing the prawns with a flashlight which allows observation of the coloration, shape, and the texture of the gonads through the exoskeleton. The observation of color changes (see Table 2) in the ovaries as a measure of female maturity is a practical method. The four stages of ovarian maturation of *P. monodon* are described as follows:

- Stage I (underdeveloped stage or spent stage): The ovaries are extremely thin, translucent, or whitish, and almost indistinguishable from the exoskeleton.
- Stage II: (developing stage): The ovaries are developing, but the outline of the rim parts are still obscure. The dorsal surfaces of the ovaries are either light yellow, yellowish brown, or yellowish green, and are slightly translucent.
- Stage III: (early ripe stage): The ovaries cover the entire dorsal part of cephalothorax and abdomen and have a light green color. The anterior and median lobes are fully developed and visible through the exoskeleton. The outline of rim parts of the ovaries is distinct and opaque.

Table 2
COLOR CHANGES IN THE OVARIES AS A DETERMINANT OF MATURITY OF FEMALE PRAWN

	Ovarian maturation stage				
Species	**I — Undeveloped stage, spent stage**	**II — Developing stage**	**III — Early ripe stage**	**IV — Ripe stage**	**Ref.**
Penaeus monodon	Translucent and whitish	Light yellow, yellowish brown, and yellowish green	Light green and green	Dark green	9, 26
P. kerathurus	Slightly greyish	Greyish	Yellowish green	Bright orange-yellow	14
P. stylirostris and *P. vannamei*	Whitish	Yellow and reddish	Golden brown	Greenish brown	15
P. plebejus	Translucent	Pink and brown	Orange and brown	Light orange and brown	18

- Stage IV (ripe stage): The ovaries become broader and can easily be observed. The shape and outline are clearly distinct and conspicuous through the exoskeleton. When the jointed area between the dorsal part of the cephalothorax and the abdomen is bent, the granular dark green colored ova can easily be observed.

These above mentioned four stages are described from external observation of live females by the naked eye. The gonadal maturation of individual prawns are evaluated in several other ways such as the Gonadal Somatic Index (GSI)[29] and by histological examination. However, these techniques are not practical.

If the development of ovaries are observed to reach stages III or IV, the females are carefully placed in 500-ℓ circular fiberglass tanks or into other suitable containers for spawning. Water temperature in the spawning tanks is maintained at 28 to 30°C. One airstone is placed in each tank to provide adequate aeration for the newly spawned eggs.

Spawning is always observed to take place between 9:00 p.m. and midnight.[19] Spawning behavior begins by active circular swimming around the tank wall just beneath the water surface. Then spawning is accomplished by violent movement of the pleopods in the middle layer of the water. The whole spawning process lasts about 1 min. Generally, regression of the unspawned ova occurs within 2 or 3 days.

The quality of maturation and spawning induced by eyestalk ablation can be determined by the following observations: spawning or no spawning, percentage of fertilized eggs, and hatching rate. Spawning follows ovary stages III or IV, except when regression occurs because females are stressed in handling. The percentage of fertilized eggs is calculated by examining the eggs under the microscope. After spawning, three types of eggs can be seen: unfertilized eggs and fertilized eggs with normal and abnormal development. The percentage of fertilized eggs can be obtained from the number of fertilized eggs divided by the total number of the eggs in the same random sample from a 1-ℓ beaker, multiplied by 100. Hatching rate (H %) is obtained by using the following formula:[19] $H\ \% = Y/X \times 100\%$; H = hatching rate, Y = total number of nauplii, and X = total number of eggs.

Although the techniques discussed in this section are presently being employed with success to induce maturation and to spawn *P. monodon* in captivity, further studies are needed on the following aspects:

1. Improvement of the ablation techniques to reduce mortality
2. Determination of the optimal age and size of prawns for ablation
3. Identification of proper nutritional diets for ovary maturation
4. Studying males to ensure successful fertilization of the ablated females
5. Seeking the optimum sex-ratio
6. Obtaining information about the optimum physical environment to ensure successful copulation
7. The determination of optimum stocking density

Once these problems are resolved, the genetic selection and hybridization of penaeids will be within our reach.

REFERENCES

1. **Liao, I. C. and Huang, T. L.,** Experiments on Propagation and Culture of Prawns in Taiwan, presented at the Symposium on Coastal Aquaculture, IPFC 14th Session, Bangkok, Thailand, November 18 to 27, 1970.
2. **Liao, I. C., Lu, Y. J., Huang, T. L., and Lin, M. C.,** Experiments on induced breeding of grey mullet, *Mugil cephalus* Linnaeus, presented at the Symposium on Coastal Aquaculture, IPFC 14th Session, Bangkok, Thailand, November 18 to 27, 1970.
3. **Liao, I. C., Huang, T. L., and Katsutani, K.,** A preliminary report on artificial propagation of *Penaeus monodon* Fabricius, *Jt. Comm. Rural Reconstr. Fish. Ser.,* 8, 67, 1969.
4. **Liao, I. C. and Chao, N. H.,** Problems to be solved for the culture of *Penaeus monodon* and *Macrobrachium rosenbergii, Harvest Farm Mag.,* 27 (23), 15, 1977.
5. **Chen, H. P.,** Report on maturation of *Penaeus monodon* Fabricius in captivity by eyestalk ablation and subsequent spawning and production of juveniles, *Chin. Fish. Mon.,* 294, 3, 1977.
6. **Chen, H. P.,** Recent progress and problems on prawn culture in Taiwan, *Chin. Fish. Mon.,* 22, 269, 1981.
7. **Liao, I. C.,** Note on the cultured spawner of red-tailed prawn, *Penaeus penicillatus* Alcock, *Jt. Comm. Rural Reconstr. Fish. Ser.,* 15, 59, 1973.
8. Southeast Asian Fisheries Development Center, Domestication of *Penaeus monodon,* in *1975 Annual Report,* Aquaculture Department, Southeast Asian Fisheries Development Center, Philippines, 1975, 5.
9. Aquacop, Reproduction in captivity and growth of *Penaeus monodon* Fabricius in Polynesia, *Proc. World Maricul. Soc.,* 8, 927, 1977.
10. **Santiago, A. C., Jr.,** Successful spawning of cultured *Penaeus monodon* Fabricius after eyestalk ablation, *Aquaculture,* 11, 185, 1977.
11. **Beard, T. W. and Wickins, J. F.,** Breeding of *Penaeus monodon* Fabricius in laboratory recirculation systems, *Aquaculture,* 20, 79, 1980.
12. **Waterman, T. H.,** *The Physiology of Crustacea,* Vol. 1, Academic Press, New York, 1960, 437.
13. **Adiyodi, K. G. and Adiyodi, R. G.,** Endocrine control of reproduction in decapod crustacea, *Biol. Rev.,* 45, 121, 1970.
14. Aquacop, Penaeid reared brood stock: closing the cycle of *P. monodon, P. stylirostris* and *P. vannamei, Proc. World Maricul. Soc.,* 10, 445, 1979.
15. Southeast Asian Fisheries Development Center, Prawn program, in *1976 Annual Report,* Aquaculture Department, Southeast Asian Fisheries Development Center, Philippines, 1976, 13.
16. **Lumare, ,** Reproduction of *Penaeus kerathurus* using eyestalk ablation, *Aquaculture,* 18, 203, 1979.
17. **Caillouet, C. W., Jr.,** Ovarian maturation induced by eyestalk ablation in pink shrimp, *Penaeus duorarum* Burkenroad, *Proc. World Maricul. Soc.,* 3, 205, 1973.
18. **Duronslet, M. J., Yudin, A. I., Wheeler, R. S., and Clark, W. H., Jr.,** Light and fine structural studies of natural and artificially induced egg growth of penaeid shrimp, *Proc. World Maricul. Soc.,* 6, 105, 1975.
19. **Chen, C. A.,** Preliminary Report on the Gonadal Development and Induced Breeding of *Penaeus monodon* Fabricius, M.S. thesis, Institute of Oceanography, National Taiwan University, Taipei, Taiwan, R.O.C., 1979, 43.
20. Aquacop, Maturation and spawning in captivity of penaeid shrimp: *Penaeus merguiensis* de Man, *Penaeus japonicus* Bate, *Penaeus aztecus* Ives, *Metapenaeus ensis* de Haan, and *Penaeus semisulcatus* de Haan, *Proc. World Maricul. Soc.,* 6, 123, 1975.
21. **Moore, D. W., Jr., Sherry, R. W., and Montanez, F.,** Maturation of *Penaeus californiensis* in captivity, *Proc. World Maricul. Soc.,* 5, 445, 1974.
22. **Conte, F. S., Duronslet, M. J., Clark, W. H., and Parker, J. C.,** Maturation of *Pendaeus stylirostris* (Stempson) and *P. setiferus* (Linn.) in hypersaline water near Corpus Christi, Texas, *Proc. World Maricul. Soc.,* 8, 327, 1977.
23. **Primavera, J. H.,** Induced maturation and spawning in five-month-old *Penaeus monodon* Fabricius by eyestalk ablation, *Aquaculture,* 13, 355, 1978.
24. **Brown, A., Jr., McVey, J., Middleditch, B. S., and Lawrence, A. L.,** Maturation of white shrimp *(Penaeus setiferus)* in captivity, *Proc. World Maricul. Soc.,* 10, 435, 1979.
25. **Lawrence, A. L., Akamine, Y., Middleditch, B. S., Chamberlain, G., and Hutchins, D.,** Maturation and reproduction of *Penaeus setiferus* in captivity, *Proc. World Maricul. Soc.,* 11, 481, 1980.
26. **Kelemec, J. A. and Smith, I. R.,** Induced ovarian development and spawning of *Penaeus plebejus* in a recirculating laboratory tank after unilateral eyestalk enucleation, *Aquaculture,* 21, 55, 1980.
27. **Brown, A., Jr., McVey, J. P., Scott, B. M., Williams, T. D., Middleditch, B. S., and Lawrence, A. L.,** The maturation and spawning of *Penaeus stylirostris* under controlled laboratory conditions, *Proc. World Maricul. Soc.,* 11, 488, 1980.
28. **Lichatowish, T., Smalley, T., and Mate, F. D.,** The natural reproduction of *Penaeus merguiensis* (de Man, 1888) in an earthen pond in Fiji, *Aquaculture,* 15, 377, 1978.
29. **Rodriguez, A.,** Growth and sexual maturation of *Penaeus kerathurus* (Forskal, 1775) and *Palaemon serratus* (Pennant) in salt ponds, *Aquaculture,* 24, 257, 1981.

HATCHERY AND GROW-OUT: PENAEID PRAWNS

I-Chiu Liao and Nai-Hsien Chao

INTRODUCTION

Many trials of propagation of penaeid prawns had been made before the first scientist succeeded in spawning parent prawns in the laboratory. It was not until 1934 that prawn larvae were reared from hatching to the mysis stage.[1] Dr. Hudinaga[2] published a well-known research paper which is considered the ''Bible'' by researchers of penaeid prawn propagation. However, his early research work was forced to a stop by World War II. It was not until 1964 that Dr. Hudinaga and colleagues[3] succeeded in the commercial production of prawn juveniles, thus initiating the large-scale hatchery system of prawns.

The artificial propagation of 14 penaeid prawn species has proved feasible up to the present time (Table 1). Among them only seven species, namely, *Penaeus aztecus, P. duorarum, P. japonicus, P. monodon, P. setiferus, P. stylirostris,* and *P. vannamei*, are able to be propagated on a commercial scale.

The pond culture of 14 species of penaeid prawn has been tried and the culture technique of 4 out of the 14 species (*P. japonicus, P. monodon, P. stylirostris,* and *P. vannamei)* is considered to be more advanced than the others. A sufficient supply of juveniles is one of the requirements to attain a successful culture of prawn. The close relationship between propagation and culture indicates that the culturable species are usually the species of successful propagation. However, species that can be propagated easily are not necessarily good for culture.

HATCHERY

The earliest propagation work was done with *P. japonicus*. The so-called ''community method''[3] was developed by Dr. Hudinaga and colleagues. By this method, the present annual production of prawn juveniles in Japan is 600 to 700 million. The normal method for juvenile production is to use 100- to 250-m^3 concrete tanks and to produce 10,000 to 15,000 P_{20} juveniles per cubic meter over a period of 30 to 40 days. The survival rate from the nauplius stage to P_{20} is 25 to 60%. This process can be repeated four times during a single spawning season.[24]

The propagation of *P. monodon* was successfully done for the first time in 1968.[13] The technique has been widely distributed to Southeast Asian countries. Among them, Taiwan (R.O.C.), Philippines, and Thailand are the leading countries where there are more progressive activities of *P. monodon* culture. The annual production of juvenile *P. monodon* from Taiwan exceeds 300 million.[14]

For the convenience of understanding the important aspects concerning juvenile production from hatcheries, a comparison between *P. japonicus* and *P. monodon*, two major cultured species, is summarized in Table 2.

GROW-OUT

The pattern of prawn culture is divided into four systems according to a variety of categories such as pond size, stocking density, inlet and outlet water system, aeration, and feeding method.

Intensive System

P. japonicus culture in Japan utilizes the intensive culture circular concrete tanks with a

Table 1
THE CURRENT STATUS OF PENAEID PRAWN CULTURES

Scientific name	Main common names	Propagation level		Production level				Major regions of prawn culture	Ref.
		Experimental	Large scale	Experimental	Extensive	Semiintensive	Intensive		
Penaeus aztecus Ives	Brown shrimp, camaron cafe		○		○			The Atlantic coast of Central and South America	4,5
P. brasiliensis Latreille	Camarao rosa, pink spotted shrimp	○		○				The Atlantic coast of Central and South America	6
P. duorarum Burkenroad	Pink shrimp, camaron rosado		○		○			The Atlantic coast of Central and South America	7,8
P. indicus H. Milne Edwards	Indian shrimp, white shrimp	○			○			Southeast Asia	9,10
P. japonicus Bate	Kuruma prawn, Japanese tiger shrimp		○				○	Japan, Taiwan, Brazil	3,11
P. merguiensis de Man	Banana prawn, white prawn	○		○				Indonesia, India	12
P. monodon Fabricius	Grass prawn, jumbo tiger prawn, sugpo		○			○		Taiwan, Philippines	13,14
P. orientalis Kishinouye	Oriental shrimp	○		○				China, Korea, Japan	10,15
P. penicillatus Alcock	Red-tailed prawn	○		○				Taiwan	16,17
P. schmitti Burkenroad	Southern white shrimp, camaron blanco, langostino blanco	○		○				The Atlantic coast of Central and South America	10,18
P. semisulcatus De Haan	Green tiger prawn, flower, bear shrimp	○		○				Taiwan, Hong Kong, Kuwait	19,20
P. setiferus (Linnaeus)	White shrimp, common shrimp		○		○			Southern U.S., Mexico	8,21
P. stylirostris Stimpson	Blue shrimp, white shrimp		○				○	The Atlantic coast of Central and South America	22,23

P. vannamei Boone	White shrimp, camaron blanco, camaron langostino	○	○	The Atlantic coast of Central and South America	22,23

Table 2
COMPARISON OF IMPORTANT ASPECTS CONCERNING JUVENILE PRODUCTION FROM HATCHERY BETWEEN *PENAEUS JAPONICUS* AND *P. MONODON* [25]

	Kuruma prawn *Penaeus japonicus*	Grass prawn *Penaeus monodon*
1. Sources of spawner	Comparatively easy to get, depend on indigenous wild 95%; Imported 5%.	Difficult to get, depend on eyestalk ablation 5—10%; Indigenous wild 55—70%; Imported wild 40—20%.
2. Number of eggs from one spawner	20—50 × 10^4	20—60 × 10^4
3. Size of rearing tank for spawning and hatching	Large concrete tank (100—250 ton)	Small fiberglass round tank (0.5—1.2 ton) — 20 ton concrete tank.
4. Sensitivity to light	Adaptable	Zoea stage: High sensitivity; Mysis stage (III): Adaptable
5. Optimal temperature and salinity	25—30°C, 27—32%	28—31°C Nauplius and zoea stage: 31—34%; after mysis stage below 30%
6. Method for rearing	Community culture method	Monoculture method or separate tank method
7. Diseases	Mysis and post larval stage: *Vibrio* infection; Post larval stage: White turbid midgut gland disease	Zoea and mysis stage: Mass mortality resulting from unknown causes; Post larval stage: *Epistylis* infection
8. Survival rate	Nauplius to P_{20} stage: 25—40%	Nauplius to $P_{12\text{-}15}$ stage: 15—20%
9. Price of juvenile	¥ 260—400/1,000 juveniles (US$ 1.2—1.9/1,000 juveniles)	NT$ 300—900/1,000 juveniles (US$ 7.9—23.7/1,000 juveniles)

diameter of 36 m and water depth of 2.5 m. A running water system is designed to maintain high dissolved oxygen. The stocking density is usually 160 juveniles per square meter and an average production of 4.5 to 24 tons/ha is obtained.[11,26] However, the highest production per unit area for intensive culture can reach 35 tons/ha.[24]

Semiintensive System

P. monodon culture in Taiwan is an example of this system. Ponds of 0.2 to 0.5 ha are used and the stocking density is 20 to 30 juveniles per square meter. The survival rate is as high as 75% or even more. Two crops each year make a total harvest ranging from 1.4 to 9.6 tons/ha and the highest harvest reached has been 12 tons/ha.[14,25]

Extensive System

Culture of *P. stylirostris* and *P. vannamei* in Ecuador is typical of the extensive system and uses large ponds bigger than 20 ha with a stocking density of 1 to 2 juveniles per square meter that are collected from nature. The harvest is usually limited to 0.1 to 0.3 tons/ha because of the limited feeding regime.

Experimental System

Research on *P. semisulcatus* and other naturally abundant prawns is now underway to develop suitable culture techniques.

For a better understanding of the possible differences between species, a comparison of growth and production between *P. japonicus* and *P. monodon* is made and shown in Table 3. In addition, a comparison of growth curves of 11 selected penaeid prawns (Figure 1) indicates that at the moment *P. monodon, P. japonicus*, and *P. vannamei* are among the best prawn species for cultivation in the world. Nevertheless, it is concluded from the study of scientists and the experience of culturists that there are some disadvantages of each of

Table 3
COMPARISON OF IMPORTANT ASPECTS CONCERNING THE GROWTH AND PRODUCTION OF *PENAEUS JAPONICUS* AND *P. MONODON*[25]

	Kuruma prawn *Penaeus japonicus*	Grass prawn *Penaeus monodon*
1. History of culture	75 years	About 300 years
2. Culture area	250 ha	2,000 ha
3. Culture method	Intensive	Semi-intensive
4. Pond size (ha)	0.03—0.1	0.2—0.5
5. Bottom type	Sand	Earth (clay-organic)
6. Aeration method	Air blower Paddy wheel	Air blower Paddy wheel
7. Optimal temperature 0°C	25—28	25—30
8. Optimal salinity %	16—34	5—25
9. Feeds used	Artificial feed plus fresh fish and shellfish	Artificial feed plus trash fish and shrimp
10. Feeding time	Night	Day and night
11. Feed conversion ratio	2.5	1.8—3.3
12. Diseases	*Vibrio* and *Fusarium* infection	*Epistylis* infection and "Red Discoloration"
13. Survival rate	70%	75%
14. Production quantity/ha	4.5—24 tons (1 crop/year)	1.4—9.6 tons (2 crops/year)
15. Production quantity (1980)	1,500 tons	5,000 tons
16. Production cost/kg	¥ 3,500—5,600 (US$ 15.2—23.4)	NT$ 184—206 (US$ 4.8—5.4)
17. Price (retail)/kg	¥ 6,000—17,000 (US$ 32—80)	NT$ 350—430 (US$ 9.2—11.3)
18. Total revenue (1980)	US$ 42,000,000	US$ 50,000,000

these three species. *P. monodon* shows a poor growth rate in culture ponds containing high salinity seawater. Attention should always be paid to a source of freshwater if culture of this species is desired. *P. japonicus* is of extremely high price but has to consume feed of high protein level which results in high culture costs. *P. vannamei* is much more euryhaline than others, however, it seldom reaches the size of the former two species as it is only a medium-sized prawn.

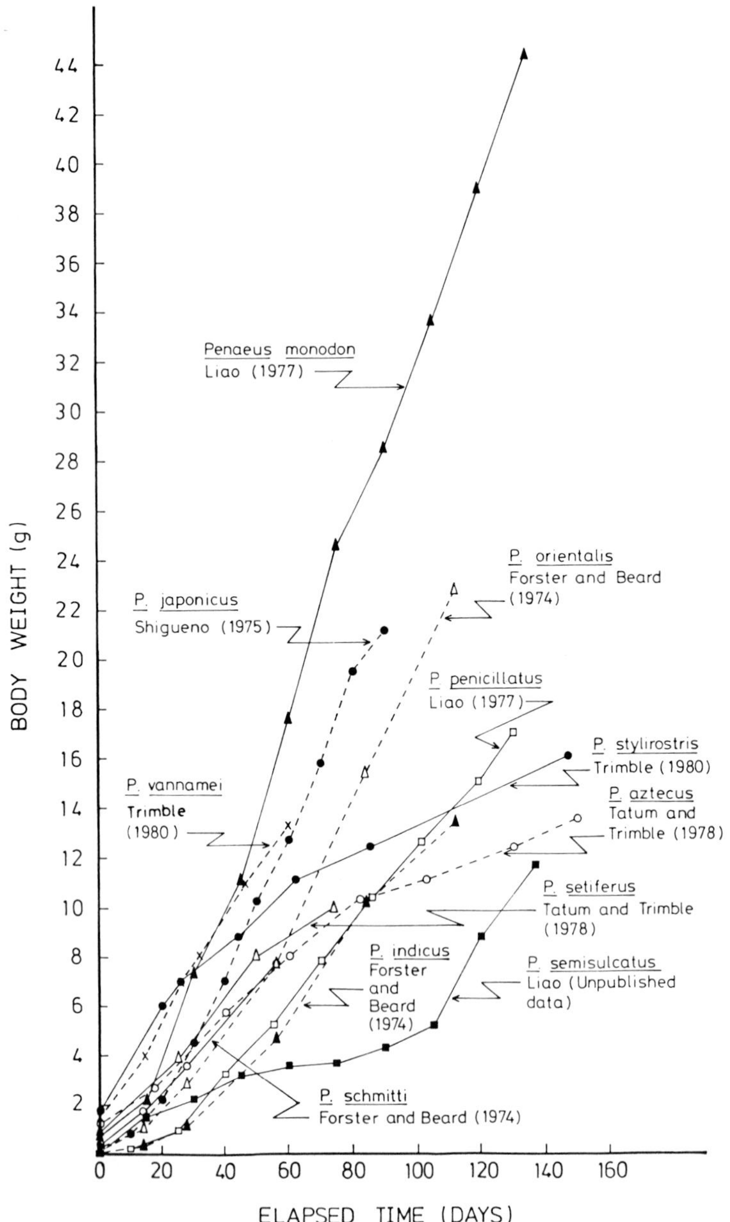

FIGURE 1. Growth curves of eleven penaeid prawns under cultivation.

REFERENCES

1. **Hudinaga, M.,** The study of *Penaeus*. I. The development of *Penaeus japonicus* Bate (1), *Rep. Hayatomo Fish. Res. Lab.*, 1, 1, 1935.
2. **Hudinaga, M.,** Reproduction, development and rearing of *Penaeus japonicus* Bate, *Jpn. J. Zool.*, 10, 305, 1942.
3. **Hudinaga, M. and Kittaka, J.,** The large scale production of the young kuruma prawn, *Penaeus japonicus* Bate, *Inf. Bull. Planktol. Jpn. Commemoration No. Dr. Y. Matsue,* p. 35, 1967.

4. **Cook, H. L. and Murphy, M. A.,** Rearing penaeid shrimp from eggs to post-larvae, *Proc. Conf. Southeast. Assoc. Game Comm.*, 19, 384, 1966.
5. **Parker, J. C. and Holcomb, H. W., Jr.,** Growth and production of brown and white shrimp (*Penaeus aztecus* and *P. setiferus*) from experimental ponds in Brazoria and Orange Counties, Texas, *Proc. World Maricul. Soc.*, 4, 215, 1973.
6. **Scelzo, M. A., Marcano, G., and Millan, J.,** Resultados sobre el crecimiento de juveniles del camaron *Penaeus brasiliensis* Latreille (Decapoda, Penaeidae), cultivados en estanques de concreto, in *I Simposio Brasileiro de Aquicultura*, Recife, Brazil, Academia Brasileira de Ciencias, Rio de Janeiro, Brazil, 1980, 397.
7. **Ewald, J. J.,** The laboratory rearing of pink shrimp, *Penaeus duorarum* Burkenroad, *Bull. Mar. Sci.*, 15(2), 436, 1965.
8. **Tatum, W. M. and Trimble, W. C.,** Monoculture and polyculture pond studies with pompano (*Trachinotus carolinus*) and penaeid shrimp (*Penaeus aztecus, P. duorarum,* and *P. setiferus*) in Alabama, 1975—1977, *Proc. World Maricul. Soc.*, 9, 433, 1978.
9. **Muthu, M. S., Pillai, N. N., and George, K. V.,** On the spawning and rearing of *Penaeus indicus* in the laboratory with a note on the eggs and larvae, *Indian J. Fish.*, 21(2), 571, 1976.
10. **Forster, J. R. M. and Beard, T. W.,** Experiment to assess the suitability of nine species of prawns for intensive cultivation, *Aquaculture*, 3, 355, 1974.
11. **Shigueno, K.,** Shrimp culture in Japan, in *Association for International Technical Promotion*, Tokyo, Japan, 1975.
12. Aquacop, Maturation and spawning in captivity of penaeid shrimp: *Penaeus merguiensis* de Man, *Penaeus japonicus* Bate, *Penaeus aztecus* Ives, *Metapenaeus ensis* de Haan, and *Penaeus semisulcatus* de Haan, *Proc. World Maricul. Soc.*, 6, 123, 1975.
13. **Liao, I. C., Huang, T. L., and Katsutani, K.,** A preliminary report on artificial propagation of *Penaeus monodon* Fabricius, *J.C.R.R. Fish. Ser.*, 8, 67, 1969.
14. **Liao, I. C.,** Status and problems of grass prawn culture in Taiwan, presented at the ROC-JAPAN Symp. on Mariculture, Taipei, Taiwan, R.O.C., December 13 to 24, 1981.
15. **Oka, M.,** Studies on *Penaeus orientalis* Kishinouye. V. Fertilization and development, *Bull. Fac. Fish. Nagasaki Univ.*, 23, 71, 1967.
16. **Liao, I. C.,** Note on the cultured spawner of red-tailed prawn, *Penaeus penicillatus* Alcock, *J.C.R.R. Fish. Ser.*, 15, 59, 1967.
17. **Liao, I. C.,** A culture study on grass prawn, *Penaeus monodon*, in Taiwan — the patterns, the problems and the prospects, *J. Fish. Soc. Taiwan*, 5(2), 11, 1977.
18. **Perez, D. P. and Suarez, M. O.,** Cultivo experimental de estadios larvales del camaron blanco *Penaeus schmitti* y del camaron acaramelado *Penaeus duorarum notialis* en laboratorio, in Proc. FAO Technical Conf. on Aquaculture, Kyoto, Japan, May 26 to June 2, 1976, 284.
19. **Liao, I. C. and Huang, T. L.,** Experiments on the propagation and culture of prawns in Taiwan, in *Coastal Aquaculture in the Indo-Pacific Region*, Pillay, T. V. R., Ed., Fishing News, London, 1973, 328.
20. **Farmer, A. S. D.,** Experimental rearing of penaeid in shrimps in Kuwait, *Proc. World Maricul. Soc.*, 10, 489, 1979.
21. **Johnson, M. C. and Fielding, J. R.,** Propagation of the white shrimp, *Penaeus setiferus* (Linn.), in captivity, *Tulane Stud. Zool.*, 4(6), 175, 1956.
22. Aquacop, Penaeid reared brood stock: closing the cycle of *P. monodon, P. stylirostris* and *P. vannamei*, *Proc. World Maricul. Soc.*, 10, 445, 1979.
23. **Trimble, W. C.,** Production trials for monoculture and polyculture of white shrimp (*Penaeus vannamei*) or blue shrimp (*P. stylirostris*) with Florida pompano (*Trachinotus carolinus*) in Alabama, 1978—1979, *Proc. World Maricul. Soc.*, 11, 44, 1980.
24. **Kurata, H., Yatsuyanagi, K., and Shigueno, K.,** Kuruma Shrimp Culture in Japan, UJNR, Aquaculture Panel 9th Joint Meeting, mimeographed, 1980.
25. **Liao, I. C. and Huang, T. L.,** Status and prospect of the culture of two important penaeid prawn in Asia, presented at 4th Simposio Latinoamericano de Acuicultura, Atlapa, Panama, January 25 to 29, 1982.
26. **Shigueno, K.,** A general view of shrimp farming in Japan, presented at the ROC-JAPAN Symp. on Mariculture, Taipei, Taiwan, R.O.C., December 13 to 24, 1981.

PENAEUS SHRIMP POND GROW-OUT IN PANAMA

Richard Pretto Malca

INTRODUCTION

The scientific rearing of shrimp in Panama was initiated in 1974, when the firm Compañia Agromarina de Panama, S.A., a subsidiary of Ralston Purina, began to operate 6 nursery ponds and 30 growout ponds covering an area of 34 ha, near the city of Aquadulce in the region of El Salado, Cocle Province. The postlarvae production laboratory was located 4 hr away in Veracruz near Panama City, Panama. At the end of 1981 Agromarina reported 650 ha of ponds under culture and a production of 12 million postlarvae per month with 70% of the postlarvae produced by maturation of captive females. This firm is considered worldwide as one of the most successful in its specialty (see Figure 1).

In 1978 operations of the second shrimp rearing firm, called Palangosta, S.A., were initiated. In contrast with Agromarina de Panama, S.A., it based its technology on the Ecuadorian system of rearing, which is the capture of postlarvae and juvenile shrimp from the wild in order to stock their ponds.

In 1979, the national government of Panama, through the National Directorate of Aquaculture of the Ministry of Agricultural Development, decided to promote the research and production of penaeid shrimp and other brackish-water organisms with the construction of the Experimental Marine Station "Ing. Enrique Enseñat".

In 1981, 12 firms dedicated to the rearing of penaeid shrimp existed in the country with all, except for Agromarina de Panama, S.A., obtaining their seed from wild stock. In 1980 the Panamanian Association of Shrimp Producers was constituted with the objective of consolidating this infant industry.

The areas where shrimp are reared in Panama are located near the sea, west of Panama City and east of Azuero Peninsula on the Pacific Coast. These lands are called salt flats (albinas) which are flooded by tides twice each lunar month during large tidal fluctuations. They are usually connected to the sea by means of estuaries. There is little or no vegetation, and the salt flats constitute an intermediate zone between the coastal margin of the mangrove areas and the lands of higher elevation covered by freshwater vegetation. In addition, there exists other lands of higher elevation (freshwater lands) and flats along the length of the Pacific Coast, especially in the Chiriqui Province, that are suitable for the rearing of shrimp.

In Panama there exists three important commercial species of white shrimp: *Penaeus occidentalis, P. stylirostris,* and *P. vannamei*. Shrimp culturists rear only *P. vannamei* and *P. stylirostris*. The culture of *P. occidentalis* is not desirable due to its slow growth and high mortality in ponds. The abundance and time of occurrence of postlarvae and juveniles in the estuaries is apparently determined by the characteristics of the rainy season. *P. occidentalis* is generally found in areas of high salinity close to where the estuary connects to the sea. *P. stylirostris* is found in intermediate salinities and *P. vannamei* is found as far as the inner estuaries, the latter constituting the species with the greatest capacity for migration within the estuary. Generally during the dry season, from January to April, only *P. occidentalis* is encountered in the estuaries. In June, July, and August, depending on when the rains begin, *P. stylirostris* begins to enter the estuaries. As rains increase, *P. vannamei* becomes more abundant and this species is predominant in October and November when rains are heaviest.

The government of Panama has negotiated a loan for 12 million dollars (U.S.) with the

FIGURE 1. Phase II of Agromarina de Panama, S.A., Cocle Province. Phase I (pilot) can be observed in the foreground.

Inter-American Development Bank (IDB) with the object of opening a line of credit through the National Bank of Panama to develop aquaculture in Panama. Part of the funds will be utilized in the development of such shrimp culture infrastructure as a closed-cycle laboratory (maturation), with an initial production of 60 million postlarvae; construction of a building including offices, a laboratory, dormitories, and storage area for the Experimental Marine Station "Ing. Enrique Enseñat". There will also be funds for biologists and extensionists with their respective vehicles and other work equipment. It is calculated that 19,600 ha of "albinas" exist and another undetermined area of freshwater soils suitable for the rearing of shrimp. Therefore, we expect significant development of shrimp aquaculture in Panama over the next 5 to 10 years.

CONSTRUCTION OF PONDS

Shrimp ponds should have a rectangular shape in order to aid water flow during water exchanges. Soil from the pond bottom is utilized to construct the dikes. Because of the soft soils, heavy machinery with "Pantaneros", extra large tracks, are used instead of tractors with conventional wheels (see Figure 2). During the construction special care is given to compacting the dikes and in giving them an adequate slope. Slopes vary from 2:1 to 4:1.

The construction of a perimeter dike constitutes the first step in building a shrimp farm to avoid wet soils during later construction.

Once the soil is sufficiently dry it is recommended to work first on both the supply and the drainage canals and finally the shaping of each individual pond. This measure permits putting the shrimp farm into operation even without having finished the development phase of a given year. One should take into account that in these areas you can work only in the dry season, which in Panama extends generally from December to March, and that the beginning of the rains totally negates the use of heavy equipment.

The pond bottoms should possess an inclination of 0.3 to 0.5% toward the drain structures so as to facilitate harvest and the correct exchange of bottom water in order to maintain optimum oxygen levels. Failure to drain the ponds completely makes harvest and competitor and predator control difficult.

FIGURE 2. Caterpillar used for the construction of ponds for the cultivation of penaeid shrimp in areas of high soil moisture.

FIGURE 3. Structure and water pumping for phase II of Agromarina de Panama, S.A.

Most shrimp farms in Panama utilize diesel pumps to supply the ponds with water. Pumping stations ought to be located close to an estuary. The pumps used are almost all of the low dynamic charge axial flow type. The motor-to-pump coupling is achieved by multiple V-belts and by gear heads. Mounting of the pumps may be vertical or inclined at 45 to 60°. The diameter of the pumps varies between 12 and 36 in., with a discharge capacity ranging from 2000 to 3000 g.p.m., at a discharge height of 2.0 to 3.5 m. The water can go directly from a pumping station (see Figure 3) to a reservoir that serves equally well either as a

FIGURE 4. Pilot project of 34 ha (DECAPASA, Los Santos Province) for the cultivation of penaeid shrimp. The relationship of reservoir and supply canal to nursery ponds and grow-out ponds can be seen. The grow-out ponds are the larger ones in the background.

sedimentation lagoon or as the principal canal supplying the ponds. The distribution canal, if constructed with a large capacity, constitutes a water reservoir that permits correcting emergency low oxygen cases in the ponds during periods of low tide (see Figure 4).

The water is filtered before entering each pond by use of screens placed at the water entrance gates of each pond. It is customary to use fine screen (1/32 in.) for the nursery ponds and larger (1/16 in.) screen for the grow-out ponds.

During high tides, shrimp have been observed entering ponds through drainage gates. This excess of larvae produces crowded conditions that cause slow growth. In addition, the excess biomass negatively affects the parameters of water quality. Therefore, it is necessary to place additional boards in the drain boxes during periods of high tides in order to avoid any entrance of water.

The ponds vary in size between 1 and 20 ha and possess independent water supply gates. The gates are constructed of concrete. A minimum of three slots are recommended for each gate although five are preferred in order to facilitate placing a series of filters of diverse sizes and, at the same time, permit the release of bottom water when bad water quality exists on the bottom and the release of surface waters when there is excessive rain or an intense algal bloom (see Figure 5). Normally, water is maintained at a depth between 70 and 90 cm.

OBTAINING SEED STOCK

In Panama the shrimp "seed" (postlarvae or small juveniles) come from two sources: the laboratory or the wild. Agromarina de Panama, S.A. possesses the only shrimp hatchery and they do not sell to other shrimp farmers. They stock the postlarvae directly in grow-out ponds at a density of $14/m^2$ or they place them in nursery or rearing ponds at high densities varying between 100 and $150/m^2$ for *P. stylirostris* and 150 and $200/m^2$ for *P. vannamei*. The pregrow-out phase lasts between 45 and 60 days after which the 0.5- to 1.0-g-size juveniles are transferred to grow-out ponds. This phase lasts some 110 days until the

FIGURE 5. Concrete gate with inverted V shape of filters in order to increase surface area. Note slots for different filters and water treatment options.

FIGURE 6. Gravity harvesting of penaeid shrimp using a cage at the back of the drainage structure.

shrimp attain a weight between 16 to 20 g. If the stocking is carried out directly into the grow-out ponds the harvest is made in 145 days (Figure 6).

Recent information suggests that it is important to transfer shrimp from the nursery ponds to the grow-out ponds before they reach 1 g, otherwise subsequent growth rates may be reduced. Agromarina prefers to transfer their shrimp at 0.75 g.

The rest of the shrimp farmers capture their seed from the wild in small lagoons (10- to

FIGURE 7. Collection of penaeid shrimp seeds by using the "chayo" in a natural shallow lagoon.

50-cm depth) formed after the highest tides of each lunar cycle. Seed size varies from 5 to 25 mm with a weight of 0.008 to 0.6 g. Seed of smaller size is considered hardier and suffers less mortality during handling. During the time of seed abundance, seeds are caught utilizing gear called "chayo" and operated by a single person (see Figure 7). The "chayo" consists of a fine conical net supported by two bamboo rods. Trawl nets are also utilized for the capture of seed. Seed collected in this way is accumulated in boxes that usually have battery powered aerators similar to those employed in aquaria. After capture, shrimp seed are transferred to transport tanks with permanent oxygen in order to be carried to the shrimp farms (Figure 8).

SEED HANDLING

The most important factors in the handling of seed that affect survival are precisely the conditions of seed transport:

1. Density during transport (1000 to 5000 postlarvae per gallon)
2. Water temperature (24 to 25°C)
3. Dissolved oxygen level (10 ppm)
4. Low water contamination by clay or organic matter in suspension
5. Time of transport

The seed ought to be fully acclimatized before being released to grow-out or rearing ponds. The seed should always be handled in water and in the coolest hours of the day or during the night. Some farmers have attempted to separate by hand the undesirable organisms such as fish larvae that come in with the shrimp seed. The treatment of shrimp seed with rotenone at a concentration of 5 ppm for 10 min is considered to be more reliable. This treatment totally eliminates the fish. It is recommended that the seed be maintained at a density of 1000/gal for 6 hr in "hapas" or cages before releasing them to nursery or grow-out ponds. These cages are made of wood with metallic mosquito mesh reinforced with 1/4-in. metal

FIGURE 8. Transportation tank with flexible tube to discharge wild penaeid shrimp seed to nursery pond.

FIGURE 9. Acclimation of seed of penaeid shrimp in a "cuna" (box made of wood with a bottom of reinforced mosquito screen).

mesh to prevent crabs from breaking into them and consuming the seed. These cages are known locally as cradles or "cunas" (see Figure 9). This practice permits reducing the large mortality from seed handling generally produced during the first hours after collection. Stocking a seed free of contaminants and of great vitality assures a greater survival, varying normally between 65 and 80%.

The quantity of seed is estimated indirectly by means of their total weight. With this method one takes a small representative sample from which the weight and number of specimens are obtained; seed coming from the laboratory are estimated by aliquots (direct method). This last method should also be used for estimating postlarvae caught from the wild as it gives a considerably lower margin of error.

Ponds receiving the seed ought to be free of competing or predator organisms. In case they cannot be completely dried the small pools of water should be treated with a chemical. A solution of 65% calcium hypochlorite at a rate of 100 g/gal of water is recommended. The advantage of this product is its small residual effect which permits stocking beginning the following day.

The nursery ponds constitute from 6 to 15% of the culture area and are stocked at densities ranging between 60 and 200 shrimp per square meter, depending on the rate of desired growth. Some farmers are accustomed to store the seed in prenursery units for 1 to 2 months, but as already stated, it may be important to make the transfer from nursery to grow-out ponds before the shrimp reach 1 g in size. The grow-out ponds are stocked at the rate of four to five juveniles per square meter. *P. vannamei* tolerates higher stocking densities than *P. stylirostris*. Ideally, for good growth with *P. stylirostris* density should not exceed $2/m^2$. Polyculture of *P. stylirostris* and *P. vannamei* at a ratio of 1:2 may be beneficial. Depending on the system utilized and the availability of seed one can harvest 1 to 2.8 times per year. Figure 10 illustrates typical growth data for *P. stylirostris* and *P. vannamei* in Panama.

FEEDING

Most farmers feed their shrimp from the time they are in the prenursery units. Among the diverse criteria that are followed, one that predominates is that based on a percentage of body weight of the number of shrimp in the pond. Initially this percentage is 25% for the small postlarvae, and is reduced to 2.5% by the time the shrimp reach market size. Other farmers use the criterion of waiting until the shrimp show a slight slowing in growth, an effect of the exhaustion of natural food; they then initiate feeding on the basis of a balanced ration. Periodic cessation of feeding, for up to three times per week, is another customary practice if shrimp are observed maintaining their stomachs full and are showing good growth and a firm exoskeleton.

To sample the population and estimate growth, cast nets are used. When the shrimp are small, small mesh seines are also used. The cast net works very well to estimate population densities of *P. stylirostris*. In the case of *P. vannamei* the results are variable, hence, there is not a reliable method to estimate the number of this species. Therefore, for *P. vannamei* it is recommended that the pond be fed on the assumption of 80% survival; this is the usual survival under normal conditions and good management.

A dry, pelleted shrimp food is produced in Panama by two commercial firms. Its protein content varies from 22 to 25%. The water stability of this pellet is very poor, and it disintegrates in a few minutes. Many believe that the pellet acts only as a fertilizer for the pond which results in the production of other organisms which are then consumed by the shrimp. This belief has motivated the intensification of studies on the use of organic fertilization and particularly on animal manures, i.e., poultry, pig, and cattle.

In initial tests organic manure has been applied uniformly over the pond bottom and water depth brought to 10 to 15 cm. An initial dose of up to 2000 kg/ha is recommended. Five days from fertilization the water depth is raised to 70 to 80 cm and stocking proceeds.

There is some evidence that the use of organic fertilizer assures a greater larval survival, up to 30%, over pelleted foods due to the improved production of natural foods. Additionally, this method can save up to 25% in costs over the use of pelleted feed.

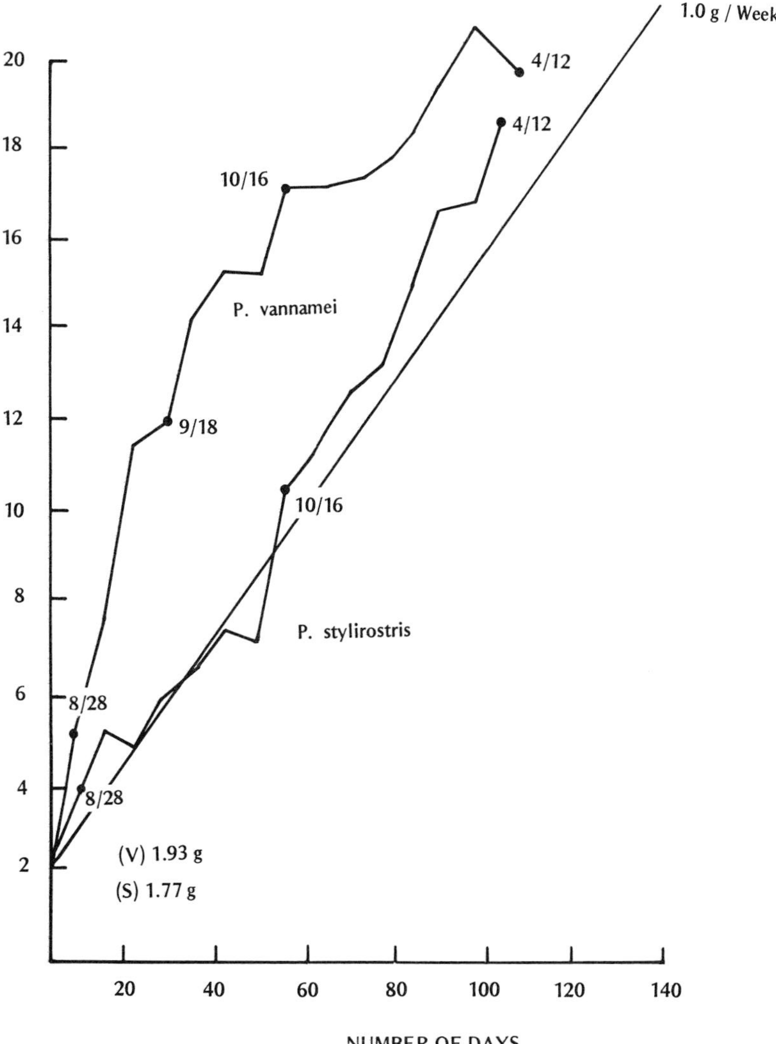

FIGURE 10. Typical growth data for *P. stylirostris* and *P. vannamei* under grow-out pond conditions in Panama. Initial stocking size was 1.93 g for *P. vannamei* and 1.77 g for *P. stylirostris*.

PUMPING AND OTHER MANAGEMENT PRACTICES

Once the shrimp are in the pond the water exchanges are daily and fluctuate between 5 and 10% of the total water volume. There are times when it is necessary to renew up to 60% of the pond water. These emergency situations are necessary when extremely low oxygen levels occur and when algal blooms establish themselves with a subsequent reduction in water quality due to self-shading.

Pumping usually is carried out daily during the high tides of each cycle. Depending on the elevation of the pumping station with respect to the tide one may pump 2 to 12 hr during each of the two daily tides.

It is fairly frequent practice not to exchange water for 10 days after postlarvae have been stocked. This procedure favors the utilization of natural food and avoids submitting the small shrimp to strong water currents which may result in their escape or damage by being forced against the screens. Thereafter, if water quality deteriorates pumping begins immediately. However, ponds that have been recently prepared for shrimp culture should have good water quality.

Unless proper screens are used at the water inflow gates, an invasion of competitive and predaceous fish may occur. Fish populations may be reduced by cast netting at the water entrance gate during pumping. Fish congregate at the gates at the commencement of pumping and are more susceptible to removal.

HARVEST AND YIELD

Harvesting can be a simple operation or a complicated and time-consuming task depending on the design of the ponds. The harvest activity ought to coincide with the cycle of low tides in order to permit draining of the ponds. If the ponds have been constructed with a good slope and exterior harvesting chamber, practically all the shrimp will be harvested at the moment of drainage on the external part of the pond. To do this it is necessary to place a bag net or a basket in the harvest chamber where the shrimp accumulate (Figure 6). If difficulty exists in draining, seines and cast nets are used, although this is not considered a recommended practice.

Before carrying out a harvest one should ascertain that the shrimp have molted and their exoskeletons hardened. The stress caused by the low water level preparatory to harvest may set off the molting mechanism. Shrimp harvested during molting lack commercial value. In this case, it is recommended to fill the pond again and postpone the harvest, taking care to maintain good water quality parameters.

If the shrimp is to be processed whole, heads on, feeding of the shrimp pellets should be suspended for at least 4 days prior to harvest. It is preferred to harvest early in the morning or during the night so as to avoid the hotter hours. When harvest is carried out at night some farmers place a light near the harvest point. Lights are said to attract the shrimp to the drain gate.

The harvested shrimp ought to be placed in ice water as soon as possible and transported to the processing plant where they are processed, graded, and packed. Some producers dehead on the farm and sell only the tails. Tails yield 54 to 66% of the total weight depending on shrimp quality (size, condition, and species).

Yields per hectare vary greatly. Agromarina de Panama, S.A. reports 2000 lb/ha per harvest; the majority of farmers obtain between 900 to 1300 lb/ha per harvest.

The factors that influence yield are many, such as quantity and quality of the pelleted feed, control of the water quality parameters, seed handling, predator and competitor control, ratio of *P. vannamei* vs. *P. stylirostris*, stocking density, survival, grow-out time, etc. All these factors may be controlled to a certain degree, some more than others.

Each year more shrimp farms are being constructed in Panama and more and more of the farms are proving successful. The shrimp culture technological package is improving everyday through our National Research Program and we expect the consolidation and expansion of this new industry in our country.

INTENSIVE LARVAL REARING IN CLEAR WATER OF *MACROBRACHIUM ROSENBERGII* (DE MAN, ANUENUE STOCK) AT THE CENTRE OCÉANOLOGIQUE DU PACIFIQUE, TAHITI

Aquacop
Centre Océanologique du Pacifique

The *Macrobrachium* program was initiated in Tahiti in 1973 as a cooperative program between CNEXO (Centre National pour l'Exploitation des Océans, a French state agency) and the French Polynesia Territory. Studies were performed in the CNEXO "Centre Océanologique du Pacifique" at Vairao, Tahiti. Until 1976 the studies were performed on an experimental scale and an original technology was developed.[1] Technical feasibility was confirmed from 1976 to 1980 in pilot-scale studies.[2] Economic feasibility was determined in the 1978 to 1981 period.[3,4] These studies involved a permanent team of four persons (one or two biologists and two or three technicians and workers) assisted by the Aquacop supporting team with specialties in nutrition, water quality control, pathology, engineering, and logistics.

Most of the larval rearing techniques utilized for *Macrobrachium* in the world are derived from the studies of Ling and Fujimura.[5,6] In these, the larval density is below 50/ℓ and the postlarval density below 30/ℓ. The Aquacop technology is derived from the Galveston technique developed for penaeid shrimp larvae:[7] the larval density is above 100/ℓ and the postlarval density above 60/ℓ. All the technology is designed so that the most important rearing parameters (temperature, light, food, water quality, diseases) are under constant and strict control and independent from the ambient environmental conditions. In Tahiti no chemical pollution has yet been encountered, but any water treatment can be easily added to the technology if needed as the needs for water are very low. Since 1980, a closed system has been investigated.

MATERIALS

Hatchery

Postlarval production has been conducted inside a 200-m^2 closed hatchery building with a north-south orientation and an opaque roof, and concrete walls and large windows on the west wide (Figure 1). The larval rearing tanks are placed along the west wall to obtain good light intensity inside the tanks. The brackish water is prepared and stored in four polyester tanks of 10 m^3: two for brackish water preparation and treatment, two for storage.

Tank Design and Accessories

The basic larval rearing tank is cylindroconical (Figure 2), and the maximum size which can be easily handled is 2 m^3. For larger sizes (5 m^3 and above), U-shaped long tanks have been successfully tested (Figure 3). The cylindroconical tanks are molded fiberglass and the U-shaped tanks are constructed with a fiberglass sheet, plywood, and galvanized iron supports. All the tanks are painted with a dark color. This color is necessary for larvae to feed well. Air stones placed at the bottom of the tank mix the water and keep an even distribution of food particles and larvae; this reduces cannibalism. The air flow rate in each 2-m^3 tank is 2.6 m^3/hr. Outlet filter meshes are in accordance with larval size. Larvae can be collected at the outlet of the tank in a 10-ℓ concentrator. For spawner and postlarval storage, cylindrical 2-m^3 tanks with flat bottom are utilized.

Recirculating System

A 5-m^3 tank is used with both mechanical and biological filtration for closed system work (Figure 3). The system consists of the following components:

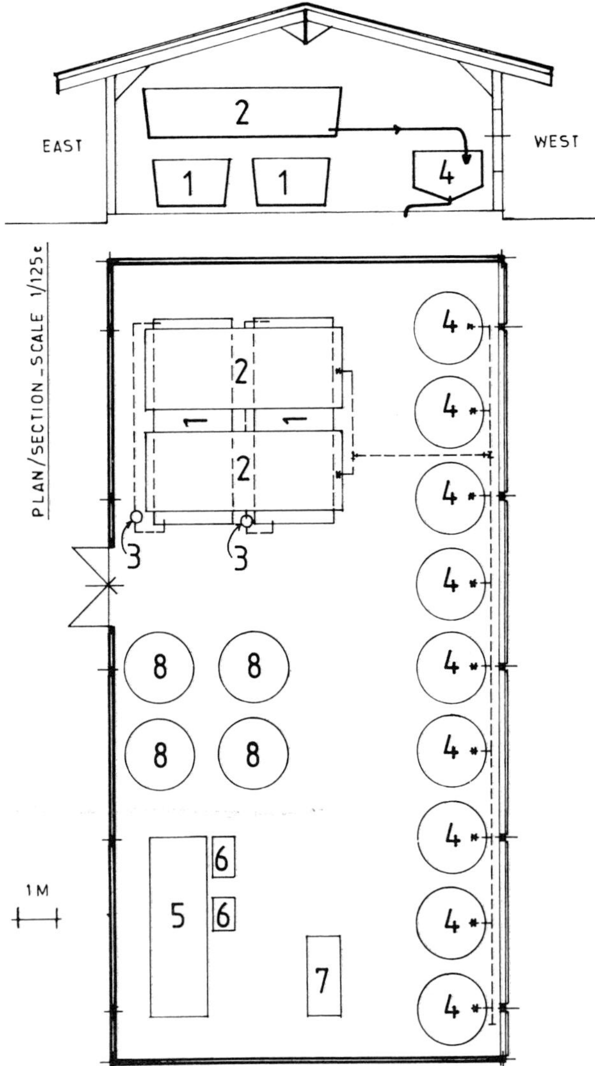

FIGURE 1. General plan of the *Macrobrachium* hatchery. (1) Tanks for the preparation and treatment of brackish water (2 × 10 m^3), (2) storage tank of brackish water (2 × 10 m^3), (3) swimming pool filters, (4) larval rearing tanks (9 × 2 m^3), (5) larval rearing tank (5 m^3), (6) biological filter for the 5-m^3 tank, (7) biological filter for four 2-m^3 tanks, (8) tanks for spawners and postlarvae storage.

1. Mechanical filter: this filter is a plywood box (1.30 × 0.70 × 0.50 m) with a 0.10-m deep sandlayer. Sand particles are 0.1 mm in diameter and a back-wash system allows a daily cleaning of the sand.
2. Biological filter: this filter is a partitioned box (1.3 × 0.65 × 0.60 m) containing a 0.15-m^3 of coral pieces (3 to 5 cm in diameter) which is utilized for bacterial treatment of metabolic wastes (Figure 4). The efficiency of this material is very good because of its great porosity and its buffer quality.
3. Water circulation: the water comes from the 5-m^3 tank by gravity. Inside the larval rearing tank, meshes of 200 to 1000 μm are placed according to the larval size. Water passes through the sand filter where it is pumped and injected into the biological filter (Figure 3).

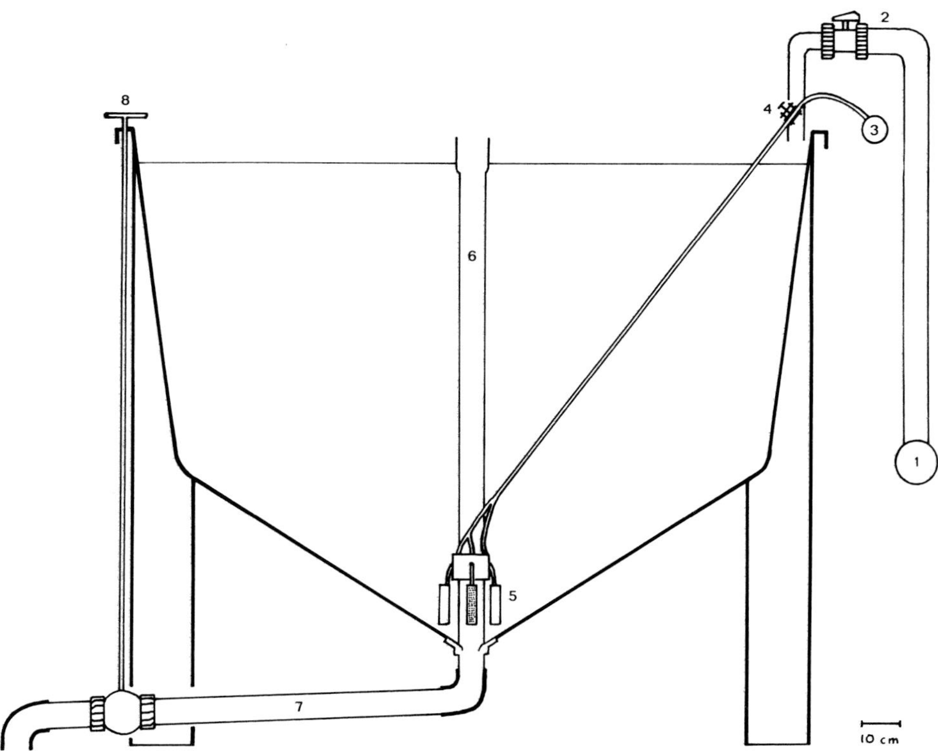

FIGURE 2. 2-m³ Larval rearing tank (static system). (1) Brackish water pipe, (2) water-gate, (3) air pipe, (4) mohr clip to adjust air flow rate, (5) air stones, (6) closing pipe, (7) draining pipe, (8) operating level of the water-gate.

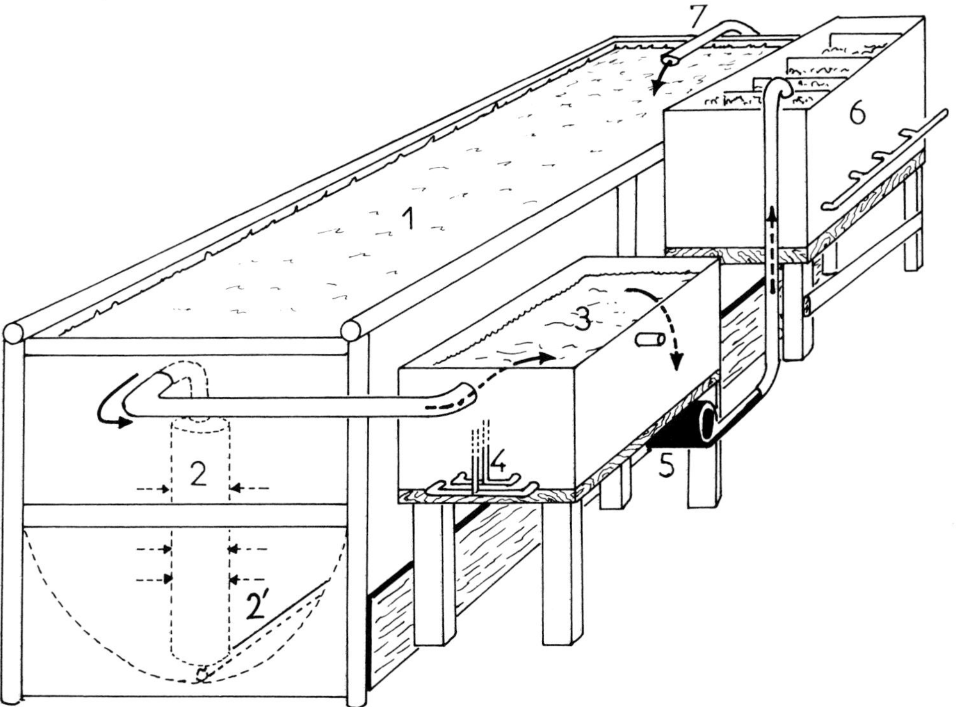

FIGURE 3. 5-m³ Larval rearing tank and recirculating system. (1) Larval rearing tank (U shape), (2) plankton mesh filter, (2′) perforated air pipe, (3) mechanical filter (sand layer), (4) air and water pipes for backwash, (5) pump 50 to 70 ℓ/mm, (6) biological filter, (7) water outlet.

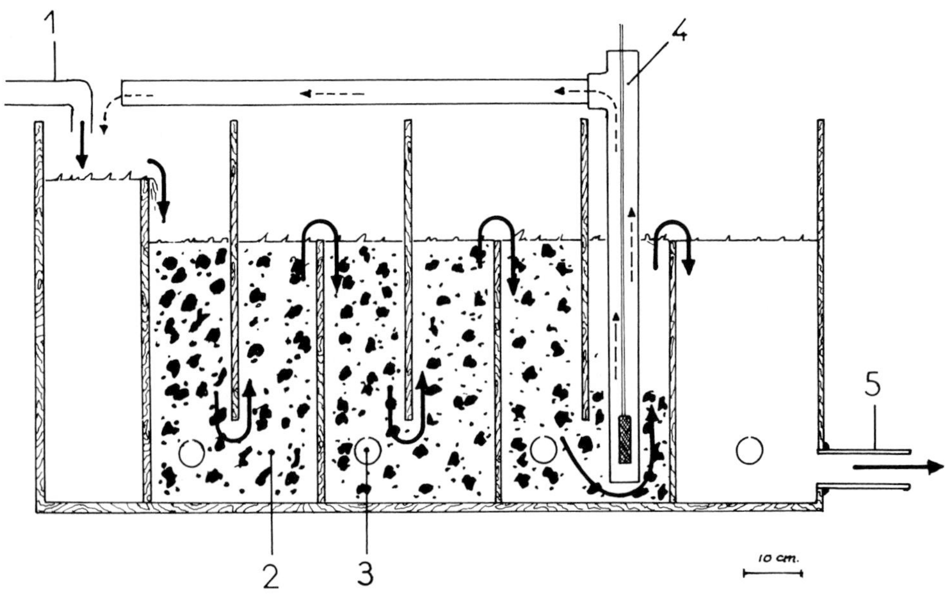

FIGURE 4. Scheme for the biological filter used with 5-m^3 larval rearing tank. (1) Water inlet, (2) coral gravels, (3) aeration pipe, (4) air lift for secondary recirculation, (5) water outlet.

WATER QUALITY

M. rosenbergii larvae are reared in brackish water at a salinity of 12 ppt. This can be done through the "green water" system, developed by Fujimura or the clear water system developed by Aquacop.[1,5] The second was selected for its greater convenience and reliability. Until 1981, the rearing system was static with a total water exchange each day. Since then, the water quality is maintained through a biological filter.

Open System

The water is renewed at the end of each day so that water quality is highest during the night when prawn larvae molt.

To obtain higher growth and survival the temperature is maintained at 30 to 31°C: hot fresh water comes from the cooling secondary circuit of generators. It is used as a fresh water source and in radiators that heat the hatchery.

The salinity is adjusted by seawater addition. The treatments of brackish water consist of chlorination (1.5 mg/ℓ free chlorine for 6 hr) to kill all bacteria and fungi, then filtration on a siliceous sand filter (50 μm) to retain organic or inorganic particles, and a dechlorination by bubbling and adsorption. In the afternoon, just before the total water exchange, the concentrations in ammonia can reach 1.5 N-NH_{3-4} mg/ℓ (Aquacop[8]). No effect has been noticed on growth and survival at this level of N-NH_{3-4}. High mortality was recorded for concentration of residual active chlorine of 0.1 mg/ℓ and larvae are stressed at much lower concentrations. It is important to make sure all free chlorine is eliminated before subjecting larvae to treated culture water.

Closed System

The turnover of water is once every hour at the maximum. The residual concentrations of ammonia and nitrite are less than 0.1 mg N per liter. At the end of the rearing period, the nitrate level is more than 5 mg N-NO_3 per liter and the change of pH is very low (7.80 inside of 8.20). No effects on larvae have been observed (Table 1).

Table 1
RESULTS OF PRODUCTION AND VARIATIONS OF WATER QUALITY IN A 5-m³ TANK WITH RECIRCULATED WATER

Time (days)	D1	D20	D35	D39	
Number of larvae	380,000	380,000	108,000	—	
Number of postlarvae	—	—	247,000	333,000	
Survival (%)	100	100	91	87	
Range of physico-chemical parameters	T°C 28.5—31	pH 7.75—8.21	$N\text{-}NH_{3\text{-}4}$ μg/ℓ 53—280	$N\text{-}NO_2$ μg/ℓ 1—112	$N\text{-}NO_3$ μg/ℓ 2—5,000

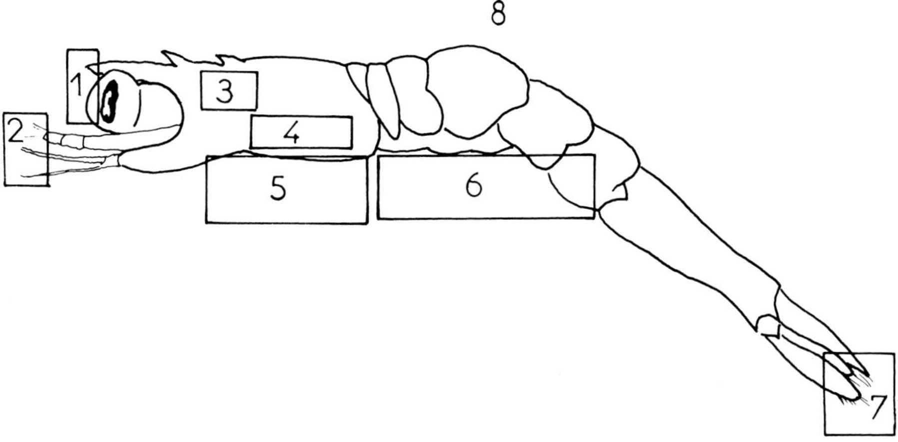

FIGURE 5. Daily check areas of larvae for sanitary and feeding controls. (1) Eye surface, (2) antennae, (3) gut content, (4) gills area, (5) pereiopods, (6) pleopods, (7) telson spines and uropod setae, (8) general pigmentation.

DISEASE CONTROL

Open System

No viral disease has been recorded. Bacterial diseases seldom occur as long as the hatchery is left empty for at least a few days between productions: the monospecific bacteria selected by the rearing conditions, which could become pathogenic, are washed out and the hatchery starts with a plurispecific, harmless, bacterial population. When one production is linked to the preceding without a waiting period, a treatment with antibiotics for 2 or 3 days, at the beginning of metamorphosis, can be useful. Selection of the antibiotic is made through antibiograms and tests on larvae. Through daily observations of the larvae in the tank (color, swimming activity, food chasing) and under the microscope (necrosis, external bacteria, missing appendages, gut content) (Figure 5), diseases can be efficiently spotted and treated.[9]

Closed System

No bacterial disease has been recorded at this time. High survival at metamorphosis is obtained without antibiotic; the bacterial population remains constant and prevents spread of pathogenic strains.

Table 2
COMPOSITION OF THE ARTIFICIAL FOOD UTILIZED IN THE LARVAL REARING OF *MACROBRACHIUM ROSENBERGII*

	Dry (%)
Formula	
Squid	27.6
Shrimp	27.6
Roe	6.9
Egg	6.9
Fish oil	14.0
Vitamins	1.0
Salts	1.0
Algin	15.0
Analysis	
Protein	54.9
Lipid	19.7
Ash	7.7

Table 3
VARIATIONS OF FOOD AMOUNT PER LARVA PER DAY DURING LARVAL REARING

Day	*Artemia* nauplii	Pellets (μg) dry weight
3	5	0
4	10	0
5—6	15	0
7	20	0
8	25	0
9	30	0
10—11	35	0
12	40	70
13—14	45	80—90
15—24	50	100—180
25—30	45	200
30+	40	200

FEEDING

Artificial food and *Artemia* nauplii are utilized. The fabrication of artificial food is derived from the technique of L'Herroux et al. developed for *Penaeus japonicus.*[10] The compound pellet is made of fresh raw materials (Table 2). The wet blend is finely ground to obtain a smooth paste. When algin is added, the blend thickens and the spaghetti-shaped blend is placed in enriched calcium chloride water which increases firmness. After drying, grinding, and sifting, dry, easily stocked, and utilizable particles are obtained. Particles have a very high water stability. This pellet, in itself, is not sufficient to provide proper growth, but allows us to save appreciably on *Artemia* nauplii.

Compound pellets are given at 8:00 a.m. and 12:00 beginning the 10th day. The amount is adjusted to the apparent consumption of larvae, and may vary widely (Table 3).

Newly hatched *Artemia* nauplii are given in the late afternoon starting on the 2nd day (Table 4). In the closed system, the recirculation of water is stopped during *Artemia* distribution. The meal schedule shown in Table 4 is due to the hatching technique for *Artemia* nauplii; the whole ration could be given at the afternoon feeding. The amount of *Artemia* nauplii (Table 3) is adjusted to get the fastest and most complete metamorphosis; any decrease gives slower metamorphosis and, further, lower survival. Food consumption, measured by the percentage of larvae with full gut, is a good index to estimate the health of larvae.

GROWTH AND SURVIVAL

Growth is measured by the evolution of the L.S.I. (Larval Stage Index) as defined by Maddox and Manzi.[11] Larvae generally take 2 days to go through one stage until L.S.I. 7 or 8, and they are generally distributed within two or three stages. The first postlarvae are observed at day 19 to 22 at L.S.I. 8.5 to 9.5; there can be as many as five larval stages present in the tank at this time.

Larval survival in the rearing tank is estimated by counting their numbers in 1-ℓ samples; in 8 to 16 samples the estimation is within 10%. Postlarval survival is estimated after cropping from the rearing tank. Postlarvae are placed in 50-ℓ buckets (5000 to 15,000 per bucket), and 6 to 18 aliquot samples of 150 mℓ each are counted; error is less than 10%. In both cases strong mixing by air stones or by hand insures accurate samples.

Table 4
DIFFERENT STEPS OF *MACROBRACHIUM* POSTLARVAE PRODUCTION, AND DAILY OPERATION OF THE REARING TANKS (STATIC SYSTEM)

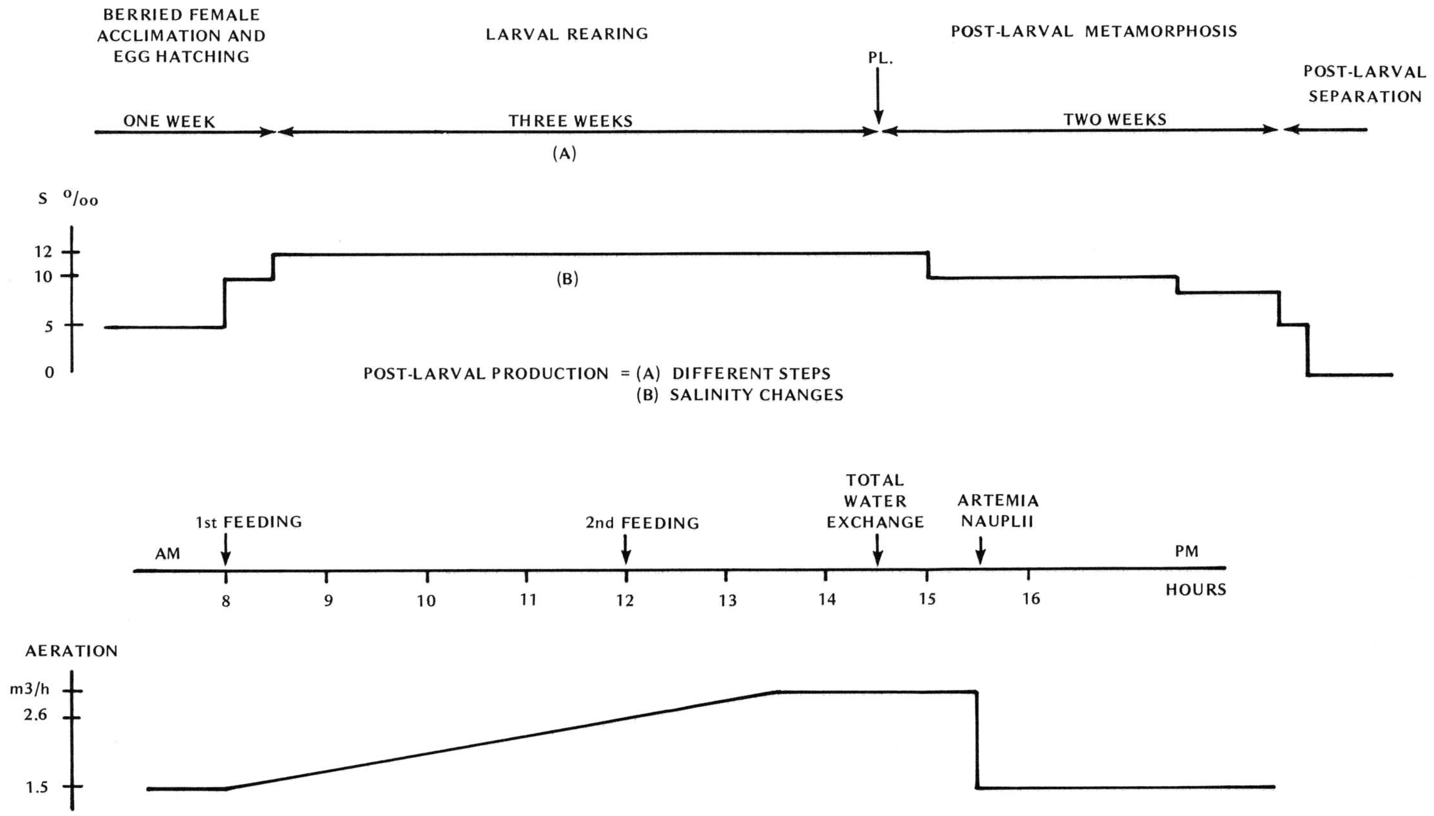

Table 5
PROJECTED COSTS FOR A HATCHERY OF *MACROBRACHIUM ROSENBERGII* PRODUCING 15 MILLION POSTLARVAE A YEAR (IN U.S. DOLLARS)

Investment depreciation	$38,000	(28.1%)
Labor (one biologist + two technicians + two workers)	65,000	(48.1%)
Artemia cysts	15,000	(11.1%)
Pellets for larvae	900	(0.7%)
Pellets for brood stock	1,600	(1.2%)
Antibiotics and treatment products	400	(0.3%)
Sea salt	600	(0.5%)
Energy — circulation pumps	5,500	(4.1%)
Energy — air blower	6,000	(4.4%)
Energy — light	500	(0.4%)
Small hardware	1,500	(1.1%)
Total	$135,000	(100%)

Note: Cost for 1,000 postlarvae: $9.00 U.S.

Usually 90% of postlarvae are obtained before day 42, but we harvest at day 30 to 35 (60 to 80% of postlarvae) for economic reasons.

Survival until first postlarvae has been 100% since 1978. Survival at metamorphosis has been above 75% since 1979. For all the productions since 1980, more than 60 postlarvae were produced per liter. The results are the same in closed system (Table 1).

POSTLARVAE CROPPING AND CONDITIONING

When postlarval density on the bottom of tank is above 2/cm^2 (100,000/2-m^3 tank), some cannibalism may occur and postlarvae must be cropped. The separation of postlarvae from larvae is made by using the planktonic behavior of larvae and the benthic behavior of postlarvae: the air bubbling is stopped and the larvae come to the surface; water is stirred by hand so that a strong circulation current is created which makes the postlarvae cling to the bottom; the larvae are scooped from the tank, and the postlarvae drained out with the water. The operation can be performed within half an hour and less than 5% of the larvae come out with the postlarvae.

The postlarvae are conditioned to freshwater within 12 hr and transferred to nursery earth ponds the day after cropping. The pH must be kept below 9 for the first weeks. Keeping postlarvae at high densities above 10,000/m^3 more than a few days leads to high mortalities, which occur generally at the next molt after transfer to the earth pond. Postlarvae are fed before shipments which take longer than a few hours.

COSTS

The economics are strictly dependent on local conditions. All the figures given in Table 5 are valid in Tahiti only. Investments include ponds for broodstock, a hatchery building, tanks for short-time storage of postlarvae, generators, air blower, all rearing tanks and accessories, *Artemia* hatching tanks, small pelletizing unit, laboratory, office, keeper house, and maintenance shop.

The three major costs are the labor, the investment depreciation, and the *Artemia* cysts. The latter will be reduced in future years. The energy cost does not include heating; the generator and solar panels can provide enough heat.

In many other countries, some costs could be lowered — mainly labor and investment.

With figures obtained from other tropical areas we can assume that the total cost could be cut by half in certain countries.

CONCLUSIONS

This technology has proven its technical reliability in Tahiti and in Martinique; but none of those hatcheries are big enough to be economically independent. In both cases the grow-out farms are still too small to make the hatchery run at full capacity. But all the available figures show that as soon as full capacity can be reached, both hatcheries will need no more financial assistance, the break-even point being approximately 7 to 10 million postlarvae per year.

REFERENCES

1. Aquacop, *Macrobrachium rosenbergii* (de Man) culture in Polynesia: progress in developing a mass intensive larval rearing in clear water, in *Proc. World Maricul. Soc. 8,* Avault, J. W., Ed., Louisiana State University, Baton Rouge, 1977, 311.
2. Aquacop, Production de masse de post-larvae de *Macrobrachium rosenbergii* (de Man) en milieu tropica l: unité pilote, in 3rd Meeting of the I.C.E.S. Working Group on Mariculture — Actes de colloques du CNEXO 4, Brest, France, May 10 to 13, 1977, 213.
3. Aquacop, Intensive larval culture of *Macrobrachium rosenbergii:* a cost study, in *Proc. World Maricul. Soc. 10,* Avault, J. W., Ed., Baton Rouge, La., 1979, 429.
4. Aquacop, Mass Production of *Macrobrachium rosenbergii* Post-Larvae in French Polynesia: Predevelopment Phase Results, presented at Symp. on Coastal Aquaculture, Marine Biological Association of India, Cochin, January 12 to 18, 1980.
5. **Fujimura, T.,** Development of a Prawn Culture Industry in Hawaii, Job Completion Report, NMFS, NOAA, and Hawaii State Division of Fish and Game, Honolulu, 1974.
6. **Hanson, J. and Goodwin, H., Eds.,** *Shrimp and Prawn Farming in the Western Hemisphere,* Dowden Hutchinson, Ross, Stroudsburg, Pa., 1977, 220.
7. **Mock, C. R. and Murphy, M. A.,** Technique for raising penaeid shrimp from eggs to post-larvae, in *Proc. World Maricul. Soc. 1,* Avault, J. W., Ed., Baton Rouge, La., 1971, 143.
8. Aquacop, *Macrobrachium rosenbergii* culture in Polynesia: water chemodynamism in an intensive larval rearing, in *Proc. World Maricul. Soc. 8,* Avault, J. W., Ed., Baton Rouge, La., 1977, 293.
9. Aquacop, Observations on diseases of Crustacean cultures in Polynesia, in *Proc. World Maricul. Soc. 8,* Avault, J. W., Ed., Baton Rouge, La., 1977, 685.
10. **L'Herroux, M., Metailler, R., and Pilvin, L.,** Remplacement des herbivores proies par des microparticules inertes: une application à l'élevage larvaire de *Penaeus japonicus,* in 3rd. Meeting of the I.C.E.S. Working Group on Mariculture — Actes de colloques du CNEXO 4, Brest, France, May 10 to 13, 1977, 147.
11. **Manzi, J. J., Maddox, M. B., and Sandifer, P. A.,** Algal supplement enhancement of *Macrobrachium rosenbergii* (de Man) larviculture, in *Proc. World Maricul. Soc. 7,* Avault, J. W., Ed., Baton Rouge, La., 1976, 677.

SEASONAL CULTURE OF FRESHWATER PRAWNS IN SOUTH CAROLINA

Paul A. Sandifer, Theodore I. J. Smith, Wallace E. Jenkins, and Alvin D. Stokes

A primary responsibility of South Carolina's Marine Resources Research Institute (MRRI) is ''to conduct research leading to the development of mariculture as a viable enterprise in South Carolina''. Currently, 5 of the institute's 16 scientists and about 25% of its 50 to 60 technical and support personnel are involved in mariculture research and development activities. Institute facilities include approximately 5500 m^2 of laboratory and office space and four research vessels (two 15-m trawlers, one 22-m trawler, and a 33-m oceanographic ship) (Figure 1). In addition, the state has recently committed $3.9 million for the construction of a Mariculture Research and Development Center which will be operated as a field station of the MRRI. This facility is expected to be operational by mid-1984. The principal purpose of the center will be to provide the large-scale facilities necessary to test and expand the results of laboratory studies into commercially viable technologies.

Since 1973, a major emphasis of the MRRI mariculture effort has been the development and demonstration of a suitable production technology for culture of the giant freshwater prawn, *Macrobrachium rosenbergii*, in South Carolina's temperate climate. Although large-scale culture of *M. rosenbergii* has centered in tropical areas where year-round production is possible, work in South Carolina, Florida, and elsewhere has indicated potential for smaller-scale, seasonal cultivation of prawns in warm temperate areas of the U.S.[1-7] The MRRI-based program has generally been recognized as the largest public sector prawn research and development program in the continental U.S. Program activities have included all phases of prawn production, from hatchery through grow-out, nutrition and feeds, breeding and genetic manipulation, behavior, systems design and engineering, economics, processing, marketing, and extension.

PRODUCTION PLAN

A basic production model for prawn farming in South Carolina and other areas of temperate climate includes four major elements: (1) a *hatchery phase* in which postlarvae are produced in indoor saltwater tanks during winter or spring; (2) a *nursery phase* in which postlarvae are reared to larger juveniles at very high population densities in specially designed nursery systems; (3) a *production phase* in which marketable prawns are produced seasonally in grow-out ponds, batch-harvested, and processed for market; and (4) a *brood stock phase* in which brood prawns are selected and maintained indoors to provide larvae for the next production cycle.[1] This chapter is limited to a discussion of the nursery and pond grow-out phases. Information on hatchery and broodstock phases is given elsewhere.[8,9]

NURSERY SYSTEMS

The climate of South Carolina permits outdoor cultivation of *M. rosenbergii* only during the warmer half of the year (May to October). Despite this restricted growing season, one crop of prawns can be produced per year from ponds in South Carolina. However, if newly metamorphosed postlarvae are stocked into ponds at the beginning of the growing season, the average size (and value) of the prawns harvested 5 to 6 months later is relatively small. If larger ''nursed'' juveniles (e.g., 0.2 to 1.0 g) are stocked instead, survival is generally higher and the mean size and value at harvest greater.[7] In temperate climates, an indoor nursery system basically provides a 1 to 3 month head start on the growing season, and the availability of nursed juveniles is likely a prerequisite for commercial culture of prawns in such areas.[1,3,6,7,10]

FIGURE 1. Aerial view of the South Carolina Marine Resources Center at Fort Johnson on Charleston Harbor. (1) Marine Resources Research Institute, (2) boat slip with research vessels, (3) headquarters of the Marine Resources Division of the S.C. Wildlife and Marine Resources Department.

Systems Design

The prawn nursery concept developed in South Carolina involves large cylindrical (2 to 6 m in diameter) or rectangular tanks equipped with artificial habitats and biological filters. Water recirculation is preferred to conserve heat, but a flow-through system may be employed where inexpensive supplies of heated water of suitable quality are available.

Cylindrical tanks are generally preferred because the circular geometry allows easy cleaning and water circulation. However, rectangular tanks make better use of floor space. Nursery tanks may be placed in a building or under a plastic roof. Natural lighting should be provided if possible, but the tanks should be shaded from direct sunlight. Holding facilities may be tanks constructed of fiberglass, concrete, plastic (e.g., plastic-lined wading pools), or small, covered earthen ponds. Prior to use, fiberglass or plastic-lined tanks should be thoroughly washed and leached to remove plastic residue and any biocides that may be incorporated in the construction materials. Similarly, the inner surfaces of concrete tanks should be painted with an epoxy paint and leached. Leaching is accomplished by filling the tank with water and letting it stand for several days. This should be repeated several times over a 2- to 4-week period. All new tanks should be bioassayed with juvenile prawns before the tanks are connected into the main nursery system.

The biological filter system may be within or outside (preferred) the nursery tank. The filter media is dolomitic gravel, and it functions primarily as a substrate for nitrifying bacteria and secondarily for particle removal and as a buffering agent.[11] Based on experience, we use 1 to 2 ℓ of gravel (~2- to 10-mm particles) per 50 ℓ of water in the system. Air-lift pumps[12] are generally placed along the perimeter of the tank and in the filter bed to circulate water, and either airlift or electric pumps are used to move water between the filter and the tank. Water is exchanged through the filter at a rate of 3 to 6 tank volumes/day, and 10 to 25% of the system water is replaced weekly.

The artificial habitat units are designed to increase the amount of surface area available to the postlarval prawns in the nursery tank. The units consist of a rigid frame constructed

FIGURE 2. Two views of an artificial prawn habitat unit constructed of plastic mesh on a rigid frame. Note the strips of window screening which serve as feeding stations on each layer.

of wood, PVC pipe, or plastic-coated wire to which layers of plastic mesh (usually 3.8 to 5.0 cm^2) are attached. An example of one habitat unit is shown in Figure 2. The rigid frame facilitates removal of the unit from the tank during cleaning activities. The wooden units have 2.5-cm slats nailed horizontally around the frame at intervals of about 2.5 cm, and the mesh layers are attached to these slats. These designs are based on a behavioral characteristic of the prawns termed the ''edge effect'' — that is, when presented with stacked, more or less solid layers, the prawns exhibited a pronounced preference for the layer edges.[13] The mesh layers markedly increase the amount of surface edges available to the prawns in both vertical and horizontal planes. Five-centimeter-wide strips of plastic window screening are placed on different layers to serve as feeding stations (i.e., food settles on these strips).

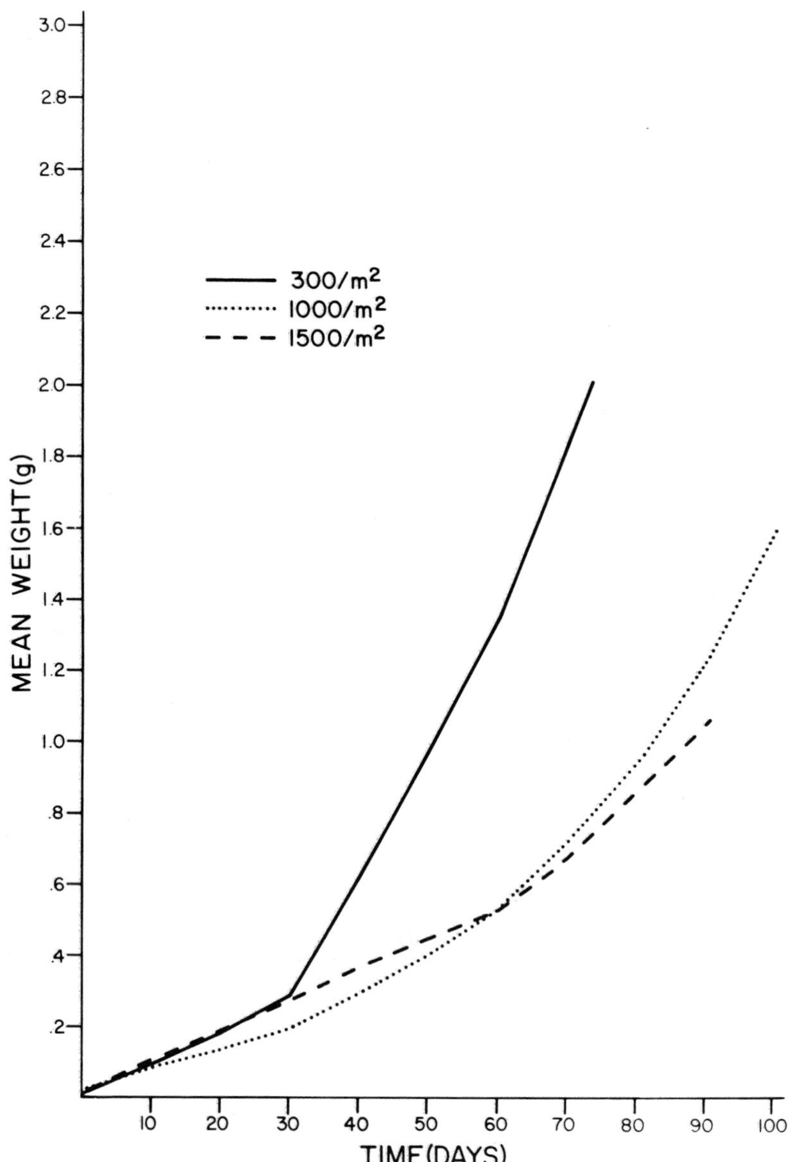

FIGURE 3. Generalized growth curves for juvenile prawns reared in indoor nursery tanks at three population densities. (From Smith, T. I. J. and Sandifer, P. A., *Proc. World Maricul. Soc.*, 10, 369, 1979. With permission.)

Stocking Density, Growth, and Survival

Depending on projected needs, postlarvae may be stocked in nursery tanks at densities from 300 to >6000/m^2 of tank floor. However, survival and growth rate are markedly density dependent (Figures 3 and 4). For short-term (e.g., 30 days) holding of postlarvae, densities of ≥2000/m^2 may be reasonable. For longer periods (e.g., 60 to 90 days) and where the desire is to produce a 0.3- to 1.0-g juvenile, stocking densities on the order of 1000 to 1500/m^2 are recommended. At these densities, postlarvae grow to approximately 0.5 g in about 60 days, with average survival rates of 92 and 84%, respectively.[14] Growth and survival may be higher at lower densities (e.g., 300 to 500/m^2), but such densities are unlikely to be economically attractive where indoor, heated nursery systems must be employed.

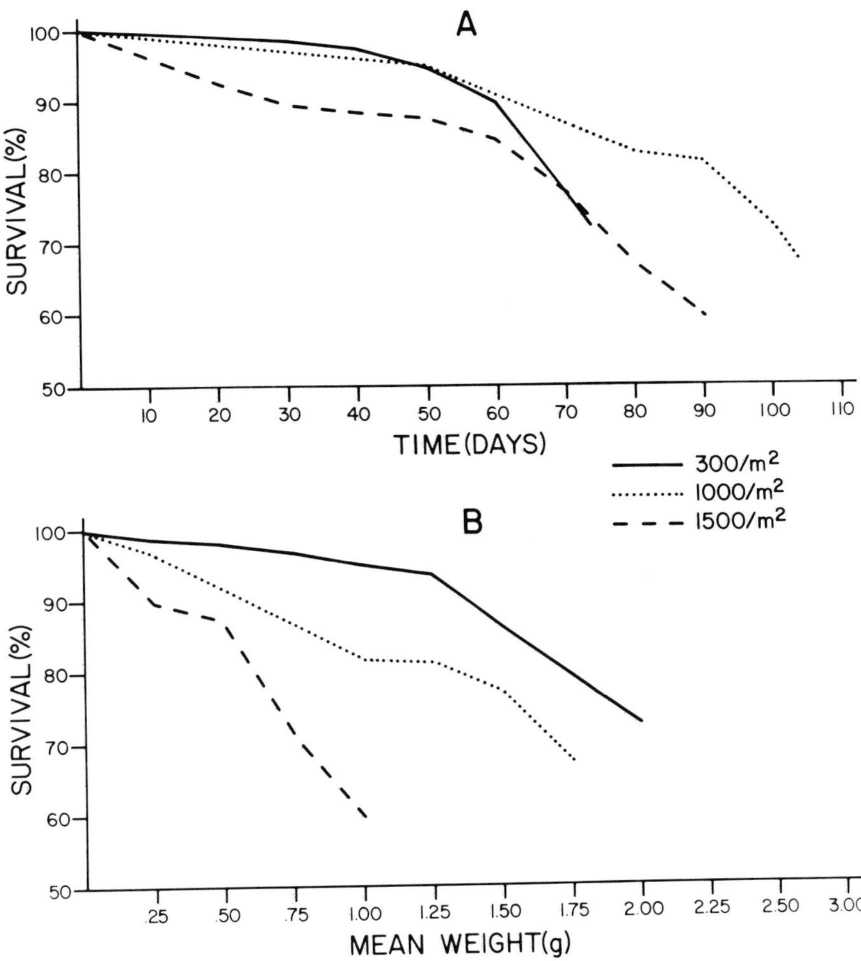

FIGURE 4. Survival in relation to time (A) and mean weight (B) for prawns reared in indoor nursery tanks at three population densities. (From Smith, T. I. J. and Sandifer, P. A., *Proc. World Maricul. Soc.*, 10, 369, 1979. With permission.)

In one large-scale nursery experiment, a total of 396,410 postlarvae was stocked into 24 laboratory tanks (cylindrical and rectangular) with an aggregate floor area of 141 m^2. Stocking densities ranged from approximately 800 to 6300/m^2 with an overall average of 2811/m^2. At these high densities, overall survival was 61.3% after 3 months (average harvest density of 1724/m^2), and mean prawn weight at harvest was 0.3 g.

Feeding

The postlarval prawns should be fed two to four times per day. In our systems Purina® Experimental Marine Ration 25 (25% protein), ground to an appropriate size for the small animals, is used as the primary food. This is regularly (at least twice per week) supplemented with natural foods, consisting of a mixture of cooked chicken eggs, fish roe (especially mullet), squid or fish flesh, and green vegetable matter (e.g., spinach). These items are chopped and sized by passing through a U.S.A. Standard Testing Sieve No. 16 (1.18-mm openings) before being fed to the prawns. The amount of food given should not exceed 15% of estimated biomass per day, and this rate can be decreased as the prawns grow. The daily feed allocation should be divided into two to four equal meals; these are broadcast over the total tank surface. Accumulation of uneaten food indicates overfeeding which may lead to serious water quality problems.

Water Quality Management

Perhaps the most critical water quality parameter in the nursery phase is temperature. The optimal range for growth is 27 to 29°C, and water temperatures should not be allowed to drop below 23°C or exceed 31°C. Other water quality parameters should be maintained as follows: pH, 7 to 8; total ammonia/nitrogen, <1 ppm; nitrite, <0.1 ppm; dissolved oxygen, ≥5 ppm.

Each nursery tank should be cleaned by siphoning off the bottom at least twice per week. This should prevent the accumulation of harmful amounts of uneaten food, fecal material, and dead prawns. Habitat units should be removed and spray cleaned about once a month. There is no need to cover the tank floor with a ''natural'' substrate such as mud, sand, or gravel, as tanks with such bottom substrates give no better results than those with bare fiberglass, plastic, or concrete floors.[14]

If a recirculating system is used, the filter bed must be large enough to accommodate the tank volume and have a well-established bacterial flora.[11] In addition, the filter flow rate should be sufficient for all the water in the nursery tank to pass through the filter at least three times per day. Plants, such as water hyacinths *(Eichhornia crassipes)* or agaria *(Agaria densa),* can be added to nursery tanks to provide additional habitat, water purification capacity, and supplemental food. In general, nursery tanks which contain plants yield somewhat better production levels than those without them.

POND CULTURE

Ponds and Pond Preparation

Most grow-out studies have been conducted in ponds of the South Carolina Wildlife and Marine Resources Department's Dennis Wildlife Center at Bonneau, S.C. These 0.25-ha ponds were constructed in 1974 primarily for the production of striped bass *(Morone saxatilis)* fingerlings. The soil texture is a sandy clay loam, and the pond levees have a 3:1 slope. Each pond is approximately 150 × 17 m and equipped with a flat-board riser water control structure and a 3 × 4 × 0.5-m-deep concrete block harvest basin (Figure 5). The pond bottoms are sloped for ease of draining, and water depth ranges from about 0.5 m at the shallow end to 1.8 m at the harvest basin (mean depth, ~1.2 m). The water inlet pipe is adjacent to the corrugated riser and allows water input directly into the harvest basin. Water for filling and flushing is obtained from an adjacent 22,000-ha reservoir, Lake Moultrie. A filter bag made of 400-μm Nitex mesh is placed over the opening of the pond inlet pipe to screen all incoming water.

Prior to filling the pond for a prawn rearing trial, all puddles remaining in the pond bottom are treated either with a chlorine solution or Rotenone to eliminate potential predators and competitors (e.g., fishes). The soil pH is also checked prior to filling. An optimal soil pH for algal production is estimated to be 6.5 to 7.0. Lime (agricultural limestone) should be spread evenly over the pond bottom if the soil pH is less than 6.0. The amount of lime needed can be calculated from data given by Boyd.[15] Addition of fertilizers may be necessary to establish a good phytoplankton bloom. Generally, if a satisfactory bloom does not develop within 1 week after filling the pond, a 20-20-10 pond fertilizer (20% nitrogen:20% phosphoric acid:10% potash) is added at the rate of ~112 kg/ha. However, we recommend that fertilizer be added at half this rate initially, with additional 56-kg/ha increments provided as needed over a 1- to 2-week period to establish a satisfactory bloom. The fertilizer should be spread as evenly as possible over the entire pond surface.

The water management structure consists of a flat-board riser and drain pipe installed in the deepest area of the pond to allow complete drainage. The drain pipe should be sufficiently sized to drain the pond within 24 hr. The riser is designed to support boards and screens which allow removal of either bottom or surface water during normal exchange. This ar-

FIGURE 5. Top: Example of 0.25-ha pond used for experimental culture of prawns in South Carolina. Bottom: Drain structure and harvest basin of above pond.

rangement reduces water stratification, permits removal of oxygen-deficient bottom water, and allows undesirable algal concentrations (e.g., blue-green blooms) to be flushed out. A semicircular corrugated aluminum riser, having two tracks on its open side and a drain pipe at the bottom, is suitable for freshwater applications; a concrete structure is preferred for brackish-water ponds. A shear gate valve may be installed in the drain pipe, but it is not essential.

Manipulation of tongue and groove boards and screens in the tracks of the riser permits selective water removal from the ponds. Typically, boards in the outside track are held ~0.3 m off the bottom with a 1.2-cm^2 mesh screen while the top board extends several centimeters above the desired water depth. The rear track is filled with boards from the bottom to exactly the desired water depth. A 1.2-cm^2 mesh screen is placed on top of the boards in the rear track to screen outflowing water and prevent the loss of prawns. This screen is covered with a finer mesh until the prawns have grown too large to pass through a 1.2-cm^2 mesh. With this arrangement, addition of water results in bottom water discharge — the water passes under the boards in the first track and over those in the second. To flush surface water, the top boards from the front track are removed prior to adding water to the pond.

If we had the opportunity to design ponds specifically for seasonal prawn culture, we would recommend the following major changes in the design described above:

1. The ponds would be larger, 0.4 to 1 ha in size, with a length-to-width ratio of 3 to 4:1.
2. The main water inflow would be installed at the end of the pond opposite the drain to improve water circulation.
3. The ponds would be somewhat deeper, with approximately 1 m of water in the shallow end and 1.8 m at the deep end.
4. The ponds would be equipped with external catch basins. Such basins are built at the end of the outflow pipe and consist of a concrete floor and three sides of cement block. The end opposite the drain pipe is left open and fitted with tracks for screens or boards. These will allow water to flow out of the basin or be retained during washing of prawns. A single such basin could serve two ponds; the drainage structures of the ponds could be located in adjacent corners, with the drain pipes emptying into the same basin.

Stocking

The major criterion for determining time of stocking is water temperature. Water temperatures should have risen to at least 22 to 23°C in bottom waters, with little chance that temperatures would drop below 20°C, before the decision is made to stock ponds. Such conditions are generally met in South Carolina by late April or early May.

Optimum stocking densities for prawn culture in South Carolina appear to range from about 4.3 to 6.5 nursed juveniles (or mixture of nursed juveniles and postlarvae) per square meter of pond surface (43,000 to 65,000/ha).[6,7,10] Such densities yield average production levels of approximately 700 to 1200 kg/ha in our 5 to 6 month growing season (Table 2). If larger juveniles (e.g., 1 to 2 g) could be stocked, yields and final mean size could be expected to increase substantially.

Accurate estimates of the number of prawns being stocked in a pond and their mean weight are essential to determine feed levels, survival, and growth. Hand counting all the prawns and individually weighing a subsample give the best results, but these methods are laborious and, thus, costly. Satisfactory estimates of mean weight can be obtained by weighing several representative subsamples of the prawns, counting the number in each subsample, and dividing the weight by the count to obtain an average. The average weight can then be divided into the total weight of the population to give an estimate of total numbers. The juvenile prawns should be weighed in small groups to prevent damage to the animals and returned to water as quickly as possible.

Juveniles are transported to the pond site in truck-mounted live tanks of 600- to 1200-ℓ capacity. The live tank should be covered to prevent loss of prawns and exposure to direct sunlight and equipped with artificial habitat units (see nursery section). An oxygen cylinder connected via a regulator to airline tubing is used to provide aeration. Care should be taken not to overstock the live tank with prawns as this may result in substantial mortalities. As a general rule, we do not exceed 15 nursed juveniles per liter of water in the live tank for transport periods of one to several hours.

Prawns are transported to the pond site in early morning or late afternoon. At the pond site, the temperature and pH of the pond water are checked and compared with that of the tank water. If the differences in these conditions are slight (≤2°C, <1 pH unit), the prawns can be acclimated to the pond conditions rapidly. This can be done easily by draining approximately half of the water from the transport tank and replacing it with pond water. The prawns are left in this mixture for 15 to 30 min and then released into the pond. If temperature and pH differ markedly between the transport tank water and the pond, more extensive acclimation is recommended.

For release into the pond, the prawns can be drained from the transport tank through a large hose or dip-netted out. In either case, we recommend that the animals be released over a window screen placed on the pond bottom in very shallow water. Any dead prawns will settle on the screen, and they can be counted to give an estimate of transport mortality. In addition, two to three other small groups of prawns (e.g., 25 to 100 prawns each) should be taken at random, counted, and placed in mesh cages in different areas of the pond. Survival of the prawns in these cages should be checked daily for 2 to 3 days to provide an estimate of stocking mortality. Then, the estimated stocking density should be adjusted for transport and stocking mortalities.

Feeding

We have used Ralston Purina® Experimental Marine Ration 25 in most of our pond culture trials. This preparation is relatively water stable, and it has given excellent food conversions when used carefully (Table 1).

The amount of feed given per day is a function of the estimated prawn biomass in the pond. Based on our experience, we recommend daily feed levels ranging from 10 to 15% of estimated biomass at the beginning of the culture period to 0.5 to 1.0% at the end (Table 2).

The amount of food to be given daily over any period can be derived by the following simple calculation:

$$N_i \times S_t \times \overline{w}_t \times FR_t = \text{Daily Ration (in g)}$$

where N_i = the initial number of prawns stocked in the pond, adjusted for transport and stocking mortalities; S_t = estimated survival at time t — a rough mortality estimate indicated by our data is 5%/month; $\overline{w}_t$ = mean prawn weight (in g) at time t. This is determined by monthly seine samples in which at least three replicate groups of 50 to 100 prawns per pond are weighed and counted to determine mean weight and variance; FR_t = the recommended feed rate for time period t (see Table 2).

As an example, let us assume that a 1-ha pond was stocked with 50,000 prawns, transport and stocking mortality amounted to 2%, and mean prawn weight at the end of July is 10 g, as determined by sampling. You now want to calculate the daily feed allotment for August: N_i = 50,000 × 0.98 = 49,000; S_t = 84%; $\overline{w}_t$ = 10 g; FR = 2% (from Table 2); 49,000 × 0.84 × 10 g × 0.02 = 8232 g feed per hectare per day.

We usually adjust feed rates at 4-week intervals, but it can be done more frequently if desired.

Feed is broadcast over as much of the pond surface as possible once daily in the late afternoon. Feeding is discontinued, reduced, or scheduled for late morning when the dissolved oxygen level drops below 2.5 ppm or is expected to do so. Similarly, the feeding rate is reduced by half near the end of the growing season when water temperatures drop below 24°C. At temperatures below 20°C feeding is stopped altogether and preparations are made to harvest the pond.

Table 1
REPRESENTATIVE STOCKING AND PRODUCTION DATA FOR PRAWNS REARED IN PONDS IN SOUTH CAROLINA

Stocking data			Harvest data				
Population description	Density (no./m²)	Mean weight (g)	Rearing period (Days)	Survival (%)	Mean weight (g)	Production (kg/ha)	Feed conversion[a]
Postlarvae + nursed juveniles (50:50 mixture)	4.3	0.4	146	73.6	20.2	674	1.1
	6.5	0.6	153	70.9	23.1	1204	2.2
Nursed juveniles only	4.3	0.4	168	86.9	25.8	995	1.2
	6.5	0.4	154	79.1	22.4	1208	1.7

[a] Number of units of feed (dry weight) needed to produce 1 unit of prawns (wet weight).

Table 2
RECOMMENDED DAILY FEEDING LEVELS FOR PRAWNS REARED IN SOUTH CAROLINA

Month	Daily feeding level (% of estimated biomass)
May	10—15
June	6—7
July	4—5
August	2—3
September	1—2
October	<1—1

Water Quality Management

In our experience, dissolved oxygen concentration, temperature, pH, secchi disk visibility, and total alkalinity are the most important water quality parameters to be monitored. Other indicators (hardness, nutrient levels, etc.) should be measured prior to stocking and checked periodically during the growing season.

Depletion of dissolved oxygen (DO) has been responsible for the loss of more prawns in South Carolina than all predators combined. Such losses can be prevented by careful monitoring and management. DO levels should be measured in bottom water samples daily (if possible) just before sunrise. At this time the oxygen concentration should be at its lowest level in the diurnal cycle. Ideally, DO concentrations ≥5 ppm should be maintained at all times; however, this is not always possible. If the DO concentration drops below 2.5 ppm, water exchange should be initiated (if the "new" water has a higher DO level or can be aerated) and supplemental aeration provided at night until the situation is remedied. Extremely high (e.g., >12 ppm) DO levels in the afternoon indicate an algal bloom which can result in nighttime DO depletion. Heavy blooms should be flushed out by water exchange and nighttime DO concentrations closely monitored. Mechanical aeration equipment should be available for use if needed.

The optimum temperature for prawn growth appears to be about 28°C, and growth decreases rapidly at temperatures <24°C and >31°C. In South Carolina, pond water temperatures <24°C occur at the beginning and end of the growing season, and there is little that can be done to improve this situation. However, at such times the source water may have a somewhat higher temperature, and water exchange may be used to increase the pond temperature slightly. In late summer pond water temperatures occasionally reach 35°C or higher, and these high temperatures are usually accompanied by depressed DO levels. Water exchange during such periods may be critical.

A preferred water pH range for prawns is 6.5 to 9.0. If the pH decreases below 6.5, it can be raised by the addition of crushed agricultural limestone. If pH rises to >9.0, heavy photosynthetic activity is likely responsible.[15] Water exchange to reduce the phytoplankton density usually results in a decrease in pH.

Secchi disk visibility provides a simple measure of light penetration in the pond. Clear water is to be avoided since it allows light to penetrate to the pond bottom and encourage the growth of vascular plants. Such plants could ultimately cover much of the pond bottom, making drain-harvesting very difficult. Highly turbid, very green water which permits little light penetration often indicates a heavy algal bloom that could cause a serious DO deficiency. In our experience, secchi disk readings in the range of 0.5 to 0.8 m are acceptable. Readings <0.5 m usually indicate a heavy algal bloom; water exchange is recommended to flush this bloom from the pond. Readings >0.8 m indicate that the water is too clear. If the reading does not begin to decrease within a couple of days, fertilizer should be added carefully to promote algal growth.

At levels <180 ppm, total alkalinity appears to have little, if any, effect on prawn growth and survival in ponds. However, in South Carolina waters with total alkalinity levels >180 ppm have occasionally been associated with significant prawn mortalities. It appears that phytoplankton blooms in such waters lead to increased pH; this may result in the formation of a calcium precipitate which settles on the prawns. It is likely that the calcium precipitate simply occludes the gills and causes mortality by suffocation. Waters with high alkalinity levels should be avoided if possible.

Predator Control

In South Carolina, predators on prawns include fish, amphibians, reptiles, birds, and mammals (Table 3). Fish predators can be controlled by prefilling eradication (i.e., poisoning any puddles in the pond bottom prior to filling) and by screening the in-flowing water from

Table 3
LIST OF KNOWN AND SUSPECTED PREDATORS OBSERVED IN AND AROUND PRAWN GROW-OUT PONDS IN SOUTH CAROLINA

Common name	Scientific name
Fishes	
American eel	*Anguilla rostrata*
Bluegill sunfish	*Lepomis macrochirus*
Largemouth bass	*Micropterus salmoides*
Pumpkinseed sunfish	*Lepomis gibbosus*
Redbreasted sunfish	*L. auritus*
Warmouth sunfish	*Chaenobryttus gulosus*
Amphibians	
Bull frog	*Rana catesbeiana*
Reptiles	
Brown water snake	*Nerodia (Natrix) taxispilota*
Cottonmouth	*Agkistrodon piscivorus*
Chicken turtle	*Deirochelys reticularia*
Florida cooter	*Chrysemys floridana*
Birds	
American anhinga	*Anhinga anhinga*
American egret	*Casmerodius albus*
Belted kingfisher	*Megaceryle alycon*
Great blue heron	*Aredea herodius*
Green heron	*Butorides virescens*
Little blue heron	*Florida caeralea*
Redbreasted merganser	*Mergus serrator*
Snowy egret	*Leucophoyx thula*
Mammals	
River otter	*Lutra canadensis*
Raccoon	*Procyon lotor*
Man	*Homo sapiens*

surface sources. Once predatory fish become established in a prawn pond, it is difficult to eliminate them without draining the pond. Wading birds can be discouraged from entering the shallow ends of ponds by stretching a clear monofilament line on stakes about 15 cm above the ground just outside the water perimeter. These birds typically light on the dike and walk into the shallow water to feed. The monofilament line serves to trip the birds, and it seems to be effective in discouraging them. Diving birds are generally not as serious predators as wading birds, but they are difficult to eliminate. Loud noises (e.g., gunshots) and dogs are effective to a limited degree. Among the reptiles, turtles, snakes, and alligators occasionally prey on prawns, but none are likely to be serious pests. Turtles and snakes can be shot if necessary. Alligators are protected; they should be trapped by a wildlife official and removed without harming them. Large frogs can be serious predators on juvenile prawns. We have found as many as nine juvenile prawns in the stomach of one bullfrog *(Rana catesbeiana).* Frogs can be eliminated by shooting or gigging. The principal mammalian predator, other than man, is the river otter. Fortunately, it has foraged in our ponds infrequently. Otters can be discouraged by a low voltage electric fence. Raccoons occasionally prey on prawns along the periphery of a pond. Such predation occurs most often when ponds are being drained, and it is probably not of much significance.

Harvesting and Processing

Preparations should be made for harvesting when water temperatures drop to 20°C in the fall, and harvesting should be completed before temperatures decline to 16°C. Typically, about half the water is drained from a pond the afternoon before the day we plan to harvest. The outflow rate is sharply restricted during the night. The following morning, the remaining water is drained and the prawns concentrated in the harvest basin. As soon as the prawns collect in the harvest basin they are removed with a small seine and hand nets. Any animals stranded on the pond bottom outside the harvest basin should be picked up and placed on ice as soon as possible. Workers should not walk on any area of the pond bottom while it is still covered with water, as each resulting footprint will retain water and provide refuge for one or more prawns. Each such depression will later have to be harvested by hand.

Within a few minutes after being removed from the pond, the prawns are washed thoroughly to remove mud, chill-killed in an ice slurry or iced brine, and weighed in tared baskets to determine yield. The washed prawns are placed in perforated plastic laundry baskets, and the baskets are set in 80- to 125-ℓ plastic trash cans filled with ice and water. The prawns in the basket are stirred to expose them thoroughly to the chilled water. After 5 to 10 min, the prawns are removed from the ice water and processed immediately or placed on ice for later handling.

Processing may involve size grading, heading, and freezing the product. Large prawns (>30 g) may be sold whole or as tails, while most smaller prawns are handled as a shrimp-tail product. Heading may be done by the farmer/processor or by the consumer. The tail yields from various size prawns are summarized in Table 4.[16]

Immediately after grading and heading, the prawns should be placed on ice or frozen. If the prawns are to be marketed fresh, they can be maintained on ice for up to 8 days.[17] However, if the product is to be frozen, it should be placed in a freezer as soon as possible after removal from the pond. We recommend freezing in a commercial blast freezer, glazing, and storage in waxed 5-lb or 2-kg institutional cartons at −20 to −40°C. Shelf life for prawns processed in this manner is estimated to be 7 months for whole prawns and 10 months for tails.[18] Further, whole prawns frozen raw generally have a longer shelf life than those frozen after cooking.[18]

Marketing

Studies in South Carolina have identified three local markets for farm-reared prawns: retail seafood markets, seafood restaurants, and direct sales to consumers.[19-21]

Prawns were readily sold as ungraded tails in retail seafood markets at prices of $8.79 to 11.01/kg.[19] The seafood retailers stated preferences for fresh prawns of medium size (78 to 110 tails per kilogram), but indicated they would accept frozen and whole product, too.

Local restaurants also successfully test-marketed prawns.[20] The restaurateurs expressed a preference for fresh product in the 35 to 65 tails-per-kilogram size range or larger. A major limitation of this market is its need for a relatively consistent supply of prawns.

The third marketing opportunity, direct sales to consumers, appears to be the best alternative for small growers.[21] Following this approach, the farmer places advertisements in local news media and contacts friends and previous customers shortly before the harvest, and then sells the entire crop at the pond site immediately after harvest. This approach greatly reduces the grower's costs for processing, packaging, storage, and transportation

Table 4
APPROXIMATE TAIL SIZES (TAIL-COUNT CATEGORY) OBTAINABLE FROM WHOLE PRAWNS OF DIFFERENT SIZES AND SEXES

Tail count[a]		Males		Females		Unsexed	
No./lb	No./kg	Weight (g)	Length[b] (mm)	Weight (g)	Length (mm)	Weight (g)	Length (mm)
<5	<11	>237.4	>200.3	>183.1	>189.4	>213.8	>195.3
6—10	12—23	237.0—113.1	200.2—160.2	182.9—92.5	189.3—153.4	213.5—104.3	195.3—159.2
11—15	24—33	112.8—72.3	160.0—139.9	92.3—61.2	153.3—135.1	104.1—67.6	157.0—137.8
16—20	34—44	72.0—52.6	139.7—127.1	61.0—45.7	135.0—123.4	67.3—49.7	137.6—125.5
21—25	45—55	52.3—41.1	126.9—117.9	45.5—36.4	123.3—115.1	49.4—39.1	125.3—116.7
26—30	56—66	40.8—33.2	117.7—110.7	36.2—30.0	114.9—108.4	38.9—31.9	116.5—109.7
31—35	67—77	33.0—28.2	110.5—105.3	29.7—25.7	108.2—103.4	31.7—27.2	109.5—104.5
36—40	78—88	27.9—24.2	105.0—100.6	25.5—22.4	103.2—99.1	26.9—23.5	104.2—100.0
41—45	89—99	23.9—21.1	100.3—96.4	22.2—19.7	98.8—95.2	23.2—20.5	99.7—96.0
46—50	100—110	20.8—18.7	96.1—93.0	19.5—17.6	94.9—92.0	20.2—18.2	95.6—92.6
51—55	111—121	18.4—16.8	92.6—90.0	17.4—15.9	91.7—89.3	18.0—16.4	92.2—89.7
56—60	122—132	16.5—15.4	89.7—87.7	15.7—14.7	88.9—87.1	16.2—15.1	89.2—87.4
61—70	133—154	15.1—12.8	87.3—83.1	14.5—12.5	86.7—82.8	14.9—12.7	87.1—83.0
71—80	155—176	12.6—11.0	82.6—79.4	12.3—10.9	82.3—79.3	12.5—11.0	82.5—79.4
81—90	177—198	10.8—9.5	78.9—75.9	10.7—9.4	78.8—76.0	10.8—9.5	78.9—76.0
>90	>198	<9.5	<75.9	<9.4	<76.0	<9.5	<76.0

[a] These tail-count categories typically represent market categories in the U.S.
[b] Length measured from orbit of eye to tip of telson.

and it is less restricted by such factors as average size, available volume, and consistency of supply.

ACKNOWLEDGMENTS

Information reported here is based on research supported by the National Sea Grant College Program, the Coastal Plains Regional Commission, and the state of South Carolina. We thank Jack Bayless for the use of facilities at the Dennis Wildlife Center and our mariculture staff for their assistance in the laboratory and field.

Reference to trade names in this paper does not imply endorsement.

REFERENCES

1. **Sandifer, P. A. and Smith, T. I. J.,** Experimental aquaculture of the Malaysian prawn, *Macrobrachium rosenbergii* (de Man), in South Carolina, U.S.A., in *Advances in Aquaculture,* Pillay, T. V. R. and Dill, W. A., Eds., Fishing News Books, Farnham, Surrey, England, 1979, 306.
2. **Smith, T. I. J., Sandifer, P. A., and Trimble, W. C.,** Pond culture of the Malaysian prawn, *Macrobrachium rosenbergii* (de Man), in South Carolina, 1974—1975, *Proc. World Maricul. Soc.,* 7, 625, 1976.
3. **Smith, T. I. J., Sandifer, P. A., and Smith, M. H.,** Population structure of Malaysian prawns, *Macrobrachium rosenbergii* (de Man), reared in earthen ponds in South Carolina, 1974—1976, *Proc. World Maricul. Soc.,* 9, 21, 1978.
4. **Willis, S. A. and Berrigan, M. E.,** Effects of stocking size and density on growth and survival of *Macrobrachium rosenbergii* (de Man) in ponds, *Proc. World Maricul. Soc.,* 8, 251, 1977.
5. **Roberts, K. J. and Bauer, L. L.,** Costs and returns for *Macrobrachium* grow-out in South Carolina, U.S.A., *Aquaculture,* 15, 383, 1978.
6. **Sandifer, P. A., Smith, T. I. J., and Bauer, L. L.,** Economic comparisons of stocking and marketing strategies for aquaculture of prawns, *Macrobrachium rosenbergii* (de Man), in South Carolina, U.S.A., in *Proc. Symp. Coastal Aquaculture, Cochin, India,* 1, 88, 1982.
7. **Smith, T. I. J. and Sandifer, P. A.,** Influence of three stocking strategies on the production of prawns, *Macrobrachium rosenbergii,* from ponds in South Carolina, U.S.A., in *Proc. Symp. Coastal Aquaculture, Cochin, India,* 1, 76, 1982.
8. **Smith, T. I. J., Sandifer, P. A., and Trimble, W. C.,** Progress in developing a recirculating synthetic seawater hatchery for rearing larvae of *Macrobrachium rosenbergii,* in *Food-Drugs from the Sea, Proc. 1974,* Webber, H. H. and Ruggieri, G. D., Eds., Marine Technology Society, Washington, D.C., 1976, 167.
9. **Sandifer, P. A. and Smith, T. I. J.,** Aquaculture of Malaysian prawns in controlled environments, *Food Technol.,* 32, 36, 1978.
10. **Smith, T. I. J., Sandifer, P. A., Jenkins, W. E., and Stokes, A. D.,** Effects of population structure at stocking and density on production and economic potential of prawn *(Macrobrachium rosenbergii)* farming in temperate climates, *Proc. World Maricul. Soc.,* 12(1), 233, 1981.
11. **Spotte, S.,** *Fish and Invertebrate Culture: Water Management in Closed Systems,* 2nd ed., John Wiley & Sons, New York, 1979, chap. 1.
12. **Castro, W. E., Zielinski, P. B., and Sandifer, P. A.,** Performance characteristics of air lift pumps of short length and small diameter, *Proc. World Maricul. Soc.,* 6, 451, 1975.
13. **Smith, T. I. J. and Sandifer, P. A.,** Observations on the behavior of the Malaysian prawn, *Macrobrachium rosenbergii* (de Man), to artificial habitats, *Mar. Behav. Physiol.,* 6, 131, 1979.
14. **Smith, T. I. J. and Sandifer, P. A.,** Development and potential of nursery systems in the farming of Malaysian prawns, *Macrobrachium rosenbergii* (de Man), *Proc. World Maricul. Soc.,* 10, 369, 1979.
15. **Boyd, C. E.,** *Water Quality in Warmwater Fish Ponds,* Agricultural Experiment Station, Auburn University, Auburn, Ala., 1979, chap. 5.
16. **Smith, T. I. J., Waltz, W., and Sandifer, P. A.,** Processing yields for Malaysian prawns and the implications, *Proc. World Maricul. Soc.,* 11, 557, 1980.

17. **Waters, M. E. and Hale, M. B.,** Quality changes during iced storage of whole freshwater prawns *(Macrobrachium rosenbergii),* in *Proc. 6th Annu. Trop. Subtrop. Fish. Technol. Conf. of the Americas,* Nickelson, R., Compiler, Texas A & M University Sea Grant College Program, College Station, 1981, 116.
18. **Hale, M. B. and Waters, M. E.,** Frozen storage stability of whole and headless freshwater prawns, *Macrobrachium rosenbergii, Mar. Fish. Rev.,* 43, 18, 1981.
19. **Liao, D. S. and Smith, T. I. J.,** Test marketing of freshwater shrimp, *Macrobrachium rosenbergii,* in South Carolina, *Aquaculture,* 23, 373, 1981.
20. **Liao, D. S., Smith, T. I. J., and Taylor, F. S.,** Restaurant Market Testing of Prawns *(Macrobrachium rosenbergii),* unpublished manuscript, 1981.
21. **Liao, D. S. and Smith, T. I. J.,** Marketing of cultured prawns, *Macrobrachium rosenbergii,* in South Carolina, *Proc. World Maricul. Soc.,* 13, in press, 1982.

COMMERCIAL SEED PRODUCTION OF THE FRESHWATER PRAWN, *MACROBRACHIUM ROSENBERGII,* IN HAWAII

Spencer Malecha

INTRODUCTION

The techniques used in Hawaiian prawn hatcheries have been previously described[1] and do not differ greatly in kind and theory from those described for other areas.[2-6] Basically, the Hawaiian system is a "semiflow-through" one whereby phytoplankton-rich, "green water" is periodically flushed through the culture tanks. Feeding is a combination of live (*Artemia* nauplii) and dead (fish flesh and/or egg custard and/or fish roe) food. Fujimura and colleagues,[12-15] following Ling's[7-11] initial success in closing the *Macrobrachium rosenbergii* life cycle, developed the mass larval rearing techniques using the "green water" system.

The purpose of this review is to: serve as an entry to published material from which more technical details can be obtained, include new material regarding broodstock selection and fecundity, and to update available information regarding practices in all the active hatcheries in Hawaii. In this regard I will concentrate on water management and feeding regimens, inasmuch as my colleagues have amply covered (elsewhere in this volume) various other nutritional aspects of prawn larval rearing.[16]

REPRODUCTIVE PERFORMANCE AND BROODSTOCK

Malecha et al.,[17] following earlier studies,[18-20] have demonstrated the ease with which *M. rosenbergii* mates and spawns in captivity — nearly all reproductively mature males and females are able to mate and spawn when paired, provided the females have undergone their prenuptial molt. Mating behavior and the structure and function of the reproductive system have been described.[20-26] Controlled spawning in captivity poses no problem to commercial culture, unlike the situation in penaeids described elsewhere in this volume.

Broodstock Selection

M. rosenbergii mates year-round in Hawaiian prawn ponds. Females extrude eggs through their gonopores into the brood chamber located under the abdomen and between the pleopods (Figure 1). Eggs undergo embryological development while attached to the female as described by Ling.[18] Early egg stages are orange due to the yolk; later stages are greyish-brown. The broodstock is currently selected by obtaining for use in hatchery larval rearing "brown-egged" gravid females similar to that shown in Figure 1, from commercial harvests. Hatchery managers can predict, by the shades of egg color, the approximate hatch or "release" date and plan the hatchery rearing cycles accordingly.

Although broodstock selection in this manner is convenient, it has two major disadvantages. First, it is inefficient because it does not take advantage of the extremely large biological fecundity of female *M. rosenbergii*. Indeed, industry practice has evolved to choosing females solely on the basis of their gravidity and availability without regard to body size. Consequently, many small (<40 g) females are needed for spawning and stocking of hatchery tanks. Inasmuch as fecundity (measured as egg mass) is an exponential function of body dimensions (the primary criteria of selectivity by the harvest nets) use of larger (>45 g) females would result in more eggs per female. Second, there is no control over the breeding value* of broodstock. Malecha et al.[27] showed that there has been little differen-

* Breeding value used in this context refers to the genetic worth of the parents of the brood used in the hatchery.

FIGURE 1. Gravid female *M. rosenbergii* collected from ponds. Left: "orange-egged", i.e., early embryological development. Right: "brown-egged" female, indicative of advanced prelarvae hatch stage. Brown-egged females like the type shown are selected by hatchery managers as broodstock for larval rearing.

tiation between the cultured Anuenue*[28] stock and the recently developed Malaysian stock**[29] and argued that the current brood practice may be selecting genetically superior as well as inferior animals, since prawns of all ages exist in undrained ponds. Doyle et al.[30] have argued that simple control of the age of broodstock may lead to genetic progress in *M. rosenbergii*. On farms which operate under the "traditional system"[31] this can be done by selecting broodstock from ponds that have been stocked only once. In "multiphased" systems[31,32] gravid females can be chosen from draindown harvests between the intermediate and final stages. Although this practice can control for age of female it cannot do so for size and fecundity, therefore, the use of broodstock ponds, separate and distinct from commercial grow-out ponds, should become a component of the industry although presently it is not a widespread practice. One farm in Hawaii is using separate brood ponds. These are usually stocked at densities (5000/0.4 acre; 0.2 ft^{-2}) much below commercial levels. Sex ratios can be altered to ratios of one male to four to five females.[36] Females can be easily raised to 100 g to increase the number of larvae released per female.

* Named by Malecha[30] after the "Anuenue" Fisheries Research Center to refer to the stock originally derived from Malaysia which has been under culture in Hawaii for 16 to 20 generations since the original import in the early 1960s.

** Second or third generation of stock imported from the same area as the original Anuenue stock. See Malecha et al.[28,32]

Table 1
EGG NUMBER (FECUNDITY), Y, FOR A HYPOTHETICAL 45-g FEMALE WHOSE LENGTH IS 152.5 mm (X) ESTIMATED BY USING THE ALLOMETRIC EQUATION, $Y = aX^b$, DERIVED IN FIVE DATA SETS

Fecundity estimated[a]	(Y) Egg no. of a 45-g female (152.5 mm)	Coefficients of $Y = aX^b$		Ref.
		a	b	
SF	41,002	0.006	3.12	(Lab spawned)[c]
PHF	36,754	0.012	2.97	(Pond spawned)[c]
LHF	20,135	0.001	3.35	19
SF	28,084	0.002	3.32	34
PHF[b]	76,236	0.079	2.74	35

[a] SF: spawning fecundity is the number of eggs that a female is biologically capable of extruding in one spawn; PHF: prehatch fecundity is the number of eggs carried by the female at any one time between spawning and larval hatch or "release"; LHF: larval hatch fecundity is the number of larvae released from the egg mass following incubation.
[b] Assumed to be PHF. Author[38] does not specify.
[c] Malecha, S. R., unpublished results.

Fecundity

In *Macrobrachium* species,[19,34-38] like other aquatics, fecundity is a linear function of body weight and an exponential function of body length. In considering practical aspects of prawn fecundity a distinction must be made between "spawning fecundity"*, "prehatch fecundity"**, and "larvae-hatched fecundity"***. The difference in numbers between these indicates attrition in egg number during incubation or loss at the time of hatching.

Wickens and Beard[19] examined larvae-hatched fecundity and other studies[34,35,38] estimated fecundity in an unspecified time between spawning and hatching (Table 1). Our recent studies[39] have shown, however, that larvae-hatched fecundity (measured as brown eggs) from females incubated in ponds is lower than females incubated in the laboratory (Figure 2). This is probably due to the continual sloughing off of dead or dying eggs due to epizootic infestation.[46] Indeed, studies on other related carideans[41-43] have shown that brooding females use the first pereiopod in cleaning and removing eggs from their incubating egg mass. The fitted regression lines in Figure 2 for "lab" and "pond" fecundity vs. body length estimate the "spawning fecundity" in the case of the "lab-spawned" animals and both "spawning" and "prehatch" fecundity in the case of the pond data. However, the pond animals represented various stages of incubation. There is about a 10% difference for a 45-g female between lab- and pond-spawned fecundity (Figure 2, Table 1). This percentage difference will increase with increasing female size because fecundity (Y) increases as a power function, $b \simeq 3$ (Table 1) of body dimension (X), according to $Y = aX^b$. Choosing gravid females on the basis of body size can result, therefore, in a large number of eggs (spawned fecundity) or larvae (larvae-hatched fecundity). For example, the need for 100 45-g (15.25 cm orbit length) females in a larval-rearing cycle can be reduced to 49 animals by choosing (or raising

* "Spawning fecundity" (SF) is the number of eggs that a female is biologically capable of extruding in one spawn.

** "Prehatch fecundity" (PHF) is the number of eggs carried by the female at any one time between spawning and larval hatch or "release".

*** "Larvae-hatched fecundity" (LHF) is the number of larvae released from the egg mass following incubation.

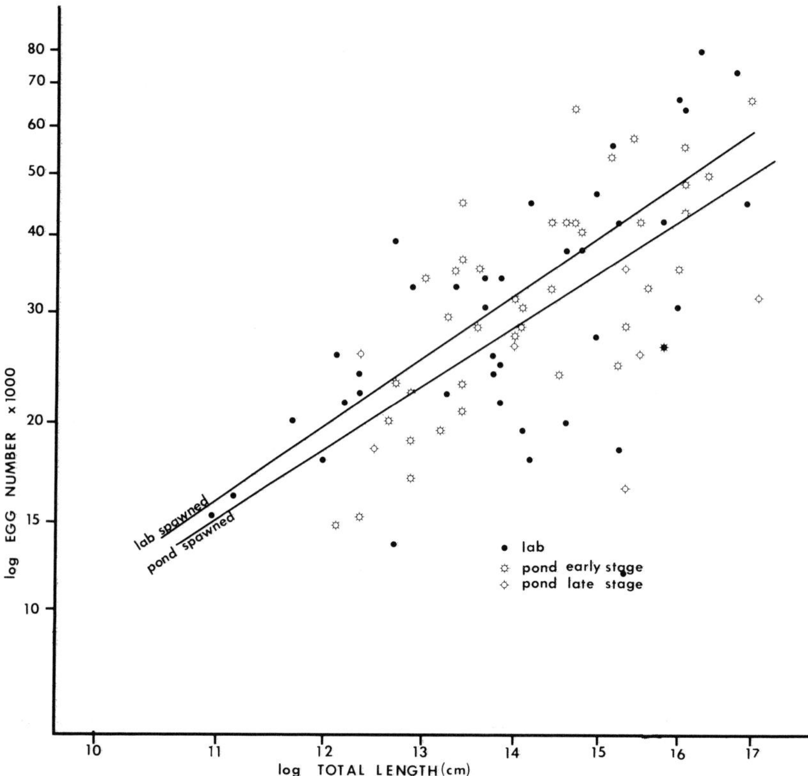

FIGURE 2. Relationship between egg number (Y) and total female length (X) for pond and laboratory spawned females. Curves are least squares regression lines fitted to pond or laboratory data sets; pond: $\log_{10} Y = -1.92 + 3.97 \log_{10} (X)$; lab: $-2.20 + 3.12 \log_{10} X$. Data represent preliminary results of a recent unpublished study and show effect of pond environment on fecundity.

in separate broodstock ponds) females that are 90 g apiece. Current industry practice is to select gravid females much less than 45 g. All in all, the several hundred females per cycle presently used could be effectively reduced to a dozen or so with the use of broodstock ponds. If laboratory spawnings are done, the number of females needed would be reduced even further because of the consistent difference between laboratory and pond-brooded egg number per female. However, caution must be exercised because of the long-term genetic implications of using very small numbers of parents in broodstock programs.[44]

LARVAL GROWTH AND DEVELOPMENT PATTERNS

Prawn larval stage number per cycle is not variable as in the case with other palaemonids.[45] Eleven larval stages can always be observed; each stage is characterized by a suite of morphological characters originally described by Ling[7-11] and later in greater detail by Uno and Kwon.[46] Although these authors give numerous criteria for each stage, abridged criteria are used as the commercial industry standard for larval identification (Figure 3). The hatch of larvae over a 96-hr period leads to an initial variance in number of stages and, in general, from two to five larval stages are present in the culture on any one day (Figure 4) following the first day. Feed density and particle size and water exchange husbandry must be adjusted periodically to accommodate this variation in larval stage and size.

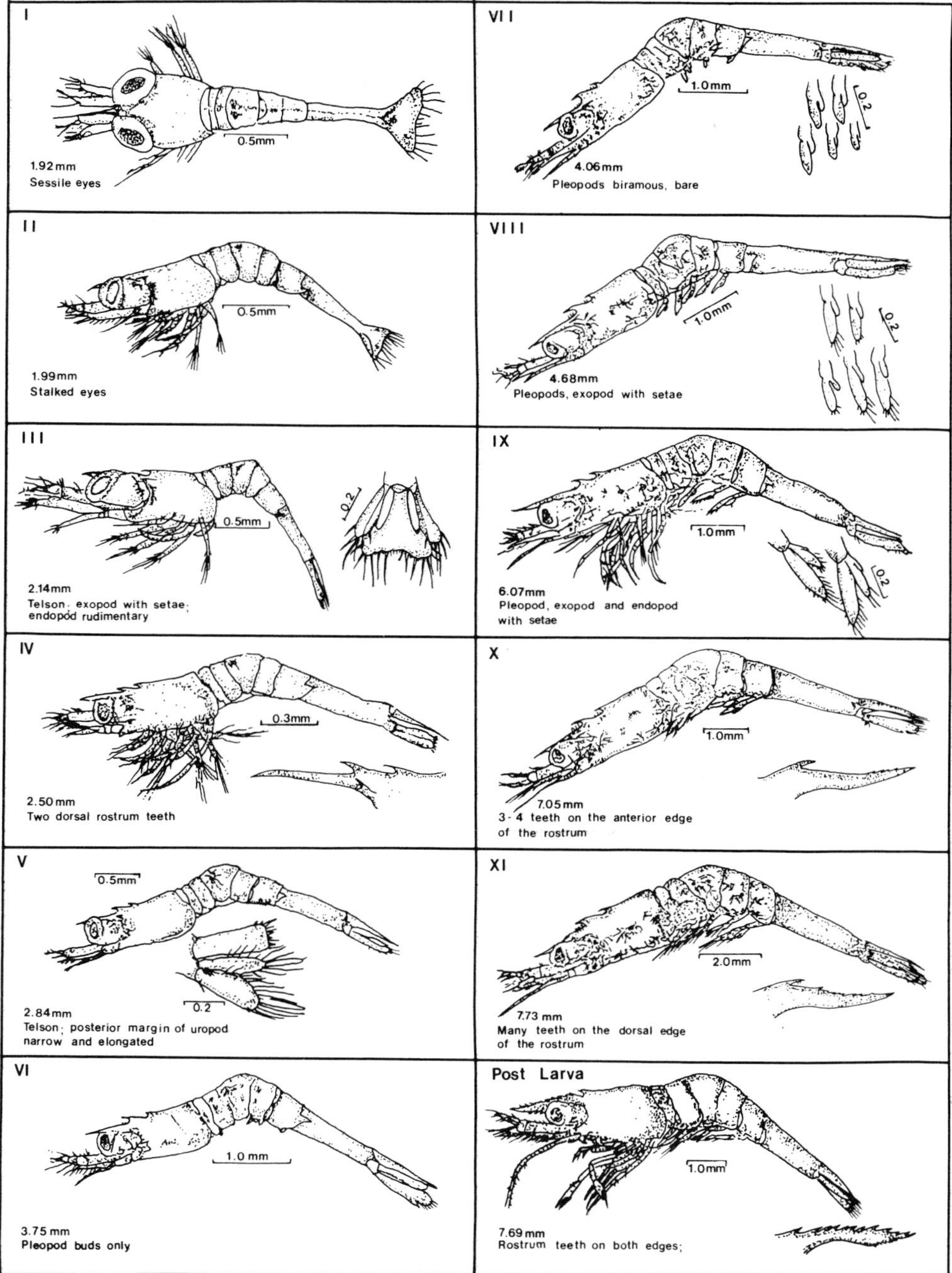

FIGURE 3. Major morphological differences between the 11 *M. rosenbergii* larval and PL stages used for larval identification under commercial and research conditions. Figure represents an abridged version of the suite of differences described by Uno and Kwon[52] for each stage and is redrawn partly from figures depicted by these authors.

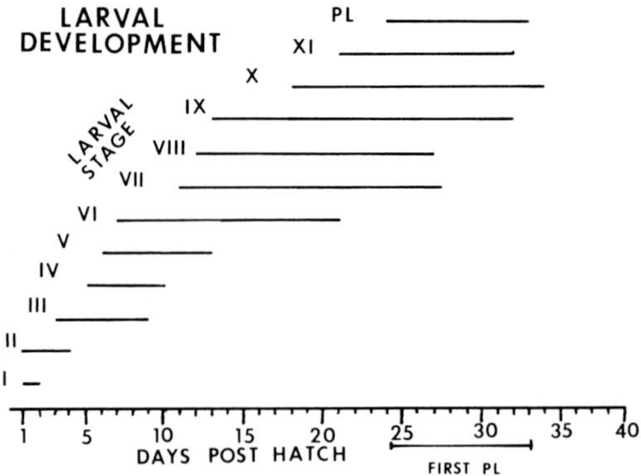

FIGURE 4. Redrawn from Malecha et al.[27] to represent larval stages present on 1 day in a "typical" 25- to 35-day rearing cycle.

As seen in Figure 4, the number of larval stages increases from two to three in the early part of the cycle (days 1 to 15). As the cycle progresses into its second half (16 to 25+ days) the range of stages increases to three to five. On all days, the culture will contain a certain small percentage (perhaps one individual!) of "scouts", i.e., larger individuals at the most advanced stage. The larval stage frequency distribution, on any one day, is skewed to the right, i.e., there will be a small percentage of larvae at an advanced stage (e.g., VII or VIII), the majority at the next two lower stages (e.g., VI and V), and a small percentage of the population in the lowest stage (e.g., IV).

Typical of invertebrate instars, growth in prawn larvae is "weight-proportional" (i.e., "type II" after von Bertalanffy[47]) as has been shown by Stephenson and Knight;[48] metabolism measured as oxygen consumption (Y) is isometric to body weight (X), i.e., $Y = aX^b$, where $b = 0.904$ and $a = 1.54$. Mammals, fishes, and invertebrate species with nonlarval forms have "surface proportional" growth: $Y = aX$,[67] i.e., metabolism as measured by oxygen consumption is a power function of body weight. Indeed, Stephenson and Knight[49] showed that $b \cong 0.7$ for juvenile prawns. One study[50] showed that $b = 1.27$ in adult prawns. Weight-proportional growth is not sigmoid since no steady state is reached because larval growth ends at metamorphosis. Figure 5 shows the growth of four cultures of prawn larvae to be nearly linear between the 10th and 25th days, although overall the growth pattern is slightly curvilinear. For all practical purposes under commercial conditions, prawn larval growth is assumed to be linear. In commercial and research culture in Hawaii, metamorphosis from larval stage XI to postlarva (PL) occurs over a 4 to 7-day period. In the hypothetical cycle discussed herein and in Table 3, the first PL appears at about day 25. This can be described as the "scout PL time" (Figure 5). After about 7 days about 90% of the late stage larvae have "PL'd" and the number of days from hatch to this occurrence is called the "90% PL drop time". It is at this time that the culture is "dropped" and all PLs are removed and the remaining larvae discarded. Under commercial conditions it is not economical to keep cultures going beyond the 90% PL drop time.

HATCHERIES AND HATCHERY PRACTICES

The following description is based upon this author's experience, observations through the years, as well as up-to-date interviews (March to April 1982) with hatchery managers of the six active commercial and research hatcheries in Hawaii, hereafter referred to as

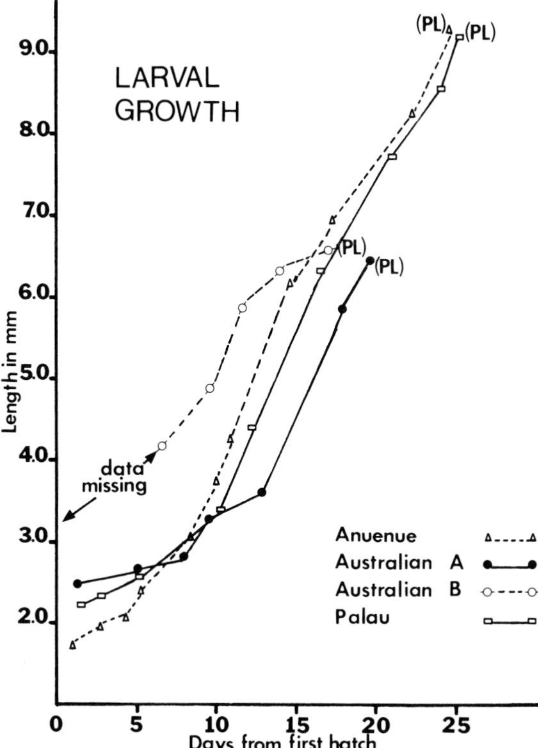

FIGURE 5. Average growth rate of four cultures of *M. rosenbergii* representing four geographic genetic groups.[29] Data points represent mean lengths of five larvae. Two stocks have abbreviated larval development.

AFRC*, Lowe**, Aquatic Farms***, Amfac†, and the PAP‡ hatcheries (Table 2, Figure 6).

The practices described herein are given in general terms of hypothetical or "typical" situations. In most cases, the specifics of a practice or a regimen will vary, depending on the particular individual engaged in the work and the specific set of environmental conditions. Indeed, it is likely that the descriptions given will not apply to any particular larval rearing cycle. All in all, the only way to successfully run a prawn hatchery is to gain the direct hands-on experience in doing so.

* Anuenue Fisheries Research Center, Hawaii State Dept. Land and Natural Resources, Sand Island, Area 4, Honolulu, Hawaii 96819. David Onizuka, Aquatic Biologist, and Francis Oishi, Hatchery Manager (Figure 6A).

** Lowe Aquafarm, P.O. Box 131, Kahuka, Hawaii 96731. Nick Carpenter, Hatchery Manager (Figure 6C). The name of this farm has recently been changed to Amoriente Aquafarm, Inc.

***Aquatic Farms, Ltd., 49-139 Kamehameha Highway, Kaneohe, Hawaii 96744. Wayne Okumura, Hatchery Manager (Figure 6B).

† AMFAC Aquatech, Hawaii Operation, P.O. Box 596, Kekaha, Kauai, Hawaii 96752. Alan Muraoka, Hatchery Manager.

‡ Prawn Aquaculture Research Program, Dept. of Animal Sciences, University of Hawaii. Hatcheries located near and within A.F.R.C.[55] hatchery. Elizabeth MacMichael, Hatchery Manager; S. R. Malecha, Principal Investigator (Figure 6D, E).

Table 2
VITAL STATISTICS FOR ACTIVE COMMERCIAL AND RESEARCH PRAWN HATCHERIES IN HAWAII AS OF 4/82, BASED ON A "HYPOTHETICAL" 25- TO 35-DAY CYCLE (SEE TEXT)

					Water reservoirs				Larval rearing tanks			Production				
Name	Year started	Sector	Exposure	PL deposition	Type[a]	Description	No.	Est. working vol (m^3)	Shape (dimensions, m)	No. in use (total)[b]	Ind. working vol (ℓ)	Est. av stocking density (ℓ^{-1})[c]	Est. PL prod. ℓ^{-1} final vol[d]	Est. av survival (%)[e]	Est. water use/cycle (m^{-3})[f]	Est. av PL prod., (m^{-3}) water used
AFRC	1973	State govt.	Enclosed greenhouse	Small farms, cooperative agreements[g]	GW	Indoor, cir., tapered bottom	10	19	Raceway (0.91 × 1.52 × 9.14)	16	9,463	160	21	55	1,136	2.8
					CW	None	—	—								
Lowe	1980	Pvt.	Indoor	Own farm, sales	GW	Outdoor, cir., enclosed, concrete	1 1	61 45	Cir., conical bottom (1.2 deep, 2.4 dia)	10	3,785	80	30	70	909	1.2
					CW	Inground cistern	1	265								
Aquatic Farms	1978	Pvt.	Outdoor	Own farm, sales	GW	Outdoor, cir., tapered bottom	7	17	Cir., flat bottom (1.0 deep, 3.7 dia)	8(20)	11,203	60	42	70	418	5.4
					CW	Outdoor, cir., tapered bottom	3	17								
Amfac	1980[h]	Pvt.	Enclosed greenhouse	Own farm	GW	Covered,	1	45	Raceway (0.74 × 1.52 × 9.14)	6(16)	7,949	94	20—25	50	421	2.5
					CW	outdoor, cir.	1	106								
PAP (A)[i]	1982	Univ. Hawaii	Outdoor	Research program	GW	Cir., outdoor	1	3.8	Rect., flat bottom (0.61 × 0.91 × 1.83)	2	852	31	18.6	60	40	1.6
					CW	None	—	—								
(B)[i]	1977	Univ. Hawaii	Enclosed greenhouse	Research program	GW	Cir., outdoor	1	19	Rect., flat bottom (0.61 × 0.91 × 1.83)	12	852	31	18.6	60	119	2.1
					CW	None	—	—	(0.53 × 0.61 × 0.91)	12	284					

[a] GW = "green water", i.e., phytoplankton culture (see text); CW = "clear water" used in flushing (see text and Table 3).
[b] Number in parenthesis indicates total number of tanks available in hatchery.
[c] Number of first stage larvae that are initially stocked before "splitting" (see text).
[d] Number of PLs obtained at end of cycle.
[e] Estimated as the difference between number of larvae stocked and number of PLs produced.
[f] Slightly conservative estimate. Estimated for only first 25 days of a 30- to 35-day cycle during flushing and refill. Estimated from individual hatchery water management regimens described in text.
[g] The State of Hawaii, through AFRC, maintains cooperative agreements with small and intermediate commercial farms to fulfill their PL stocking needs (see Table 1 in Rference 32.
[h] Hatchery was previously owned by another company no longer in prawn culture. (Note added in proof: the Amfac farm went out of business in late 1982.)
[i] Hatchery A located outdoors, near AFRC's; Hatchery B located within the AFRC hatchery building (Figure 6).

6

FIGURE 6. Examples of *M. rosenbergii* hatcheries in Hawaii. (A) AFRC,[57] (B) Aquatic Farms,[59] (C) Lowe Aquafarms, (D, E) PAP hatcheries, (F) ''trickle flow-through'' using drip irrigation tubing in PAP hatchery shown in frame D. GWR = green water reservoirs, LRT = larval rearing tanks, BS = brine shrimp hatching tanks, BSHT = broodstock holding tanks. The Amfac hatchery is not shown but utilizes raceway larval rearing tanks similar to those in A.

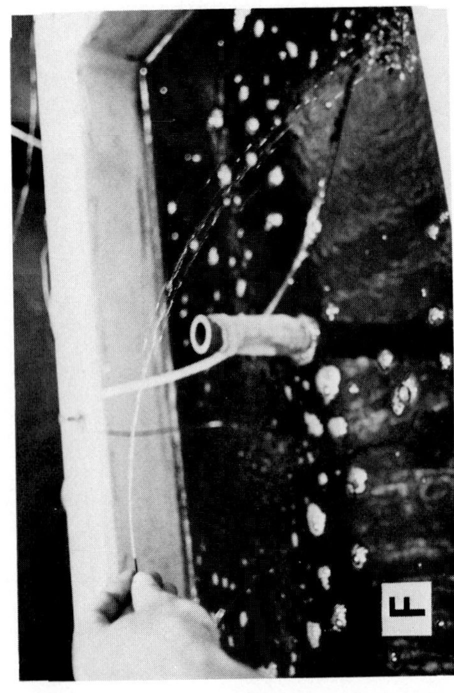

FIGURE 6F.

FIGURE 6E.

Physical Plant

Table 2 presents vital statistics regarding, and Figure 6 shows the overall layout of, Hawaiian prawn hatcheries. Physical facilities are basically divided into larval rearing units and reservoirs for phytoplankton ("green water") culture and "clear" water; both are used in water exchange husbandry. Hawaiian hatcheries are either indoors or fully outdoors. Three hatcheries (Amfac, AFRC, and PAP) are enclosed in a greenhouse structure. The Lowe hatchery is fully enclosed in a large building with minimal solar incidence on the larval rearing tanks. The Lowe green water and clear water reservoirs are outside and covered. In addition, the Lowe hatchery water is heated by solar exchange on the roof of one of the reservoirs and one on the roof of the hatchery building itself. The Aquatic Farms hatchery has heating coils located in the concrete foundation "pad" of the larval rearing tanks. The other hatcheries do not have supplemental heating other than direct solar incidence.

Water reservoir shapes and sizes matter little in prawn larval rearing as long as enough water is available. The Lowe hatchery uses a large in-ground cistern which was in place at the site before the hatchery was built. So not counting the Lowe hatchery, the average total reservoir to larval rearing unit volume is 1.21:1, i.e., reservoir volume is about 25% greater than larval culture tank volume. This ratio for each hatchery is 1.25:1 for AFRC; 11.41:1 for Lowe; 1.89:1 and 0.75:1 for 8 and 20 larval rearing tanks for Aquatic Farms; 3.16:1 for the 6 larval culture tanks currently in use and 1.58 for the 16 total larvae tanks for Amfac; and 1.11:1 and 1.38:1 for the PAP hatcheries A and B, respectively (Table 2).

Incubation and Larval Hatch Husbandry

Handling and husbandry of gravid females in the hatchery prior to release of their larvae varies slightly among the active hatcheries. As mentioned before, only one commercial hatchery (Amfac) and, of course, the research hatchery (PAP) utilize separate broodstock sectors to the production scheme. In any case, "brown-egged" females are brought to the hatchery and either incubated in the larval rearing tanks themselves (Amfac) or in separate tanks (Lowe, Aquatic Farms, AFRC, PAP). In the latter, larvae are collected in buckets by siphoning, counted, and dispersed into the larval rearing units. The Amfac hatchery places the gravid females in wire cages within the larval rearing tanks. Larval release is into 11‰. The Lowe, Aquatic Farms, PAP, and AFRC hatcheries allow incubation and larvae hatch into 14, 13, 15, and 12 to 15‰, respectively.

Larval release is usually over a 4-day period with a peak release between 24 to 72 hr as described by Wickens and Beard.[19] Consequently, females are left in the incubation and larval rearing tanks for up to 4 days.

Density Management and Mortality

Sick and Beaty[3] have shown that 40 ind. ℓ^{-1} is an optimum density under husbandry conditions similar to that described herein. Indeed, the final production (ℓ^{-1}) in Hawaiian hatcheries seems to confirm this (Table 1). However, this 40-ℓ^{-1} density is equivalent to a final "standing crop"; higher densities can be tolerated at the younger (and smaller) earlier stages. Indeed, resources (e.g., space, water, food) are optimized when larvae are initially stocked at high densities and then "split" at midcycle. This follows from the stock division manipulation that is practiced in culture of juvenile-adult fish[62] and is analogous to a "two-phase" growout of adult prawns as discussed elsewhere in this volume.[52] The Lowe, Amfac, and AFRC hatcheries stock larvae at 80, 94, and 160 ind. ℓ^{-1}, respectively, and split the culture roughly at day 12 (stages V to VII). Depending on the conditions, Lowe splits 1:2 ($80 \rightarrow 2 \times 40$ ind. ℓ^{-1}); Amfac splits 1:3 ($94 \rightarrow 3 \times 31.5$ ind. ℓ^{-1}); and AFRC splits 1:4 ($160 \rightarrow 4 \times 40$ ind. ℓ^{-1}). Realistically, there is mortality between day 1 and 12 but operationally hatchery managers rarely budget for "significant" mortality in the first half of the rearing cycle. Survival can be 80 to 90% up to stages IX, X, and XII and then drop to 50 to 70%. Consequently, it is this "late stage mortality" that is most important in typical, healthy commercial cultures. This is different from acute or chronic disease-related mortalities.[40,53]

The Aquatic Farms and PAP hatcheries do not split their larvae cultures, the latter because of research design and the fact that experience has shown that the small rearing tanks can be effectively utilized without a stock division. The Aquatic Farms hatchery stocks at about 60 ind. ℓ^{-1} and maintains the stock of this density (minus attrition) until postlarvae are removed from culture.

Water Exchange Husbandry and Water Quality

Water management varies among the hatcheries in Hawaii in terms of the total water used and regimen of its use (Table 3). Two commercial and one research water management regimens are practiced: "green water toilet flush", "clear water flow flush", and "trickle flow-through". On the commercial scale only AFRC practices exclusively a daily green water toilet flush regimen — water level in the culture tank is dropped to a level (50%) and refilled with green water. This is done before and after the midcycle 1:4 split. The Lowe, Amfac, and Aquatic Farms hatcheries practice the green water toilet flush on certain days in the early part of the cycle and then switch to a clear water flow flush for the remaining part of the cycle. The Amfac and Lowe regimens are similar and differ from the Aquatic Farms regimen. In the Lowe hatchery, between days 3 and 6 a 30% green water toilet flush is conducted (i.e., water is lowered by 30% and refilled by that percentage) between days 7 and 9 there is a similar toilet flush of 40%. Between days 10 and 13, flow flushing is practiced — culture volume is lowered by 50% and the remaining 50% is then flushed with its entire volume of clear water. The culture tank is then brought up to full volume with phytoplankton-rich green water. On day 14, after the 1:2 split, the culture water is dropped to a 50% level and 100% of the total tank volume is flushed with clear water. This represents a 200% exchange of the remaining volume containing the larval culture. This flushing regimen continues until the PLs are removed.

The Amfac water management regimen is similar to the Lowe regimen. On days 3 and 5 (every other day), only a 50% toilet flush is conducted. From days 7 to 13 there is a 50% drop and then a 50% (of total tank volume) clear water flow flush representing, as in the Lowe case, a 100% exchange of the remaining volume. On day 14, following the 1:3 stock split, the flow flush rate is increased to 100% of tank volume or 200% of remaining larval culture water volume. Amfac's culture tanks are 11203 ℓ and this volume can pass through, and exchange with, the larval culture water volume in about 1.5 hr at about 30 gpm (114 ℓpm). The Aquatic Farms regimen uses the flow flush method but differs from the two procedures just described in not only the percentage of lowered volume but in the exchange volume during the flow flush. At days 4 and 5, there is a 25% toilet flush with green water. On days 6 and 7 this is increased to 67% and, on days 8 to 10, to 75%. From day 11 to day 25 the tank volume is lowered to 75% and the remaining 25% is exchanged with an equal flow flush volume. This represents only 25% of tank volume and 100% of the remaining volume flow flush compared with a 200% remaining volume flow flush in the Amfac and Lowe hatcheries.

The PAP research hatchery A located within the AFRC hatchery building practices the simplest water management regimen — no flush on day 1, a 50% green water toilet flush on day 2, and between 3 and the last day of the cycle, a 95% green water toilet flush. Research hatchery B, on the other hand, uses trickle flow-through. Drip irrigation tubing is used to flow a "trickle" of water (Figure 6F) continuously through the culture tanks. Pump flow and drip irrigation emitters are set for a flow of water equal in volume to the culture tank working volume. However, this "new-water-for-old-water" exchange rate will be less than 100% since new and old water mix before leaving the tank through a standpipe drain. We estimate that about 50% day^{-1} of the old water is exchanged for new water.

Maintenance of good water quality per se, i.e., prophylaxis against toxins, is carried out through water exchange, siphoning, and the use of phytoplankton-rich water left in the culture overnight. Inorganic toxins, pesticides, and heavy metals are not monitored unless an acute problem needing immediate diagnosis develops. Similarly, the major metabolic nutrient toxins, ammonia and nitrite, are not usually monitored on a continual basis because of the short residence time (24 hr or less) of the culture water due to flushing which keeps the total ammonia levels (both ionized and deionized) well below the 10 $mg\ell^{-1}$ toxic levels suggested by Armstrong et al.[54] Similarly, nitrite is kept below the 2 to 3 $mg\ell^{-1}$ danger levels.[55] Along with flushing, the phytoplankton culture keeps the larval rearing pH area within the tolerable and optimum range for decapod crustacea.[56]

The green water cultures are maintained according to the general methods outlined elsewhere in this volume.[16] Cohen et al.[57] have shown that prawn larvae do not digest, (i.e., derive nutritional benefit from) the green water phytoplankton. Larvae passively ingest phytoplankton cells to a very small degree and can come to have phytoplankton in their digestive tract by way of the artemia which are phytoplanktivorous. Manzi et al.[58] have shown that green water culture reduces ammonia levels in contrast to the data of Cohen et al.[57] Also green water may aid in food capture or increase production in ways other than water quality maintenance, i.e., ammonia, nitrite reduction. In any case the "green water" method seems to increase production by 10 to 20% in comparable systems not using green water although clear water culture has been found to be viable.[59]

Foods, Feeding, Husbandry

All hatcheries in Hawaii feed a combination of live (newly hatched brine shrimp nauplii) and dead (fish flesh, roe, or egg custard) food, but differ in feeding regimens. The amount of food per unit culture volume or larval density is done to a large degree by "eye" and varies among hatchery managers. Table 3 describes feeding, water management, and siphoning regimens in the active commercial and research hatcheries in Hawaii. Corbin et al.[16] should be consulted for additional detail regarding the AFRC hatchery and for additional information regarding larvae nutrition.

Brine Shrimp

Commercially available brine shrimp cysts are hatched according to standard methods and added *ad libitum* to the culture at the end of the working day (~5 p.m.). This usually constitutes the fifth feeding of the day (Table 3). Feeding rates are adjusted daily based on a visual assessment of overnight consumption. (See Corbin et al.[16] for details.)

Fish Flesh

All hatcheries use fish flesh prepared in essentially the same way (Figure 7). Two commercially available species are most widely used — tuna ("aku", *Katsuwonus pelamis*) used fresh and pollack *(Theragra calcogramma)* used from frozen blocks.

Feed preparation involves forcing (with water jets) fillets of fish through stainless steel mesh screens (Figure 7A, B). Four screen sizes are utilized which result in slurries of fish flesh in three sizes: fine, medium, and coarse. The screen sizes and their corresponding number of holes per centimeter are coarse — 7.88, medium — 15.75, fine — 23.62, and a catch screen of 31.50 holes per centimeter. The water jet breaks apart the muscle into small segments and forces them through the mesh creating a slurry of fish flesh particles (whose size roughly corresponds to the mesh diameter) which collect on the collecting screen. The slurry is usually placed in a container and an equal volume of water is then added. The mixture is refrigerated until use. Food is made up every 2 to 3 days in normal practice.

Table 3
QUALITATIVE FEEDING, WATER EXCHANGE, AND SIPHON REGIMENS FOR THE FIRST 20 DAYS OF A HYPOTHETICAL SUMMER (JUNE TO SEPTEMBER) LARVAL REARING CYCLE IN THE ACTIVE COMMERCIAL AND RESEARCH HATCHERIES IN HAWAII[a]

			Feeding[d]					Flushing		
Day	Approx. larval stages[b]	Hatchery[c]	1	2	3	4	5	Lower vol/flushed vol	Type[e]	Siphon[f]
1	I	AFRC	FFM	FFM	FFM	FFM	0	0	—	0
		LOWE	0	0	0	0	BS	0	—	0
		AMFAC	EC	0	EC	EC	BS	0	—	0
		AF	0	0	0	0	BS	0	—	0
		PAP[g]	0	0	0	0	0	0	—	0
2	I, II	AFRC	FFM	FFM	FFM	FFM	BS	0	—	0
		LOWE	0	0	0	0	BS	0	—	0
		AMFAC	EC	0	EC	EC	BS	0	—	0
		AF	0	0	0	0	BS	0	—	+
		PAP	FFM, BS	FFM, BS	FFM, BS	BS	0	50/—	TF[h]	0
3	I, II	AFRC	FFM	FFM	FFM	FFM	BS	50/—	TF	+
		LOWE	0	FFM[i]	0	0	BS	25/—	TF	0
		AMFAC	EC	0	EC	0	BS	50/—	TF	+
		AF	0	0	0	0	BS	0	—	0
		PAP	FFM, BS	FFM, BS	FFM, BS	BS	0	95/—[j]	TF[j]	+
4	I, II, III	AFRC	FFM	FFM	FFM	FFM	BS	50/—	TF	0
		LOWE	0	FFM[i]	0	0	BS	30/—	TF	0
		AMFAC	EC	0	EC	EC	BS	0	TF	0
		AF	FFM	0	FFM	0	BS	25/—	TF	+
		PAP	FFM, BS	FFM, BS	FFM, BS	BS	0	99/—	TF	0
5	I, II, III, IV	AFRC	FFM	FFM	FFM	FFM	BS	50/—	TF	+
		LOWE	FFM	0	FFM	0	0	30/—	TF	+
		AMFAC	EC	EC	EC	EC	BS	50/—	TF	+
		AF	FFM	FFM[k]	FFM	0	BS	25/—	TF	0
		PAP	FFM, BS	FFM, BS	FFM, BS	BS	0	95/—	TF	+

6	II, III, IV, V	AFRC	FFM	FFM	FFM	FFM	BS	50/—	TF	0
		LOWE	FFM	0	FFM, BS	0	0	30/—	TF	0
		AMFAC	EC	EC	EC	EC	BS	0	TF	0
		AF	FFM	FFM	FFM	0	BS	67/100	FF[k]	+
		PAP	FFM, BS	FFM, BS	FFM, BS	BS	0	95/—	TF	0
7	II, III, IV, V	AFRC	FFM	FFM	FFM	FFM	BS	50/—	TF	+
		LOWE	FFM	0	FFM, BS	0	0	40/40	FF	+
		AMFAC	EC	EC	EC	EC	BS	50/50	FF, 1/2 hr	+
		AF	FFM	FFM	FFM	0	BS	67/100	FF, 1 hr	0
		PAP	FFC, BS	FFC, BS	FFC, BS	BS	0	95/—	TF	+
8	III, IV, V	AFRC	FFM	FFM	FFM	FFM	BS	50/—	TF	0
		LOWE	FFF	FFM	FFM	BS	0	40/40	FF	0
		AMFAC	EC	EC	EC	EC	BS	50/50	FF	0
		AF	FFM	FFM	FFM, FR[k]	FFM	BS	75/75	FF, 1 hr	+
		PAP	FFC, BS	FFC, BS	FFC, BS	FFC,[k] BS	0	95/—	TF	0
9	III, IV, V, VI	AFRC	FFM	FFM	FFM	FFM	BS	50/—	TF	+
		LOWE	FFM	FFM	FFM	BS	0	40/100	FF	+
		AMFAC	EC	EC	EC	EC	BS	50/100	FF, 1 hr	–
		AF	FFM	FFM	FFM, FR[k]	FFM	BS	75/75	FF	0
		PAP	FFC, BS	FFC, BS	FFC, BS	FFC,[k] BS	0	95/—	TF	+
10	IV, V, VI	AFRC	FFM	FFM	FFM	FFM	BS	50/—	TF	0
		LOWE	FFM	FR	FFM	BS	0	50/100	FF	0
		AMFAC	EC	EC	EC	EC	BS	50/00	FF	0
		AF	FFC	FFC	FR	FFC, EC	BS	75/75	FF	+
		PAP	FFC, BS	FFC, BS	FFC, BS	FFC, BS	0	95/—	TF	0
11	V, VI, VII	AFRC	FFM	FFM	FFM	FFM	BS	50/—	TF	+
		LOWE	FFC	FR	FFC	BS	0	50/100	FF	+
		AMFAC	EC	EC	EC	EC	BS	50/100	FF	+
		AF	FFC	FFC	FR	FFC, EC	BS	75/100	FF	0
		PAP	FFC, BS	FFC, BS	FFC, BS	FFC, BS	0	95/—	TF	+
12[l]	V, VI, VII	AFRC	FFC	FFC	FFC	FFC	BS	50/—	TF	0
		LOWE	FFC	FR	FFC	BS	0	50/100	FF	0
		AMFAC	EC	EC	EC	EC	BS	75/100	FF	+
		AF	FFC	FFC	FR	FFC	BS	75/100	FF	+
		PAP	FFC, BS	FFC, BS	FFC, BS	FFC, BS	0	95/—	TF	0

Table 3 (continued)
QUALITATIVE FEEDING, WATER EXCHANGE, AND SIPHON REGIMENS FOR THE FIRST 20 DAYS OF A HYPOTHETICAL SUMMER (JUNE TO SEPTEMBER) LARVAL REARING CYCLE IN THE ACTIVE COMMERCIAL AND RESEARCH HATCHERIES IN HAWAII[a]

			Feeding[d]					Flushing		
Day	Approx. larval stages[b]	Hatchery[c]	1	2	3	4	5	Lower vol/flushed vol	Type[e]	Siphon[f]
13	V, VI, VII, VIII	AFRC	FF	FF	FF	FF	BS	50/—	TF	+
		LOWE	FR	FFC	FR	BS	0	50/200	FF	+
		AMFAC	EC	EC	EC	EC	BS	50/200	FF	+
		AF	FFC	FFC	FR	FFC, EC	BS	75/100	FF	0
		PAP	FFC, BS	FFC, BS	FFC, BS	FFC, BS	0	95/—	TF	+
14	VI, VII, VIII, IX	AFRC	FF	FF	FF	FF	BS	50/—	TF	0
		LOWE	FR	FFC	FR	BS	0	50/200	FF	0
		AMFAC	EC	EC	EC	EC	BS	50/200	FF	0
		AF	FFC	FFC	FR	FFC, EC	BS	75/100	FF	+
		PAP	FFC, BS	FFC, BS	FFC, BS	FFC, BS	0	95/—	TF	0
15	VI, VII, VIII, IX	AFRC	FF	FF	FF	FF	BS	50/—	TF	+
		LOWE	FR	FFC	FR	BS	0	50/200	FF	+
		AMFAC	EC	EC	EC	EC	BS	50/200	FF	+
		AF	FFC	FFC	FR	FFC, EC	BS	75/100	FF	0
		PAP	FFC, BS	FFC, BS	FFC, BS	FFC, BS	0	95/—	TF	+
16	VI, VII, VIII, IX	AFRC	FF	FF	FF	FF	BS	50/—	TF	0
		LOWE	FR	FFC	FR	FFC	BS	50/200	FF	0
		AMFAC	EC	EC	EC	EC	BS	50/200	FF	0
		AF	FFC	FFC	FR	FFC, EC	BS	75/100	FF	+
		PAP	FFC, BS	FFC, BS	FFC, BS	FFC, BS	0	95/—	TF	0
17	VI, VII, VIII, IX	AFRC	FF	FF	FF	FF	BS	50/—	TF	+
		LOWE	FR	FFC	FR	FFC	BS	50/200	FF	+
		AMFAC	EC	EC	EC	EC	BS	50/200	FF	+
		AF	FFC	FFC	FR	FFC, EC	BS	75/100	FF	0
		PAP	FFC, BS	FFC, BS	FFC, BS	FFC, BS	0	95/—	TF	+

18	VI, VII, VIII, IX, X	AFRC	FF	FF	FF	FF	BS	50/—	TF	0
		LOWE	FR	FFC	FR	FFC	BS	50/—	FF	0
		AMFAC	EC	EC	EC	EC	BS	50/—	FF	0
		AF	FFC	FFC	FR	FFC, EC	BS	100/—	FF	+
		PAP	FFC, BS	FFC, BS	FFC, BS	FFC, BS	0	95/—	TF	0
19	VI, VII, VIII, IX, X	AFRC	FF	FF	FF	FF	BS	50/—	TF	+
		LOWE	FR	FFC	FR	FFC	BS	50/200	FF	+
		AMFAC	EC	EC	EC	EC	BS	75/100	FF	+
		AF	FFC	FFC	FR	FFC, EC	BS	75/100	FF	0
		PAP	FFC, BS	FFC, BS	FFC, BS	FFC, BS	0	5/—	TF	+
20	VII, VIII, IX, X	AFRC	FF	FF	FF	FF	BS	50/—	TF	0
		LOWE	FFC	FFC	FR	FFC	BS	50/200	FF	0
		AMFAC	EC, FR	EC, FR	EC, FR	EC, FR	BS	50/200	FF	0
		AF	FFC	FFC	FR	FFC, EC	BS	75/100	FF	+
		PAP	FFC, BS	FFC, BS	FFC, BS	FFC, BS	0	95/—	TF	0

[a] Data obtained through observations and interviews with hatchery managers and reflect subjective assessments for a hypothetical but "typical" summer cycle when water temperatures are high enough (26 to 30°C) for rapid larval growth and healthy phytoplankton cultures.

Anuenue Fisheries Research Center, Hawaii State Dept. Land and Natural Resources, Sand Island, Area 4, Honolulu, Hawaii 96819. David Onizuka, Aquatic Biologist and Francis Oishi, Hatchery Manager.

Lowe Aquafarm, P.O. Box 131, Kahuku, Hawaii 96731. Nick Carpenter, Hatchery Manager.This farm has recently been renamed Amoriente Aquafarm, Inc.

AMFAC Aquatech. Hawaii Operation, P.O. Box 596, Kekaha, Kauai, Hawaii 96752. Alan Muraoka, Hatchery Manager.

Aquatic Farms, Ltd., 49-139 Kamehameha Highway, Kaneohe, Hawaii 96744. Wayne Okumura, Hatchery Manager.

Prawn Aquaculture Research Program, Dept. of Animal Sciences, University of Hawaii, Hatchery Location near and within A.F.R.C. hatchery. Elizabeth MacMichael, Hatchery Manager; S. R. Malecha, Principal Investigator.

[b] After Uno and Kwon[46] and Figures 3 and 4; actual situations are variable.

[c] See References 57 to 61.

[d] 1:0700 to 0800 hr; 2:1000 to 1200 hr; 3:1300 to 1400 hr; 4:1500 to 1600 hr; and 5:before close of business ~1700 to 1800 hr.

[e] See text for detail. A 50:100 lower vol/flush vol ratio means tank lowered to 50% volume and flushed with clear water equal to 100% of the remaining volume; a 50:200 ratio indicates a clear water flush equal to 200% of remaining volume.

Table 3 (continued)
QUALITATIVE FEEDING, WATER EXCHANGE, AND SIPHON REGIMENS FOR THE FIRST 20 DAYS OF A HYPOTHETICAL SUMMER (JUNE TO SEPTEMBER) LARVAL REARING CYCLE IN THE ACTIVE COMMERCIAL AND RESEARCH HATCHERIES IN HAWAII[a]

[f] Usually every other day.
[g] Hatchery A, Table 2.
[h] TF = green water toilet flush, i.e., lower volume and refill; FF = flow flush of volume remaining after total tank volume has been lowered.
[i] Time is variable.
[j] PAP actually flows green water through the 5% remaining volume for 20 to 30 min.
[k] Considered optional depending on conditions and feeding rates by hatchery managers.
[l] Approximate day of density adjustment by splitting for Lowe, Amfac, and AFRC (see text).

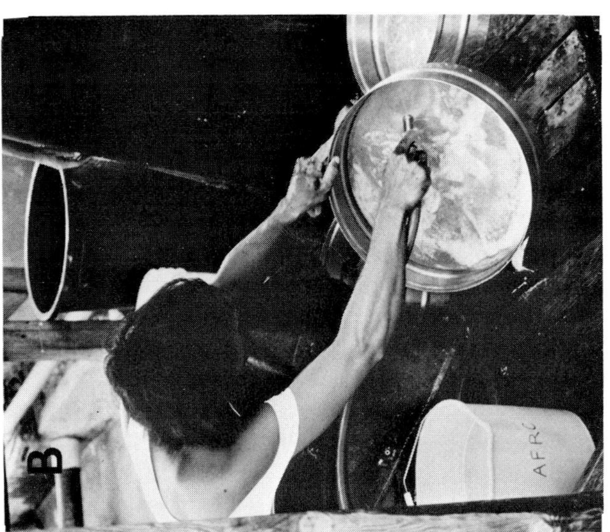

FIGURE 7. Preparation of fish flesh for use in prawn larval culture. (A) High pressure water jet minces bulk fillet through stainless steel screens of varying mesh sizes (see text); (B) fish flesh slurry that collects on screen after passing through screen of larger dimension; (C) applying fish flesh slurry to larval rearing tank; (D) coarse screen: 7.88 holes cm^{-1} (coin is 19 mm); (E) medium screen: 15.75 holes cm^{-1}; (F) fine screen: 23.62 holes cm^{-1}.

FIGURE 7C.

FIGURE 7D.

FIGURE 7E.

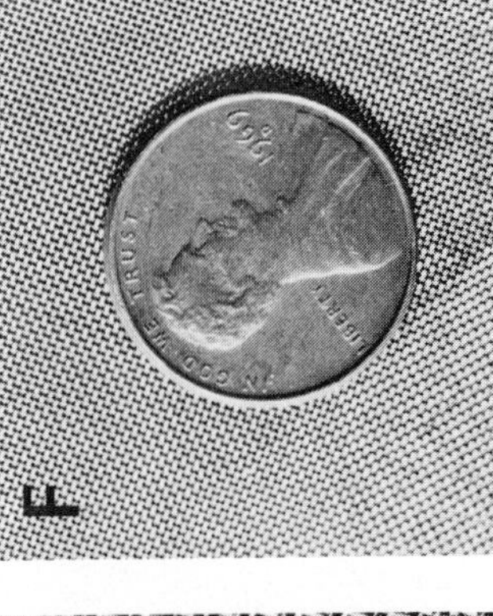

FIGURE 7F.

The fish species and its preparation (fresh, frozen) that is used greatly affects the consistency and texture of the fish particles in the slurry. For example, fine particles are difficult to obtain with pollack so some hatcheries begin with the medium screen size. Application of fish slurry is done by hand (Figure 7) in order to achieve a uniform suspension of fish particles in the culture water. Moller[60] has shown that *M. rosenbergii* feed by means of chance encounters with food particles but larvae may move to a particle that is in the immediate vicinity. All in all, food particles have to be "brought to" the larvae which, in practice, means not only maintaining an optimum ratio of food particles to larvae but keeping the food in suspension by means of aeration currents. Of course, there is limit to the amount of food particles that can be applied at any one time in order to avoid deterioration of uneaten food. Once food is grasped prawn larvae use their chemotoxic sense to "taste" the particle, rejecting unsuitable food types or conditions. Over the years, hatchery practice has evolved into feeding four times a day roughly at 0730, 1030, 1230, 1600 hr, named the first, second, third, and fourth feedings in Table 3. Some hatcheries feed fish flesh only once per day in the first part of the rearing cycle or alternate with egg custard or fish roe as indicated in Table 3.

Egg Custard

"Egg custard" (EC, Table 3) can be prepared a variety of ways. One common method is simply to break and scramble whole, raw chicken eggs (whites and yolks) into a bowl and then microwave the mixture until cooked. This is followed by passing the cooked egg mass through one of the medium- or fine-mesh screens used in fish flesh preparation. Egg custard can be stored for 2 to 3 days and is versatile because it can be a vehicle for supplemental ingredients such as vitamins, lipids, and, if necessary in the future, antibiotics.[3] Egg custard without live food is not suitable for commercial culture as shown by Sick et al.[3]

Feeding Regimens

Although feeding regimens vary among the six hatcheries, certain general trends are apparent (Table 3). First, light feeding or no feeding at all is practiced in the first 1 to 3 days of the rearing cycle, since embryological food reservoirs are still being utilized by the newly hatched first-stage (I) larvae in addition to the fact that the larvae are extremely small. In general, fish flesh is not fed in great amounts in the early stages.

Pollack fish flesh was being used by all hatcheries at the time of the data collection (4/82) for the Table 3. This fish does not give a very good "fine" (FFF, Table 3) grade. Beginning on day 3 to 5, some hatcheries begin feeding fish flesh more often, depending upon environmental conditions and larval appetite. By the 8th day (stages III to V) all hatcheries have their cultures on at least four feedings a day. It is around this time that the hatcheries that use it begin testing the acceptance of fish roe (FR, Table 3). Generally, mullet (*Mugil* sp.) roe is used. It is so oily feeding is done before the daily water change. Also, depending upon specific conditions, the fish flesh particle size may be changed to coarse (FFC, Table 3). By the 12th day (stages V to VII) all hatcheries that use it have their cultures on the coarse fish flesh particle size. Two hatcheries, Lowe and Aquatic Farms, may have one feeding consisting entirely of fish roe. Also, Aquatic Farms may begin consistently using egg custard (EC, Table 3) as a supplement in the fifth feeding. On the 13th day or so, the Lowe hatchery may begin alternating feeding fish roe (FR) and coarse fish flesh (FFC) for the first four feedings followed by a brine shrimp (BS) feeding in the sequence FR-FFC-FC-FFC-BS. Aquatic Farms, on the other hand, prefers to apply egg custard along with fish flesh at the fourth feeding. From the 17th day (stages VII to IX) onward, there is usually minimal change in the feeding regimen, although the quantities fed will be altered according to the specific culture conditions which are assessed continually by the hatchery manager and staff.

There has been only one quantitative study on the relationship between food particle density and larval production which showed that as little as five *Artemia* nauplii $m\ell^{-1}$ is suitable for a maintenance diet. As pointed out by Corbin et al.,[16] we can assume that culture practices described herein use nauplii concentrations between 5 to 15 ind. $m\ell^{-1}$ even though quantitative assessments have not been made.[61]

Fish flesh or fish roe particle density has not been quantified. Corbin et al.[16] say the AFRC procedure is to apply from about 100 to 500 mℓ per tank per feeding of a 50:50 fish flesh slurry (the volume of residual slurry left on screen diluted with equal volume of water) during the first phase of the cycle before the split. This is readjusted to 100 mℓ per tank per feeding after the split. This 100 to 500 mℓ of fish flesh slurry amounts to a total of 0.042 to 0.21 mℓ of "food" ℓ^{-1}day^{-1} (four feedings a day). These feeding rates are more or less followed by other hatcheries that use the fish flesh.

Production

Estimated average production for the active Hawaiian prawn hatcheries are given in Table 2. It must be emphasized that Table 2 does not reflect a "whose best" situation; locale-specific, historic, economic, and environmental factors all make each hatchery unique. Some general conclusions can be made, but, of course, these observations will have to be confirmed and experimentally tested. First, the overall average production final vol^{-1} of ~30 ℓ^{-1} in Hawaii hatcheries compares very favorably with that reported in a survey of prawn hatcheries[4] (18 ℓ^{-1}) and the production recently reported (1.3 to 20 ℓ^{-1}).[5,6] Second, survival is high (>50%) as compared to other published studies.[5,6] Third, Hawaiian hatcheries use a lot of water — 1 m^3 for every 100 to 500 PLs produced. Fourth, there appears to be a positive relationship between production and total water use. Fifth, clear water flow flushing seems to increase survival. It is plausible that since some mortality is due to biotic agents whose doubling time is less than the resident time of the culture water, then daily flushing, i.e., replacing "old" water with "new" water, will serve to keep the numerical census of these agents in check. Lastly, tank splitting does not seem to be biologically necessary, although clearly it may be an economic necessity, i.e., it is easier, and, therefore, more economical, for a given labor force to care for a minimum number of tanks in the first half of the rearing cycle.

SUMMARY

1. Three of the four commercial hatcheries in Hawaii utilize gravid females collected from commercial harvests for hatchery broodstock.
2. Females are collected on the basis of their gravidity and not on the basis of size (or their egg number).
3. In female *M. rosenbergii* from Hawaiian ponds, fecundity (egg number, Y) is a power function of length and is about 10% lower than the fecundity for a 45-g (15.25-cm) animal spawned in the laboratory.
4. One commercial farm maintains separate broodstock ponds in order to raise large broodstock females so the number of eggs (and larvae hatched) per female is increased.
5. Separate broodstock ponds should be instituted where possible to take advantage of the favorable exponential relationship between female size and fecundity and also to allow for the possibility of genetic progress due to pedigree and age class control.
6. All hatcheries in Hawaii utilize the "green water" semiflow-through system of larval rearing which incorporates live (*Artemia* nauplii) and dead (fish flesh and/or egg custard and/or fish roe) in their husbandry.
7. Feeding husbandry regimens differ slightly among Hawaiian hatcheries in that food items are applied to larval cultures at different times and on different days.

8. Water management husbandry regimens differ among Hawaiian hatcheries; no two hatcheries have the same regimen. Hatcheries either exclusively use the "green water toilet flush" or a combination of this and a "clear water flow flush". The latter, in turn, varies among the hatcheries in the amount of water "flushed" through larval culture tanks following a reduction in its volume. The reduction is also variable among hatcheries. The research hatchery uses a "trickle flow through" using continuous flow of water (trickle) through drip irrigation tubing.
9. Larvae culture tank cleaning by siphoning husbandry differs little among the hatcheries — siphoning is carried out every other day from about the second or third day to the end of the cycle.
10. Hawaiian hatcheries use a lot of water — about 1 m^3 for every 100 to 500 PLs produced although production based on final culture water volume is good ($\sim$30 ind. ℓ^{-1}), which compares favorably with published data.

ACKNOWLEDGMENTS

Sincere thanks go to Amy Dicksion and Deb Ciampa for their editorial, clerical, and graphic assistance. Thanks also go to Deborah Ishihara for graphic help. Special thanks go to the hatchery managers: Nick Carpenter of Amoriente Aquafarms, Inc., Alan Muraoka of Amfac Aquatech, Wayne Okamura of Aquatic Farms, and David Onizuka of Anuenue Fisheries Research Center, who patiently answered (and reanswered) my questions regarding feeding and water use regimens. Thanks also go to Scott Masuno and Bruce Cyr who worked on the fecundity data. Finally, the expert assistance of Dr. Gideon Hulata in correcting the galley proofs is gratefully acknowledged.

The preparation of this report was funded in part by the Aquaculture Development Program, State of Hawaii; by Prawn Tech, 41-052 Hihimanu Street, Waimanalo, Hawaii 96795; and by the University of Hawaii Sea Grant College Program under Institutional Grant No. NA-81-AA-D-00070. This is Sea Grant Publication No. UNIHI-Sea Grant-MC-82-01.

REFERENCES

1. **Malecha, S. R.,** Aquaculture of the freshwater prawn, *Macrobrachium rosenbergii* in Hawaii: history, present status and application to other areas, in Simposio Brasileiro De Aquicultura, Recife, Academia Brasileira De Ciencias Rio de Janeiro, RJ, July 1980.
2. **Dugan, C. C., Hagood, R. W., and Frakes, T. A.,** Development of spawning and mass larval rearing techniques of brackish-freshwater shrimps of the genus *Macrobrachium* (Decapoda Palaemonidae), *Fla. Dep. Nat. Resour. Mar. Res. Lab. Publ.*, No. 12, 1975.
3. **Sick, L. V. and Beaty, H.,** Culture Techniques and Nutritional Studies for Larval Stages of the Giant Prawn, *Macrobrachium rosenbergii,* Georgia Mar. Sci. Ctr., University of Georgia Technical Report, Athens, Ser. No. 74-5, 1974.
4. **Sandifer, P. A., Hopkins, J. S., and Smith, T. I. J.,** Production of juveniles, in *Shrimp and Prawn Farming in the Western Hemisphere,* Hanson, H. L. and Goodwin, J. A., Eds., Dowden, Hutchinson and Ross, Stroudsburg, Pa., 1977, 220.
5. **New, M. B., Ed.,** Growing the larvae — hatchery management, in *Giant Prawn Farming,* Elsevier, New York, 1982, 115.
6. **Lee, C. L.,** Progress in developing standardized systems for production of juvenile *M. rosenbergii* (de Man) at Mardi, Malaca, in *Giant Prawn Farming,* New, M. B., Ed., Elsevier, New York, 1982, 129.
7. **Ling, S. W. and Merican, A. B. O.,** Notes on the life and habits of the adults and larval stages of *Macrobrachium rosenbergii* (de Man), in Proc. 9th F.A.O. Indo-Pacific Fisheries Council, Section II, 1961, 55.

8. **Ling, S. W.,** Studies on the Rearing of Larval and Juveniles and Culturing of Adults of *Macrobrachium rosenbergii,* F.A.O. Indo-Pacific Fisheries Council, Current Affairs Bulletin No. 35, 1962.
9. **Ling, S. W.,** A General Account on the Biology of the Giant Freshwater Prawn, *Macrobrachium rosenbergii,* and Methods for its Rearing and Culturing, F.A.O. Indo-Pacific Fisheries Council, Contribution Paper No. 40, 1964.
10. **Ling, S. W.,** The general biology and development of *Macrobrachium rosenbergii* (de Man), *F.A.O. Fish. Rep.,* 57(3), 589, 1969.
11. **Ling, S. W.,** Method of rearing and culturing *Macrobrachium rosenbergii* (de Man), *F.A.O. Fish. Rep.,* 57(3), 607, 1969.
12. **Fujimura, T.,** former chief of the Anuenue Fisheries Research Center, Honolulu, chronicled the development of mass rearing techniques in a series of unpublished quarterly and annual reports to his grant agencies between 1966 and 1974; each of these reports are listed in the book cited in Reference 4.
13. **Fujimura, T.,** Notes on the Development of a Practical Mass Culturing Technique of the Giant Prawn, *Macrobrachium rosenbergii,* in Proc. 12th Session, F.A.O., Indo-Pacific Fisheries Council, (IPFC/C66/WP47), 1966.
14. **Fujimura, T. and Okamoto, H.,** Notes on Progress Made in Developing a Mass Culture Technique for *Macrobrachium rosenbergii* in Hawaii, in Proc. 14th Session, F.A.O., Indo-Pacific Fisheries Council, 1970, 17pp.
15. **Fujimura, T. and Okamoto, H.,** Notes on progress made in developing a mass culture technique for *Macrobrachium rosenbergii* in Hawaii, in *Coastal Aquaculture in the Indo-Pacific Region,* Pillay, T. V. R., Ed., Fishery News Books, Farnham, Surrey, England, 1972.
16. **Corbin, J. S., Fujimoto, M. M., and Iwai, T. Y., Jr.,** Feeding practices and nutritional considerations for *Macrobrachium rosenbergii* culture in Hawaii, in *CRC Handbook of Mariculture,* McVey, J., Ed., CRC Press, Boca Raton, Fla., 1983.
17. **Malecha, S. R., Masuno, S., and Onizuka, D.,** Feasibility of measuring the genetic control of growth pattern variation in the cultured freshwater prawn, *Macrobrachium rosenbergii* (de Man): juvenile growth, *Aquaculture,* in review, 1982.
18. **Ling, S. W.,** The general biology and development of *Macrobrachium rosenbergii* (de Man), *F.A.O. Fish. Rep.,* 3(57), 589, 1969.
19. **Wickens, J. F. and Beard, T. W.,** Observations on the breeding and growth of the giant freshwater prawn, *Macrobrachium rosenbergii* (de Man) in the laboratory, *Aquaculture,* 3, 159, 1974.
20. **Rao, R. M.,** Breeding behavior in *Macrobrachium rosenbergii* (de Man), *Fish. Tech. (India),* 2(1), 19, 1965.
21. **Chow, S., Ogasawara, Y., and Taki, Y.,** Male reproductive system and fertilization of the palaemonid shrimp, *Macrobrachium rosenbergii, Bull. Jpn. Soc. Sci. Fish.,* 48(2), 177, 1982.
22. **Sandifer, P. A. and Smith, T. I. J.,** A method for artificial insemination of *Macrobrachium* prawns and its potential use in inheritance and hybridization studies, in *Proc. World Maricul. Soc.,* 10, 403, 1979.
23. **Nagamine, C., Knight, C. W., Maggenti, A., and Paxman, G.,** Effects of androgenic gland ablation on male primary and secondary sexual characteristics in the Malaysian prawn, *Macrobrachium rosenbergii* (de Man) (Decapoda, Palaemonidae), with first evidence of induced feminization in a nonhermaphroditic decapod, *Gen. Comp. Endo.,* 41, 423, 1980.
24. **Nagamine, C., Knight, A. W., Maggenti, A., and Paxman, G.,** Masculinization of female *Macrobrachium rosenbergii* (de Man) (Decapoda, Palaemonidae) by androgenic gland implantation, *Gen. Comp. Endo.,* 41, 442, 1980.
25. **Sreekumar, S., Adiyodi, R. G., and Adiyodi, K. G.,** Aspects of semen production in *Macrobrachium* sp., in *Giant Prawn Farming,* New, M. B., Ed., Elsevier, New York, 1982, 83.
26. **Sandifer, P. A. and Lynn, J. W.,** Artificial insemination of caridean shrimp, in *Advances in Invertebrate Reproduction,* Clark, W. H., Jr. and Adams, T. S., Eds., Elsevier, Amsterdam, 1980, 271.
27. **Malecha, S., Sarver, D., and Onizuka, D.,** Approaches to the study of domestication in the freshwater prawn, *M. rosenbergii,* with special emphasis on the Anuenue and Malaysian stocks, in *Proc. World Maricul. Soc.,* 11, 500, 1980.
28. **Malecha, S. R.,** Genetics and selective breeding of *Macrobrachium rosenbergii,* in *Shrimp and Prawn Farming in the Western Hemisphere,* Hanson, J. A. and Goodwin, H. L., Eds., Dowden, Hutchinson and Ross, Stroudsburg, Pa., 1977, 328.
29. **Malecha, S. R.,** Development and general characterization of genetic stocks of *Macrobrachium rosenbergii* and their hybrids for domestication, *Univ. Hawaii Sea Grant Q. News,* 2 (4), 6, 1980.
30. **Doyle, R., Singholka, S., and New, M.,** Indirect selection for genetic change: a quantitative analysis illustrated with *Macrobrachium rosenbergii, Aquaculture,* 1982, in press.
31. **Malecha, S. R.,** Commercial pond production of the freshwater prawn in Hawaii, in *CRC Handbook of Mariculture,* McVey, J., Ed., CRC Press, Boca Raton, Fla., 1983.

32. **Malecha, S. R., Polovina, J., and Moav, R.,** Multi-Stage Rotational Stocking and Harvesting System for Year-Round Culture of the Freshwater Prawn, *Macrobrachium rosenbergii,* University Sea Grant Technical Report, TR-81-01, Honolulu, September 1981.
33. **Gibson, R.,** Amafac Aquatech, personal communication.
34. **Rajyalakmi, T.,** Studies on maturation and breeding in some estuarine palaemonid prawns, *Proc. Nat. Inst. Sci. India,* 27B(4), 179, 1961.
35. **Patra, R. W. R.,** The fecundity of *Macrobrachium rosenbergii* de Man, *Bangladesh J. Zoo.,* 4(2), 1, 1976.
36. **Shakuntala, K.,** The relationship between body size and number of eggs in the freshwater prawn, *Macrobrachium lamarrei* (H. Milne Edwards) (Decapoda, Caridea), *Crustaceana,* 33(1), 17, 1977.
37. **Koshy, M. and Tiwari, K. K.,** Clutch size and its relation to female size in two species of freshwater shrimps of the genus *Macrobrachium* Bate, 1868 (Crustacea: Caridea: Palaemonidae) from Calcutta, *J. Inland Fish. Soc. India,* 7, 109, 1975.
38. **Shakuntala, K.,** A note on the changes in egg weight during the early development of *Macrobrachium rude* (Heller), *J. Island Fish. Soc. India,* 8, 109, 1976.
39. **Malecha, S. R., Cyr, B., and Masuno, S.,** The effect of pond rearing on fecundity of *Macrobrachium rosenbergii,* in preparation.
40. **Brock, J. A.,** Diseases (infectious and non-infectious), metazoan parasites, predators and public health considerations in *Macrobrachium* culture and fisheries, in *CRC Handbook of Mariculture,* McVey, J., Ed., CRC Press, Boca Raton, Fla., 1983.
41. **Bauer, R. T.,** Anti-fouling adaptations of caridean shrimps: cleaning of the antennal flagellum and general body grooming, *Mar. Biol.,* 49, 69, 1982.
42. **Bauer, R. T.,** Anti-fouling adaptations of marine shrimp (Decapoda: caredea): gill cleaning mechanisms and grooming of brooded embryos, *Zoo. J. Linn. Soc.,* 65, 281, 1979.
43. **Bauer, R. T.,** Grooming behavior and morphology in the decapod crustacea, *J. Crust. Biol.,* 1, 153, 1981.
44. **F.A.O.,** Conservation of the genetic resources of fish: problems and recommendations, *F.A.O. Fisheries Tech. Pap.,* no. 217, 1981.
45. **Sandifer, P. A. and Smith, T. I. J.,** Possible significance of variation in the larval development of palamonid shrimp, *J. Exp. Mar. Biol. Ecol.,* 39, 55, 1979.
46. **Uno, Y. and Kwon, C. S.,** Larval development of *Macrobrachium rosenbergii* (de Man) reared in the laboratory, *J. Tokyo Univ. Fish.,* 55(2), 179, 1969.
47. **von Bertalanffy, L.,** Metabolic types and growth types, *Am. Nat.,* 85(821), 111, 1951.
48. **Stephenson, M. J. and Knight, A. W.,** Growth, respiration and caloric content of larvae of the prawn, *Macrobrachium rosenbergii, Comp. Biochem. Physiol.,* 66A, 385, 1980.
49. **Stephenson, M. J. and Knight, A. W.,** The effect of temperature and salinity on oxygen consumption of post-larvae of *Macrobrachium rosenbergii* (de Man) (Crustacea: Palaemonidae), *Comp. Biochem. Physiol.,* 67A, 699, 1980.
50. **Iwai, T. Y., Jr.,** Energy Transformation and Nutrient Assimilation by the Freshwater Prawn *Macrobrachium rosenbergii* under Controlled Laboratory Conditions, M.S. thesis, University of Hawaii, Honolulu, 1976.
51. **Snow, J. R.,** Increasing the Yield of Channel Catfish Rearing Ponds by Periodic Division of the Stock, presented at the 13th Ann. Conf. Southeast Assoc. Game and Fish Comm., 1976, 239.
52. **Malecha, S. R.,** Commercial pond production of the freshwater prawn in Hawaii, in *CRC Handbook of Aquaculture,* McVey, J., Ed., CRC Press, Boca Raton, Fla., 1983.
53. **Nakamura, R., Akita, G., Miyamoto, G., Fujimoto, M., Oishi, F., Sumikawa, D., Onizuka, D., and Brock, J.,** Epizootiologic study of mid-cycle disease of larval *Macrobrachium rosenbergii,* in *Proc. World Maricul. Soc.,* 13, 1983, in press.
54. **Armstrong, D. A., Chippendale, D., Knight, A. W., and Colt, J. E.,** Interaction of ionized and unionized ammonia on short-term survival and growth of prawn larvae, *Macrobrachium rosenbergii, Biol. Bull.,* 154, 14, 1978.
55. **Armstrong, D. A., Stephenson, M. J., and Knight, A. W.,** Acute toxicity of nitrite to larvae of the giant Malaysian prawn, *M. rosenbergii, Aquaculture,* 9, 39, 1976.
56. **Colt, T., Mitchell, S., Tchobanoglous, G., and Knight, A.,** The Use and Potential of Aquatic Species for Freshwater Treatment. Appendix C: The Environmental Requirements of Crustaceans, Publication No. 65, California State Water Resources Control Board, Sacramento, Calif., 1979.
57. **Cohen, D., Finkel, A., and Sussman, M.,** On the role of algae in larviculture of *Macrobrachium rosenbergii, Aquaculture,* 8, 199, 1976.
58. **Manzi, J. J., Madox, M. B., and Sandifer, P. A.,** Algal supplement enchancement of *Macrobrachium rosenbergii* (de Man) larviculture, in *Proc. World Maricul. Soc.,* 8, 201, 1977.
59. Aquacop, *Macrobrachium rosenbergii* (de Man) culture in Polynesia: progress in developing a mass intensive larval rearing technique in clear water, in *Proc. World Maricul. Soc.,* 8, 311, 1977.

60. **Moller, T. H.,** Feeding behavior of larvae and postlarvae of *M. rosenbergii* (de Man) (Crustacea: Palamonidae), *J. Exp. Mar. Biol. Ecol.,* 35, 251, 1978.
61. **Manzi, J. J. and Maddox, M. B.,** Requirements for *Artemia* nauplii in *Macrobrachium rosenbergii* (de Man) larviculture, in *The Brine Shrimp Artemia,* Vol. 3, Persoone, G., Sorgeloos, P., Roels, O., and Jaspers, E., Eds., Universa Press, Wetteren, Belgium, 1980, 313.

COMMERCIAL POND PRODUCTION OF THE FRESHWATER PRAWN, *MACROBRACHIUM ROSENBERGII*, IN HAWAII

Spencer R. Malecha

INTRODUCTION

The main purpose in the following will be to provide a general review of pond culture practices in Hawaii and their theoretical background, provide a guide to the literature, point out highlights of the systems, and emphasize certain topics — notably, PL poststocking mortality, microclimatic factors in site selection, inventory control, and the usefulness of production management systems developed for warm water fish pond culture in prawn pond aquaculture. Space limitations do not permit a thorough treatment, in a ''how-to-recipe-book'' fashion, of the technical details of freshwater prawn pond culture in Hawaii. In this regard, careful examination of the published material referred to herein will provide the reader with sufficient details.

GENERAL

Macrobrachium rosenbergii is a large, decapod caridean crustacean found throughout the Indo-Pacific region.[1-4] It readily completes its life cycle in captivity and its general biology and suitability for culture have been reviewed.[5-13] Specific aspects of prawn culture in year-round growing seasons like Hawaii have been discussed;[14] a technical manual[15] and a comprehensive treatment of the prawn's biology and culture[16] are in preparation. Management principles developed for warm water fish culture[17-20] apply to a large degree to prawn pond culture.

THE HAWAIIAN PRAWN INDUSTRY

The status and development of traditional Hawaiian prawn industry practices have been described[21-29] and should be compared with that given for temperate[30-32] and other tropical areas.[33,34] All the active prawn ponds in Hawaii, with one exception, are in the private sector (Table 1). Most farms are located on the windward coasts and are concentrated on the Island of Oahu (Figure 1). The large- and intermediate-sized farms are designed and operated to provide the sole source of income for the owner company. All the smaller farms provide a second income to the owners or parent companies who are engaged in other primary businesses. Most commercial ponds are managed in the traditional way, described herein, but nontraditional practices utilizing batch harvesting with pond draindown (with or without traditional cull and seine), nursery culture (both high and intermediate density), and size grading are being instituted.

Site Selection

Usually this topic is discussed together with pond construction. Inasmuch as ponds used in the Hawaiian industry are either the nondraindown traditional type or the draindown nontraditional type, a review of the physical descriptions of the two types will be included separately when the two systems are discussed.

The physical topography, soil condition, and water availability and quality and their costs are the main factors in site selection. In general, the site selection criteria for ponds differ little among warm water species and the general criteria adopted for warm water fish culture[17-19,35] can be followed for prawns. Soil should be clay-loam or silty clay with 85%

Table 1
VITAL STATISTICS FOR HAWAIIAN PRAWN FARMS AS OF 4/82. ORGANIZATION IDENTIFICATION NUMBERS CORRESPOND TO THOSE GIVEN IN FIGURE 1

Organization	Business start date	Water surface (acres/ha)	No. ponds	Type[a] (innovation[b])	Av water use[c] (ℓ/m/ha)
Large sized					
1	5/80	100/40.5	118	1 Phase	150—168
2	9/81	40/16.2	20	3, 4 Phase	94—140
Intermediate sized					
3	7/77	27/10.9	22	Traditional (N, MH)	187
4	11/77	24.4/9.9	12	Traditional	196—271
5	6/73	20.3/8.2	7	Traditional	140
Small sized					
6	7/80	8.0/3.2	8	Traditional	—
7	11/76	7.7/3.1	12	Traditional	
8	10/76	5.8/2.3	11	Traditional	140
9	8/80	5.3/2.1	3	Traditional	—
10	7/81	5.0/2.0	3	Traditional (MH, R)	—
11	10/76	4.0/1.6	6	Traditional	—
12	10/81	1.0/4	21	Research[d]	230
Very small sized					
13[e]	9/73	1.5/0.6	2	Traditional	—
14	7/81	1.5/0.6	5	Traditional	—
15	8/80	0.9/0.4	6	Traditional	—
16	7/77	0.7/0.3	3	Traditional	—
17	10/79	0.6/0.2	1	Traditional	—
18	1/75	0.6/0.2	3	Traditional	—
19	8/78	0.5/0.2	1	Traditional	—
20	8/81	0.5/0.2	1	Traditional	—
21	7/73	0.3/0.1	4	Traditional	—

[a] See Figure 2. All 3 and 4 phases have a nursery and use pond draindown.
[b] Used in part of operation: N = nursery, MH = mechanical harvesting, R = raceway ponds.
[c] Based on survey of pond managers.
[d] Prawn Aquaculture Research Program, University of Hawaii, experimental ponds (Figure 6).
[e] Gentaro Ota farm, Punaluu, Hawaii, used by Fujimura (see in-text footnote on Fujimura) in developing traditional system. Ponds represented in Figure 4.

water retention. Bedrock should be greater than 1.5 m (5 ft) deep. The seepage rate should be between 1.60 and 5 cm hr^{-1} (0.63 to 2 in. hr^{-1}). Slopes of land area should be 0 to 3°. Water and soil analysis profiles considered representative of Hawaii conditions are given by Fujimoto et al.[22]

It is good practice to have site soils tested for residual pesticides and heavy metals. Furthermore, the history of the site should be checked out by examining government and private records and interviewing neighbors, especially on land previously used for other agriculture practices.

The microclimatic aspects of a site have not been given the careful consideration they demand — specifically, cloud cover, wind speed and direction, and certain topographic relief features (e.g., hills), all of which can affect pond water temperature. Indeed, temperature, one of the two major variables affecting production, can only be cost effectively "managed" through site selection. Cloud and high wind areas should be examined closely because, even though high dissolved oxygen concentrations are easier to maintain under these conditions, they produce low pond water temperatures due to convective heat loss across, and low solar incidence on, the pond surface. It is probably better to have lower

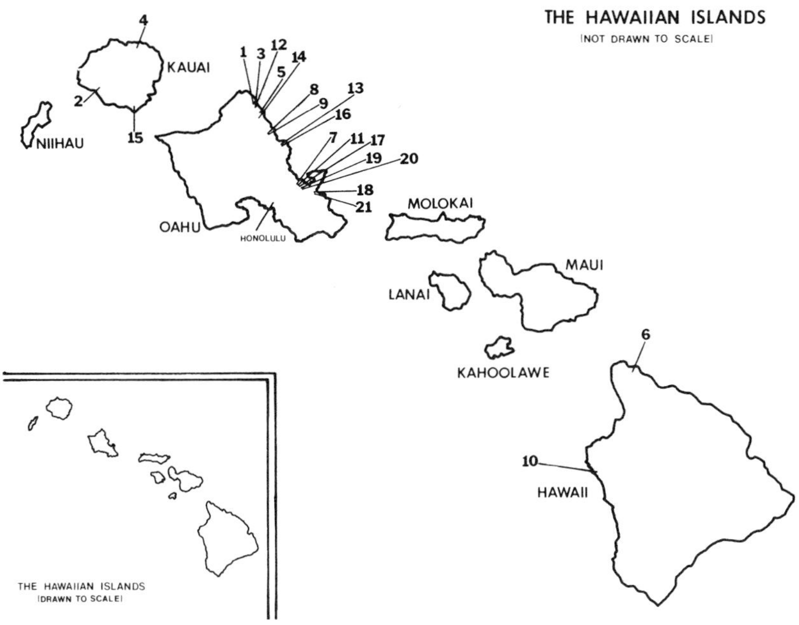

FIGURE 1. Map of Hawaii showing locations of active commercial and research prawn ponds. Numbers are the same as in Table 1 and identify separate operations.

wind velocities and higher solar radiation and be faced with managing DO crises periodically than it is to be faced with chronic slow growth rates and thermal shock at stocking.

Environmental Requirements

The environmental requirements in traditional and nontraditional culture systems are about the same — two important variables are dissolved oxygen and temperature.

Dissolved Oxygen (DO)

Dissolved oxygen is a well-known and important state variable in the prawn pond. The extension emphasis[36-38] and the pond studies of this variable and its management which have been conducted[39-42] and reviewed[43,44] for catfish culture apply to prawn ponds in Hawaii and elsewhere. On-farm practices of measuring DO varies — large and intermediate farms take daily a.m. and/or p.m. DO readings. Studies have shown that oxygen consumption in prawns is an allometric (log) function of body weight[45-47] but nearly a linear function of oxygen concentrations (mmHg) below about 55 mmHg (6.45 ppm or 75% saturation at 22°C).[48] The latter study showed that prawns will consume oxygen at concentrations as low as 10 mmHg (1.17 ppm or 11.7% saturation at 22°C). Our studies have shown that, in general, the prawn is well adapted to hypoxia.[48,49] In those cases where low oxygen levels result in death it is at an oxygen level which is well below the tolerance of most species of crustaceans.[63] The lowest oxygen level where the animal can be maintained without stress is between 25 and 30% saturation or about 2.25 to 2.75 ppm in the range of 25 to 30°C. Consequently, on-farm practice should be directed, at the very least, to maintaining DO concentrations above 3 ppm. This is about the level set by others[50,51] as the minimum acceptable DO concentration. Indeed, an on-farm rule of thumb is that, at 24 to 28°C, 1 ppm is the tolerance level, but levels between 3 and 5 ppm are considered dangerous so pond managers like to maintain the DO between 6 and 8 ppm. DO concentration fluctuation is seasonal since it is a function of temperature and phytoplankton biomass. Losordo[50] has furnished guidelines for pond DO and temperature management and has shown that water column oxygen demand was between 48 and 87% of total pond oxygen loss at night.

Prawn ponds in Hawaii do stratify and it is good practice to monitor both bottom and near-surface DO levels. Likewise, concentrations of DO in large ponds vary with location within the pond. Tables 2 and 3 illustrate a yearly profile of DO concentration from selected grow-out and nursery ponds from an intermediate-sized prawn farm. Danger levels are obviously in the summer months when solar radiation is maximum and wind velocity (which is responsible for DO exchange between the water column and the atmosphere) is reduced.

In general, constant vigilance of meteorological and pond conditions is warranted to prevent DO crises. Usually prophylaxis for low DO is by means of pond flushing, i.e., increasing the outflow of nutrients and phytoplankton biomass. This requires a considerable amount of water something which is perhaps lacking in some areas. Peak demand during crises flushing is 3.74 m^3ha^{-1} (400 gpm $acre^{-1}$) for 24 hr.

Temperature and Its "Management"

There have been no published experimental data on the quantitative relationship between pond temperature and postlarval, juvenile, or adult growth and production. However, several laboratory studies,[52-55] published in English and one in Chinese,[56] point to the optimum larval rearing temperature to be around 28 to 30°C. Moreover, experimental pond and tank nursery studies have noted the relationship between growth and temperature, the latter a nontreatment effect in these studies.

There is no question that prawns grow faster at "higher" temperatures (26 to 28°C) and slower at ones lower than these.[53-55] Inasmuch as these temperature regimens develop in semitropical areas such as Hawaii, they should be a major factor in site selection, therefore, temperature management.

Tables 2 and 3 show a profile of temperature fluctuations under commercial conditions (farm 3, Table 1, Figure 1). These data compare favorably with published data[22] and are typical of the area which contains a concentration of prawn farms (Figure 1).

Nutrient and Mineral Water Quality Parameters

Like temperature, there have been no published studies on the quantitative relationship between prawn growth and production and various nutrients, toxicants, pollutants, etc., with the exception of the econometric analysis of Morita.[52] Therefore, rules of thumb for managing these variables follow those developed for warm water fish culture.

Residual pesticides should be tested for during site selection and obviously acute (or chronic) exposures avoided. One study[58] found a difference in molting patterns and fivefold differences in growth, between adult prawns reared in water containing 0.65 to 5.00 $\mu M\ell^{-1}$ Ca (65 to 500 ppm $CaCO_3$). This study noted that slow growth and encrustation of *Bryozoa* sp. and *Epistylus* sp. had been observed in a commercial prawn farm whose pond water had Ca^{++} concentrations of 305 to 638 ppm. All in all, prawn farm water sources should have calcium levels below 100 ppm. The calcium requirements for molting have been given.[59]

A short published summary[60] and longer unpublished studies[61,62] give specific water quality data from commercial production ponds and (inasmuch as these ponds are commercially viable) these data can be consulted for more detailed baseline information (Table 4). In general, water hardness, pH, and alkalinity beyond the limits established for decapod crustacea[63] are to be avoided. Nitrogenous nutrients (NH_4, NO_2, NO_3) and the phosphorus nutrients (PO_4) do not pose any serious problem in commercial pond situations because their levels are kept low by their continual demand and metabolism by fish or phytoplankton. The latter, of course, should be managed as a water quality parameter because of its effect on dissolved oxygen concentration.

Phytoplankton

Five studies[50,61,62,64,65] have provided data on phytoplankton biomass under Hawaiian

Table 2
AVERAGE WEEKLY TEMPERATURE, SECCHI DISK DEPTH, AND DISSOLVED OXYGEN (DO) READINGS FOR ONE CALENDAR YEAR (MAY 1981 TO MAY 1982) FROM THREE PONDS ON AN INTERMEDIATE-SIZED PRAWN FARM (NO. 3, TABLE 1, FIGURE 2)[a]

Week of		Temp[b] (°C)	Secchi disk depth (cm)	DO (ppm)[c] Top	DO (ppm)[c] Bottom
			1981		
May	11	24.0	24.5	6.0	5.7
	18	24.7	26.8	5.0	4.8
	25	25.9	25.6	—	—
	1	25.8	24.4	4.7	4.5
June	8	25.9	22.1	4.9	4.8
	15	25.8	24.1	4.1	3.9
	22	26.3	24.9	4.2	4.1
	29	25.9	27.4	5.0	4.8
July	6	26.2	25.7	4.5	4.4
	13	26.0	22.3	4.4	4.3
	20	25.7	24.3	5.3	5.1
	27	25.9	24.7	5.1	4.9
Aug	3	25.9	26.2	4.9	4.7
	10	26.3	25.6	4.3	4.1
	17	25.9	25.9	5.4	5.2
	24	26.9	27.9	5.4	5.2
	31	26.0	26.9	5.0	4.9
Sept	7	27.4	26.0	5.5	5.3
	14	27.0	22.8	4.3	4.1
	21	26.4	21.3	4.2	4.1
	28	25.1	22.1	4.6	4.4
Oct	5	24.9	20.7	5.0	4.9
	12	25.4	23.5	5.4	5.3
	19	24.0	22.1	5.3	5.2
	26	24.0	15.3	5.3	5.1
Nov	2	24.0	19.0	5.8	5.7
	9	23.5	25.5	6.5	6.4
	16	23.2	22.2	6.9	6.7
	23	21.2	23.9	7.0	6.9
	30	21.4	23.6	6.4	6.0
Dec	7	19.6	28.1	7.3	7.2
	14	23.1	27.2	6.1	6.0
	21	21.9	23.5	5.9	5.8
	28	22.0	29.3	6.9	6.8
			1982		
Jan	4	22.6	21.2	5.3	5.1
	11	22.6	18.4	4.8	4.7
	18	21.5	23.8	6.5	6.4
	25	20.0	27.0	7.2	7.1
Feb	1	20.0	29.5	7.6	7.5
	8	21.0	29.7	6.6	6.4
	15	23.0	25.8	4.8	4.7
	22	22.0	21.8	4.1	4.0

Table 2 (continued)
AVERAGE WEEKLY TEMPERATURE, SECCHI DISK DEPTH, AND DISSOLVED OXYGEN (DO) READINGS FOR ONE CALENDAR YEAR (MAY 1981 TO MAY 1982) FROM THREE PONDS ON AN INTERMEDIATE-SIZED PRAWN FARM (NO. 3, TABLE 1, FIGURE 2)[a]

Week of		Temp[b] (°C)	Secchi disk depth (cm)	DO (ppm)[c] Top	DO (ppm)[c] Bottom
			1982		
Mar	1	—	—	—	—
	8	—	—	—	—
	15	23.0	19.3	3.2	3.0
	22	19.9	8.1	6.5	6.3
	29	21.0	17.9	6.0	5.9
Apr	5	21.0	19.5	5.3	5.1
	12	21.7	20.7	6.1	6.0
	19	21.4	18.3	5.6	5.6
	26	21.0	17.8	6.1	6.0
May	3	23.0	19.2	5.2	5.0

[a] Table data is arithmetic average of daily readings made on three ponds on a commercial prawn farm (No. 3, Table 1, Figure 2). Data average of calendar weeks (Monday through Friday); readings made between 0700 and 1000 hr.

[b] Temperature data from one pond only.

[c] Readings made with YSI oxygen meter (Yellow Springs Instrument Co., Yellow Springs, Ohio) just below pond surface ("top") and just off pond bottom ("bottom").

conditions. Two studies[61,62] summarized by Iwai[60] reported chlorophyll *a* pigment levels, measured at 1300 hr, ranging from 19.2 to 170.8 $\mu g \ell^{-1}$. Laws and Malecha[64] showed that phytoplankton growth is light-limited and that phytoplankton should be managed at chlorophyll *a* levels between 150 and 400 $\mu g \ell^{-1}$ to prevent hypoxia. Losordo[50] measured total suspended solids (the majority of which would be phytoplankton) in production ponds and recommended guidelines for pond management because of the strong association between Secchi disk depth and oxygen demand. Normal "on-farm" measurement of phytoplankton density is done with Secchi disk readings. Table 2 gives a typical profile of Secchi disk readings from a commercial prawn farm. If we assume the relationship between Secchi disk readings and phytoplankton biomass to be linear and using the data in Table 2, then we can see that, on the average, commercial pond phytoplankton biomass levels should be managed to give Secchi disk readings of between 15 and 35 cm as recommended.[50] Small nursery ponds have considerably higher phytoplankton levels (Table 3).

Malecha et al.[66] have demonstrated the feasibility of commercial prawn-fish polyculture. Unpublished studies[65] in preparation by our group show that fish biomass stabilizes the pond ecosystem and dampens out large oscillations of phytoplankton biomass and the concomitant dissolved oxygen fluctuations. Numerous other studies[67-73] on the effect of fish on nutrient cycling and materials balance in "pond-like" aquatic systems argue strongly for the institution of prawn-fish polyculture systems where possible and economically feasible. All

Table 3
AVERAGE BIWEEKLY TEMPERATURE, DISSOLVED OXYGEN, AND SECCHI DISK DEPTH READINGS FOR HIGH DENSITY NURSERY AND RESEARCH PONDS (NO. 12, TABLE 1, FIGURES 1 AND 8) LOCATED ON AN INTERMEDIATE-SIZED COMMERCIAL FARM (NO. 3, TABLE 1, FIGURES 1 AND 8)

	Temp (°C)		DO (ppm)		Secchi disk depth (cm)	
1981—1982	**Min**	**Max**	**a.m.**	**p.m.**	**a.m.**	**p.m.**
March 1	23.27	29.25	7.71	10.48	64.86	59.44
2	21.86	27.46	9.32	10.90	64.90	62.49
April 1	22.99	28.33	8.80	10.23	59.25	58.12
2	23.19	28.99	9.02	10.80	58.06	54.11
May 1	22.59	28.95	8.67	10.66	59.40	57.95
2	24.03	29.81	9.12	9.70	58.94	57.28
June 1	24.54	30.48	8.01	10.20	62.82	61.41
2	24.58	30.73	7.42	9.62	65.38	64.67
July 1	24.42	30.80	7.76	10.03	74.47	72.63
2	24.22	32.06	6.32	13.15	—	—
Aug 1	24.77	31.02	7.27	11.02	—	—
2	25.11	31.62	7.25	9.49	—	—
Sept 1	25.00	31.74	5.91	10.82	—	—
2	24.64	29.94	6.52	10.34	—	—
Oct 1	24.04	29.30	6.26	9.81	—	—
2	23.30	28.88	6.40	8.84	—	—
Nov 1	23.09	28.68	6.18	9.26	—	—
2	21.78	26.25	6.75	9.39	—	—
Dec 1	20.45	24.89	6.26	10.07	—	55.29
2	22.26	27.16	5.55	9.84	—	50.82
Jan 1	—	—	4.86	9.91	—	—
2	20.49	25.00	7.14	9.95	—	49.60
Feb 1	20.72	26.04	6.78	11.56	—	50.12
2	23.04	28.23	8.73	12.91	—	47.50

Note: Twenty-one total ponds on site, but not all were measured on any given day. Data taken daily but averaged into biweekly periods.

commercial pond prawn culture in Hawaii is a monoculture because of historical reasons as well as the lack of development of a production and marketing system for freshwater fishes. Hawaiian prawn farmers, therefore, manage dissolved oxygen with either flushing or emergency aeration. However, in other areas where polyculture of prawns and fishes is feasible, this practice is recommended for pond ecosystem stability.

Salinity

Several studies[74-79] have been directed at the effect of salinity on prawn osmoregulation and growth. Although prawns tolerate salinities that are higher than their iso-osmotic point (18‰)[75] optimum growth conditions are at fresh or slightly brackish (0 to 4‰) concentrations. Two studies[46,79] have demonstrated that salinity affects respiration and metabolic rates at the range of temperatures (20 to 30°C) likely to be encountered by prawns in nature or under culture. Interestingly, Nelson et al.[79] showed a significant difference in metabolic rate between 0 to 7‰ (the optimum culture range[74,75]) at 27°C, an average Hawaiian prawn pond

Table 4
SUMMARY OF SELECTED WATER QUALITY PARAMETERS REPORTED FOR HAWAIIAN PRAWN PONDS

Variable	Ref.
Temperature °C	
19.8—27.8	22, 60
pH	
7.21—8.84	60
5.20—8.20	22
Secchi disk depth (cm)	
25—35	50
Dissolved oxygen (ppm or $\mu g\ell^{-1}$)	
5.4—11.6	60
5.07—8.35 (bottom)	50
9.92—11.50 (surface)	50
Chlorophyll *a* ($\mu g\ell^{-1}$)	
19.2—170.8	60
140—400	64
60.91—494 (Manure)	65
13.73—95.79 (Conventional feeds)	65
Nitrogen (ppm or $mgN\ell^{-1}$)	
0.1172—0.2204 (Total dissolved)	60
0.0039—0.473 (Inorganic)	60
0.017—0.35 (NH_4)	22
Phosphorus (P)	
0—0.0189 $mgP\ell^{-1}$ (Dissolved organic)	60
0.0003—0.0065 $mgP\ell^{-1}$ (Dissolved inorganic)	60
2.2—9.8 μM (Total P, conventional feed)	65
2.3—68.9 μM (Total P, manure)	65
Alkalinity	
34.3—76.1 mg $CaCO_3\ell^{-1}$	60

Note: See original sources for sample technique and other details.

temperature. Indeed, the one Hawaiian prawn farm which uses slightly brackish water (0.25 to 0.75 ppt) has reported very good growth rates.[80] All in all, culture of prawns in slightly brackish water is biologically viable.

The Traditional System

Background

The traditional pond grow-out system reviewed in References 21 to 29 was instituted by Fujimura* after he developed mass larval rearing techniques (described elsewhere in this volume) following the initial successes of Ling[81] in closing the prawn's life cycle. However, readers desiring a thorough understanding of the system should study the description of its development described in a series of (unfortunately) unpublished quarterly and annual reports** written by Fujimura between 1966 and 1972. More widely available summaries by Fujimura and colleagues of the work contained in these reports appeared in 1966,[82] 1970,[83] 1972,[21] and 1977.[22]

* T. Fujimura, former chief, Anuenue Fisheries Research Center, State of Hawaii, Department of Land and Natural Resources, Sand Island, Area 4, Honolulu, Hawaii 96819.

** Takuji Fujimura wrote a series of quarterly and final reports covering the period 1966 to 1972. Another report (job completion) covered the period July 1969 to June 30, 1972, and along with References 21 to 29 will give the reader an adequate summary of the development of the traditional system. The book cited by Fujimoto et al.[22] (pages 407 to 408) should be consulted for a complete citation of the reports.

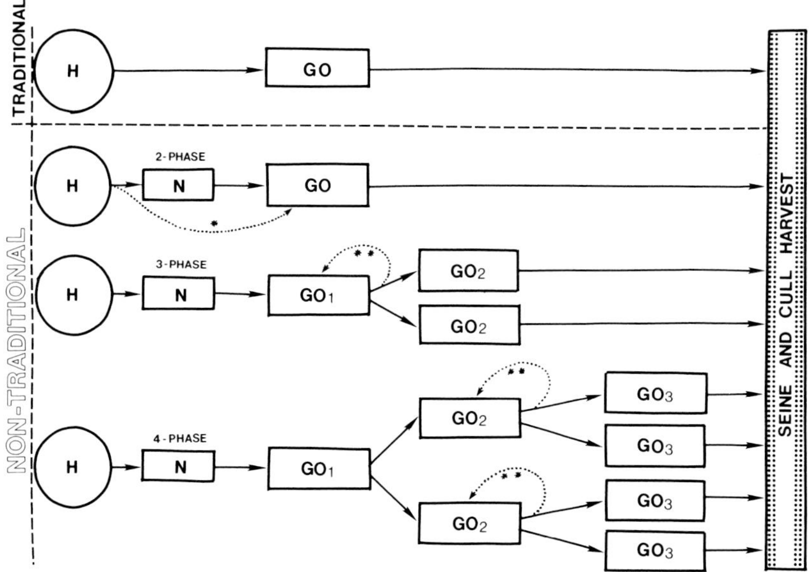

FIGURE 2. Representation of production phases in traditional, nontraditional, and multiphased systems practiced in commercial prawn farming in Hawaii. H = hatchery; N = nursery ponds, GO = grow-out ponds 1, 2, 3 in multiphased systems. * Indicates the partial use, on traditional, single-phased farms, of nursery phases. ** Indicates that the divided prawn stock, "drain harvested" (see text) from a particular grow-out phase pond, may be placed back into the same pond and not necessarily split among two (or more) different ponds. Either operation achieves the same result of stock division for inventory and density control. All phases use a traditional "seine and cull" harvest in their terminal segments, although pond draindown is practiced between grow-out phases (GO1, GO2) in order to harvest and/or rotate stock.

The traditional practice encompasses only a hatchery and pond grow-out phase (Figure 2) and can be described as a "continuous stocking and harvesting system".[84] Postlarvae (PLs) obtained from local hatcheries are stocked at least once, but when available, twice or three times per year after studies[85] have shown the advantages of multiple stocking. Ponds contain cohorts from previous stockings; selective harvesting is done by "seine and cull" — pulling a seine net through the water column which selectively retains certain sizes (>30 g). Ponds are rarely (if ever) drained in routine husbandry. Draining is sometimes done to repair the pond itself or to remove sediments.

Under the traditional system a population of mixed stocking classes develops in the pond which results in a wide range of size classes present in the standing crop (Figure 3). Fujimura* felt that the best way to manage the wide size ranges was to periodically cull (harvest) the large market-sized animals (30 to 45 g) to "make room" for the smaller sized classes to grow into the size class vacated by the harvested animals. This is shown in Figure 4 which represents two cohorts which distinguished themselves as the modes (size modes are shaded differently in Figure 4) in a bimodal size frequency distribution; periodic selective harvesting of the higher mode reduced its census as shown. In practice, it is impossible to distinguish stocking cohort size classes in overlapping regions (Figure 4). This is especially true in ponds that have been subjected to a number of stockings and harvests. Therefore, it is possible that very old, small, residual populations remain in the pond, a major drawback with the traditional system.[86]

"Batch" harvesting was not instituted as a viable alternative in Hawaii because the wide size ranges lowered the overall product value; prices paid for prawns are a function of their size.[88] The periodic seine harvesting is designed to continually produce a uniform, large-

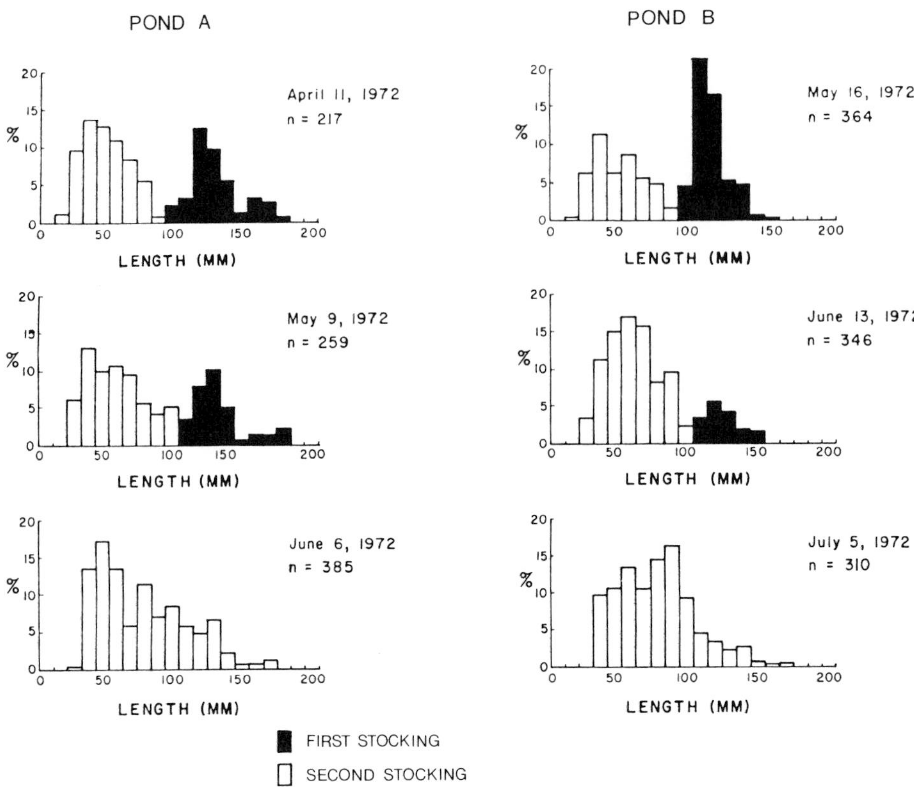

FIGURE 3. Size frequency distribution in two commercial ponds of cohorts representing two pond stockings following the last three cull harvests before the first stocking cohort is "harvested out" of the population. Harvesting began for pond A in February 1971; for pond B in March 1971, and continued monthly through dates shown. Pond A was first stocked in August 1970 and then in May 1971. Pond B was first stocked in September 1970 and then in November 1971. Diagrams represent the pioneering "in-pond" studies by Fujimura (see in-text footnote on Fujimura) which led, in part, to the development of the traditional "stock-'em-and-seine-'em" method. Figure reproduced with permission from job completion report by Fujimura.

sized (and, therefore, valuable) product throughout the year analogous to fruit harvest of perennial plants whereby new growth emerging from a basic stock is picked when ripe.

In the populations represented in Figure 4, production was 3314 kg $ha^{-1}year^{-1}$. The production did not "grow" from 1700 kg (it's assumed standing crop [see later]) to 3314 kg in a traditional sense; harvested production was due to faster growing animals that accelerate their growth to replace the larger ones that are removed from the population. One study,[87] which directly compared seine and cull harvests followed by a draindown harvest in one set of ponds and a batch harvest in another set, showed an increased yield in the ponds that had been selectively harvested. Surprisingly, the culled biomass represented only a small percentage of the difference between the two methods indicating that selective seine and cull somehow induces the growth and production of the unculled (and smaller) animals. This point has been discussed by Malecha et al.[84]

The continual harvest of ponds referred to in Figure 4 yielded 3314 kg $ha^{-1}year^{-1}$ or 276 kg $ha^{-1}month^{-1}$, roughly the amount per harvest. The Hawaiian industry began with the expectation (following the pioneering studies of Fujimura [see preceding footnote]) of a 276-kg $ha^{-1}harvest^{-1}$ (~3000 lb $acre^{-1}year^{-1}$).

Production

Pond production varies with pond age, size, and location. In general, the smaller and

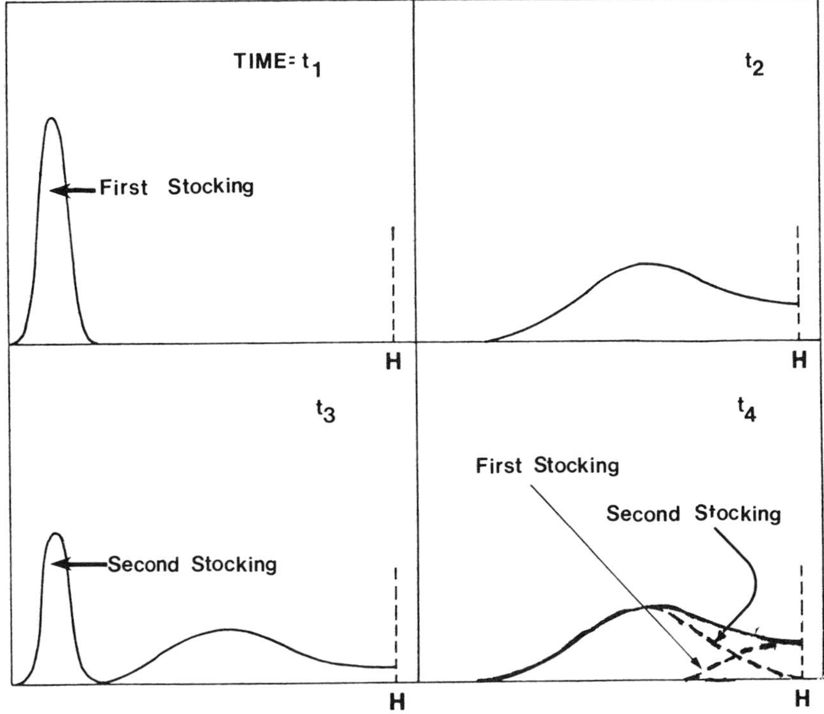

FIGURE 4. Hypothetical relationship between first and second stocking cohorts in commercial prawn pond using the traditional "stock-'em-and-seine-'em" method. H = harvest cull and is assumed to be 100% efficient as shown. (Figure is reproduced [with lettering additions and modifications] from Polovina, J. and Brown, J., *Proc. World Maricul. Soc.*, 9, 393, 1978. With permission.)

older the pond, the higher the production. These rules of thumb, however, must be viewed cautiously (Tables 5 and 6).

In subtropical areas like Hawaii, production in traditional systems can be "predicted" only by relying on actual production records. However, first stocking growth (no previous stocking) can be "predicted" using the pond growth in the summer in one-season, one-stocking studies.[88-90]

Table 5 and 6 display production data from selected prawn farms and ponds in Hawaii. Table 5 shows the average yield per harvest (i.e., per month) per hectare for the years 1974 to 1981 for certain small and intermediate farms using traditional practices. The observed yield per harvest is less than expected based on initial trials (276 kg $ha^{-1}harvest^{-1}$) (see preceding footnote on Fujimura). Several things may account for this. First, there is wide variability in pond management sizes and practices. Second, it may be that the traditional "stock-'em-and-seine-'em" system is "good" for only two or three stockings. Indeed, the pilot pond studies (Figure 4) that led to the institution of the traditional system were based on stocking and seine and cull production practices over only a 2-year period; the two ponds had only two stockings (Figure 3, shaded and unshaded modes). It was assumed that the same system (periodic stocking and periodic seine and cull) could sustain production (~250 kg $ha^{-1}harvest^{-1}$) through the years of stocking and harvest. Tables 5 and 6 clearly show this not to be the case. A reason for this may be that the stocking cohorts do not clearly differentiate into modal classes and that through time a "mixed bag" of year classes results. Newly stocked cohorts must "come up through" these residents with, perhaps, significant lowering of production. Prawns with long resident times may be slow growing. Malecha et

Table 5
AVERAGE YIELD PER HARVEST PER POND FOR CERTAIN SMALL AND INTERMEDIATE-SIZED PRAWN FARMS (TABLE 1) BETWEEN 1974 AND 1981[a]

Calendar month	Yield (kg ha^{-1})	Pond months[b]	Hectare months[c]
January	148.5	169	97.1
February	131.8	168	97.7
March	136.7	180	100.7
April	144.1	189	103.6
May	148.8	199	112.9
June	158.4	205	117.4
July	171.4	216	125.4
August	170.3	222	126.6
September	159.1	224	132.2
October	163.9	236	137.3
November	153.5	227	132.6
December	166.6	231	134.9

[a] Table assumes one harvest per month. Monthly harvest data was obtained from computerized data base of the Prawn Aquaculture Research Program which was compiled from commercial prawn farms under cooperative agreements with the Anuenue Fisheries Research Center, State of Hawaii[81] for the calendar years, 1974 to 1981. Yield, Y_i, was calculated as follows:

$$Y_i = \frac{\Sigma_j \Sigma_k P_{ijk}}{\Sigma_j \Sigma_k h_{ijk}}$$

where j = 1974 to 1981; k = 1 . . . n; P_{ijk} is the total production in kg for the kth pond in the ith month of the jth year; h_{ijk} is the hectarage that produced P_{ijk}. If the P_{ijk} for a pond was zero for a month and year, then its hectarage was not added to the h_{ijk} and, thus, the data are inflated to some degree. Idle ponds bring down the overall estimates. The table can be interpreted as follows: the average production in January between, and including, 1974 and 1981 is 148.5 kg ha^{-1} harvest^{-1} pond^{-1}.

[b] Pond months: total number of ponds reported in each month between, and including, the years 1974 to 1981 summed over these years (e.g., 169 pond months means that there were 169 pond-specific harvest reports for January between 1974 and 1981).

[c] Hectare months: the total water surface hectarage represented by each month calculation.

al. have argued this point in detail.[84] Perhaps what the results represented in Figure 3 and the actual production of the industry (Tables 5 and 6) which has utilized the system beyond the range [2 years] of the pilot studies tell us is that the traditional seine and cull method should be confined to definite segments punctuated by pond draindown. This is basically what is practiced in the terminal phase of a multiphased growth system (see Figure 2).

It must be pointed out that some small ponds are able to maintain expected production

Table 6
AVERAGE PRODUCTION (kg ha^{-1} year^{-1}) IN CERTAIN TRADITIONAL SMALL- AND INTERMEDIATE-SIZED PRAWN FARMS BETWEEN AND INCLUDING THE YEARS 1974 TO 1981

Production year[a]	Farm size (ha)			
	0.405	0.405—0.809	0.809	Av
1	1915	1700	1396	1670
	(21)	(24)	(17)	(62)
2	2152	2206	1647	2002
	(18)	(22)	(15)	(55)
3+	2159	2093	1988	2080
	(36)	(39)	(25)	(100)
Average	2075	2000	1677	1917
	(75)	(85)	(57)	(217)

Note: Data grouped by "production" (not calendar) year[a] and by farm size. Numbers in parenthesis denote number of farms in data set.

[a] Data obtained from computer data base (Prawn Aquaculture Research Program) compiled from monthly harvest reports from farms under cooperative agreement with the Anuenue Fisheries Research Center.[81] The first production year for a pond was defined as the first 12-month period beginning with the date of the first reported harvest. If there were less than 6 months of production reported within the 12-month period, the pond's data was not entered into the calculations. If there were less than 12 but more than 6 months of production reported, the production was prorated for 12 months. The second production year was defined as the 12-month period beginning at the end of the first production year. Subsequent production years were defined in like manner.

levels, and studies[28] show that prawn farms can be economically viable with production levels below those originally obtained by Fujimura.

Practices

Pond culture practices in Hawaii have been described (see preceding footnote on Fujimura)[21-29,82-85] and technical information can be assembled in great detail from these sources. This section will, therefore, summarize "hypothetical" traditional prawn farming practices in Hawaii.

Pond Construction

Traditional rectangular pond design specifications are featured in detail by Shang.[28] Figure 5 (from Shang[28]) illustrates cross-section views of a traditional type prawn pond. Ponds are generally rectangular with a depth of about 0.93 m at the shallow end sloping to 1.09 m at the deeper end. Overflow pipes or sluice gates (maka'ha' in Hawaiian) are located at the deep end. The "monk" drainage system described for fish culture[19] is not used. Berms between ponds are designed to accommodate vehicular traffic — typically a 3.10 m berm would be built for a one 0.41-ha (1 acre) pond. Berms should be at least 0.31 m above the water surface. Ponds generally have a 3:1 slope but this varies, especially among the smaller farm ponds which are constructed according to the local topography ("lay of the land"). Shang[28] has provided details on the volume and costs of soil movement in pond construction. Figure 6 shows the layout of a typical medium-sized prawn farm utilizing nursery ponds.

Cross-Sectional View of Embankments

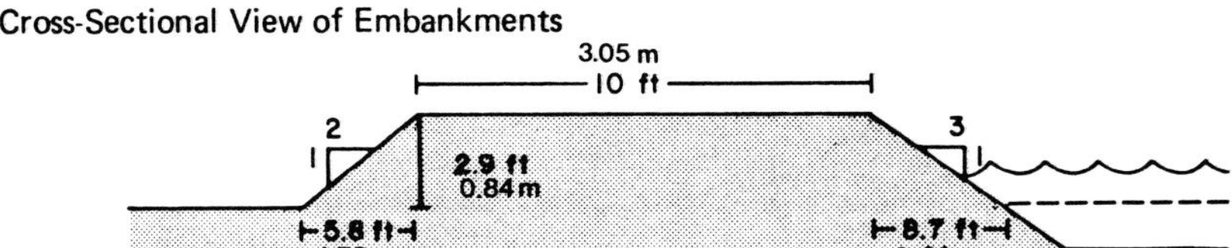

Cross-Sectional View of the Width

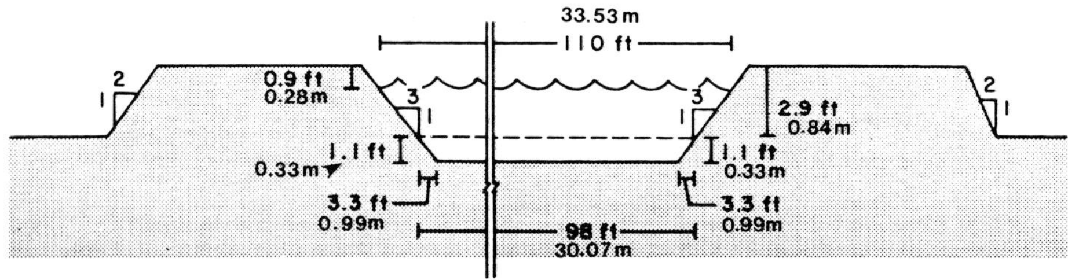

Cross-Sectional View of the Length

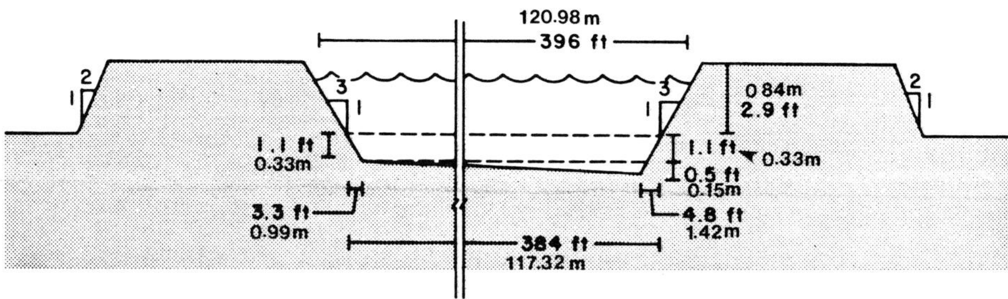

FIGURE 5. Cross-sectional dimensions of a conventional 0.40-ha (1-acre) prawn pond. Metric dimensions added. (Reproduced from Shang, Y. C., Fresh-water Prawn *(Macrobrachium rosenbergii)* Production in Hawaii: Practices and Economics, UNIHI-SEAGRANT-R-81-07, Sea Grant Miscellaneous Report, 1981. With permission.)

Pond Stocking

PLs are stocked directly from the hatchery into the production ponds at a rate which varies around 16 PL m^{-2} (1.5 ft^{-2}) to allow for a 50% mortality and pond standing crop population census of 11 ind. m^{-2}. However, 72-hr stocking mortality is highly variable from pond to pond[91] so some farms practice a post stocking "watch" by stocking the postlarvae in areas of the pond sectioned off by nets and then releasing the PLs after a day or so after visual inspection has been made of abundance and overall health. A count is not made. This method is not as efficient or effective as the 72-hr bag test described below.

Until now pond stocking in the traditional mode has been to directly place postlarvae from hatcheries into ponds. However, on-farm experience and empirical data[91,92,94] have shown that some prestocking acclimation period is needed. Culturists are urged to be conservative in their hatchery-to-pond transfer practices — PLs should be acclimated to fresh water in the hatchery and transferred to an intermediate (not necessarily nursery stage as described below) holding phase for 1 week or so and then stocked in ponds with pH and temperature acclimitization. PLs should not be stocked in water that is less than 20°C — probably a safer lever would be 22°C.

Poststocking predation can be a problem and, although empirical data is lacking, casual

FIGURE 6. Intermediate-sized (10.9 ha) commercial prawn farm in Hawaii (no. 3, Table 1; Figure 1) showing small, 0.016-ha ponds (foreground) used for high density nursery culture. Ponds are also part of experimental pond complex of the University of Hawaii Prawn Aquaculture Program (no. 12, Table 1; Figure 1).

observations have demonstrated that carnivorous fishes, frogs, insect instars, birds, and even larger prawns can prey on newly stocked postlarvae. Mosquito fish should not be placed in ponds contrary to original recommendations (see preceding footnote on Fujimura).

In nature, adaptation to postlarval mortality is through high fecundity whose cost is subsumed in the overall species survival. Under culture, PL mortality ''cost'' is in ''cash'' and is assumed by either the farmer or government subsidy. Attention has been called to postlarval costs in studies done in temperate climates[93-102] because, at first glance, the importance of these costs are more acute in these areas. Indeed, PLs have to be nursed in intensive systems to compress the growing season into the confines of a temperate climate spring and winter. In Hawaii, PL biology, husbandry, and costs have not been emphasized until recently. To some degree, the concept that ''PLs are free'' developed, perhaps because the traditional system was instituted by a government program. Also, Hawaii has been considered tropical with a year-round growing season. The latter is true when compared to temperate climates but growth and stocking conditions are by no means constant throughout the year. Indeed, Tables 2 and 3 show the temperature fluctuations possible over a year's time. Several of these temperatures would not be conducive for PL stocking or optimum growth.

Juvenile to Adult Growth, Inventory Control

There are a few published studies that have given growth and production vs. time data for prawn farming in Hawaii (see preceding footnote on Fujimura).[85,93-95] Overall, these studies show that the traditional system is designed to manage a prawn population to reach, and remain at, a certain standing crop sometime after the second stocking and a series of seine and cull harvests. Precise estimates of standing crop for Hawaiian ponds are not available simply due to the fact that ponds are never drained and enumerated. Indirect estimates of 200 gm^{-2} [85] and 270 gm^{-2} [93,94] are probably upper estimates. Malecha et al.,[65] using data from other studies[88-90] as well as their own, estimated the standing crop for one season's growth under monoculture to be between 145 and 207 gm^{-2}, or about 1700 kg ha^{-1} (1521 lb $acre^{-1}$).

Total pond growth rates under the traditional system have no meaning once the population has reached the standing crop; one must, therefore, focus on the growth rate of the stocking cohorts (the modal classes shown in Figure 3). Three studies have developed simulation models to predict pond growth and harvest yield.[85,93-95]

In the traditional system, the population profiles (census, size frequency distribution, and standing crop) and economic forecasting derived can only be crudely estimated from periodically sampling the pond. Because of this, precise economic forecasting is not possible. In Hawaii, sampling practice is scattered. Some farms, usually the larger ones and those that are not a second income generator, maintain a routine, constant sampling procedure so that economic forecasting can be accomplished. Sampling is often done with a 4-mm mesh net (dimensions: 1.22 m high and 27.45 m long). Usually, one end is anchored to one bank, the other extended out to the full net length, and then brought around in a semicircle to another location on the bank. Pond corners can, on occasion, be sectioned off with the procedure. A number of ''passes'', usually three, are conducted for each pond. Quite obviously there are several drawbacks to this procedure: (1) although the mesh size is small enough to capture juveniles, in-pond conditions prevent an assessment of small-size classes; (2) the procedure is not a random sample of the population — many size classes go undetected; and (3) the procedure cannot be used for accurate numerical estimation. Because of the latter, a prawn farmer practicing the traditional ''stock 'em and seine 'em'' system never really knows the pond inventory. This is a serious drawback to the traditional system in larger farms where standard economic forecasting is an economic necessity. Smaller, backyard, second income farmers do not need, nor do they attempt, to achieve tight control or knowledge of their prawn inventory.

Harvesting

Harvesting is done by seining under the traditional system. The seine net used throughout the Hawaiian industry has been described in detail.[22] Basically, it is a monofilament double-knotted mesh with a 5.1-cm ''eye'' that is designed to retain animals greater than 10.5-cm orbit length (~30 g). In practice, smaller animals unable to escape around or through the advancing net because of congestion are also retained. The float line (top border) is made from polypropylene, the ''sinker line'' (bottom border) from nylon.

The net position in the pond during a hypothetical (but typical) harvesting operation is shown in Figures 7, 8A and 8C. The practice is to secure one end of the seine net (wing section) on the bank and slowly work the leading edge (other wing section) along the pond bank. Eventually the entire net is stretched across the pond width and then slowly pulled to one end. In some cases, the anchored end stays put, in others it is moved along the pond banks corresponding with the movement of the other end. Eventually, the two free wing ends meet and both wing sections are pulled on shore. Care is taken to keep the sinker line on the bottom. The final step is ''bagging'' (Figure 7, frame 9), where the netted prawns are confined to the bag portion, removed using dip nets (Figure 8C), and transferred to a live haul transport tank. This usually consists of a 1135- to 1890-ℓ tank equipped with aeration located on the back of a truck. In some cases, two or three wire mesh baskets are placed in the tank for easy prawn removal at market destination. Live haul transports can be equipped with ''stand alone'' diesel or gasoline-powered aerators, back-up oxygen ''bottles'', and be located on a trailer hitched to tractors or trucks. Small farmers use the tank-on-a-truck system; larger farms use live haul systems on trailers. Almost all farms, regardless of size, use some form of live haul to transport live prawns to markets, to processing facilities, or to another pond in multiphased culture (see later). A scale able to handle up to approximately 136 kg can be attached to the live haul and is utilized to weigh ''net fulls'' of prawns as they are netted from the seine net bag.

The seine harvesting operation in the traditional system is very labor intensive, slow, and inefficient. Overall, it takes four persons about 1 hour to harvest a 1/2-ha (~1 acre) pond

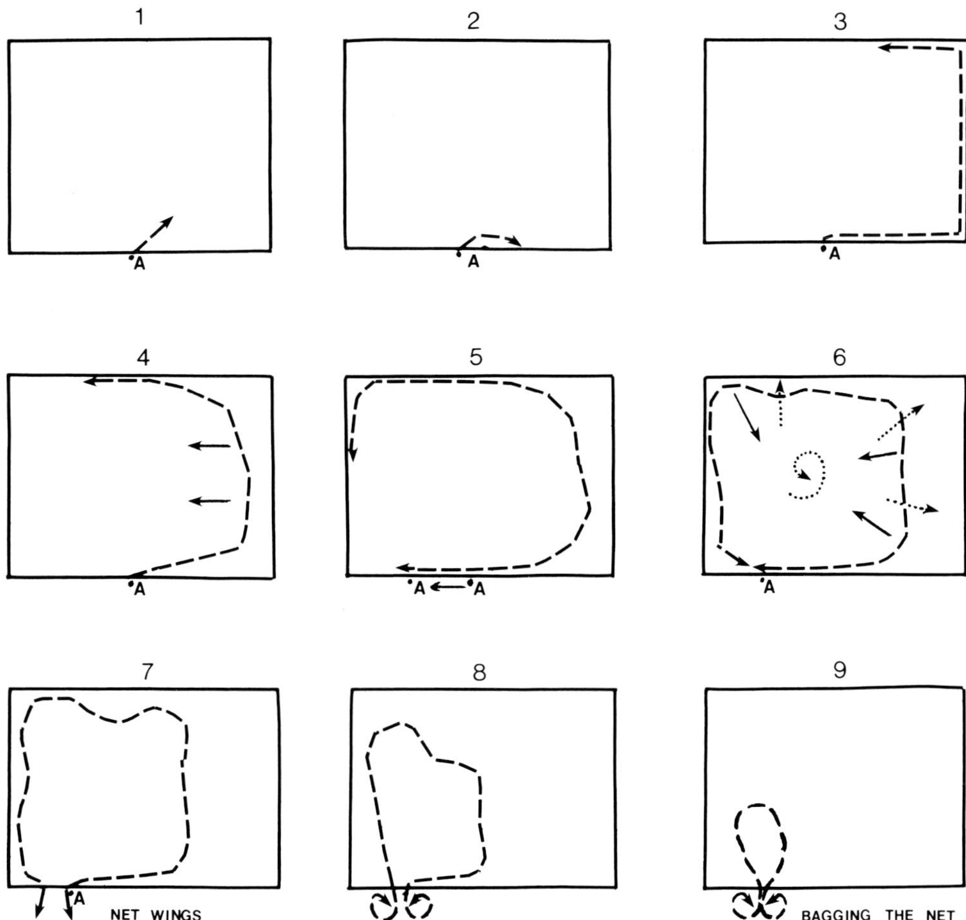

FIGURE 7. Stylized representation of the position of seine net during a traditional "seine and cull" harvest operation. Frame 4: solid arrows indicate direction of net advance. Frame 4 is equivalent to A in Figure 8; frame 9 (bagging) is equivalent to frame C in Figure 8. A = bank "anchor" position for stationary wing which can be moved (frame 5) to a new location. In frame 6 dotted lines indicate escaping under harvest-sized prawns; solid lines indicate harvest-sized animals. Not drawn to scale.

including the time the truck containing the harvest crew and net pulls up to the pond side until the time the truck leaves. Harvest efficiency is variable making the operation more of a "hunt", in the sense of a capture fishery, than a complete harvest of all new growth. This can be seen in Figure 3. The shaded mode contains animals larger than 10 cm (orbit length). Complete harvest efficiency should have captured all animals when the ponds were first harvested but 16 to 18 months of harvesting were needed to remove all the market-sized animals from the two ponds illustrated in the figure. Size frequency distribution profiles for the last 2 months are shown. If we add the 9 months between stocking and final harvest, then some animals were in the pond for 27 months! As pointed out previously, the degree of overlap between the stocking cohorts is not known — the shaded and nonshaded areas represent the author's (see preceding footnote on Fujimura) interpretation of differences between the stocking cohorts. An actual relationship between the cohorts is more like that shown in Figure 4 from Polovina and Brown.[85]

The Nontraditional System

Background

Some new farms have instituted new designs in their physical plant and new stock man-

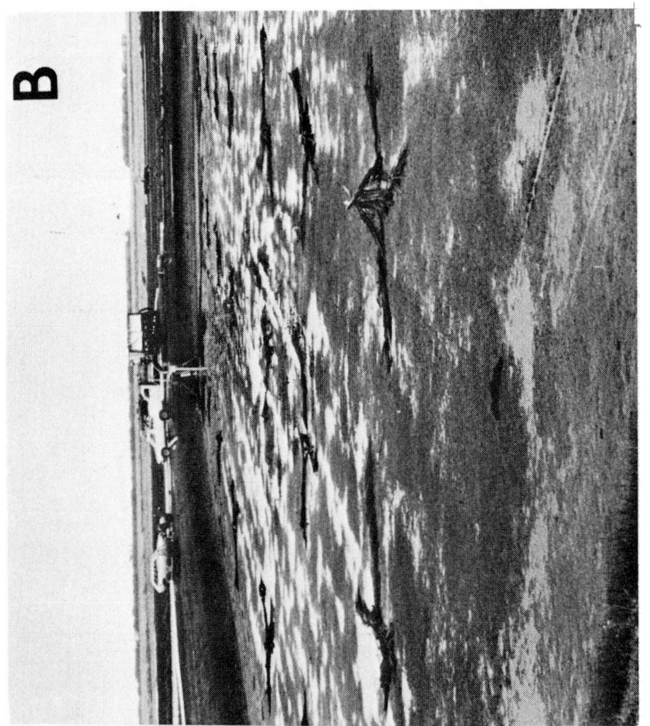

FIGURE 8. Comparison of analogous traditional (A, C) and nontraditional (B, D) operations in mid (A, B) and terminal (C, D) stages of harvesting in commercial prawn farms in Hawaii. (A) Seine and cull method of animal capture, (B) pond draindown method (pond drained away from camera to sump area near trucks, (C) "bagging" the seine net containing harvest-sized animals (see Figure 6, frame 9); (D) sump area in draindown harvest. PL = pump line, BFL = back flush line, S = screen between sump area and outflow, DT = deceleration tower (Figure 9), P = pump, SU = sump, SA = sump area (see Figure 10). Operator is "working" back flow line to flush additional water into sump area to keep animals in "suspension" so they can be pumped and/or to keep them from hanging up in the mud beyond water line not drawn in figure.

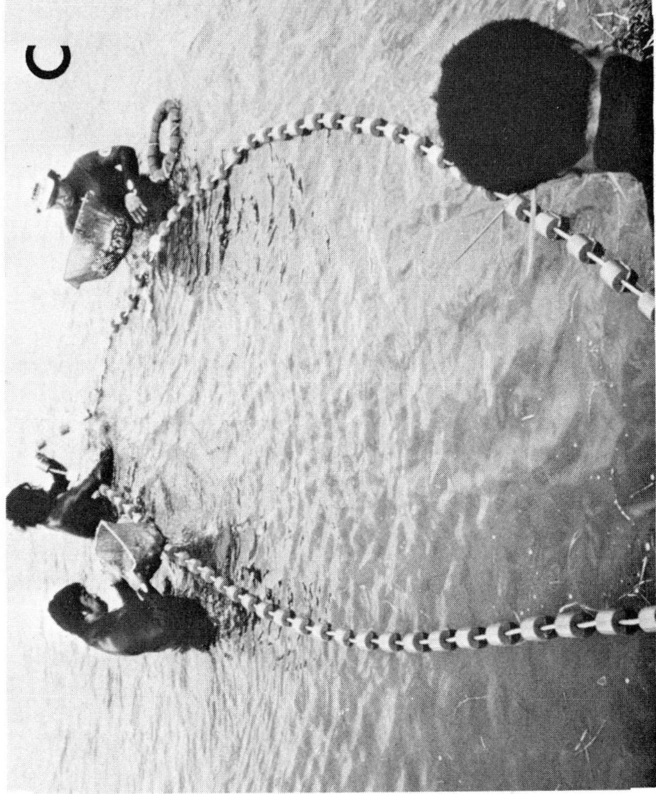

FIGURE 8C.

FIGURE 8D.

agement procedures as the result of their own recognition of the drawbacks of the traditional system and the alternatives pointed out.[84] In addition, traditional farms have instituted new "nontraditional" components, notably a nursery, into their production schemes.

Nursery

A nursery phase has no direct climatic necessity in Hawaii, the necessity for, and feasibility of, nursery grow-out systems have been discussed for temperate areas,[96-100] as well as tropical ones.[101,102] The fact that PL survival among commercial prawn ponds is variable[91] and the need for inventory control (accountability) in economic planning stimulated the present use by three commercial farms of a nursery phase in at least part of their operations. Since laboratory studies[96-102] show that it is feasible to grow PLs stocked at rates of up to 100 ind. m^{-2} for 60 days in nursery systems, one farm is developing high density (>700 m^{-2}) nursery culture in very small ponds (150 m^2) constructed specifically for this purpose (Figure 6). On other farms the nursery ponds are slightly smaller (0.8 to 1.4 ha) than the production ponds. Density varies between 100 and 200 ind. ha^{-1} with final survival averaging 80%.[80]

Pond Draindown and Batch Harvesting

Two farms presently practice pond draining to remove stock from either nursery ponds or intermediate grow-out ponds (Figures 8B and 8D). In both cases, pond water is drained down to a corner or section of the pond containing a depression or "sump". In the case illustrated in Figure 8 the prawns (in heavy concentration in the sump and the area surrounding it, hereafter referred to as the "sump area") are pumped from the sump into a live haul transport. The pump, illustrated in Figure 8D, is basically a vaneless food or fish pump. Mortality is minimal, but there are several practical considerations to draining and transferring prawns in a pond draindown. First, ponds have to be specifically constructed to facilitate efficient water drainage and concentration of prawns in sump and surrounding area; second, heavy concentrations of prawns during the latter phases cause increased oxygen demand in the water mass; third, the relatively thin layer of water that is in the sump area will heat up rapidly in direct sunlight causing heat stress and increased oxygen demand; fourth, prawns can "hang up" to some degree in soft mud in the sump area.

All these considerations can be managed. Since many fish species are harvested by the draindown batch system, prawn pond construction can be similar to those used for these species.[19] Hypoxia and thermal stress can be avoided by rapid operation in the draining process, conducting the operation at hours with minimal solar radiation or on cloudy days, and "backflushing" water from a point source near the sump itself into the surrounding area. This not only serves to reduce oxygen demand per water volume unit but also to free prawns that are hung up in the mud. The latter can also be accomplished by pumping water from in front of the sump area towards the sump itself from which there is an outflow. The effect achieved is a temporary flow-through system in the sump and sump area during the pumping of prawns to the live haul. Flow patterns of incoming water are regulated to keep the sump area resident prawn population in suspension and readily capable of coming into the immediate vicinity of the pump intake pipe (Figures 8D and 9).

Draindown operators try to establish a certain ratio of water-to-prawn volume and number and a pattern distribution in order to achieve a "cushioning" effect in the pump and its intake and outflow pipes (Figure 10 shows three highly stylized cases).

The live haul transport used in draindown harvests are modified with a decceleration tower to reduce the velocity of the water entering the live haul. This device is basically an inverted triangular shaped box with the apex fitted to the pump outflow. Water and prawns spill over the top and into the live haul or grader (Figure 11).

Usually subsamples are taken from the live haul to assess mean size, size frequency distribution, and biomass of the prawns removed by draindown. This is an important operation

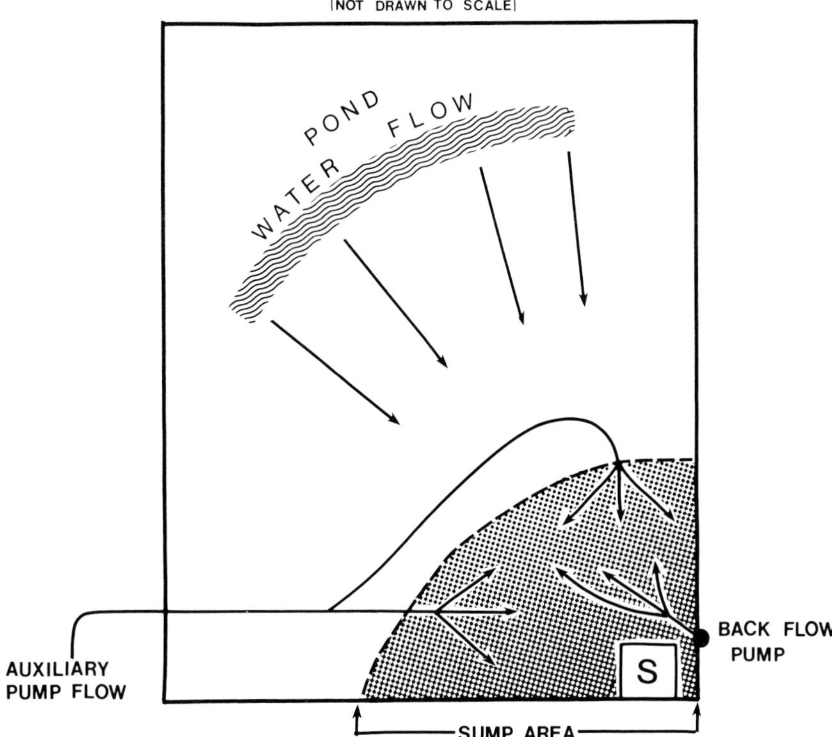

FIGURE 9. Water flow patterns in hypothetical drain harvest. S = sump. Compare this figure with frames C and D in Figure 8. Auxiliary pump flow used to free animals "hung up" in mud on sump area periphery. Back flush pump flow also aids in this regard and keeps water volume in sump area at certain level.

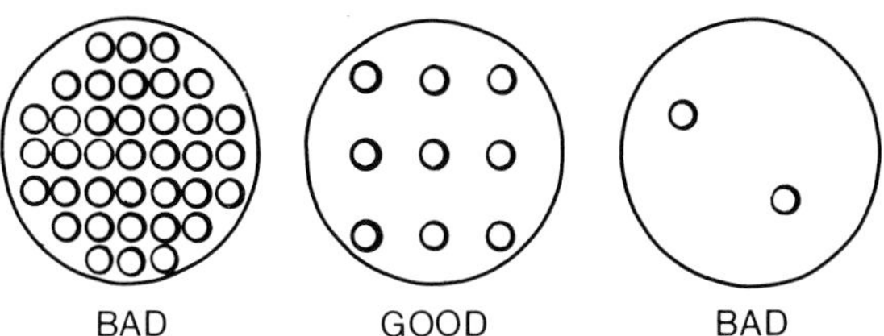

FIGURE 10. Diagrammatic representation of a cross section of a pump line in drain harvest operation illustrating the need to maintain a certain density and distribution of prawns (open circles) to achieve a lateral cushioning effect as prawns are pumped from sump into deceleration tower.

in inventory control. One farm can grade prawns into three size classes before transfer to the live haul.

Multiphased Grow-out Systems

Malecha et al.[84] have pointed out the advantages of phased grow-out systems in prawn aquaculture and similar systems have been shown to be effective in fish culture.[103,104] Two farms practice multiple phase grow-out (Figure 2). One farm uses a two-phase (hatchery → nursery → grow-out) system and the other farm is investigating both a three-phase and four-phase (hatchery → nursery → grow-out I → grow-out II → grow-out III). Animals are harvested from the nursery and grow-out ponds by draindown.

FIGURE 11. Top view of water flow from deceleration tower connected to pump shown in Figure 8. Prawns ''spill'' through tower opening either to holding box as shown and then to live haul or directly to live haul, or into size grader and then to a live haul.

A main purpose of a multiphase system is to establish inventory control so that precise economic forecasting can be done. In addition, management practices such as those involving feeding rates can be more precisely adjusted to the actual standing crop when the latter is known with some precision. Size grading and sex ratio management would be impossible without a multiphased system.[84] Lastly, data based management systems, especially those using on-farm microcomputers,*[105] can easily be implemented in such a system. Transfer mortality does not appear to be a problem; the actual movement of prawns, however rapid, through the pump system, has a minimal effect per se on mortality since *M. rosenbergii* is naturally adapted to such water movement in nature. Thermal and anoxia stress, obviously, should be avoided. There appears to be an actual reduction in overall mortality in the phased system although this remains to be experimentally confirmed.[80]

The last grow-out phase (Figure 3) is harvested largely by the traditional seine and cull method. However, the ponds can be drained, inventoried, and harvested at any time so that the full advantageous tradition of the seine and cull method (uniform harvest size) can be realized without the full impact of the disadvantages (long, unmanaged prawn resident time, poor or no inventory control, and reduced production due to intrapopulational behavioral interaction[84]). Although it is too soon to say for sure (May 1982), the combination of seine and cull and batch harvest-transfer by draindown may prove to be the best pond management strategy especially for intermediate and corporate level prawn farms in areas with year-round growing seasons.

Other Innovations

The 72-Hr Bag Test

Our group[91] has shown that postlarval (PL) stocking mortality within 72 hr is highly

* The Prawn Aquaculture Research Program, University of Hawaii, is developing a microcomputer data based management system patterned after a system developed for its own research for a multiphased farm in Hawaii.

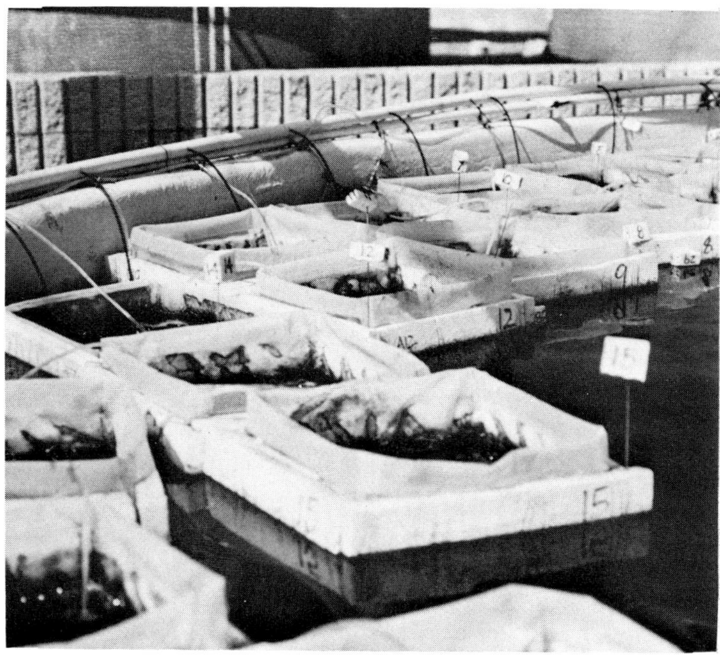

FIGURE 12. "Floating bags" of the type used in ponds to conduct the "72-hr bag test" (see text). Bags in figure are floating in a large tank as part of another study.

variable among outwardly normal prawn ponds. Consequently, it is recommended that initial mortality be assessed in all PL and juvenile pond stockings. This can be done in what has come to be called in Hawaii the "72-hr bag test". The test bags (60 × 60 × 90 cm, 324 ℓ) are illustrated in Figure 12. They are made of some small mesh (window screen is cheap but not durable; nitex is expensive but durable) and floated (anchored) in the pond. About 300 PLs are added to the bags, fed daily, and counted at the end of 72 hr. Simultaneously with this, stocklings are put into the ponds and their expected production can be adjusted by the bag mortality. It has been estimated[80] that between 5 and 10% of overall mortality in any one phase is due to acute stocking mortality. The bag test does not assess predator-related mortality since predators are kept out of the bags.

Mechanical Harvesting

Williamson and Wang[107] have described a mechanical harvesting system that has been designed and tested under commercial conditions. This system consists of modifications to the traditional seine haul net described above and by Fujimoto et al.[22] and uses a tractor or truck in place of humans to pull the net through the water column. Harvesting efficiency is at least as good as "hand" harvesting[107] and, of course, the seining time is shorter due to the power assistance. However, the "bag time" remains the same, i.e., once the seine is bagged (Figures 7, frame 9, and 8C), it requires the same amount of time to remove animals to the live haul in both the mechanical and hand harvesting methods. Three Hawaii farms utilize mechanical harvesting to some degree.

Pond Design

Designs for the nontraditional pond systems must accommodate draindown and/or ease of mechanical harvesting. Draindown has been accommodated by designing ponds with bottoms which slope to sump areas similar to those used in fish culture.[17-19] Gibson and

Wang[108] have argued for the use of ''canal'' type ponds to ease management and harvesting. Indeed, in Hawaii the ratio of bank dimension to water volume has been decreasing as new ponds are constructed. This indicates that ponds are becoming more narrow and rectangular rather than square.

SUMMARY

1. Site selection and water quality management for prawn ponds in Hawaii follow criteria developed for warm water fish culture.
2. Hawaiian pond management practices fall into the traditional ''stock-'em-and-seine-'em'' and nontraditional ''multiphased'' systems.
3. The traditional system is confined to small and intermediate farms and utilizes only a hatchery and grow-out phase whereby PLs are stocked at least once a year and harvesting is done by the ''seine and cull'' method.
4. Nontraditional systems are used on larger farms consisting of at least a two-phased grow-out system (hatchery → nursery → grow-out) and involve pond draindown between phases.
5. In the multiphased systems traditional seine and cull methods are practiced in the terminal phase so that the advantages of both systems are realized.
6. The traditional system is advantaged by harvested product size uniformity and disadvantaged by poor inventory control and long prawn resident time which may reduce expected production.
7. The multiphased system is advantaged by more precise inventory control, a *sine qua non* for accurate economic forecasting, and possibly higher overall survival rates.
8. Nursery systems, both high and intermediate density, are being incorporated into both traditional and nontraditional systems in order to improve PL survival and to improve grow-out inventory control.

CONCLUSION

Small farms (Table 1) will stay traditional since it is economically feasible to do so. Large, corporate level farms and some intermediate-sized farms will evolve into a combination of nontraditional ''multiphased'' and traditional ''seine and cull'' systems to utilize the advantages of both. The economic picture should continue to improve as culturists become more and more familiar with techniques which reduce net costs. In this regard, tighter control of product inventory will be exercised and economic forecasting will be utilized more and more, especially those based on on-farm computerized data based management systems. The use of nursery systems should continue to rise as culturists try to improve their PL survivability and control their grow-out inventories. Improvements in harvesting, grading, feeds, and feeding[112] will continue and serve as short-term economic inducement, but ultimately prawn production will be optimized only by changing the prawn's biological requirements through domestication.[2,110,111,114]

ACKNOWLEDGMENTS

Sincere thanks go to Amy Dicksion and Deb Ciampa for assistance in preparation of this manuscript. Also the graphics assistance of Deborah Ishihara is gratefully acknowledged. Thanks go to the following prawn farmers who kindly provided information for this report: R. Hanohano, S. and C. Katsutani, A. Rietow, R. Santos, L. Wong; R. Gibson of Amfac Aquatech; A. Kuljis, C. Greenwald, and E. Scura of Aquatic Farms; R. Kido of Brigham

Young University Aquaculture Farm; and J. Peterson, J. Murphy, and J. Wallace of Amoriente Aquafarms, Inc. Also, M. Fujimoto, D. Onizuka, T. Iwai, F. Oishi, and D. Sumikawa of the Anuenue Fisheries Research Center, Division of Aquatic Resources, Department of Land and Natural Resources, State of Hawaii, provided information. Lastly, thanks go to Scott Masuno who prepared the data for Tables 2, 3, 5, and 6 and to Rick Gibson of Amfac Aquatech for permission to use the pictures of drain harvesting. Finally the expert assistance of Dr. Gideon Hulata in correcting the galley proofs is gratefully acknowledged.

Preparation of this manuscript was funded in part by the Aquaculture Development Program, State of Hawaii; by Prawn Tech., 41-052 Hihimanu Street, Waimanalo, Hawaii 96795; and by the University of Hawaii Sea Grant College Program under Institutional Grant No. NA-81-AA-D-00070. This is Sea Grant Publication No. UNIHI-Sea Grant-MC-82-02.

REFERENCES

1. **Malecha, S. R.,** Genetics and selective breeding of *Macrobrachium rosenbergii,* in *Shrimp and Prawn Farming in the Western Hemisphere,* Hanson, J. A. and Goodwin, H. L., Eds., Dowden, Hutchinson and Ross, Stroudsburg, Pa., 1977, chap. 14.
2. **Malecha, S. R.,** Development and general characterization of genetic stocks of *Macrobrachium rosenbergii* and their hybrids for domestication, *Univ. Hawaii Sea Grant Q. News,* 2(4), 1980.
3. **Hedgecock, D., Stelmach, D. J., Nelson, K., Lindenfelser, M. E., and Malecha, S. R.,** Genetic divergence and biogeography of natural populations of *Macrobrachium rosenbergii, Proc. World Maricul. Soc.,* 10, 873, 1979.
4. **Johnson, D. S.,** Sub-specific and infra-specific variation in some freshwater prawns of the Indo-Pacific region, in *Proc. Centenary and Bicentenary Congress in Biology,* Purchon, R. D., Ed., Singapore, 1960, 259.
5. **Miyajima, L. S.,** About *Macrobrachium* species, in *Shrimp and Prawn Farming in the Western Hemisphere,* Hanson, J. A. and Goodwin, H. L., Eds., Dowden, Hutchinson and Ross, Stroudsburg, Pa., 1977, chap. 2.
6. **John, M. C.,** Bionomics and life history of *Macrobrachium rosenbergii* (de Man), *Bull. Cent. Res. Inst. Univ. Kerala Ser. C (Nat. Sci.),* 5(1), 93, 1957.
7. **Ling, S. W. and Merican, A. B. O.,** Notes on the life and habits of the adults and larval stages of *Macrobrachium rosenbergii* (de Man), in *Proc. Indo-Pac. Fish. Counc.,* 9(2), 55, 1961.
8. **George, M. J.,** Genus *Macrobrachium* Bate 1868, *Cent. Mar. Fish. Res. Inst. Bull.,* 14, 179, 1969.
9. **Ling, S. W.,** The general biology and development of *Macrobrachium rosenbergii* (de Man), *F.A.O. Fish. Rep.,* 3(57), 589, 1969.
10. **Wickens, J. F. and Beard, T. W.,** Observations on the breeding and growth of the giant freshwater prawn *Macrobrachium rosenbergii* (de Man) in the laboratory, *Aquaculture,* 3, 159, 1974.
11. **Wickens, J. F.,** Prawn biology and culture, *Oceanogr. Mar. Biol. Rev.,* 14, 435, 1976.
12. **Ling, S. W. and Costello, T. J.,** The culture of freshwater prawns: a review, in *Advances in Aquaculture,* Pillary, T. V. R. and Dill, W. A., Eds., Fishing News Books, Farnham, Surrey, England, 1976, 299.
13. **Forster, J. R. M. and Beard, T. W.,** Experiments to assess the suitability of nine species of prawns for intensive cultivation, *Aquaculture,* 3, 355, 1974.
14. **New, M. B., Ed.,** *Giant Prawn Farming,* Elsevier, New York, 1982.
15. **New, M. B. and Singholka, S.,** *A Field Manual for the Culture of Macrobrachium rosenbergii,* Food and Agriculture Organization, Rome, Italy, in press.
16. **Malecha, S. R.,** *The Biology and Culture of the Freshwater Prawn, Macrobrachium rosenbergii,* in preparation.
17. **Huet, M.,** *Textbook of Fish Culture,* Fishing News Books, Farnham, Surrey, England, 1979.
18. **Stickney, R. R.,** *Principles of Warmwater Aquaculture,* John Wiley & Sons, New York, 1979.
19. **Hepher, B. and Pruginin, Y.,** *Commercial Fish Farming with Special Reference to Fish Culture in Israel,* John Wiley & Sons, New York, 1981.
20. **Pillay, T. V. R., Ed.,** Proc. F.A.O. world symposium on warm water fish culture, *F.A.O. Fish. Rep.,* 3 and 4 (44), 1967.

21. **Fujimura, T. and Okamoto, H.,** Notes on progress made in developing a mass culturing technique for *Macrobrachium rosenbergii* in Hawaii, in *Coastal Aquaculture in the Indo-Pacific Region,* Pillay, T. V. R., Ed., Fishing News Books, Farnham, Surrey, England 1972, 313.
22. **Fujimoto, M., Fujimura, T., and Kato, K.,** Pond growout systems, in *Shrimp and Prawn Farming in the Western Hemisphere,* Hanson, J. A. and Goodwin, H. L., Eds., Dowden, Hutchinson and Ross, Stroudsburg, Pa., 1977, chap. 6.
23. **Malecha, S. R.,** Development of prawn aquaculture in Hawaii, *Commer. Fish Farmer,* March 14, 1978.
24. **Malecha, S. R.,** Aquaculture of the Freshwater Prawn, *Macrobrachium rosenbergii,* in Hawaii: History, Present Status, and Application to Other Areas, presented at 1st Brazilian Symp. on Aquaculture, Recife, Brazil, July 24 to 28, 1978.
25. **Malecha, S. R.,** Research and Development in Freshwater Prawn *(Macrobrachium rosenbergii)* Culture in the U.S.: Current Status and Biological Constraints with Emphasis on Breeding and Domestication, presented at 9th Joint Meeting U.S./Japan Aquaculture Panel, Kyoto, Japan, May 26 and 27, 1980.
26. **Shang, Y. C.,** Economic Feasibility of Freshwater Prawn Farming in Hawaii, University of Hawaii, Economic Research Center, 1974.
27. **Shang, Y. C. and Fujimura, T.,** The production economics of freshwater prawn (*Macrobrachium rosenbergii*) farming in Hawaii, *Aquaculture,* 11, 99, 1977.
28. **Shang, Y. C.,** Freshwater Prawn *(Macrobrachium rosenbergii)* Production in Hawaii: Practices and Economics, UNIHI-SEAGRANT-MR-81-07, Sea Grant Miscellaneous Report, Honolulu, 1981.
29. **Lee, S. R.,** The Hawaiian Prawn Industry: A Profile, Aquaculture Development Program, Department of Planning and Economic Development, State of Hawaii, Honolulu, 1979.
30. **Roberts, K. J. and Bauer, L. L.,** Costs and returns for *Macrobrachium* grow-out in South Carolina, U.S.A., *Aquaculture,* 15, 383, 1978.
31. **Brody, T., Cohen, D., Barnes, A., and Spector, A.,** Yield characteristics of the prawn *M. rosenbergii* in temperate zone aquaculture, *Aquaculture,* 21, 375, 1980.
32. **Smith, T. I. J., Sandifer, P. A., Jenkins, W. E., and Stokes, A. D.,** Effect of population structure at stocking and density on production and economic potential of prawn *(M. rosenbergii)* farming in temperate climates, in *Proc. World Maricul. Soc.,* 12, 1982, in press.
33. **Vejkaran, K. and Uraironk, P.,** Economic analysis of freshwater prawn farming in Thailand, in *Giant Prawn 1980,* International Foundation for Science, ISBN: 91-857, 98-08-8, Stockholm, Sweden, 1980.
34. **Shang, Y. C.,** Comparison of freshwater prawn farming in Hawaii and Thailand: culture practices and economics, *Proc. World Maricul. Soc.,* 12, 1982, in press.
35. Soil Conservation Service, Ponds for Water Supply and Recreation, Agriculture Handbook 387, Washington, D.C., 1971.
36. **Boyd, C. E. and Lichtoppler, F.,** Water Quality Management in Pond Fish Culture, R & D Series No. 22, Agriculture Exp. Station, Auburn University, Auburn, Ala., 1979.
37. **Avault, J., Jr.,** More on oxygen depletion, *Aquaculture Mag.,* 5(6), 33, 1979.
38. **Conte, F. S.,** Oxygen and Water — Emergency Aeration, California Aquaculture Newsletter, Cooperative Extension Service and Sea Grant College Program, University of California, Berkeley, February 1981.
39. **Boyd, C. E.,** The chemical oxygen demand of waters and biological materials from ponds, *Trans. Am. Fish. Soc.,* 102(3), 606, 1973.
40. **Boyd, C. E., Romaire, R. P., and Johnston, E.,** Predicting early morning dissolved oxygen concentration in channel catfish ponds, *Trans. Am. Fish. Soc.,* 107(3), 484, 1978.
41. **Boyd, C. E. and Tucker, C. S.,** Emergency aeration of fish ponds, *Trans. Am. Fish. Soc.,* 108, 299, 1979.
42. **Romaire, R. P. and Boyd, C. E.,** Effects of solar radiation on the dynamics of dissolved oxygen in channel catfish ponds, *Trans. Am. Fish. Soc.,* 108, 473, 1979.
43. **Boyd, C. E.,** *Water Quality in Warmwater Fish Ponds,* Auburn Experiment Station, Auburn University, Auburn, Ala., 1979.
44. **Boyd, C. E.,** *Water Quality Management for Pond Fish Culture,* Elsevier, New York, 1982.
45. **Iwai, T.,** A Preliminary Investigation on Oxygen Consumption of *Macrobrachium rosenbergii,* Working Paper No. 31, Sea Grant, University of Hawaii, May 1978.
46. **Stephenson, M. and Knight, A. W.,** The effect of temperature and salinity on oxygen consumption of post-larvae of *Macrobrachium rosenbergii* (de Man) (Crustacea: Palemonidae), *Comp. Biochem. Physiol.,* 67A, 699, 1980.
47. **Stephenson, M. J. and Knight, A. W.,** Growth, respiration and caloric content of larvae of the prawn, *Macrobrachium rosenbergii, Comp. Biochem. Physiol.,* 66A, 385, 1980.
48. **Mauro, N. and Malecha, S. R.,** Anoxia tolerance among geographic stocks of the freshwater prawn, *Macrobrachium rosenbergii,* 1982, in preparation.
49. **Mauro, N. and Malecha, S. R.,** The effects of hypoxia on acid base balance and lactate production in *Macrobrachium rosenbergii,* 1982, in preparation.

50. **Losordo, T.,** An Investigation of the Oxygen Demand Materials of the Water Column in Prawn Grow-Out Ponds, M.S. thesis, Department of Agriculture Engineering, University of Hawaii, Honolulu, 1980.
51. **Spotts, D. G.,** Low oxygen stress in the freshwater prawn, *Macrobrachium rosenbergii, Proc. World Maricul. Soc.,* 13, in review.
52. **Uno, Y., Bejie, A. B., and Igarashi, Y.,** Effects of temperature on the activity of *Macrobrachium rosenbergii* La Mer, *Bull. Soc. Fr. Jpn. Oceanogr.,* 13(3), 150, 1975.
53. **Crowell, S. K. and Nakamura, R. M.,** Effects of thermal stress on body weight in the freshwater prawn, *Macrobrachium rosenbergii, Proc. West. Sec. Am. Soc. An. Sci.,* 31, 31, 1980.
54. **Corwell, S. K.,** The Effects of Water Temperature on the Growth Rate of the Freshwater Prawn, *Macrobrachium rosenbergii,* M.S. thesis, University of Hawaii, Honolulu, 1981.
55. **Farmanfarmian, A. and Moore, A.,** Diseasonal thermal aquaculture. I. Effect of temperature and dissolved oxygen on survival and growth of *Macrobrachium rosenbergii, Proc. World Maricul. Soc.,* 9, 55, 1978.
56. **Cheng, F. F. C.,** Studies on Nutrition, Cannibalism and Cultural Management of *Macrobrachium rosenbergii,* (in Chinese with English tables and charts) Final Report to DAPD, National Taiwan College of Marine Science and Technology, August 8, 1979.
57. **Morita, S. K.,** An econometric model of prawn pond production, *Proc. World Maricul. Soc.,* 8, 741, 1977.
58. **Cripps, M. C. and Nakamura, R. M.,** Inhibition of growth of *Macrobrachium rosenbergii, Proc. World Maricul. Soc.,* 10, 575, 1979.
59. **Fieber, L. A. and Lutz, P. L.,** Calcium requirements for molting in *Macrobrachium rosenbergii, Proc. World Maricul. Soc.,* 13, 1982, in review.
60. **Iwai, T.,** A seven-month study of 24-hour fluctuations of selected water quality parameters of a prawn (*Macrobrachium rosenbergii*) rearing pond, Oahu, Hawaii, in *Shrimp and Prawn Farming in the Western Hemisphere,* Hanson, J. A. and Goodwin, H. L., Eds., Dowden, Hutchinson and Ross, Stroudsburg, Pa., 1977, 247.
61. **Leary, D. and Iwai, T.,** A Five Month Study of Selected Water Quality Parameters in Freshwater Prawn *(Macrobrachium rosenbergii)* Rearing Pond, Oahu, Hawaii, unpublished manuscript, Anuenue Fisheries Research Center, Area 4, Sand Island, Honolulu, 1975.
62. **Iwai, T.,** A Liminological Investigation of Selected Water Quality Parameters from a Freshwater Prawn *(Macrobrachium rosenbergii)* Rearing Pond, Oahu, Hawaii, unpublished manuscript, Anuenue Fisheries Research Center, Area 4, Sand Island, Honolulu, 1979.
63. **Colt, T., Mitchell, S., Tchobanoglous, G., and Knight, A.,** The Use and Potential of Aquatic Species for Freshwater Treatment. Appendix C: The Environmental Requirements of Crustaceans, Publication No. 65, California State Water Resources Control Board, Sacramento, Calif., 1979.
64. **Laws, E. and Malecha, S. R.,** Application of a nutrient-saturated growth model to phytoplankton management in freshwater prawn *(Macrobrachium rosenbergii)* ponds in Hawaii, *Aquaculture,* 24, 91, 1981.
65. **Pierce, B., Craven, D. B., and Laws, E. A.,** Water Quality Effects of Water Hyacinths and Fish Polycultures in Hawaiian Prawn *(Macrobrachium rosenbergii)* Aquaculture Ponds, unpublished manuscript, Prawn Aquaculture Research Program, University of Hawaii, Honolulu.
66. **Malecha, S. R., Buck, D. H., Baur, R. J., and Onizuka, D. R.,** Polyculture of the freshwater prawn, *M. rosenbergii,* Chinese and common carps in ponds enriched with swine manure. I. Initial trials, *Aquaculture,* 25, 101, 1981.
67. **Herbacek, J., Dvorakova, M., Korinek, V., and Prochozkova, L.,** Demonstration of the effect of the fish stock on the species composition of zooplankton and the intensity of metabolism of the whole plankton association, *Verh. Int. Verein. Limnol.,* 14, 192, 1961.
68. **Kitchell, J. F., Koonce, J. F., and Tennis, P. S.,** Phosphorus flux through fishes, *Verh. Int. Verein. Limnol.,* 19, 2478, 1975.
69. **Hurlbert, S. H., Zedler, S., and Fairbanks, D.,** Ecosystem alteration by mosquito fish (*Gambusia affinis*) predation, *Science,* 175, 639, 1972.
70. **Keen, W. H. and Gagliardi, J.,** Effect of brown bullheads on release of phosphorus in sediment and water systems, *Prog. Fish Cult.,* 43(4), 183, 1981.
71. **Andersson, G., Berggren, H., Cronberg, G., and Gelin, C.,** Effects of planktivorous and benthivorous fish on organisms and water chemistry in eutrophic lakes, *Hydrobiologia,* 59(1), 9, 1978.
72. **Lammarra, V. A.,** Digestive activities of carp as a major contributor to the nutrient loading of lakes, *Verh. Int. Verein. Limnol.,* 19, 2461, 1975.
73. **Sinha, V. R. P., Khan, H. A., Chakraborty, D. P., and Gupta, M. V.,** Preliminary observations on the nitrogen balance of some ponds under composite fish culture, *Aqua. Hung. (Szarvas),* 2, 105, 1980.
74. **Perdue, J. A. and Nakamura, R.,** The effect of salinity on the growth of *Macrobrachium rosenbergii, Proc. World Maricul. Soc.,* 7, 647, 1976.
75. **Sandifer, P. A., Hopkins, J. S., and Smith, T. I. J.,** Observations on salinity tolerance and osmoregulation in laboratory-reared *Macrobrachium rosenbergii* post larvae (Crustacea: Caridea), *Aquaculture,* 6, 103, 1975.

76. **Singh, T.,** The isosmotic concept in relation to the aquaculture of the giant prawn, *Macrobrachium rosenbergii, Aquaculture,* 20, 251, 1980.
77. **Chatry, M. and Huner, J. V.,** Growth and survival of *Macrobrachium rosenbergii* in brackish water (10—20 ppt) ponds, *Proc. World Maricul. Soc.,* 13, 1982, in review.
78. **Silverthorn, S. V. and Reese, A. M.,** Cold tolerance at three salinities in post-larval prawns, *Macrobrachium rosenbergii* (de Man), *Aquaculture,* 15, 249, 1978.
79. **Nelson, S. G., Armstrong, D. A., Knight, A. W., and Li, H. W.,** The effects of temperature and salinity on the metabolic rate of juvenile *Macrobrachium rosenbergii* (Crustacea: Palaemonidae), *Comp. Biochem. Physiol.,* 56A, 533, 1977.
80. **Gibson, R. T.,** Amfac-Aquatech, Hawaii Operation, P. O. Box 596, Kekaha, Kauai 96752, personal communication, 1982. (Note added in proof: The Amfac-Aquatech farm went out of business in late 1982.)
81. **Ling, S. W.,** The general biology and development of *Macrobrachium rosenbergii* (de Man), *F.A.O. Fish. Rep.,* 3(57), 589, 1969.
82. **Fujimura, T.,** Notes on the Development of a Practical Mass Culturing Technique on the Giant Prawn, *Macrobrachium rosenbergii,* IPFC/C66/WP47, Indo-Pacific Fish. Council, 12th session, Honolulu, Hawaii, 1966.
83. **Fujimura, T. and Okamoto, H.,** Notes on Progress Made in Developing a Mass Culturing Technique for *Macrobrachium rosenbergii* in Hawaii, Paper IPFC/C70/Sym 53, Proc. Indo-Pacific Fish. Council, 14th session, Bangkok, Thailand, 1970.
84. **Malecha, S. R., Polovina, J., and Moav, R.,** Multi-Stage Rotational Stocking and Harvesting System for Year-Round Culture of the Freshwater Prawn, *Macrobrachium rosenbergii,* University Hawaii Sea Grant Technical Report TR-81-01, Honolulu, September 1981.
85. **Polovina, J. and Brown, J.,** A population dynamics model for prawn aquaculture, *Proc. World Maricul. Soc.,* 9, 393, 1978.
86. **Smith, T. I. J., Waltz, W., and Sandifer, P. A.,** Processing yields for Malaysian prawns and the implications, *Proc. World Maricul. Soc.,* 11, 557, 1980.
87. **Willis, S. A. and Berrigan, M. E.,** Effects of Fertilization and Selective Harvest on Pond Culture of *M. rosenbergii* in Central Florida, Completion Report for U.S. Department of Commerce, N.O.A.A., N.M.F.S., PL 88-309, No. 2-298-R-a, Job 3B, Gainesville, Fla., 1978.
88. **Willis, S. A. and Berrigan, M. E.,** Effects of stocking size and density on growth and survival of *M. rosenbergii* (de Man) in ponds, *Proc. World Maricul. Soc.,* 8, 251, 1977.
89. **Smith, T. I. J., Sandifer, P. A., and Trimble, W. C.,** Pond culture of the Malaysian prawn, *Macrobrachium rosenbergii* (de Man), in South Carolina, 1974—1975, *Proc. World Maricul. Soc.,* 8, 625, 1977.
90. **Smith, T. I. J., Sandifer, P. A., Jenkins, W. E., and Stokes, A. D.,** Effect of population structure at stocking and density on production and economic potential of prawn *(M. rosenbergii)* farming in temperate climates, *Proc. World Maricul. Soc.,* 12, 1982, in press.
91. **Sarver, D., Malecha, S., and Onizuka, D.,** Possible sources of variability in stocking mortality in post larval *Macrobrachium rosenbergii,* in *Giant Prawn Farming,* New, M. B., Ed., Elsevier, New York, 1982, 99.
92. **Harrison, K. E.,** The changing salinity requirements of *Macrobrachium rosenbergii* from larval to adult: a review of old and new findings, *Proc. World Maricul. Soc.,* 13, in review.
93. **Gibson, R. T.,** A Strategy for Earthen Pond Prawn Production Management *(Macrobrachium rosenbergii),* M.S. thesis, Department of Agriculture Engineering, University of Hawaii, Honolulu, 1979.
94. **Gibson, R. T. and Wang, J. K.,** A prawn population management model, *Trans. Am. Soc. Agric. Eng.,* 22(1), 207, 1979.
95. **Huang, W.-Y., Wang, J. K., and Fujimura, T.,** A model for estimating prawn population in ponds, *Aquaculture,* 8, 57, 1976.
96. **Smith, T. I. J. and Sandifer, P. A.,** Development and potential of nursery systems in the farming of Malaysian prawns, *M. rosenbergii* (de Man), *Proc. World Maricul. Soc.,* 10, 369, 1979.
97. **Sandifer, P. A. and Smith, T. I. J.,** Effects of population density on growth and survival of *M. rosenbergii* reared in recirculating water management systems, *Proc. World Maricul. Soc.,* 6, 43, 1975.
98. **Willis, S. A., Hagwood, R. W., and Eliason, G. T.,** Effects of four stocking densities and three diets on growth and survival of postlarval *M. rosenbergii* and *M. acanthurus, Proc. World Maricul. Soc.,* 7, 655, 1976.
99. **Eble, A. F., Evans, M. C., Deblois, N., and Stolpe, N. E.,** Maintenance of broodstock, larval rearing and nursery techniques used to grow *M. rosenbergii* in waste-heat discharge waters of an electric generating station in New Jersey (U.S.A.), *Actes Colloq. C.N.E.X.O.,* 4, 233, 1977.
100. **Mancebo, V. J.,** Growth in tank-reared populations of the Malaysian prawn, *Macrobrachium rosenbergii* (de Man), *Proc. World Maricul. Soc.,* 9, 83, 1978.
101. **Kneale, D. C. and Wang, J. K.,** A laboratory investigation of *Macrobrachium rosenbergii* nursery production, *Proc. World Maricul. Soc.,* 10, 359, 1979.

102. **Piyatiratitivokul, S. and Menasveta, P.,** A comparative study on nursery techniques of the giant freshwater prawn post-larvae (*M. rosenbergii* de Man), in *Giant Prawn 1980,* International Foundation for Science, ISBN:91-857, 98-08-8, Stockholm, Sweden, 1980.
103. **Snow, J. R.,** Increasing the Yield of Channel Catfish Rearing Ponds by Periodic Division of the Stock, presented at 13th Annu. Conf. South East Assn. Game Fish Comm., 1976.
104. **Bardach, J. E., Ryther, J. H., and McLarney, W. O.,** Milkfish culture, in *Aquaculture: The Farming and Husbandry of Freshwater and Marine Organisms,* John Wiley & Sons, New York, 1972, chap. 17.
105. **Stamp, N. H. E.,** Computer technology and farm management economics in shrimp farming, *Proc. World Maricul. Soc.,* 9, 383, 1978.
106. **Williamson, M. R. and Wang, J. K.,** An improved harvesting net for freshwater prawns, *Aqua. Eng.,* 1, 81, 1982.
107. **Wang, J. K.,** Department Agriculture Engineering, University of Hawaii, Honolulu, personal communication.
108. **Gibson, R. T. and Wang, J. K.,** An Alternative Prawn Production Systems Design in Hawaii, University Hawaii Sea Grant Tech. Report, TR-77-05, Honolulu, May 1977.
109. **Corbin, J. S., Fujimoto, M. M., and Iwai, T., Jr.,** Feeding practices and nutritional considerations for *Macrobrachium rosenbergii* culture in Hawaii, *CRC Handbook of Mariculture,* McVey, J., Ed., CRC Press, Boca Raton, Fla., 1983, in press.
110. **Malecha, S. R.,** Genetics and selective breeding of *Macrobrachium rosenbergii*, in *Shrimp and Prawn Farming in the Western Hemisphere,* Hanson, J. A. and Goodwin, H. L., Eds., Dowden, Hutchinson and Ross, Stroudsburg, Pa., 1977, 328.
111. **Malecha, S. R., Sarver, D., and Onizuka, D.,** Approaches to the study of domestication in the freshwater prawn, *M. rosenbergii,* with special emphasis on the Anuenue and Malaysian stocks, *Proc. World Maricul. Soc.,* 11, 500, 1980.

MATURATION AND SPAWNING OF THE AMERICAN LOBSTER, *HOMARUS AMERICANUS*

D. Hedgecock

The scale at which farming of the lobster *Homarus americanus* will likely prove economical within the U.S. — 450 metric tons annual production — implies year-round harvesting from sophisticated, three-dimensional, automated, intensive aquaculture systems.[1,2] Such levels of production will only be attained if sufficient numbers of lobster larvae are continuously available for stocking into these systems. Presently, techniques for truly continuous production of larvae are not available although significant progress in their development is being made.

In their description of hatchery methods for the production of *Homarus* juveniles, Schuur and co-workers[3] gave three means of acquiring ovigerous female lobsters: (1) berried females may be obtained from fishermen specially licensed to trap such females for the Massachusetts State Lobster Hatchery or for scientific research purposes; (2) nonovigerous, but reproductively mature females may be obtained from the natural fishery in August or September. (A weight of 0.8 kg or larger was stated as the criterion of maturity, but assessing female maturity may be considerably more difficult.[4]) "Sixty to eighty percent of these females will extrude eggs in the next one to seven months. The variation in time of extrusion is the result of interplay between the state of development of the ovary at the time of capture and *the temperature at which the female is held subsequent to capture*"[3] (italics added; more recent evidence presented below suggests that temperature is *not* the factor controlling oviposition). (3) Finally, the complete reproductive process — mating of newly molted females, oviposition, and hatching — may be carried out in the laboratory.

Only this last means of producing seed stock is suitable as a foundation for commercial lobster aquaculture because this industry cannot depend on obtaining females from nature. Acquisition of berried lobsters conflicts with public policy on fisheries conservation,* and utilization of fall-caught, nonovigerous females from the natural fishery would, at best, yield larval production during only part of the year. Complete control over reproduction, on the other hand, provides the potential for both larval production on a year-round basis and for improvement of stock through selective breeding. Unfortunately, while aspects of lobster reproduction have been observed in several laboratories,[3,5,8] the rate of extrusion and hatching success from controlled matings has been lower outside of the Martha's Vineyard hatchery. Prior to 1976, for example, only five extrusions had been documented at the University of California's Bodega Marine Laboratory (BML) and only one of these egg clutches was carried to successful hatching.[8] Thus, supply of seed lobster fry remains one of the major biological constraints upon lobster aquaculture.

A major focus of the lobster research carried out at BML since 1972 under funding from the California Sea Grant College Program has been control of reproduction. Over the last 5 years, significant advances have been made in understanding ovarian maturation and oviposition, but a number of problems remain to be solved. This chapter reviews progress in controlling the lobster ovarian cycle and proposes a broodstock management scheme designed to yield year-round egg production. In discussing outstanding problems, in particular egg loss from ovigerous females, we emphasize that a still more complete understanding

* All 11 coastal states bordering on the lobster's natural range have enacted legislation forbidding the capture of ovigerous (egg-bearing or "berried") females or the removal of eggs from such females (American Lobster Fishery Management Plan, 1978). Moreover, Maine law demands that females laying eggs while being held in lobster pounds be notched in the telson, returned to sea, and protected from capture while the tail notch remains visible.

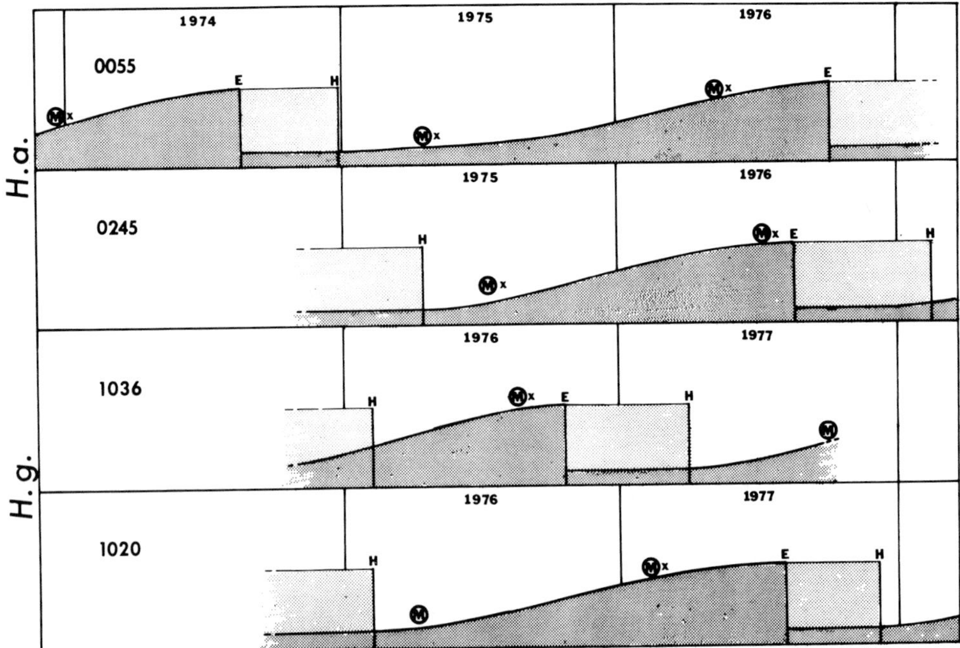

FIGURE 1. Representative reproductive and molt cycles in laboratory-held American (animals 0055 and 0245) and European (1036 and 1020) lobsters *(Homarus)*. E = extrusion, H = hatch, M = molt, "x" by molt indicates laboratory mating. (From Hedgecock, D., Moffett, W. L., Borgeson, W., and Nelson, K., *Proc. World Maricul. Soc.*, 9, 497, 1978. With permission.)

of the reproductive biology of both sexes is required to achieve seed supply independent of natural productivity and sufficient for commercial lobster farming.

THE BIENNIAL REPRODUCTIVE CYCLE OF *HOMARUS*

In nature, clawed lobsters lay eggs every other year.[9] This pattern, also seen in the majority of female lobsters held in the laboratory,[5,9] frequently features the alternation of barren and fertile molt cycles, at least in the size range of 80- to 110-mm carapace length (CL) (Figure 1). The majority (80% or greater) of *H. americanus* females in nature conforms to a seasonal reproductive pattern with oviposition occurring chiefly in July and August and hatching mainly in the following May, June, and July.[9,10] A minority of females lay eggs in fall or winter, but whether these eggs are ready to hatch the following summer is uncertain. Reports cited by Herrick[9] concerning the reproductive cycle of the European lobster, *H. gammarus*, also support a biennial mode although oviposition in *consecutive* years has been frequently observed. At the Bodega Marine Laboratory, *H. gammarus* females display an annual ovarian cycle more often than not (Figure 1[8,11]). In any case, as in *H. americanus*, egg laying and hatching in natural populations of the European clawed lobster are seasonal, supporting the suggestion that temperature or photoperiod might influence the reproductive cycle of *Homarus*.

EFFECTS OF TEMPERATURE AND PHOTOPERIOD ON OVARIAN DEVELOPMENT

It has been reported that oviposition is suppressed and molting accelerated for female lobsters held at 15°C or above.[5,8] These reports demonstrate the antagonistic or competitive relationship between ovarian maturation and somatic growth in the female lobster. Since female lobsters do not develop eggs and oviposit at the high temperatures (21°C) used to

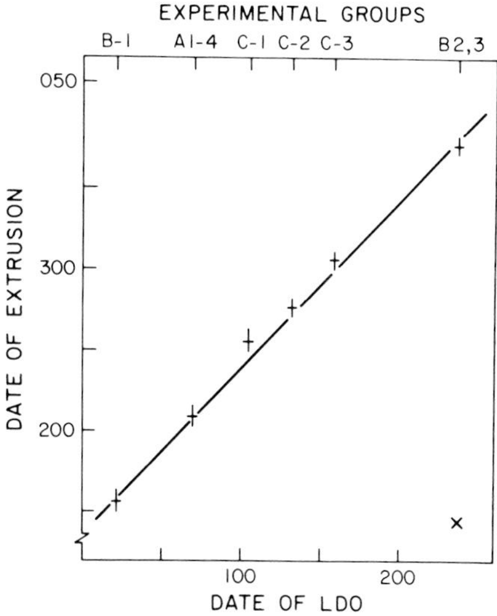

FIGURE 2. Correlation between date of long-day onset (LDO, see text) and date of subsequent extrusion in ten experimental groups of lobsters. (From Nelson, K., Hedgcock, D., and Borgeson, W., *Can. J. Fish. Aqua Sci.*, submitted.

optimize growth rate in aquaculture systems, lobster broodstock will require separate, cooler seawater systems after reproductive maturity is reached.*

When mature lobster females are held at seawater temperatures between 10 to 15°C, changing photoperiod from short days (8hL:16hD) to long days (16hL:8hD) — termed long-day onset or LDO in the following discussion — synchronizes egg development and promotes extrusion in an average of 55% of experimental subjects.[11] These experiments mimicking seasonal changes in photoperiod utilized mainly recently molted (i.e., shiny-shelled) mature female lobsters captured in the fall; according to Schuur and co-workers,[3] 60 to 80% of these should extrude eggs. That a switch from short- to long-day lengths promotes this high rate of egg extrusion provides a likely explanation of why fall-caught female lobsters have routinely extruded at the Martha's Vineyard Hatchery where broodstock are exposed to ambient photoperiods but have not extruded in the past at the Bodega Marine Laboratory where broodstock were on uncontrolled photoperiods prior to 1975.

These experiments now raise the question of how short the period of short days can be to ready the ovary for final vitellogenesis. Four months is, thus far, the shortest period of 8hL:16hD photoperiods to have yielded a high rate of oviposition following LDO. A recent experiment in which paired groups of 18 fall-caught females were exposed, one for 90 days and the other for 135 days, to short daylengths prior to LDO yielded 13 and 33% ovipositions, respectively, suggesting a lower limit to the "winter" daylength period (unpublished data). We do not know why the overall rate of extrusion was low in this particular experiment, and clearly more data are needed on the minimum short-day period required to set the stage for final vitellogenesis.

More striking than the *rates* of extrusion in experimental groups of female lobsters exposed to a change from short days to long days is the high *correlation* between date of LDO and date of oviposition (Figure 2). Notice that egg extrusion follows LDO by the same number

* A successful criterion of female maturity is a ratio of second abdominal segment width to carapace length (AW/CL) greater than 0.65. Differentiation of the wider female abdomen is noticeable at CLs of about 55 cm with some variation among geographic stocks.[4]

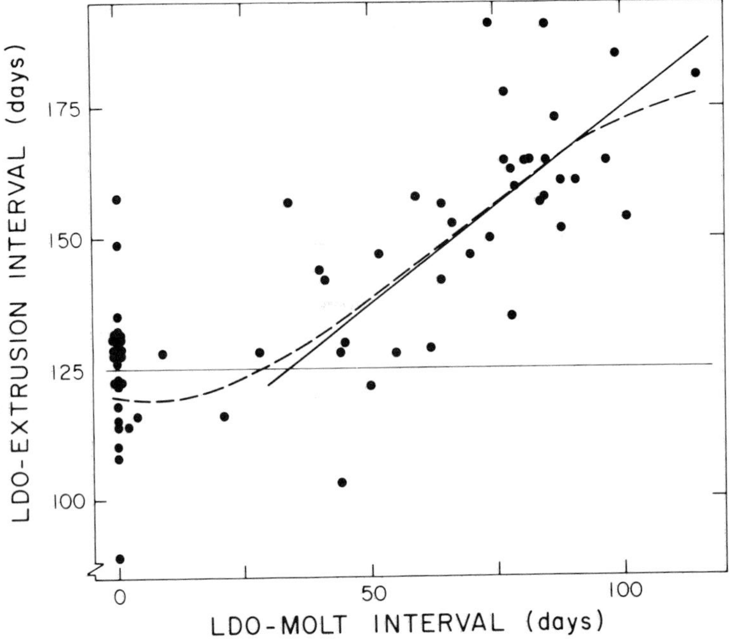

FIGURE 3. Influence of molting, expressed as the interval between long-day onset (LDO) and molt, upon extrusion, expressed as the interval between LDO and extrusion. Dashed line, polynomial fit to all data with LDO molt interval, $x > 0$; $y = 119.4 - 0.25x + 0.014x^2 - 10^4 x^3$. Solid linear regression line, $y = 99.7 + 0.748x$ for $y > x > 35$.

of days for LDOs ranging from day 026 to day 240. Photoperiod manipulation appears to be an effective (and inexpensive) means of regulating the time of oviposition in *Homarus;* controlled short daylengths can suppress vitellogenic activity far beyond the vernal equinox (day 80), which probably represents the natural photoperiod trigger for egg laying in midsummer.[11] While an artificial extension of winter lighting conditions allows the scheduling of egg laying for the otherwise barren months of late fall and winter of the following year, we propose an alternate management practice below.

The correlation between LDO and oviposition is somewhat more complicated than indicated in Figure 2, because of the influence of molting upon final vitellogenesis. Consider the relationship between LDO and molt vs. LDO and extrusion (Figure 3). If the molt previous to egg extrusion falls before LDO (represented by the cluster of points above zero on the abscissa), oviposition occurs about 125 days after LDO. If, on the other hand, the molt occurs after LDO, and especially after about 45 or 50 days post-LDO, final vitellogenesis and oviposition are delayed. This observation again emphasizes the antagonism between growth and reproductive processes, and points to the necessity of avoiding this antagonism, if possible, through appropriate broodstock husbandry.

AN OPERATIONAL HYPOTHESIS FOR LOBSTER BROODSTOCK MANAGEMENT

The chief requirements of a lobster broodstock management scheme are (1) control over growth and reproductive events, (2) avoidance of conflict between these antagonistic processes, (3) elimination of alternating barren and fertile molt cycles, (4) production of eggs in virtually all months of the year, and, ultimately, (5) elimination of seasonality in hatching.

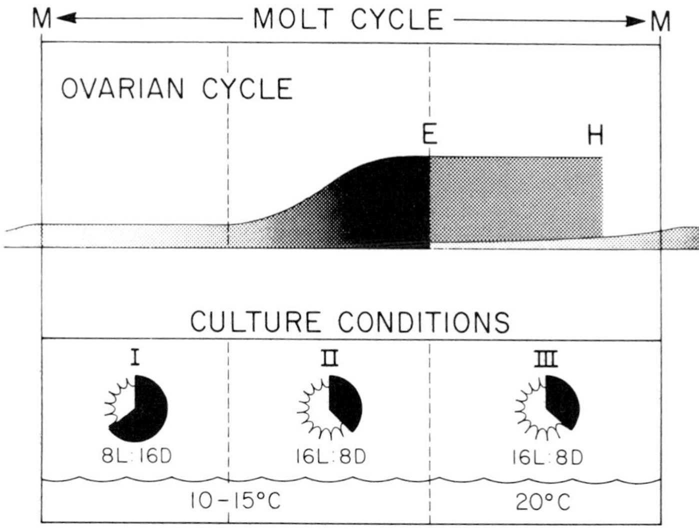

FIGURE 4. Working hypothesis for successful management of *Homarus* broodstock. Cycling of female lobsters through three holding systems with indicated temperatures and photoperiods should result in egg extrusion and hatching within each intermolt interval (see text).

We have reviewed the evidence for photoperiod control over final vitellogenesis and oviposition in *Homarus;* these reproductive events can be controlled and scheduled with remarkable accuracy. The molt cycle of laboratory-held lobsters and the effect of temperature upon growth have been described mathematically[12] so that molting is also a predictable event.

We believe that the alternation of fertile and barren molt cycles observed in lobsters in nature is due, not to an intrinsic biennial ovarian cycle, but to the interplay between somatic and ovarian growth in an environment with great seasonal fluctuations in temperature. We hypothesize that conflict between somatic and ovarian growth can be avoided by manipulating temperature and photoperiod such that a complete ovarian cycle, including brooding and hatching, is contained within each molt cycle. Our operational broodstock management scheme (Figure 4) should result in consecutive fertile molts in female lobsters.

Three distinct seawater systems are used to implement the scheme shown in Figure 4. The first system is housed in a light-tight room with timer-controlled short daylengths and seawater temperatures of 10 to 15°C. (Ambient seawater temperature in the Aquaculture Facility at BML averages a convenient 12 ± 3°C.) The second system is also maintained at near ambient seawater temperatures, but is subject to the long daylengths (15hL:9hD) maintained in the Aquaculture Facility's large wetlab. Finally, a third system on the same photoperiod as system II is maintained at 20°C by a thermostat-controlled, heat exchange loop.

All three culture systems are semirecirculating systems with UV treatment, sand filtration, and make-up flow rates amounting to nearly two complete replacement volumes per day.[13] The holding tanks themselves are stacked, rack-mounted, injection-molded plastic troughs (36 × 152 × 26 cm) subdivided into as many as ten compartments by perforated PVC partitions.[13] Culture water enters at both ends of each trough and exits from a central standpipe. Chopped squid, mussels, fish, or shrimp is provided daily at a rate of about 2% of the adult lobster's body weight per day. Animal care and systems maintenance requires 6 man-hours/day for a broodstock of about 250 lobsters.

Broodstock are cycled through the three systems according to the following "typical" schedule (Figure 4): (1) Soon after a nuptial molt the female lobster is placed in system I (cold water, short daylengths); she is held there for 120 days, the minimum short-day length period known to ready the ovary for stimulation of final vitellogenesis; (2) after 120 days in system I, the female lobster is transferred to system II (cold water, long daylengths) where she undergoes final vitellogenesis and extrudes eggs after 125 days; (3) upon extrusion the female is transferred to system III where warm water and long daylengths accelerate embryonic development and result in hatching after approximately 4 months.[3,14] As hatch approaches the female is temporarily held in a special compartment of the warm-water larval rearing system designed to capture the newly hatched planktonic larvae (see this volume "Lobster (*Homarus*) Hatchery Techniques"), (4) after hatching the female is retained in system III until her next molt; (5) the cycle is then reinitiated by transferring the female back to system I.

So far, 39 females have been taken through step 3; 23 have extruded eggs for an extrusion rate of nearly 60%. However, extrusion rate appears to depend on how quickly step 1 is taken following a nuptial molt. Only 4 of 15 females (27%) placed in system I after 21 days postmolt extruded while 19 of 24 females (79%) placed in system I within 21 days postmolt subsequently laid eggs. Judging from the duration of intermolt substages (see Aiken's Table I[15]), the ovarian cycle is less readily initiated if the molt cycle is allowed to proceed beyond stage C_1. This may also explain why some 40% of fall-caught, nonovigerous females exposed to short days, then LDO, fail to extrude in the following spring. It will be of interest to correlate rate of extrusion with molt cycle stage at the time of transfer to system I.

There are two control points in this management scheme for delaying or advancing egg production and hatching from a particular female lobster: (1) as already mentioned, final vitellogenesis may be delayed many months if females are merely kept in system I; on the other hand, there appears to be a lower limit of about 4 months for step 1 to be effective. (2) Rate of embryo development and the length of the brooding period are temperature dependent and may range from 4 months at 20°C to 9 or 10 months at 10°C.[14] However, scheduling of egg production and hatching in all months of the year may be more easily achieved by introducing cohorts of broodstock into the typical management cycle in each month. This is readily accomplished because laboratory-held lobsters gradually lose seasonality of molting,[11] and because the duration of the molt cycle is size dependent, being about 9 months in the first adult year and thereafter increasing to well over a year over the size range of 80 to 120 mm CL. Movement of various-sized broodstock through this husbandry system, however, will be complex so that scheduling of lobster fry production will probably require computer assistance. A broodstock inventory program is currently under development at the Bodega Marine Laboratory.

CURRENT PROBLEMS IN LOBSTER BROODSTOCK DEVELOPMENT

Since 1978 most of our laboratory-mated broodstocks have been subjected to photoperiod regimes that were more or less appropriate for the stimulation of ovarian maturation and oviposition. How have these crosses fared? Of a total 109 laboratory matings 70 (64%) subsequently extruded. Comparing this to an overall extrusion rate of 11.1% for the period of December 1973 to March 1977,[8] we can say that there has been significant progress towards controlling lobster reproduction (Table 1). However, of the 70 females that successfully oviposited only 12 had hatched and 22 were still berried in January 1982. Thus, only about half of the females that successfully lay eggs in the laboratory carry those eggs until hatch, and in many of these cases, clutch size is reduced to a few hundred eggs. Egg loss has now become the major problem in achieving control of lobster reproduction in the laboratory.

There may be several reasons for this egg loss. Perkins[16] estimates that even in nature up

Table 1
COMPARISON OF (A) RESULTS OF *HOMARUS* MATINGS AT THE BODEGA MARINE LABORATORY BEFORE PHOTOPERIOD CONTROL OF OVULATION WAS DISCOVERED (DECEMBER 1973 TO MARCH 1977 FROM HEDGECOCK ET AL.[8]) WITH (B) RESULTS OF NATURAL OR LABORATORY MATINGS FOLLOWED BY A CHANGE IN PHOTOPERIOD FROM 8L:16D TO 16L:8D

	Extrusions		Hatches	
Number of matings	Number	Percent of crosses	Number	Percent of extrusions
A—135	15	11.1	6	40.0
B—109	70	64.2	34[a]	48.6

[a] Actually, 22 of these were still brooding as of January 1982.

to 36% of the egg clutch is lost before hatch; the causes of this attrition are unknown. In the laboratory egg loss may occur immediately after extrusion or over such a long period of time that it is difficult to detect on a daily or weekly basis. In the latter cases, development of the remaining eggs proceeds normally until a very few larvae actually hatch.

One explanation for early egg loss may be that these females have not been inseminated when paired with a male at their nuptial molt. Loss of unfertilized eggs in lobsters has been reported previously,[4] and Cheung's[17] suggestion that the cement for egg attachment arises from the egg after fertilization and is not a secretion from the mother's abdominal glands would certainly explain these occurrences. Unfertilized eggs, however, do form attachments to pleopod setae that appear normal, but these eggs deteriorate and are eventually shed.[20]

Because female lobsters secrete a sex pheromone upon molting (see literature cited by Aiken and Waddy[4]), male lobsters are introduced into the tanks of newly molted female lobsters and left either until courtship and copulation have transpired or overnight if courtship is proceeding too slowly to permit observation. When the male is removed from the female's tank, the presence or absence of a spermatophore on the female's annulus or sperm recepticle is noted. However, presence or absence of a spermatophore is not a failsafe indication of insemination since successful hatches have resulted from matings in which no spermatophore was visible. Nevertheless, a contingency chi-square test of association between early or late (or no) egg loss and whether a spermatophore was or was not observed in a total of 43 matings shows a significant tendency for visible spermatophores to be associated with late loss or hatching rather than early loss (Table 2). Examination of individual male mating records indicates that some males consistently fail to sire progeny while others are consistently successful.[21] Moreover, direct microscopic examination of testes from male laboratory brood-stock and from males caught in Maine waters in October of 1981 revealed little meiotic activity.[22] As there are no published reports concerning the annual reproductive cycle of the male *H. americanus,* we do not know to what extent such a cycle might also be seasonal as in females.

Causes for late egg loss are probably numerous.[4] Perhaps some males have reduced fertility so that only a portion of the clutch is fertilized and retained with normal development. Egg attachment may be abnormal due to nutritional or other environmental stresses on the extruding female. Torsion of the last walking leg and groups of setae that form "cleaning brushes" especially equip the female lobster to clean her abdomen before extrusion and

Table 2
CONTINGENCY TEST FOR ASSOCIATION BETWEEN OBSERVATIONS OF SPERMATOPHORE DEPOSITION FOLLOWING COPULATION AND FATE OF SUBSEQUENTLY EXTRUDED EGGS

	Egg loss		
Spermatophore	**Early, total**	**Late or incomplete**	**Total**
Observed	4	9	13
Not observed	19	11	30
	—	—	—
	23	20	43
Total	$\chi^2 = 3.86, p < 0.05$		

while brooding. Aberrant cleaning behavior, however, may be triggered by certain stresses of laboratory holding conditions. Herrick[9] describes two cases of aquaria-held females picking and scratching off all eggs within a few days of their laying.

Clearly, while we have progressed in our understanding of what factors control egg maturation, spawning, and hatching, we have much to learn about the reproductive biology of both sexes of lobsters before reliable broodstocks for aquaculture are a reality.

SIZE OF BROODSTOCK FOR COMMERCIAL CULTURE

Despite problems that remain in controlling lobster reproduction, it may be useful to estimate on the basis of our research experience the size of a broodstock population required to meet anticipated commercial production levels. As mentioned at the outset, computer modeling studies suggest commercial facilities must produce on the order of 80,000 1-lb lobsters per month. In order to quantify the size of broodstock needed to provide seed for such production levels, the following assumptions are made:

1. An average female of 85-mm carapace length will produce approximately 8500 eggs based on the equation

$$\log_{10} Y = -1.6017 + 2.8647 \log_{10} X$$ [18]

   where Y = fecundity (number of eggs), X = size (carapace length in mm).*
2. About 70% of females in the broodstock management cycle successfully extrude eggs.
3. Presently half of the clutches produced by broodstock are lost while attrition of the remaining clutches appears to be high, but is not yet well quantified. If we accept the natural attrition rate of 36% suggested by the data of Perkins,[16] then the overall probability of egg loss per clutch is 0.50 + (0.36) (0.50) = 0.68.
4. Mortalities suffered during the larval rearing period using standard laboratory methods and optimal densities,[3,19] will be approximately 20%.
5. Losses occurring during the movement of fifth-stage juveniles into individual rearing containers are 5%.[23]
6. Losses during the juvenile to market-size rearing period will be 25%. This figure is

* Other estimates of fecundity give somewhat higher values; for example, Herrick's[9] regression equation yields a fecundity of 10,500 eggs for an 85-mm CL female (see Reference 4 for discussion). However, we wish to be conservative in our estimates of fecundity.

high considering estimates available for mortality but reasonable if one considers the need for culling slow growing lobsters at the first juvenile transfer.

Working backward, then, an estimate of the number of extruded females needed on a monthly basis can be calculated.

- 80,000 450-g lobsters = 75% effective survival of 106,667 loaded juveniles
- 106,677 loaded juveniles = 95% survival of 112,281 newly stocked fifth-stage postlarvae
- 112,281 fifth-stage postlarvae = 80% survival of 140,351 first-stage larvae
- 140,351 first-stage larvae = 30% survival of 467,836 extruded eggs
- 467,836 extruded eggs ÷ 8500 eggs per female = 55 ovipositing females per month
- 55 berried females per month = 70% of a cohort of 79 females subjected to the broodstock management cycle

Thus, under the above assumptions, roughly 950 reproductive female adult lobsters would be needed for larval production on a yearly cycle. Although these simplified calculations assume no greater control of the reproductive cycle than is presently possible, the estimated size of a commercial broodstock is only about five times as large as the existing BML broodstock population.

Thus, scaling up broodstock to commercial production levels entails much less uncertainty than presently exists for lobster *grow-out* operations, which will have to be scaled up several orders of magnitude beyond current laboratory production levels.

CONCLUSIONS

The great strides that have been made in the past 7 years in understanding ovarian maturation, oviposition, and the complex interactions between somatic and ovarian growth processes are cause for optimism that complete reproductive control and efficient broodstock husbandry will be achieved in the near future for *Homarus*. The seed supply problem in lobster aquaculture has been solved at least to the extent that proof of concept demonstrations at larger than laboratory scale may be conducted. Even with state-of-the-art husbandry techniques, seed supply for a full commercial farming operation (450 metric tons per year) appears attainable since year-round egg production and hatching have been accomplished and maintenance of the estimated total number of broodstock required, 950, would consume only a fraction of total operating costs. By preventing egg loss, which currently amounts to nearly 70% of eggs extruded by female lobsters, the efficiency of commercial broodstocks will be greatly improved. The causes of egg loss are only vaguely understood but one prime candidate, low or seasonal fertility of male lobsters, should be amenable to similar research methods as have solved many of the riddles of female lobster reproduction. With the seed supply problem solved, a program aimed at the genetic improvement of lobsters for culture can grow together with this nascent aquaculture industry.

ACKNOWLEDGMENTS

This work is a result of research sponsored in part by NOAA, Office of Sea Grant, Department of Commerce, under Grant No. 04-8-MO1-189 R/A 28. The U.S. government is authorized to produce and distribute reprints for governmental purposes notwithstanding any copyright notation that may appear hereon.

REFERENCES

1. **Allen, P. G. and Johnston, W. E.,** Research direction and economic feasibility: an example of systems analysis for lobster aquaculture, *Aquaculture,* 9, 144, 1976.
2. **Johnston, W. E. and Botsford, L. W.,** Systems analysis for lobster aquaculture, EIFAC/80/Symp.: E/56, Symp. on New Developments in the Utilization of Heated Effluents and of Recirculation Systems for Intensive Aquaculture, Stavanger, Norway, 1980.
3. **Schuur, A., Fisher, W. S., Van Olst, J. C., Carlberg, J. M., Hughes, J. T., Shleser, R. A., and Ford, R. F.,** Hatchery methods for the production of juvenile lobsters *(Homarus americanus),* Sea Grant Publication No. 48, University of California, Institute of Marine Resources, La Jolla, 1976.
4. **Aiken, D. E. and Waddy, S. L.,** Reproductive biology, in *The Biology and Management of Lobsters,* Vol. 1, Cobb, J. S. and Phillips, B. F., Eds., Academic Press, New York, 1980, 215.
5. **Aiken, D. E. and Waddy, S. L.,** Controlling growth and reproduction in the American lobster, *Proc. World Maricul. Soc.,* 7, 415, 1976.
6. **Hughes, J. T. and Matthiessen, G. C.,** Observations on the biology of the American lobster, *Homarus americanus, Limnol. Oceanogr.,* 7, 414, 1962.
7. **Carlberg, J. M., Van Olst, J. C., and Ford, R. F.,** A comparison of larval and juvenile stages of the lobsters *Homarus americanus, Homarus gammarus,* and their hybrid, *Proc World Maricul. Soc.,* 9, 109, 1978.
8. **Hedgecock, D., Moffett, W. L., Borgeson, W., and Nelson, K.,** Progress and problems in lobster broodstock development, *Proc. World Maricul. Soc.,* 9, 497, 1978.
9. **Herrick, F. H.,** Natural history of the American lobster, *Bull. U.S. Bur. Fish.,* 29, 149, 1909.
10. **Herrick, F. H.,** The reproduction of the lobster, *Zool. Anz.,* 18, 226, 1895.
11. **Nelson, K., Hedgecock, D., and Borgeson, W.,** Photoperiodic and ecdysial control of vitellogenesis in lobsters *(Homarus),* (Decapoda, Nephropidae), *Can. J. Fish Aqua. Sci.,* submitted.
12. **Botsford, L. W., Rauch, H. E., and Shleser, R. A.,** Optimal temperature control of a lobster plant, *I.E.E.E. Trans. Autom. Control,* ac-19(5), 541, 1974.
13. **Hand, C.,** Development of aquaculture systems, Sea Grant Publication No. 58, University of California, Institute of Marine Resources, La Jolla, 1977.
14. **Perkins, H. C.,** Development rates at various temperatures of embryos of the northern lobster *Homarus americanus* Milne-Edwards, *Fish. Bull.,* 70, 95, 1972.
15. **Aiken, D. E.,** Molting and growth, in *The Biology and Management of Lobsters,* Vol. 1, Cobb, J. S. and Phillips, B. F., Eds., Academic Press, New York, 1980, 91.
16. **Perkins, H. C.,** Egg loss during incubation from offshore northern lobsters (Decapoda: Homaridae), *Fish. Bull.,* 69, 451, 1971.
17. **Cheung, T. S.,** The development of egg membranes and egg attachment in the shore crab, *Carcinus maenas,* and some related decapods, *J. Mar. Biol. Assoc. U. K.,* 46, 373, 1966.
18. **Saila, S. B., Flowers, J. M., and Hughes, J. T.,** Fecundity of the American lobster, *Homarus americanus, Trans. Am. Fish. Soc.,* 98, 537, 1969.
19. **Hughes, J. T., Shleser, R. A., and Tchobanoglous, G.,** A rearing tank for lobster larvae and other aquatic species, *Prog. Fish. Cult.,* 36, 129, 1974.
20. **Talbot, P.,** UC Riverside, personal communication.
21. **Wilson, P.,** unpublished data, Aquaculture Enterprises, personal communication.
22. **Roberts, F.,** University of Maine, Orono, and Hedgecock, D., personal observations.
23. **Conklin, D.,** personal observation.

LOBSTER *(HOMARUS)* HATCHERY TECHNIQUES

Ernest S. Chang and Douglas E. Conklin

The relatively brief larval development period (approximately 12 days at 20°C)[1] of *Homarus* has been one of the attractive features stimulating the continued interest in *lobster aquaculture*.

At the University of California Bodega Marine Laboratory, the berried female lobster is monitored for size of clutch and embryonic development subsequent to egg extrusion and fertilization. Approximately two weeks prior to hatching, as estimated by the eyespot index (for *H. americanus*;[2] for *H. gammarus*[3]), berried females are placed into a hatching system. In these extruded plastic holding containers (75 × 25 × 30 cm, l × h × w) filtered and ultraviolet-sterilized water, heated to 20° C, flows into the animal compartment, over a dam, and through a catch basket before entering the drain. The catch basket is made of plexiglas and approximately 1 mm mesh fiberglass screening and will prevent loss of the newly hatched larvae. Hatching females tend to release their larvae in the early morning hours.[4] Each morning, the removable catch baskets are examined for larvae and, if present, they are rinsed into a larval rearing system.

In the larval rearing system, the pelagic larvae go through a series of four larval stages (Figure 1) before taking on the morphology of adults. As the larvae are cannibalistic from the time of hatching, larval rearing systems have been designed with the idea of providing sufficient circulation to maintain an even distribution of larvae and to minimize larval interactions. Many researchers utilize a "kreisel"[5] system (Figure 2) developed at the Massachusetts State Lobster Hatchery.[6] In this kreisel, seawater is forced out at the bottom at a rate of 5 to 6 ℓpm through a series of jets producing a spiral upwelling pattern. In order to obtain the maximum survival rate of 70%, the 40-ℓ kreisel is stocked with approximately 1500 larvae from a single day's hatch.[7] Relating this to a hypothetical one million pound-per-year lobster production facility,[8] it would appear that this type of larval rearing system could easily be scaled up in that 60 of the present kreisels, each handling two batches of larvae per month, would meet the production demands of such a facility (see Maturation and Spawning page 261). Although the dimensions of the individual kreisels could be scaled up, there are advantages for retaining numerous smaller-sized culture containers in order to adequately deal with progeny from individual females as well as being able to isolate various groups in case of disease.

Lobster larvae will feed on a variety of diets: live brine shrimp, various species of zooplankton, chopped fish and mollusks, and artificial preparations. Due to their cannibalistic nature they will also readily feed on each other. Large age differences among the larvae will result in increased cannibalism, thus, the hatches from no more than 2 to 3 succeeding days are put into a single kreisel. A diet of live adult brine shrimp has been considered superior to other diets in promoting maximum survival[7] and is used almost exclusively at the Bodega Marine Laboratory.

Satisfactory survival rates are also obtained with frozen adult brine shrimp. If a constant supply of live brine shrimp is unavailable, or to insure against unforeseen interruptions in regular delivery schedules, it is advisable to stockpile frozen brine shrimp in convenient packets so that they need be thawed only once. This is done by straining out the required amount of live brine shrimp from a vigorous batch and placing them into aluminum foil dishes. These dishes are then rapidly frozen and stored until use. Commercially available brine shrimp that have been previously frozen are often unacceptable for successful larval culture. We have also successfully raised lobster larvae on brine shrimp reared on ricebran in a modified raceway system.[9]

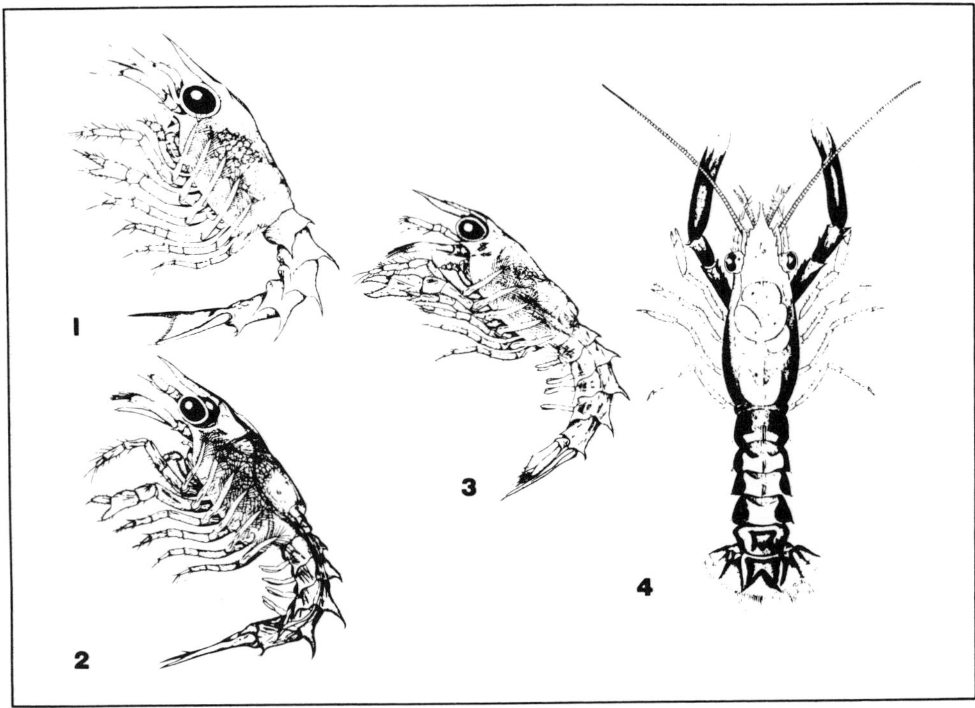

FIGURE 1. The four larval (free swimming) stages of *Homarus*. The second stage may be distinguished from the first by the presence of the biramous abdominal pleopods. The third stage may be distinguished from the second by the presence of the uropods. In the fourth stage, the larval lobster displays the juvenile morphology with the straightened abdomen and forward pointing chelipeds. (Modified from Herrick, F. H., *Bull. Bur. Fish.*, 29, 149, 1909. With permission.)

At the Bodega Marine Laboratory, lobster larvae are fed adult brine shrimp once a day to excess as determined by visual inspection. Other researchers have suggested the use of an automatic feeding system,[10] particularly if frozen brine shrimp are used. In the kreisel system it has been suggested that a ratio of four adult brine shrimp per larva should be maintained throughout the culture period.[7] Other investigators found that in culturing individual larvae at 19°C, maximum survival and growth were obtained when feeding eight adult brine shrimp per larva per day.[11] How analogous this is to the feeding levels in the kreisel system is somewhat difficult to ascertain in that even at this feeding level the cumulative percentage survival was only 45% in the absence of any possible cannibalism. "Equivalent survival rates" have also been reported[12] utilizing a much simpler static culture system. Rectangular aquaria (holding 50 to 100 ℓ) and equipped with air stones for circulation were used with periodic (every 2 to 3 days) changing of the water. The larvae were fed freshly hatched brine shrimp nauplii in excess.

At the Bodega Marine Laboratory, two larval rearing systems are maintained, each utilizing eight kreisels. These larval systems are kept isolated from juvenile and adult holding systems. Although the larvae are generally considered more susceptible to disease and fouling organisms than adults, this has not been a major problem in our experience. The larval stages are maintained at 20°C with recirculating seawater, 2.4 ℓpm to each kriesel, which has been treated by ultraviolet irradiation and filtration (Figure 3). At 20°C, larvae reach the fourth stage in as little as 10 days while decreasing temperatures result in increasing lengths of time spent in the larval stages. Temperature is maintained by either immersion heaters (Vycor®, Corning) or heat exchanger systems regulated by temperature monitors (Yellow Springs Instruments).

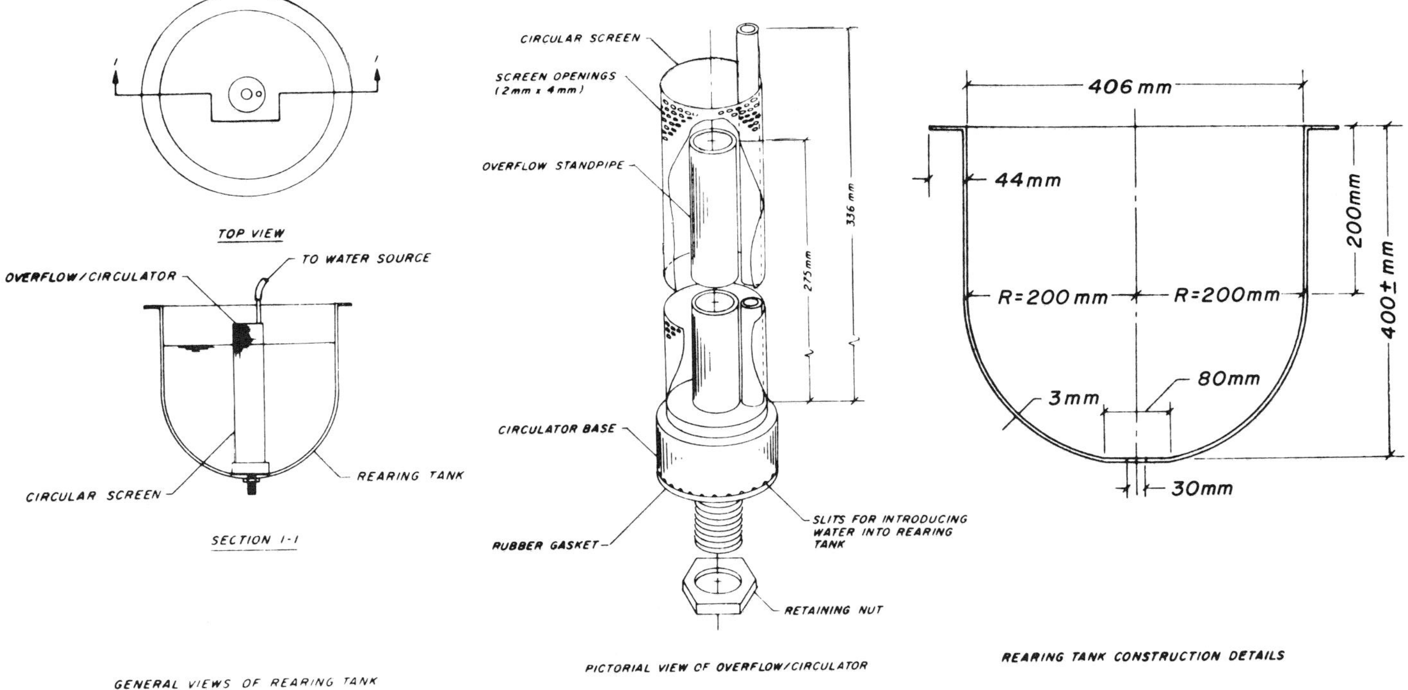

FIGURE 2. The kreisel or rearing tank developed by Hughes.[6] (Modified from Schuur, A., Fisher, W. S., Van Olst, J. C., Carlberg, J. M., Hughes, J. T., Schleser, R. A., and Ford, R. F., Sea Grant Publication No. 48, University of California, Institute of Marine Resources, La Jolla, 1976. With permission.)

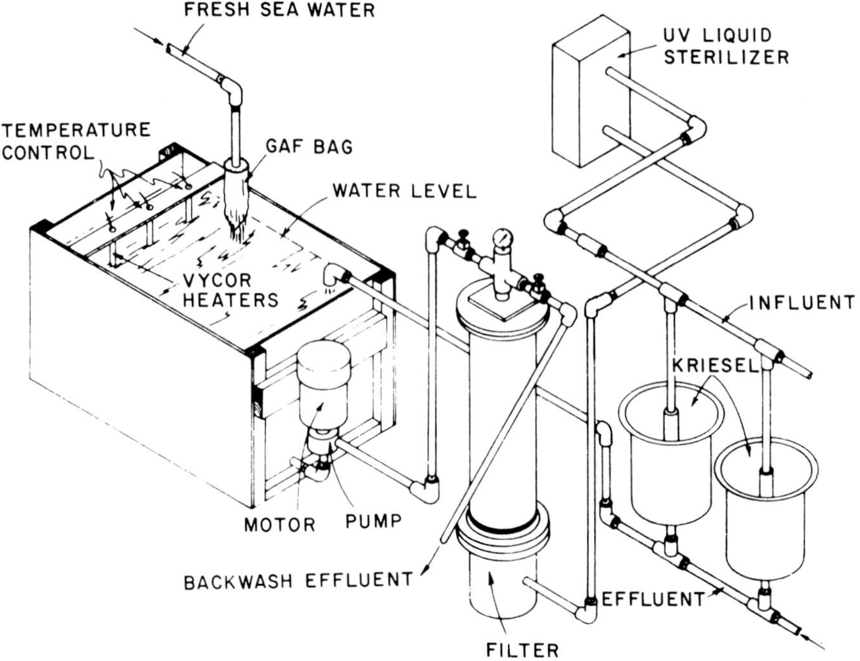

FIGURE 3. Semiclosed system utilized at the Bodega Marine Laboratory for the culture of larval lobsters. See text for details. (Modified from Schuur, A., Fisher, W. S., Van Olst, J. C., Carlberg, J. M., Hughes, J. T., Schleser, R. A., and Ford, R. F., Sea Grant Publication No. 48, University of California, Institute of Marine Resources, La Jolla, 1976. With permission.)

The constant input of fresh seawater (20 ℓpm into a 700-ℓ sump) prevents the accumulation of any metabolic waste. Although closed larval systems based on utilization of kreisels have been described,[10] it would appear that the static culture system described earlier may be simpler for those facilities with a limited seawater supply.

ACKNOWLEDGMENTS

This work is a result of research sponsored in part by NOAA, Office of Sea Grant, Department of Commerce, under Grant No. 04-8-MO1-189 R/A 28. The U.S. government is authorized to produce and distribute reprints for governmental purposes notwithstanding any copyright notation that may appear hereon.

REFERENCES

1. **Hughes, J. T. and Matthiessen, G. C.,** Observations on the biology of the American lobster, *Homarus americanus, Limnol. Oceanogr.*, 7, 414, 1962.
2. **Perkins, H. S.,** Development rates at various temperatures of embryos of the northern lobster *(Homarus americanus), Fish. Bull.*, 70, 95, 1972.
3. **Richards, P. R. and Wickins, J. F.,** Lobster Culture Research, Laboratory Leaflet No. 47, Ministry of Agriculture, Fisheries, and Food, Directorate of Fisheries Research, Lowestoft, U. K., 1979.
4. **Ennis, G. P.,** Observations on hatching and larval release in the lobster *Homarus americanus, J. Fish. Res. Board Can.*, 32, 2210, 1975.

5. **Greve, W.,** The "plankton kreisel", a new device for culturing zooplankton, *Mar. Biol.,* 1, 201, 1968.
6. **Hughes, J. T., Shleser, R. A. and Tchobanoglous, G.,** A rearing tank for lobster larvae and other aquatic species, *Prog. Fish Cult.,* 36, 129, 1974.
7. **Schuur, A., Fisher, W. S., Van Olst, J. C., Carlberg, J. M., Hughes, J. T., Shleser, R. A., and Ford, R. F.,** Hatchery methods for the production of juvenile lobsters *(Homarus americanus),* Sea Grant Publication No. 48, University of California, Institute of Marine Resources, La Jolla, 1976.
8. **Johnston, W. E. and Bostford, L. W.,** Systems analysis for lobster aquaculture, EIFAC/80/Symp.: E/56, Symp. on New Developments in the Utilization of Heated Effluents and of Recirculation Systems for Intensive Aquaculture, Stavanger, Norway, 1980.
9. **Sorgeloos, P.,** personal communication.
10. **Serfling, S. A., Van Olst, J. C., and Ford, R. F.,** A recirculating culture system for larvae of the American lobster, *Homarus americanus, Aquaculture,* 3, 303, 1974.
11. **Carlberg, J. M. and Van Olst, J. C.,** Brine shrimp *(Artemia salina)* consumption by the larval stages of the American lobster *(Homarus americanus)* in relation to food chemistry and water temperature, *Proc. World Maricul. Soc.,* 7, 379, 1976.
12. **Stewart, J. E. and Castell, J. D.,** Various aspects of culturing the American lobster, *Homarus americanus,* in *Advances in Aquaculture,* Pillay, T. V. R. and Dills, W. A., Eds., Fishing News Books, Farnham, Surrey, England, 1979, 314.
13. **Herrick, F. H.,** Natural history of the American Lobster, *Bull. Bur. Fish.,* 29, 149, 1909.

GROW-OUT TECHNIQUES FOR THE AMERICAN LOBSTER, *HOMARUS AMERICANUS*

Douglas E. Conklin and Ernest S. Chang

Provision of the necessary biological data base for commercial lobster aquaculture has been the focus of the multidisciplinary research program at the University of California Bodega Marine Laboratory since the early 1970s under funding from the California Sea Grant College Program. As the smaller juvenile lobsters were the most useful for these fundamental studies, grow-out, in many respects, has been a somewhat neglected phase of lobster culture at our laboratory. The principal features of the grow-out phase are individual rearing compartments, continuous partial seawater replacement, and the provision of food. With a minimum of attention to detail, individual lobsters, which are extremely hardy, can easily be maintained and grown from juveniles to reproductive adults.

This ease of culture, which continues to stimulate interest in commercial exploitation, also somewhat obscures the necessity to clearly understand the mechanisms relating growth to the culture environment. Thus, there are a number of factors, as of yet still inadequately addressed, which significantly affect the economics of rearing large numbers of juvenile lobsters to market size. This section will review the culture techniques commonly used in the laboratory and pinpoint several important aspects of the lobster's biology which still need to be understood.

LABORATORY CULTURE TECHNIQUES

Researchers working with lobsters have developed a variety of culture systems designed to conveniently hold large numbers of individually housed lobsters.[1-8] The two types of grow-out systems presently used at the Bodega Marine Laboratory (see Figures 1 and 2) are similar to other research systems in that ready and easy access to all animals is essential. Additionally, these systems are all based on the concept of having a number of individually isolated lobsters located in a common container so that, in the case of small juveniles, several hundred individuals being used for a particular experiment will experience the same environmental conditions. In order to achieve this goal each unit has a large sump (see Figure 1) from which the seawater is pumped to the rearing containers. In the sump, water returning from the rearing containers is mixed with new incoming sand-filtered seawater from the main laboratory system. Water temperature (typically 20°C) is maintained by means of a temperature-sensitive relay that controls a separate pump that circulates water from the sump through a heat exchanger. Also located in the sump is a standpipe connected to a drain in order to maintain the proper water level.

Seawater pumped from the sump is treated by cartridge filtration and ultraviolet irradiation (laboratory fabricated units based on the design of Loosanoff and Davis[9]) before entering the distribution pipes of the rearing units. In the case of the fiberglass-coated plywood water tables (1.2 m × 2.4 m × 18 cm deep) used for smaller juveniles, water is distributed from underneath the rearing containers either through perimeter pipes[10] or in the newer table systems through rotating arms.[11] The rotating arm distribution system (shown in Figure 1) was developed because of concern for slight table location effects and to inhibit the accumulation of uneaten food particles below the rearing trays as observed in nutrition experiments utilizing formulated rations. The 36 × 152 × 26-cm-deep molded plastic trough rearing troughs (see Figure 3) which are used for larger juveniles (above approximately 50 g) and adults have simple water input hoses at both ends. The seawater then flows through the perforated PVC partitions toward the drain standpipe located at the center. These troughs can be stacked for maximum utilization of floor space (see Figure 2).

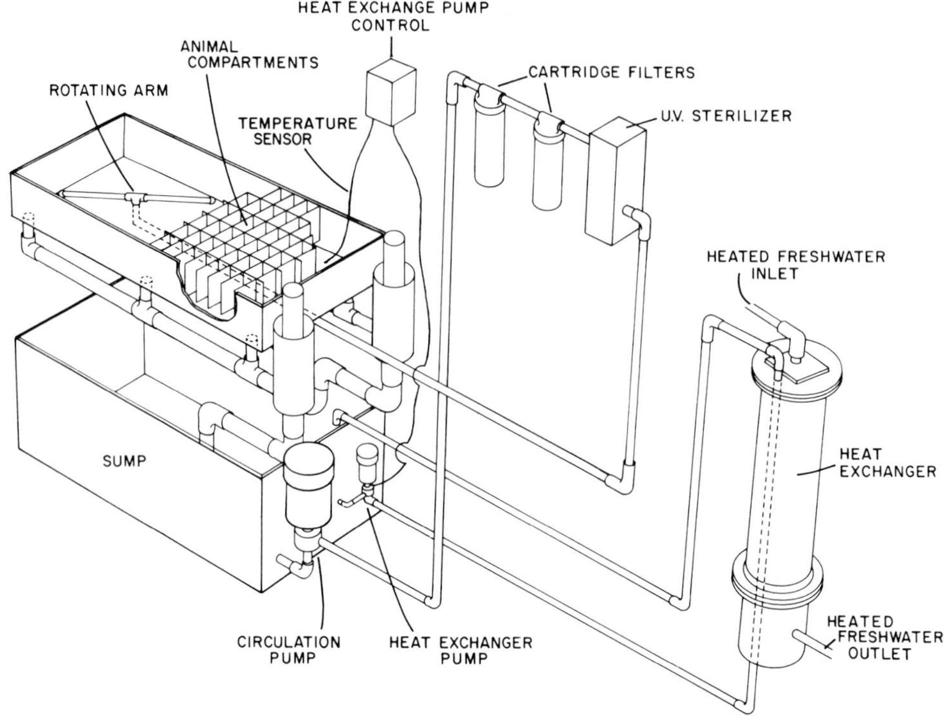

FIGURE 1. Schematic diagram of juvenile grow-out system for lobsters. This recent design incorporates a rotating arm[11] for water circulation.

Typically, fourth-stage larval lobsters are moved into the individual compartments of the juvenile rearing systems even though this lobster stage still maintains a pelagic existence. It is not until the following molt (fifth stage) that lobsters begin to spend the majority of their time moving around on the bottom of the rearing containers. Although a dipper is used in the transfer of the larvae so that they always remain suspended in water,[12] a number of deaths invariably occur, presumably resulting from the stress of transfer. However, further losses after the first few days are unusual provided that the lobsters are fed an adequate diet and sufficient water quality is maintained.

Plastic trays (60 × 120 cm) typically containing 72 identical compartments about 10 × 10 × 15 cm deep are used to rear individual lobsters on the water tables. Each water table holds four of these trays. Water level in the compartments is maintained at 10 cm. The individual compartments have a 1-mm mesh fiberglass screen floor to allow water exchange. The screen is supported by a sheet (60 × 120 cm) of styrene louvre (1.27 × 1.27 × 1.27 cm) made for fluorescent lighting fixtures. The top of the trays protrude about 2 cm above the side of the rearing table so that in the case of an overflow the lobsters are not flooded from their compartments. For those water tables with the perimeter distribution system the trays rest on the pipes; for the newer water tables with the rotating arms the trays are suspended from the sides by means of "S"-shaped hooks. The trays can be built with larger individual compartments for medium-sized juveniles (15 to 50 g), however, this size of lobster frequently tears the fiberglass mesh at the bottom of the container. For these medium-sized juveniles a tray built of perforated plastic is being evaluated along with an automatic siphon device used for lobster culture at the Fisheries Experiment Station at Conwy.[7] The siphon device is placed over the present table standpipe and insures continuous water exchange in each of the individual compartments.

FIGURE 2. Stacked troughs for large juvenile and adult lobster grow-out.

The major parameters for optimal water quality for lobster culture are known and are summarized in Table 1, which was adapted from data presented by Van Olst and co-authors[13] on *Homarus americanus*. From the culture information available on *H. gammarus*[7] it would appear that the two species have similar water quality requirements. The Bodega Marine Laboratory is located on an exposed coastal site (approximately 60 mi north of San Francisco, Calif.) providing access to high quality ocean water. Large roof tanks supplying all of the laboratory facilities are kept filled automatically. Gravity-fed incoming new seawater to the rearing tank sumps is added at approximately 16% of the rate of which water from the sump is being pumped to the rearing containers. Consequently, water quality is of minimal concern due to the continuous dilution of nitrogenous waste metabolites and the stabilization of the other parameters such as salinity and pH. Water quality tests for nitrogenous wastes and bacterial levels are carried out periodically during experiments as a check for unusual conditions. All of the rearing systems are also periodically cleaned; the water tables roughly

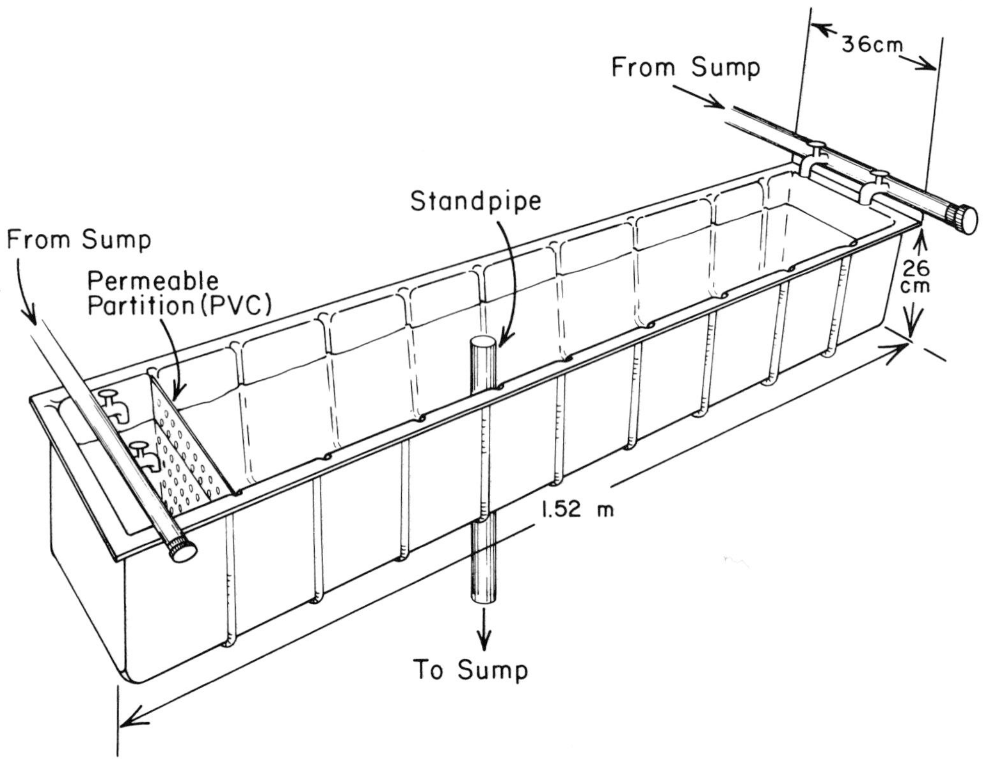

FIGURE 3. Plastic trough used for adult lobster grow-out. Water enters at either end and drains in the center compartment.

Table 1
WATER QUALITY REQUIREMENTS FOR LOBSTER CULTURE

	Optimal	Lethal
Temperature (°C)	20—22	<0, >31
Salinity (0/00)	30	<8, >45
Dissolved oxygen (mg/ℓ)	6.4	<1, >saturation
pH	8.0	<5, >9
NH_3-N	<9.4	94 (48 hr LC_{50})
NO_3-N	<50	>5000
NO_2-N	<10	>100
Chlorine residual (mg/ℓ)	<0.01	0.69 (48 hr LC_{50})

Adapted from Hedgecock, D., Nelson, K., and Shleser, R. A., *Proc. World Maricul. Soc.*, 7, 347, 1976. With permission.

on monthly intervals while the trough systems are cleaned less frequently depending on experimental conditions. Both types of systems are easily cleaned by removing the animals and then the rearing trays or perforated partitions and draining all the water by removing the standpipes.

Successful laboratory culture of small juvenile lobsters at the Bodega Marine Laboratory traditionally has depended on the feeding of live adult brine shrimp, *Artemia salina*, which are purchased from a commercial supplier in the San Francisco area. Frozen brine shrimp can be successfully used, however, juvenile growth rates are approximately 40% less than those achieved with the live diet.[14] Other food items such as chopped mussels, chopped crabs, and shrimp can be used but again growth is not always comparable to those achieved with live brine shrimp. Recently a formulated laboratory ration has been developed for rearing juvenile lobsters.[15] A few lobsters have been grown to maturity on the diet. While survival was good, the growth rate was approximately 35% less than what is achieved on a diet of live brine shrimp. Although this formulated diet is being used extensively as the basis for nutritional experiments, most of the small juveniles cultured in the laboratory are still maintained on brine shrimp. Larger juveniles and adults are fed chopped squid, mussels, fish, or shrimp. Lobsters in the laboratory are generally fed once a day in the morning after any uneaten food from the previous day has been siphoned out. Depending on experimental conditions, the lobsters are fed either a specific amount, in excess, or what is judged by experience to satiety.

COMMERCIAL GROW-OUT

While maintaining the image of certain profitability, the concept of lobster rearing has evolved from that of a simple cottage industry[16] to one of large-scale, highly specialized, automated production facilities. Two facets of lobster biology, its cannibalistic nature and its response to elevated temperature, have strongly influenced this evolution. Though limited phases of communal rearing may be involved, the majority of the grow-out period is anticipated to involve individual containment to prevent cannibalism. This necessity to provide individual rearing containers led towards the development of large-scale capital-intensive facilities. The accelerated growth rate of lobsters observed at elevated culture temperatures[17] was originally seen as a boon to culture in that time to marketability could be cut to a fourth of what it was in nature. However, during the last decade in which the present concepts of lobster farming were evolving, energy costs were undergoing even more dramatic changes. In response to energy costs and the consequent lessening of potential profitability, the rearing concepts were further intensified and enlarged to take advantage of the economies of scale.

Presently being envisioned are facilities producing a constant stream of marketable lobsters on the order of 80,000/month or an annual production of 450 metric tons. In order to keep space and labor costs within bounds these facilities are conceived to be highly intensive and automated. A variety of schemes have been proposed to arrange the large numbers of individual containers into integrated systems. Van Olst and co-authors have discussed the advantages and disadvantages of most of these designs[4,13] which, in the final analysis, will depend to a great extent on the specific site used.

A common theme of all of the proposed production schemes is the set of physiological requirements of the lobsters and the economics of providing for these needs. The magnitude of the various cost factors have been estimated using a mathematical model of a generalized production facility.[18] This model is based on a number of assumed relationships between the known or probable responses of the lobster to environmental variables and estimated engineering costs. A discussion of the development of the model along with an updating of the data base was recently presented by Botsford and Johnston.[19] A summary of the cost analysis resulting from their paper is presented in Table 2.

As this model is based on a wide variety of factors, a number of which are probable but as of yet unproven or ill defined, it should only be considered as an approximation. However, while the model may suffer from numerous imperfections, the general allocation and rankings of cost are thought to be valid and should be useful in focusing research aimed at improving the economics of lobster aquaculture.

An examination of Table 2 shows that by far the greatest single cost is associated with energy. In order to avoid using the prohibitively expensive fossil fuels, most observers have concluded that lobster aquaculture facilities will have to be sited where either ambient

Table 2
HYPOTHETICAL LOBSTER PRODUCTION COST ANALYSIS

	With heat	Without heat	
Cost factors	$/Animal	$/Animal	% Of total
Heat	6.48	—	
Space	1.62	1.62	48
Food	1.02	1.02	30
Labor	0.44	0.44	13
Pumping	0.09	0.09	3
Waste treatment	0.10	0.10	3
Aeration	0.06	0.06	2
Larvae	0.05	0.05	1

Adapted from Johnston, W. E. and Botsford, L. W., *Symp. on New Developments in the Utilization of Heated Effluents and of Recirculation Systems for Intensive Aquaculture,* EIFAC/80/Symp.: E/56, FAO European Inland Fisheries Advisory Commission 11th Session, Stavanger, Norway, 1980. With permission.

temperatures are favorable or seawater is available at an elevated temperature as an inexpensive by-product of other industrial activities such as power generation. The utilization of cooling water from coastal electrical generating stations has been extensively examined by researchers at San Diego State University (for a review see Van Olst et al.[13]) and appears to be an attractive option. Without the cost of heat the two most appropriate areas for research aimed at improving the economics of lobster aquaculture would be the space and food requirements of the lobster. In the updated version of the mathematical model these two factors combined account for 78% of the production costs (see Table 2).

Space

The space requirements of lobsters have been examined several times (for reviews see Aiken[20] and Van Olst et al.[13]) but the physiological response of the animal is still not clearly understood. Additionally, there are two recent publications which suggest it is not the perception by the lobster of limited space per se which inhibits growth but possibly some other factor. Aiken and Waddy[21] found that similar growth rates were achieved at a specific density of lobsters irregardless of whether the animals were separated by physical barriers or banded but free to roam within the total container. More recently Nelson and co-authors[22] have found that a density-dependent growth inhibition factor is produced by juvenile lobsters and suggest that this factor may have influenced the earlier "space limitation" experiments. As the space required by each individual lobster for optimal growth will have a major effect on the overall cost of space in a commercial facility, an understanding of the mechanisms involved might lead to significant cost savings. One key to understanding this phenomenon of "space limitation" would be defining whether the lobsters initial response is a simple reduction of food intake or whether the response is more complicated and involves metabolic changes leading to inefficient utilization of feed. These types of measurements might provide a more instantaneous measure of "space limitation", thereby enhancing the gathering of data as opposed to depending solely on longer-term growth data.

Food

The lack of adequate formulated feeds is the major remaining problem for commercial grow-out of lobsters. Available diets do not promote as rapid rates of growth as are apparently possible. As mentioned previously, the best growth rates of small juveniles in the laboratory have been produced using a diet of live adult brine shrimp.

The actual rate of growth achieved by feeding brine shrimp will vary with temperature and other culture parameters as well as with different progeny. As a useful index, a rate of growth equal to 0.1-mm/day increase in carapace length has been suggested[23] to be representative of ideal culture conditions. A lobster growing at this rate will reach 80 mm in carapace length, weighing 454 g (the traditional 1 lb minimum market size) in a little over 2 years. It should be noted that lobsters cannot actually be grown in this fashion on the brine shrimp diet. After about 6 months of juvenile growth the now relatively small size of an individual brine shrimp limits its effectiveness to support the maximum growth rate in lobsters. Additionally, as Richards and Wikens[7] note, this 2-year growth rate has rarely been achieved. Thus, the brine shrimp standard growth rate is basically an extrapolation of an idealized index; nevertheless, it has come to represent the goal of lobster aquaculture.

When compared to the goal, the recently developed formulated ration[15] (see also Lobster Nutrition page 413) is seen as less than satisfactory. A few lobsters have been grown to market size on this diet with good survival but 3 years were required. While rapid improvements can be expected in the laboratory diets, a suitable formulated commercial diet is thought to be a number of years away. To fill this gap and allow large-scale pilot plant testing to proceed, a combination of a formulated diet with a periodic supplement of fresh natural food items should be examined. This approach would ease the burden (see Lobster Nutrition page 413) on the commercial producer of having to rely completely on fresh food items.

As food is the second most expensive production factor, additional information is also needed on the feeding by the lobster in culture. For example, feeding everyday has been found to be necessary for maximum growth.[7,24,25] However, the very slight reduction in growth rate noted when the feeding is halved, i.e., every other day,[7] suggests that a careful examination of lobster feeding behavior, periodicity, and conversion efficiencies under differing regimes may again result in significant economic gains.

Hormonal Manipulations

Even further improvements in growth rate and economics may ultimately be possible through hormonal manipulation of growth. Since the initial ablative and reimplantation experiments on decapod crustaceans,[26-28] it has been repeatedly demonstrated in a number of crustacean species that molting is controlled by a humoral factor originating in the eyestalks. These experiments have resulted in the postulation of a molt-inhibiting hormone (MIH) that is secreted by neurosecretory neurons in the eyestalk which comprise a gland called the X-organ[29] with endings that terminate in a neurohemal organ called the sinus gland.[30] The action of MIH is thought to be upon the synthesis and release of the molting hormone by the thoracic molting gland, known as the Y-organ (see Kleinholz and Keller[31] and Aiken[20] for reviews). Since crustaceans must periodically shed their confining exoskeletons before growth can be manifested, the control of this process is of importance when considering maximization of growth for aquaculture purposes.

The observed acceleration of the molt cycle as a result of eyestalk ablation has been observed in lobsters as well[32-34] (see Figure 4). We have observed that ablated juvenile lobsters (approximately 4-cm carapace length) have molt cycles that last about 25% of the time required by their control siblings. In addition, Castell et al.[35] have observed significant weight gains following the molt of ablated vs. control animals.

Experiments have also been conducted on molt acceleration in lobsters by means of molting hormone (ecdysteroid) injections. Although large, single administrations of 20-hydroxyecdysone, the active form of the molting hormone,[36] were able to accelerate proecdysial preparations for molt, most of the animals died at ecdysis.[32] If, however, hormone was

FIGURE 4. Effects of eyestalk ablation on lobster growth. The larger animal was bilaterally ablated approximately 2 months after hatching. The sibling control was unoperated. Both animals were subsequently kept in the same system for approximately 1 year and fed excess amounts of food.

injected in a slow-release form (as an acetate derivative in an oil emulsion), the lobsters molted more successfully.[37] Thus, on the laboratory level at least, acceleration of the molt cycle by means of manipulations of the endocrine system appear to have some potential in applied aquaculture.

As mentioned previously, however, there is still a large void in the information concerning the application of laboratory-scale lobster grow-out techniques to large production practices. Since ablated lobsters eat much more than their intact controls, it is not certain whether the accelerated ecdyses and weight gains are economically feasible since metabolic rates and food conversion efficiencies have not yet been determined for ablated animals. Using previously published data,[33] Botsford[38] did show, by use of his theoretical model, that the culture costs could be significantly decreased if the metabolic rate remained constant or increased by no more than a factor of two.

The decreased amount of time required for growth certainly makes attractive the continued research in the area of hormonal manipulations. The benefits are a faster stock turnover and diminished chances for the occurrence of unpredictable events, such as pump failures, labor strikes, or natural disasters. These resulting benefits must be weighed, on the other hand, against the increased labor costs involved in the animal manipulations and the greater mortalities produced by injections or surgery.

As alluded to previously, these uncertainties regarding the transfer to commercial production of laboratory-scale lobster grow-out techniques can only be adequately addressed with further research at the pilot plant scale.

ACKNOWLEDGMENTS

This work is a result of research sponsored in part by NOAA, Office of Sea Grant,

Department of Commerce, under Grant No. 04-8-MO1-189 R/A 28. The U.S. government is authorized to produce and distribute reprints for governmental purposes notwithstanding any copyright notation that may appear hereon.

REFERENCES

1. **Chanley, M. H. and Terry, O. W.,** Inexpensive modular habitats for juvenile lobsters *(Homarus americanus), Aquaculture,* 4, 89, 1974.
2. **Lang, F.,** A simple culture system for juvenile lobsters, *Aquaculture,* 6, 389, 1975.
3. **Sastry, A. N.,** An experimental culture-research facility for the American lobster, *Homarus americanus, Proc. Eur. Mar. Biol. Symp.,* 10, 419, 1975.
4. **Van Olst, J. C., Carlberg, J. M., and Ford, R. F.,** A description of intensive culture systems for the American lobster *(Homarus americanus)* and other cannibalistic crustaceans, *Proc. World Maricul. Soc.,* 8, 271, 1977.
5. **Hand, C.,** Development of aquaculture systems, Sea Grant Publication No. 58, University of California, Institute of Marine Resources, La Jolla, 1977.
6. **Boghen, A. D. and Castell, J. D.,** A recirculating system for small-scale experimental work on juvenile lobsters *(Homarus americanus), Aquaculture,* 18, 383, 1979.
7. **Richards, P. R. and Wickins, J. F.,** Lobster culture research, *Laboratory Leaflet, MAFF Direct. Fish. Res.,* Lowestoft 47, 1979.
8. **Stewart, J. E. and Castell, J. D.,** Various aspects of culturing the American lobster, *Homarus americanus,* in *Advances in Aquaculture,* Pillay, T. V. R. and Dill, W. A., Eds. Fishing News Book, Farnham, Surrey, England, 1979, 314.
9. **Loosanoff, V. L. and Davis, H. C.,** Rearing of bivalve mollusks, in *Advances in Marine Biology,* Vol. 1, Russell, F. S., Ed., Academic Press, New York, 1963, 1.
10. **Conklin, D. E., Devers, K., and Shleser, R. A.,** Initial development of artificial diets for the lobster, *Homarus americanus, Proc. World Maricul. Soc.,* 6, 237, 1975.
11. **Conklin, D. E., Bordner, C. E., Garrett, R. E., and Coffelt, R. J.,** Improved facilities for experimental culture of lobsters, *Proc. World Maricul. Soc.,* (in press).
12. **Hedgecock, D., Nelson, K., and Shleser, R. A.,** Growth differences among families of the lobster, *Homarus americanus, Proc. World Maricul. Soc.,* 7, 347, 1976.
13. **Van Olst, J. C., Carlberg, J. M., and Hughes, J. T.,** Aquaculture, in *The Biology and Management of Lobsters,* Vol. 2, Cobb, J. S. and Phillips, B. F., Eds., Academic Press, New York, 1980, 333.
14. **Van Olst, J. C., Ford, R. F., Carlberg, J. M., and Dorband, W. R.,** Use of thermal effluent in culturing the American lobster, in *Power Plant Waste Heat Utilization in Aquaculture,* Workshop 1, PSE & G, Newark, N.J., 1976, 71.
15. **Conklin, D. E., D'Abramo, L. R., Bordner, C. E., and Baum, N. A.,** A successful diet for the culture of juvenile lobsters, *Aquaculture,* 21, 243, 1980.
16. **Lord, R. A.,** *Crab, Shrimp, and Lobster Lore,* George Routledge and Sons, London, 1867, 122.
17. **Hughes, J. T., Sullivan, J. J., and Shleser, R.,** Enhancement of lobster growth, *Science,* 177, 1110, 1972.
18. **Allen, P. G. and Johnston, W. E.,** Research direction and economic feasibility: an example of systems analysis for lobster aquaculture, *Aquaculture,* 9, 155, 1977.
19. **Johnston, W. E. and Botsford, L. W.,** Systems analysis for lobster culture, in *Symp. on New Developments in the Utilization of Heated Effluents and of Recirculation Systems for Intensive Aquaculture,* EIFAC/80/ Symp.: E/56, FAO European Inland Fisheries Advisory Commission 11th Session, Stavanger, Norway, 1980.
20. **Aiken, D. E.,** Molting and growth, in *The Biology and Management of Lobsters,* Vol. 1, Cobb, J. S. and Phillips, B. F., Eds., Academic Press, New York, 1980, 91.
21. **Aiken, D. E. and Waddy, S. L.,** Space, density and growth of the lobster *(Homarus americanus), Proc. World Maricul. Soc.,* 9, 461, 1978.
22. **Nelson, K., Hedgecock, D., Borgeson, W., Johnson, E., Daggett, R., and Aronstein, D.,** Density-dependent growth inhibition in lobsters, *Homarus* (Decapoda Nephropidae), *Biol. Bull.,* 159, 162, 1980.

23. **Conklin, D. E., Devers, K., and Bordner, C.,** Development of artificial diets for the lobster, *Homarus americanus, Proc. World Maricul. Soc.,* 8, 841, 1977.
24. **Shleser, R. A.,** The effects of feeding frequency and space on the growth of the American, *Homarus americanus, Proc. World Maricul. Soc.,* 5, 149, 1974.
25. **Bordner, C. E. and Conklin, D. E.,** Food consumption and growth of juvenile lobsters, *Aquaculture,* 24, 285, 1981.
26. **Brown, F. and Cunningham, O.,** Influence of the sinus gland of crustaceans on normal viability and ecdysis, *Biol. Bull.,* 77, 104, 1939.
27. **Abramowitz, R. K. and Abramowitz, A. A.,** Moulting, growth, and survival after eyestalk removal in *Uca pugilator, Biol. Bull.,* 798, 179, 1940.
28. **Kleinholz, L. H. and Bourquin, E.,** Effects of eye-stalk removal on decapod crustaceans, *Proc. Natl. Acad. Sci. U.S.A.,* 27, 145, 1941.
29. **Knowles, F. G. W. and Carlisle, D. B.,** Endocrine control in the Crustacea, *Biol. Rev.,* 31, 396, 1956.
30. **Bliss, D. E. and Welsh, J. H.,** The neurosecretory system of brachyuran Crustacea, *Biol. Bull.,* 103, 169, 1952.
31. **Kleinholz, L. H. and Keller, R.,** Endocrine regulation in Crustacea, in *Hormones and Evolution,* Vol. 1, Barrington, E. J. W., Ed., Academic Press, New York, 1979, 159.
32. **Rao, K., Fingerman, S., and Fingerman, M.,** Effects of exogenous ecdysones on the molt cycles of fourth and fifth stage American lobsters, *H. americanus, Comp. Biochem. Physiol.,* 44, 1105, 1973.
33. **Mauviot, J. C. and Castell, J. D.,** Molt- and growth-enhancing effects of bilateral eyestalk ablation on juvenile and adult American lobsters *(Homarus americanus), J. Fish. Res. Bd. Can.,* 33, 1922, 1976.
34. **Chang, E. S. and Bruce, M. J.,** Ecdysteroid titers of juvenile lobsters following molt induction, *J. Exp. Zool.,* 214, 157, 1980.
35. **Castell, J. D., Covey, J. F., Aiken, D. E., and Waddy, S. L.,** The potential for eyestalk ablation as a technique for accelerating growth of lobsters, *(Homarus americanus)* for commercial culture, *Proc. World Maricul. Soc.,* 8, 895, 1977.
36. **Chang, E. S., Sage, B. A., and O'Connor, J. D.,** The qualitative and quantitative determinations of ecdysones in tissues of the crab, *Pachygrapsus crassipes,* following molt induction, *Gen. Comp. Endocrinol.,* 30, 21, 1976.
37. **Gilgan, M. W. and Burns, B. G.,** The successful induction of molting in the adult male lobster *(Homarus americanus)* with a slow-release form of ecdysterone, *Steroids,* 27, 571, 1976.
38. **Botsford, L. W.,** Current economic status of lobster culture research, *Proc. World Maricul. Soc.,* 8, 723, 1977.

Section III
Pathology and Disease Treatments for Crustaceans

DISEASES OF CULTURED PENAEID SHRIMP

Donald V. Lightner

INTRODUCTION

The development of experimental and commercial culture of penaeid shrimp has been accompanied by the occurrence of diseases of infectious and noninfectious etiologies. Two general methods of shrimp culture are practiced in the world.[1] The culture of shrimp in high density, intensively managed pens, ponds, tanks, and raceways is defined as intensive culture, while extensive culture is the culture of shrimp in low density ponds or tidal enclosures in which little or no management is exercised or possible. It is in the former method of culture that the recognition, prevention, and treatment of disease is possible. Furthermore, except for certain parasitic diseases, it is the very nature of intensive culture systems (i.e., high density per unit volume of water used) that encourages the development and transmission of many diseases. The economic incentives of using intensive culture systems also require that disease be understood and controlled. This review of the diseases of cultured penaeids reflects the growth and development of shrimp culture and the importance of disease recognition and control in its development. The majority of the literature on the diseases of shrimp and other cultured crustaceans has been published in the past few years, and virtually all of it in the past 10 years.

This review emphasizes those diseases that cause significant losses, both in terms of the number of shrimp lost and in terms of revenue lost. However, a number of diseases are discussed that are either of common occurrence, but of little economic significance, and those diseases that have caused significant losses, but are of unknown or poorly understood etiology. Not included in this review are parasitic diseases due to the metazoan parasites, as these organisms seldom cause significant problems in shrimp culture. However, the reader is referred to the excellent reviews on these parasites by Kruse,[2] Hutton,[3] Overstreet,[4,5] and Couch.[6]

INFECTIOUS DISEASES

Viruses, Rickettsia, and Chlamydia

Three virus diseases of cultured penaeid shrimp are known and have been described,[7-9] and a fourth is suspected[10] (Tables 1 and 2). Including the viruses in penaeids, nearly 20 virus diseases are now known to occur in the crustacea.[11-14] Major groups of viruses represented in the crustacea include the Reoviridae, Picornaviridae, Baculoviridae, Paramyxoviridae, Rhabdoviridae, and Iridoviridae.[12,14,15] Hence, as the culture of penaeid shrimp increases, the discovery of other virus diseases from these groups (and possibly others) seems highly probable.

While no rickettsia or chlamydia have yet been described in penaeid shrimp, they are known to occur in other crustaceans. Rickettsia or rickettsial-like organisms are known to occur in crabs in Europe[16] and in amphipods in Florida.[17] A chlamydial-like organism has been found recently infecting the Dungeness crab, *Cancer magister*, in the Pacific Northwest.[18] It also seems highly likely that representatives of these organisms will also be found to occur in cultured penaeids.

Penaeid Baculoviruses

The known penaeid baculoviruses infect the epithelial cells of the hepatopancreas and, less commonly, the anterior midgut. The diseases that they cause typically result in high

Table 1
REPORTED HOSTS AND KNOWN GEOGRAPHIC RANGE OF VIRUS-CAUSED DISEASES IN CULTURED PENAEID SHRIMP

Virus	Known hosts	Known geographic range	Ref.
Baculovirus penaei (BP)	*P. duorarum, P. aztecus, P. setifeus*	Florida, Mississippi	5, 6
	P. vannamei, P. stylirostris	Panama, Costa Rica, Equador	99, 100
Monodon baculovirus (MBV)	*P. monodon*	Philippines, Taiwan	8
		Tahiti, Hawaii, Mexico	10
Baculoviral midgut gland necrosis virus (BMNV)	*P. japonicus*	Southern Japan	9
Infectious hypodermal and hematopoietic necrosis virus (IHHNV)	*P. stylirostris*	Hawaii	19
	P. monodon	Guam	19

mortality in the postlarval or early juvenile stages,[9,10,14] but may be present in and cause mortalities in older animals.[7,8] The virus and the pathological changes that they cause are similar in each of these diseases. In every case, the hepatopancreatic tubule epithelium (and, in very young postlarval or in larval animals, the anterior midgut epithelium) is affected (Table 2). Death apparently results from destruction of the epithelium of these organs.

Baculovirus penaei (BP), known to occur in several penaeid species, cultured on the northern coast of the Gulf of Mexico and the Pacific coast of Central America, is diagnosed by the demonstration of tetrahedral inclusion bodies in the greatly hypertrophied nuclei of affected cells (Figure 1). Demonstration of polyhedra may be accomplished in unstained squashes of the hepatopancreas or in stained histological preparations of the organ.Monodon baculovirus (MBV)[8,10] is diagnosed in the same manner as BP disease, except that the polyhedral inclusion bodies tend to be multiple and spherical within affected host cell nuclei (Figure 2). The baculovirus occurring in *Penaeus japonicus* (BMN) does not form polyhedral inclusion bodies in the nucleus, but, nevertheless, it may be diagnosed in its normal host species by the histological demonstration of greatly hypertrophied nuclei within hepatopancreatic epithelial cells that are undergoing necrosis (Figure 3).

Other Virus Diseases

Infectious hypodermal and hematopoietic necrosis (IHHN) is a recently discovered disease of cultured blue shrimp *(P. stylirostris)* of viral etiology.[19] Small particles (16 to 28 nm) of cubic symmetry have been demonstrated in affected tissues of *P. stylirostris* with IHHN (Figure 4A). A similar, if not identical, disease is known to occur in *P. monodon,*[19] and similar virus-like particles have been observed in apparently healthy *P. aztecus.*[135] The shape and small size of IHHN virus suggest that it may be a picornavirus or a parvovirus. The presence of virus-like particles in aggregates and lattice structures in the cytoplasm of some affected cells suggests it to be a picornavirus. However, the intranuclear eosinophilic inclusion bodies (Figures 4C to 4F) are typical to parvovirus infections in insects and mammals. Hence, the classification of IHHN virus will depend upon determination of its nucleic acid type. Shrimp dying with acute IHHN show a massive destruction of the cuticular hypodermis and often of the hematopoietic organs, of glial cells in the nerve cord, and of loose connective tissues such as the subcutis and the gut serosa (Figure 4B). To date, little is known about the disease, but so far only populations of *P. stylirostris* or *P. monodon* within the size range of 0.05 to 1.0 g have been observed to have epizootics from the disease that result

Table 2
SUMMARY OF THE KNOWN HOSTS, THE ORGANS AFFECTED, AND THE VIRUS MORPHOLOGY IN PENAEID DISEASES OF KNOWN OR SUSPECTED VIRAL ETIOLOGY

Disease	Known hosts	Life stage affected	Target organ(s)[a]	Virus type	Morphology: Shape	Morphology: Virion	Inclusions or polyhedra	Ref.
Baculovirus penaei (BP)	*P. duorarum*	All	HP	Baculovirus	Rod	74 × ~ 270 nm	Eosinophilic, intranuclear, pyramidal (single or multiple)	5, 6
	P. setiferus, P. aztecus	All	HP, AMG					
	P. vannamei	Larval,	HP, AMG					99, 100, 128
	P. stylirostris	Postlarval	HP, AMG					
Monodon baculovirus (MBV)	*P. monodon*	All	HP, AMG	Baculovirus	Rod	69 × ~ 275 nm	Eosinophilic, intranuclear, spherical (single or multiple)	8
Baculoviral midgut gland[b] necrosis (BMN)	*P. japonicus*	Larval, postlarval	HP, AMG	Baculovirus	Rod	72 × ~ 310 nm	None	9
Infectious hypodermal and hematopoietic necrosis (IHHN)	*P. stylirostris P. monodon*	Postlarval, juvenile	CH, HE, HEO, CT	Picornavirus or parvovirus	Icosahedron	16—28 nm	Eosinophilic, intranuclear, single, and basophilic cytoplasmic	19
Not named	*P. aztecus*	Juvenile	Heart (?)	Picornavirus or parvovirus	Icosahedron	23 nm	Cytoplasmic in fixed phagocytes of heart	135

[a] Abbreviations: HP = hepatopancreatic epithelium, AMG = anterior midgut epithelium, CH = cuticular hypodermis, HE = hemocytes, HEO = hematopoietic organs, and CI = connective tissue.
[b] Midgut gland = hepatopancreas.

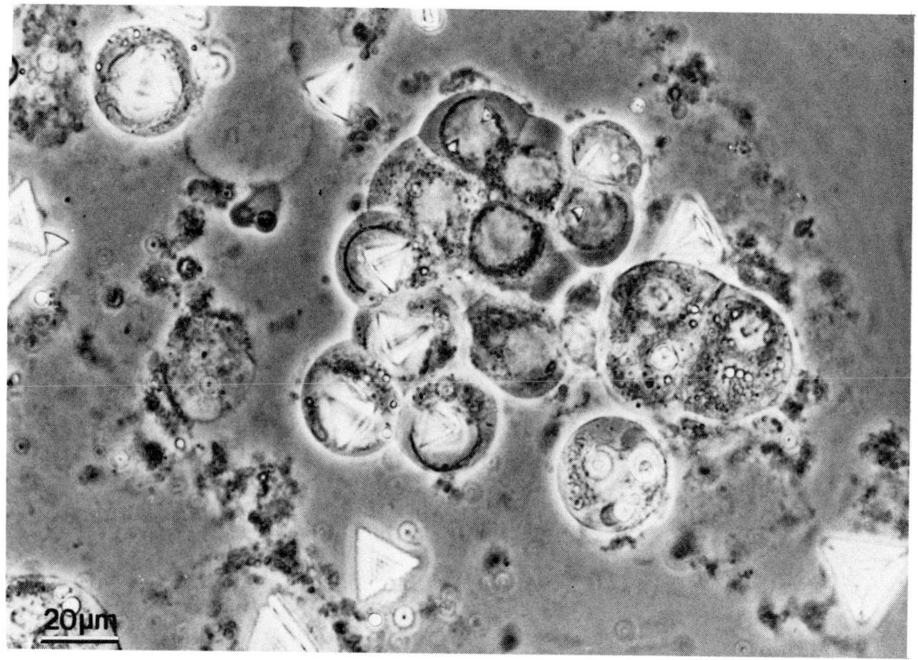

A

B

FIGURE 1. (A) Intranuclear polyhedral bodies of *Baculovirus penaei* in a fresh squash of the hepatopancreas of a pink shrimp *Penaeus duorarum*. (No stain.) (Photo courtesy of J. A. Couch, U.S.E.P.A., Gulf Breeze, Fla.) (B) Electron micrograph of a *B. penaei* (BP) polyhedral body within the hypertrophied nucleus of hepatopancreatic epithelial cell. Rod-shaped particles of BP virus (arrows) are present free and occluded within the polyhedron. (Lead citrate and uranyl acetate.) (Photo courtesy of J. A. Couch, U.S.E.P.A., Gulf Breeze, Fla.)

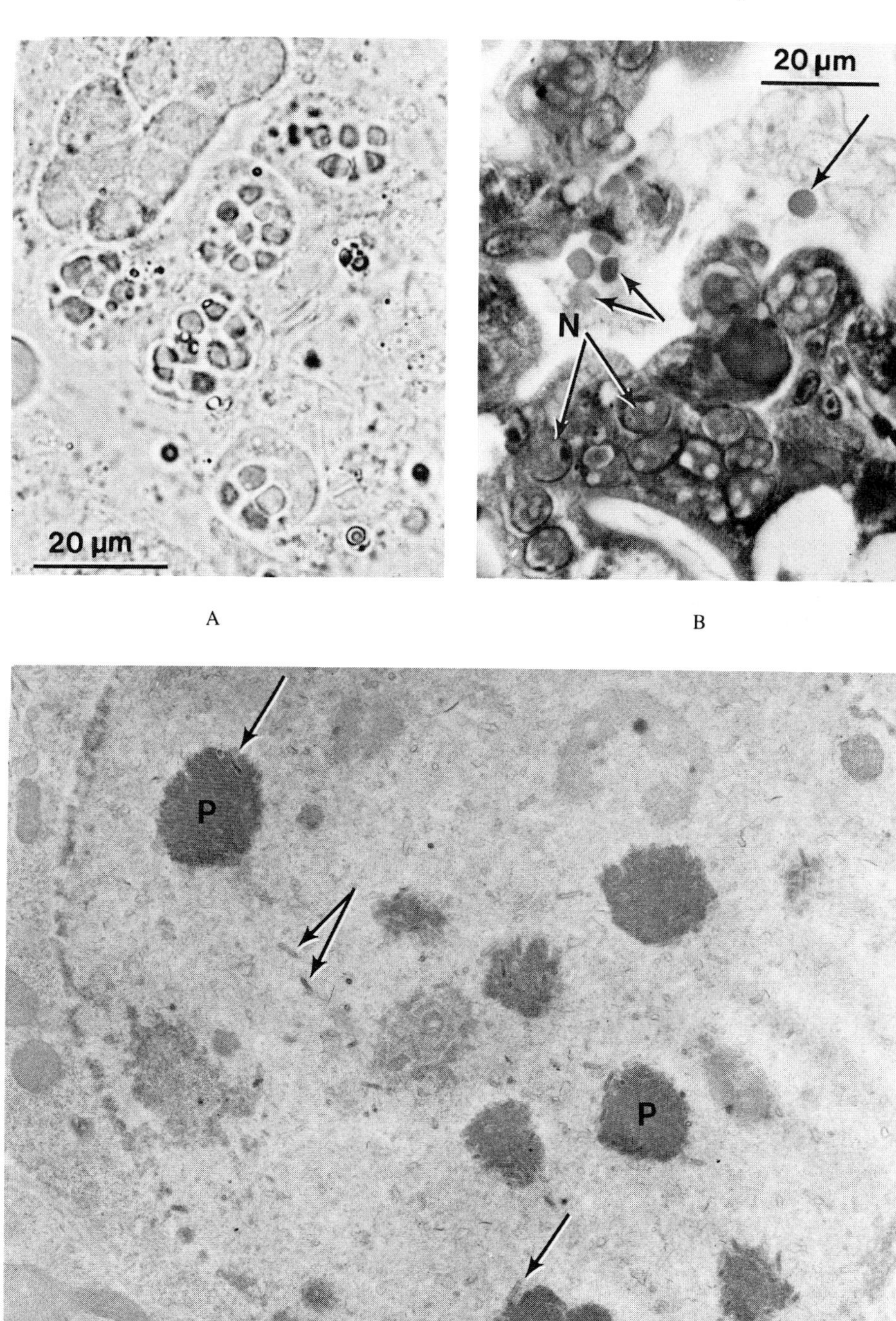

FIGURE 2(A) Intranuclear baculovirus polyhedral bodies in a fresh squash of the hepatopancreas from a postlarval *Penaeus monodon*. (Squash stained with 0.05% aqueous malachite green.) (B) Light photomicrograph of a histological section of the hepatopancreas of *P. monodon*. A number of hepatopancreatic tubule epithelial cells contain multiple polyhedra within hypertrophied nuclei (N). Several polyhedra (arrows) have been released into the tubule lumen by lysis of tubule epithelial cells. (Hematoxylin and eosin.) (C) Electron micrograph of a hypertrophied nucleus in an hepatopancreatic epithelial cell from a postlarval *P. monodon*. Free and occluded baculovirus particles (arrows) are present in the nucleoplasm and in the developing polyhedral bodies (P). (Lead citrate and uranyl acetate.)

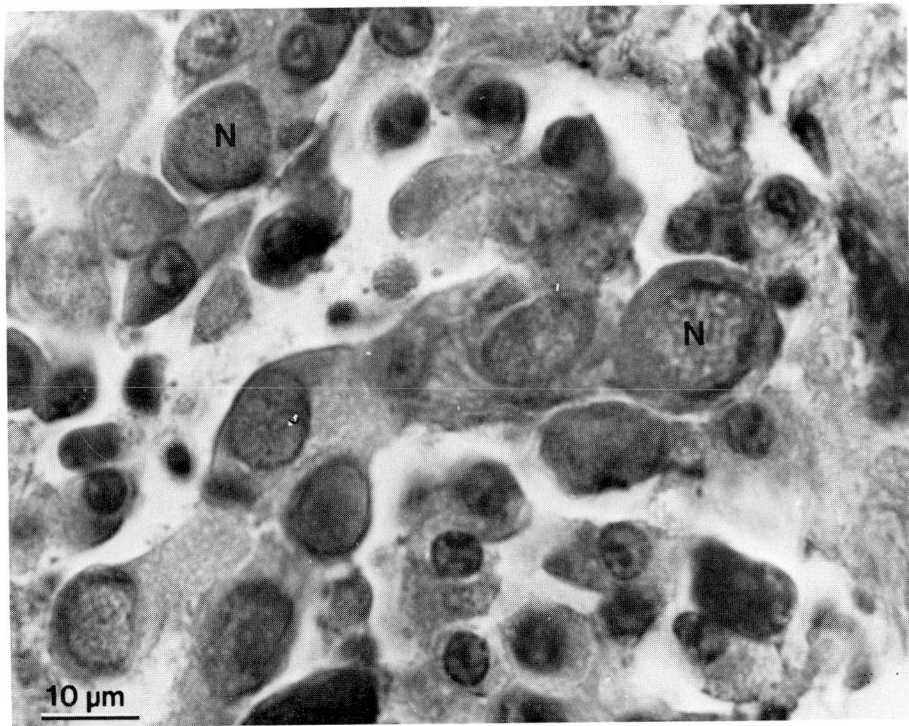

A

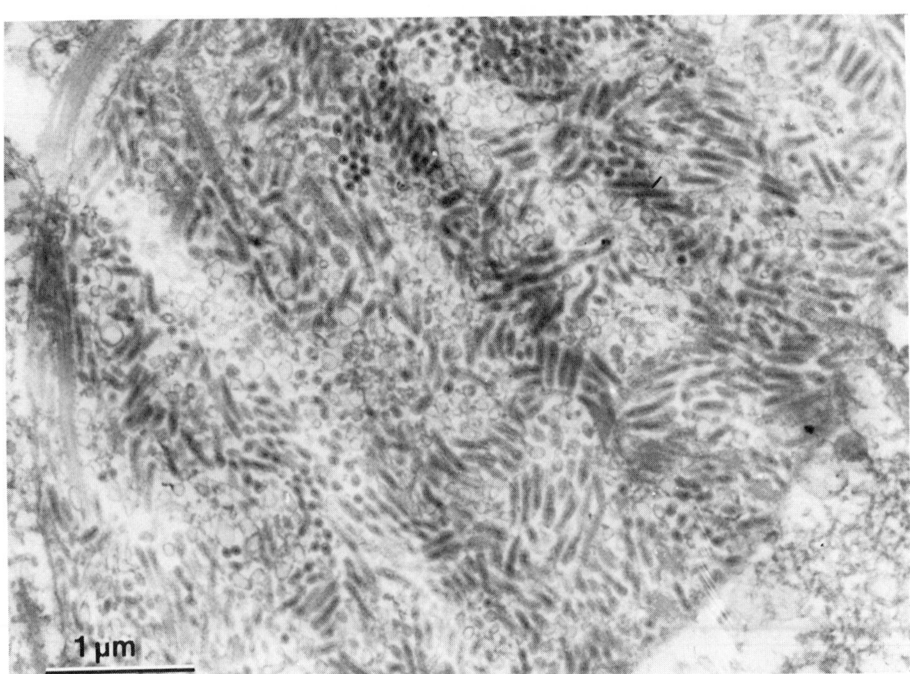

B

FIGURE 3. (A) Histologic section of the hepatopancreas of a postlarval *P. japonicus* with baculoviral midgut gland (= hepatopancreas) necrosis disease. Numerous hypertrophied nuclei (N) of the hepatopancreatic tubule epithelium are shown. Polyhedral bodies are not known to occur in this disease. (Hematoxylin and eosin.) (B) Electron micrograph of one of the hypertrophied nuclei shown in 3A. Baculovirus particles nearly fill the nucleus. (Lead citrate and uranyl acetate.)

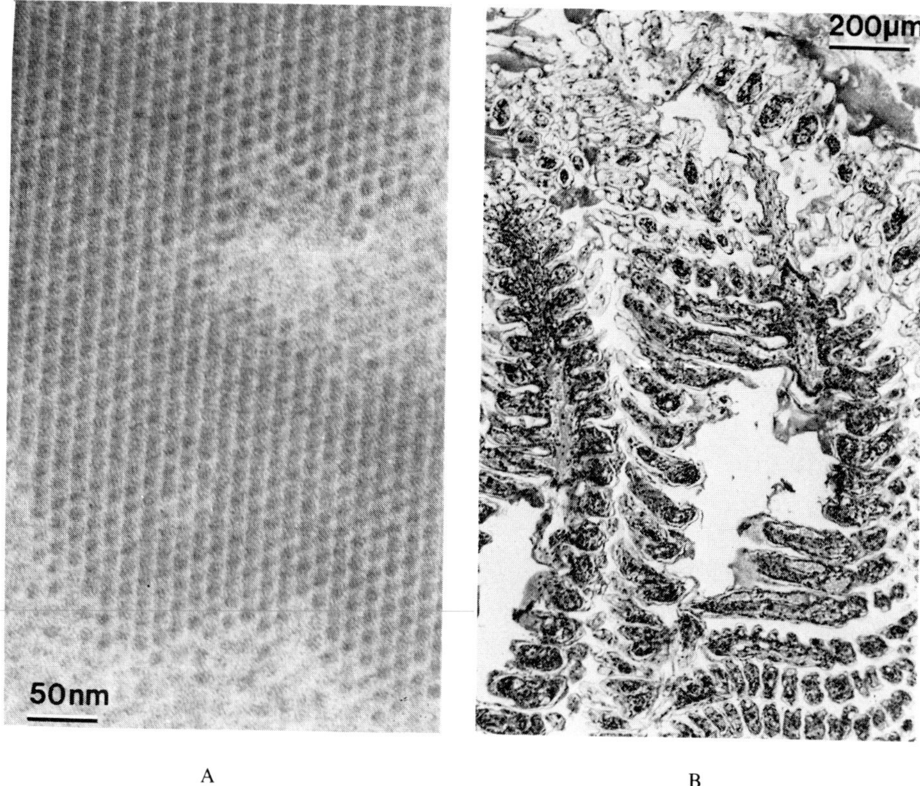

FIGURE 4. (A) Electron micrograph of a lattice structure of IHHN virus, a probable picornavirus from *P. stylirostris*. This structure was part of a larger cytoplasmic inclusion body in a subhypodermal connective tissue cell. (Lead citrate and uranyl acetate.) (B) Histological section of the gills from a juvenile *P. stylirostris* with an advanced IHHN infection. Necrosis of gill rachi and lamellae is more advanced apically (top). (Hematoxylin and eosin.) (C,D,E, and F) Examples of the common eosinophilic inclusion bodies (arrows) within hypertrophied nuclei that occur in certain target organs of *P. stylirostris* during IHHN infections. (Davidson's fixation; hematoxylin and eosin.)

in massive mortalities (often 80 to 90% of the population within 2 weeks of onset). Older, larger shrimp of these species apparently may contract the disease, but high mortalities in larger shrimp with the disease have not been observed.

IHHN disease is diagnosed by the histological demonstration of large eosinophilic inclusion bodies (Figures 4C to 4F) within the nuclei of cuticular hypodermal, hematopoietic or connective tissue cells in tissues showing other signs of necrosis from moribund shrimp.

Bacterial Diseases

A number of diseases caused by bacteria have been reported from penaeid shrimp. While bacterial diseases of a primary bacterial etiology have been reported from penaeid shrimp,[20,21] the majority are of a secondary etiology.[6,22] The only true primary bacterial disease in cultured decapod crustacea is gaffkemia disease that occurs in lobsters of the genus *Homarus*. That disease is caused by a Gram-positive tetrad-forming bacterium that was originally described as *Gaffkya homari*[23,24] and is now known as *Aerococcus viridans* variety *homari*.[25]

In every reported case of bacterial infections in penaeid shrimp, motile, Gram-negative, oxidase-positive, fermentative rods have been isolated[22,26-31] (Table 3). Most isolates have been *Vibrio* species, usually *V. alginolyticus*, *V. parahaemolyticus*, or *V. anguillarum*. Certain other Gram-negative rods including *Pseudomonas* spp. and *Aeromonas* spp. may occasionally be involved in bacterial disease syndromes in penaeid shrimp (Table 4). Most,

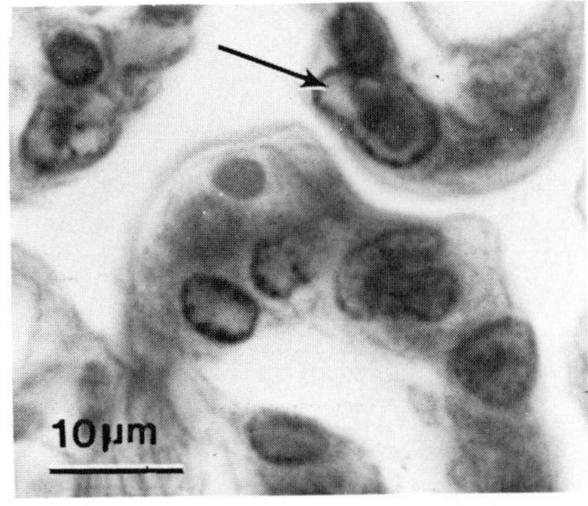

FIGURE 4C.

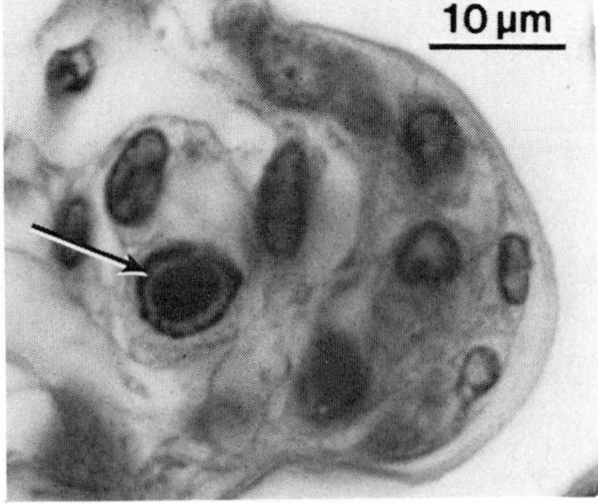

FIGURE 4D.

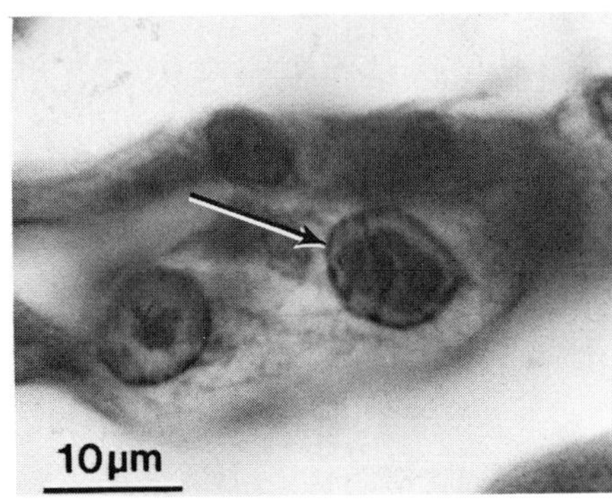

FIGURE 4E.

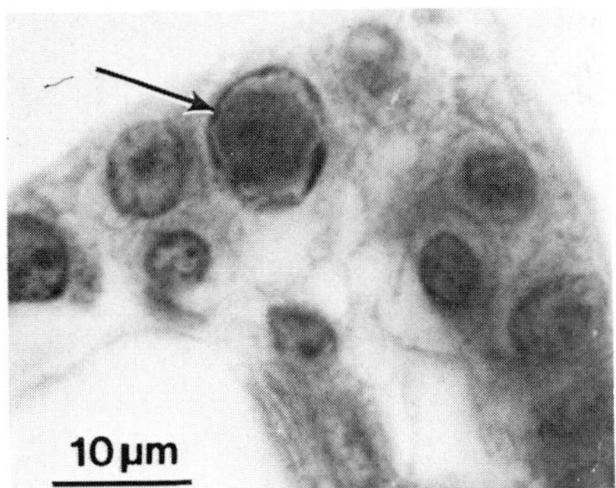

FIGURE 4F.

if not all, of these organisms are part of the normal microflora of these animals.[32-35] Some bacterial taxonomists maintain that certain chitinase-positive vibrios and pseudomonads should be classified as *Beneckea* sp. However, as *Beneckea* is not recognized in the most recent edition of *Bergey's Manual,*[36] that classification will not be used here.

Bacterial infections in shrimp may take two forms: localized pits in the cuticle (shell disease) or localized infections within the body and generalized septicemias (Table 4). All life stages may be affected.

In shell disease,[20,37,38] the causative organisms cause erosions of the cuticle by the production of the enzyme chitinase. If such lesions are not successfully resolved by the host's inflammatory response, septicemia and death will result. Other bacterial diseases may occur as localized, well-encapsulated abscesses in the shrimp viscera, muscle, gills, etc., or as generalized septicemias. These latter types of infections, as well as shell disease, typically are secondary, resulting from some primary lesion due to another infectious organism, a parasite, to mechanical trauma or to nutritional, chemical, or physical stress.[22]

Diagnosis of bacterial infections is made by the demonstration of Gram-negative motile rods in the hemolymph or tissues of shrimp with advanced shell disease, localized septic lesions, or generalized septicemias. Successful therapy of bacterial infections in penaeid shrimp (Tables 5 and 6) has been reported by the direct addition of certain antibiotics to the culture tank water,[30,39,40] which is practical only in shrimp hatching, larval rearing, and nursery tanks, or by the incorporation of antibiotics in the ration for larger shrimp.[22,29,39,41-43] Drying, cleaning, and disinfection of spawning, hatching, larval rearing, and nursery tanks between uses is effective in reducing bacterial diseases in larvae and postlarvae.[22,30,41]

Table 3
KEY CHARACTERISTICS OF BACTERIA COMMONLY ISOLATED FROM SHRIMP[a]

Reaction	*Vibrio alginolyticus*	*V. parahaemolyticus*	*Pseudomonas* sp.	*Aeromonas* sp.	*Flavobacterium* sp.
Gram stain	−	−	−	−	−
MacConkey (growth)	+	+	+	+	+
Oxidase	+	+	+	+	+
Motility	+	+	+	+	−
Glucose (fermented)	+	+	−	+	−
Glucose (oxidized)	+	+	+	+	+
Lactose (acid)	−	−	+/−	+/−	+/−
Glucose (acid)	+	+	+/−	+	−
Sucrose (acid)	+	−	+/−	+/−	−
Indole	+	+	−	+	+/−
Nitrate	+/−	+/−	+/−	+	+
Arginine	−	−	+/−	+	−
Citrate	+/−	+/−	+	+/−	+
Pigment	−	−	Fluorescent green, pink	−	Yellow, orange, red
TSI	A/A	K/A	K/K	K(A)/A	K/N to K(A)
Requires NaCl	+	+	−	−	−
TCBS	Yellow	Blue	()[b]	()	()

[a] Using Analytab Products, Inc. system plus conventional tests.
[b] () = Information not available or not tested.

Table 4
RELATIVE FREQUENCY OF ISOLATION OF CERTAIN BACTERIA[a] FROM THE HEMOLYMPH OR TISSUES OF CULTURED PENAEID SHRIMP WITH VARIOUS DISEASE SYNDROMES

	Disease[b]				
Organism	SD	HE	BD	BSX	Isolates (%)
Vibrio alginolyticus	16	20	4	1	42
V. parahaemolyticus	7	0	0	0	7
Vibrio spp.[c]	8	5	1	1	15
Pseudomonas sp.	6	0	0	6	12
Aeromonas sp.	1	1	1	2	5
Flavobacterium	2	0	0	5	7
Pasteurella sp.	3	0	0	0	3
Moraxella sp.	0	0	0	4	4
Achromobacter sp.	0	0	0	2	2
Acinetobacter sp.	0	0	1	0	1
Yersinia sp.	0	0	1	0	1

[a] Identifications made from culture by API 20E[R] system (Analytab Products, 200 Express St., Plainview, N.Y. 11803).
[b] SD = shell disease consisting of melanized erosions of cuticle on general body surface, gills, or appendages; HE = hemocytic enteritis; BD = black death (ascorbic acid deficiency syndrome); BSX = blue shrimp syndrome.
[c] Includes several *Vibrio* species, including *V. anguillarum*.

Table 5
CAUSATIVE ORGANISM, LESION TYPE, AND PREFERRED METHOD OF TREATMENT FOR THE MOST COMMON BACTERIAL DISEASES OF CULTURED PENAEID SHRIMP

Bacterial disease	Most common causative organism(s)	Reported treatment(s)[a]	Ref.
Hatchery			
Shell disease, appendage rot, septicemias	*Vibrio alginolyticus, V. parahaemolyticus, V. anguillarum, Pseudomonas* spp., *Aeromonas* spp.	Determine and alleviate stress factors; dry and clean rearing tanks between uses	22, 30, 56
		Add antibiotic to culture tank water	77, 101
		Furacin, Furanace (1 mg/ℓ)	22, 102
		Chloramphenicol (1—10 mg/ℓ)	30, 103
		Oxytetracycline (60—250 mg/ℓ)	40
Juvenile and adult			
Septicemias, shell disease, brown spot disease, infected wounds	*Vibrio alginolyticus, V. parahaemolyticus, V. anguillarum, Pseudomonas* spp., *Aeromonas* spp.	Add antibiotic to feed	101
		Oxytetracycline	42, 43
		(500—1000 mg/kg feed)	41
		Furacin, furanace	41, 79
		NF-180 (100—500 mg/kg feed)	39

[a] FDA approval for this usage may be required.

Table 6
SENSITIVITY OF *VIBRIO* SPP. ISOLATED FROM CULTURED PENAEID SHRIMP

Antibiotic	Isolates sensitive	Sensitive (%)
Aureomycin	18	95
Bacitracin	0	0
Colistin	4	21
Erythromycin	2	11
Furacin	19	100
Furoxone	11	58
Nalidixic acid	19	100
Neomycin	1	5
Nitrofurantoin	16	84
Novobiocin	4	21
Oleandomycin	0	0
Penicillin	0	0
Polymyxin B	9	47
Streptomycin	0	0
Sulfonamides	0	0
Terramycin	19	100
Tetracycline	16	84

Fungus-Caused Diseases

Two general types of fungus diseases occur in cultured penaeid shrimp. These are the systemic noninflammatory mycoses of the larval and postlarval stages, and the generally localized mycoses of the juvenile and adult stages that are typically accompanied by an inflammatory response.

Table 7
FUNGI REPORTED TO CAUSE DISEASE IN CULTURED PENAEID SHRIMP OR IN CLOSELY RELATED SPECIES

Fungus	Host species	Geographic location	Ref.
Class: Phycomycetes			
Lagenidium callinectes	All penaeids	Ubiquitous	22, 44, 47, 48, 104
Siropidium sp.	*P. monodon, P. merguiensis, P. japonicus, P. aztecus*	Tahiti, Texas	22
Pythium sp.	*Palaemon serratus*	Great Britian	37, 105
Leptolegnia marina	*P. monodon*	India	106
Saprolegnia parasitica	*P. monodon*	India	106
	Palaemonetes kadiakensis	Ohio	107
Achlya flagellata	*P. kadiakensis*	Ohio	107
Saprolegnia sp., *Achlya* sp., *Aphanomyces* sp., *Phythium* sp., and *Leptomitus* sp.	*Macrobrachium rosenbergii*	India	108
Atkinsiella dubia	*P. aztecus*	Texas	47
Haliphthoros milfordensis	*P. duorarum, P. setiferus*	North Carolina	109
Class: Fungi Imperfecti			
Fusarium solani	*P. japonicus*	Japan, Tahiti, Hawaii	30, 47, 53, 110
	P. aztecus,	Texas	54
	P. duorarum, P. setiferus	North Carolina	22
	P. californiensis, P. stylirostris, P. vannamei	Mexico, Hawaii, Ecuador, Peru	47, 128

The systemic mycosis of larval and postlarval penaeids have been the cause of serious mortalities in penaeid hatcheries throughout the world.[22,30,44-46] The reported fungi that cause this group of diseases are all phycomycetes (Table 7). Two genera of phycomycetes, *Lagenidium* and *Siropidium,* are the best known of this group of shrimp pathogens and apparently are ubiquitous. Figure 5A shows the typical appearance of a larval penaeid shrimp which is infected with a phycomycete. Identification of the causative fungus may sometimes be made by demonstration of the method of sporogenesis. The most common phycomycetous infections are caused by *L. callinectes* and *Siropidium* sp. and may be distinguished by the method of sporogenesis (Figures 5B and 5C).

Larval penaeids apparently lack the ability to mount a significant inflammatory response to phycomycetous fungi once an infection is established. Hence, these fungi grow unrestricted within infected larvae, and in a matter of only a few hours after initiation, the infection is well established. An infection is established when a fungal zoospore attaches to, and encysts upon, the cuticle of an egg or larval shrimp. The encysted spore then germinates, with the germ tube penetrating the cuticle. The mycelium grows from this point of entry, replacing most of the host's muscle and other soft tissues, until the larva is literally filled with hyphae. Infected individuals become immobile and will settle to the bottom of the tank if aeration or circulation of the tank water is interrupted. Sporogenesis begins when the host's nutrients have been depleted, with the formation of sporangia within the body of the host, the formation of discharge tubes, and eventual release of zoospores.[44,47]

The source of the fungi infecting larval shrimp in larval-rearing tanks may be from the

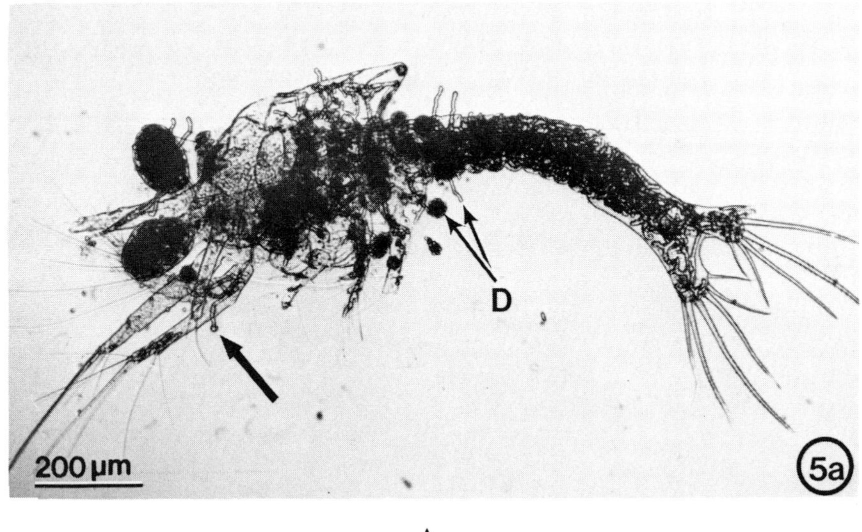

A

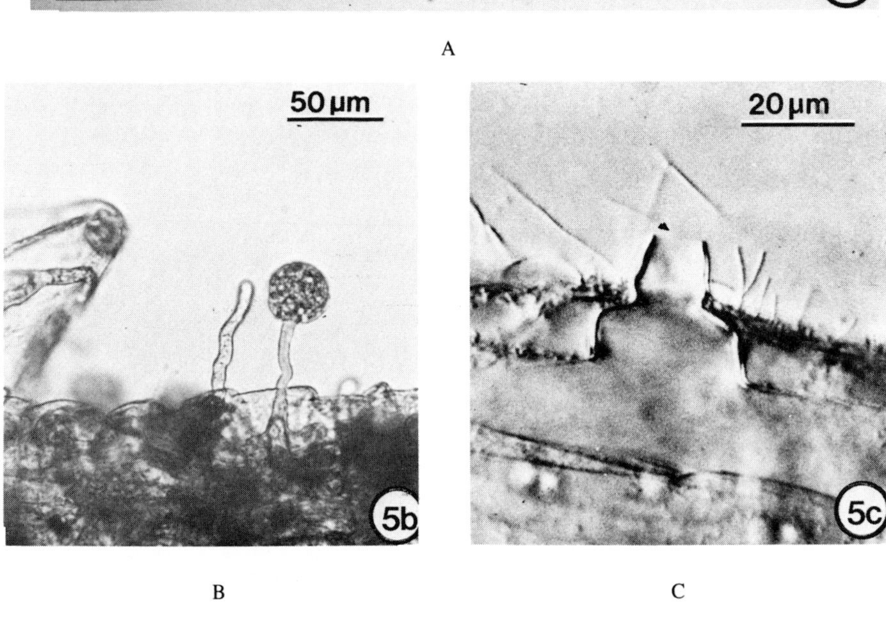

B C

FIGURE 5. (A) A larval *P. setiferus* with an advanced infection of *Lagenidium callinectes*. Hyphae are visible in the abdomen and cephalothorax, and specialized discharge tubes (D) that are used in sporogenesis are shown protruding from the shrimp. Also shown is a germinated zoospore (arrow) from which the infection may have been initiated. (No stain.) (B) A higher magnification of a discharge tube and terminal vesicle of *Lagenidium*. (No stain.) (C) Wet mount of a sporangium and discharge tube of *Siropidium* sp. in an appendage of a larval penaeid. (No stain.) (Photo courtesy of C. E. Bland, East Carolina University, Greenville, N.C.)

parent broodstock or from the presence of carrier hosts in the seawater supply. Besides penaeids, the eggs and larvae of various species of crabs and barnacles have also been found to be infected by *Lagenidium* spp.[48,49] and a variety of other phycomycetes.[46,50]

Several methods of chemotherapy for larval crustacean mycoses have been reported,[30,51,52] and two chemicals have shown promise in controlling the fungus. Malachite green oxylate at 0.006 ppm (static) was reported as effective in arresting *Lagenidium* epizootics in larval rearing tanks or in preventing epizootics, if added prior to the time the epizootic becomes established.[22,46,52] Other investigators[30,51] found Treflan® (trifuralin, Elanco Products Co.) to be effective in preventing *Lagenidium* and *Siropidium* epizootics in the culture of larval crabs and larval penaeid shrimp. For control of these fungi in shrimp, Aquacop[30] found that a single application of 0.01 ppm trifuralin was enough to kill *Lagenidium* and *Siropidium* zoospores, but had no effect on hyphae. Aquacop[30] further reported that multiple 6-hr-duration applications of Treflan® in the parts-per-billion range are effective in preventing the disease.

Larval mycoses may be prevented in some facilities without the use of chemicals, provided that the seawater used in rearing the larvae has been pretreated or filtered to insure the absence of zoospores. This has been accomplished by collection of spawned eggs on a fine-meshed nylon screen and then rinsing the eggs with several changes of clean seawater or seawater containing malachite green or trifuralin.[30,41] Other facilities separate healthy nauplii from spawning debris, unhatched eggs, and diseased nauplii by their phototaxic response to point light source at the surface of the hatchery tank.

Penaeid juvenile and adult life stages are reported to be infected by a variety of phycomycetous fungi and a single genera of the imperfect fungi (Table 7). To date, only infections caused by the imperfect fungus *Fusarium solani* have been reported to result in serious losses in aquaculture; hence, only *F. solani* will be discussed in detail. *F. solani* has been reported from a wide variety of freshwater and marine cultured and wild decapod crustaceans, and its distribution appears to be worldwide.[22,47,53,54] It has been found to cause a serious disease in several species of penaeids, with *P. japonicus* and *P. californiensis* being especially susceptible. Other species are less susceptible (e.g., *P. stylirostris* and *P. vannamei*) while others (e.g., *P. monodon*) appear to be highly resistant.[47,55]

F. solani is present in soils and detritus, and, hence, may be introduced into aquaculture systems from the pond bottom mud and detritus or from the seawater source. Spores of *F. solani* have been found in seawater from wells placed in beach sands.[55] The fungus is an opportunistic pathogen and, in highly susceptible species, the slightest wound or abrasion may become infected.[56] Well-developed *F. solani*-caused lesions are typically darkly melanized (Figure 6A). In penaeids,[47,53,57] *F. solani*-caused lesions occur in the gills (causing the black gill disease of *P. japonicus*), at the bases of the appendages, or on the cuticle.

Diagnosis of fusarium disease is made by the examination (at 100 ×) of a wet mount of material scraped from a suspect lesion. Demonstration of the "boat-shaped" macroconidia (Figure 6B), which is the characteristic spore for members of the genus *Fusarium*, provides a presumptive diagnosis.

No effective chemotherapy or preventative measures are known for fusarium disease,[55,58] and mortalities due to fusarium disease in highly susceptible species may approach 100%, especially in high density culture systems. Rearing of susceptible species in low density, or the rearing of *F. solani*-resistant species in high density culture systems is, at the present time, the only known method of preventing fusarium disease.[55]

NONINFECTIOUS DISEASES

Diseases Caused by Epicommensals

Among the more serious diseases of penaeids are those caused by noninfectious epicommensal organisms. All life stages may be affected, but the most serious losses are encountered in juvenile and adult stages when the gills of the host become fouled by heavy infestations of epicommensal organisms such as *Leucothrix mucor, Zoothamnium* sp., and *Epistylis* sp. Table 8 lists the more commonly observed epicommensal organisms that alone, or with other epicommensals, cause disease in cultured penaeids. The more important diseases are discussed here.

Leucothrix Disease

L. mucor is a common ubiquitous estuarine and marine bacterium.[59,60] It attaches to living

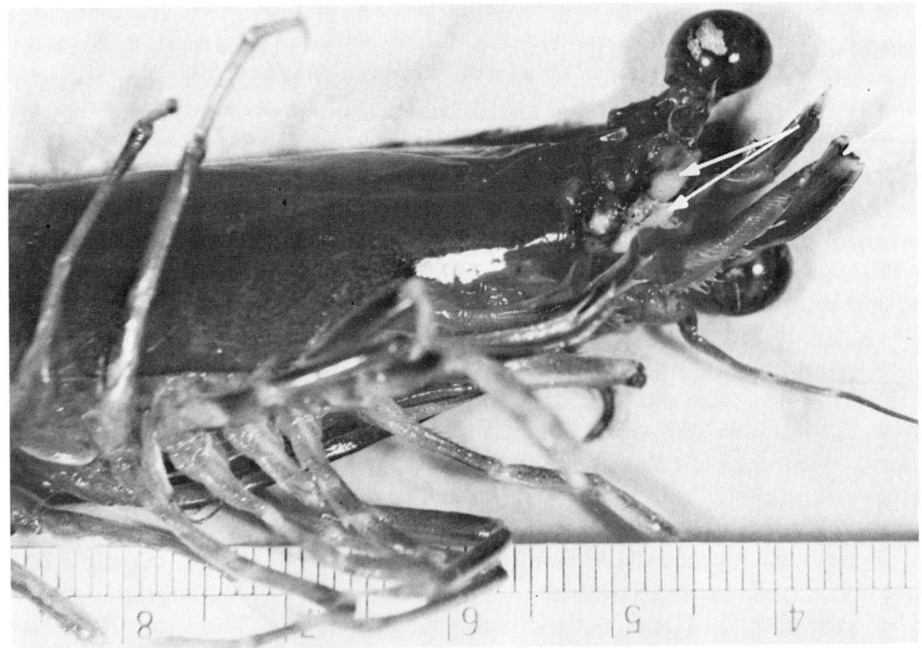

A

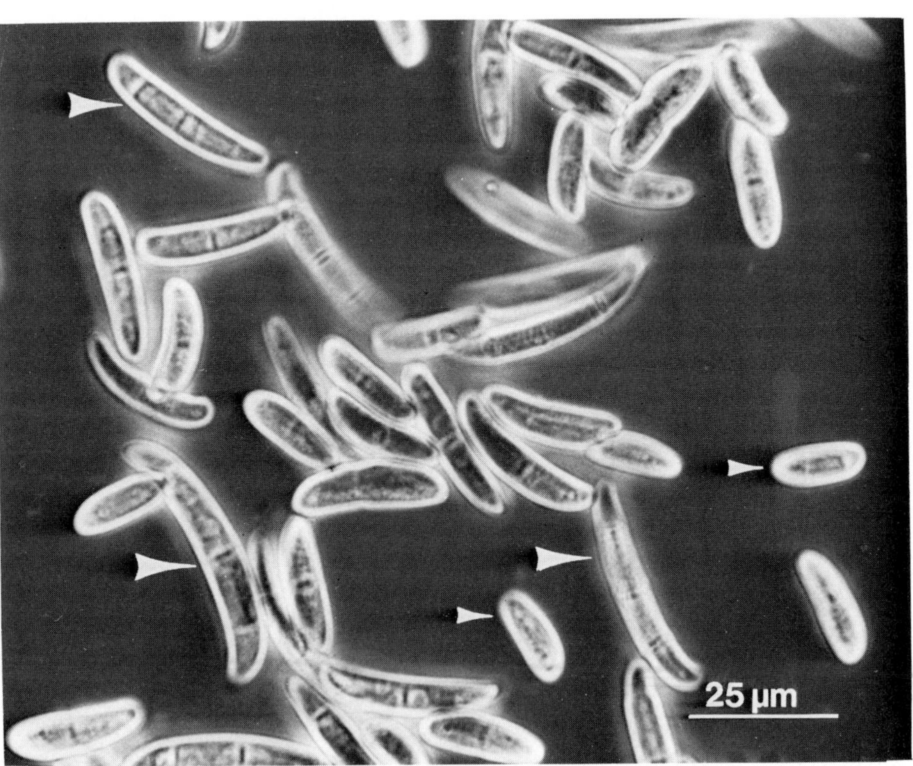

B

FIGURE 6. (A) A lesion (arrow) due to *Fusarium solani* on the head appendages of a juvenile *P. californiensis*. (B) Wet mount of conidiospores of *F. solani*. Both macroconidia (large arrows) and microconidia (small arrows) are shown. (No stain.)

Table 8
EPICOMMENSAL ORGANISMS OBSERVED OR REPORTED TO CAUSE DISEASE IN CULTURED PENAEID SHRIMP

Classification	Ref.
Bacteria	
Leucothrix mucor	22, 41, 59, 62, 68
Thiothrix sp.	
Cytophaga sp.	
Flexibacter sp.	
Blue-green algae (Oscillatoriaceae)	
Spirulina subsalsa	97
Schizothrix calcicola	4
Ciliates (Protozoa)	
Epistylis sp.	4, 22, 25, 41, 65, 67, 73
Zoothamnium sp.	
Vorticella sp.	
Lagenophrys sp.	73
Apostome ciliate	6
Suctoria (Protozoa)	
Acineta sp.	73
Diatoms	
Amphora sp.	96, 111
Nitzschia sp.	
Achanthes sp.	

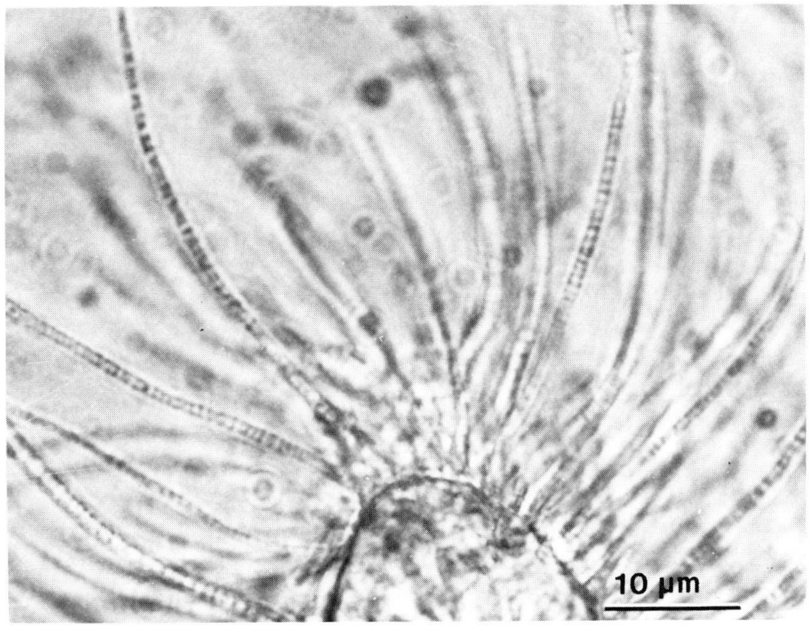

A

FIGURE 7. (A) *Leucothrix mucor* infestation of a gill lamellus from *P. californiensis*. (No stain.) (B) Scanning electron micrograph of gill lamellus from a *P. stylirostris* with a heavy infestation of *L. mucor*.

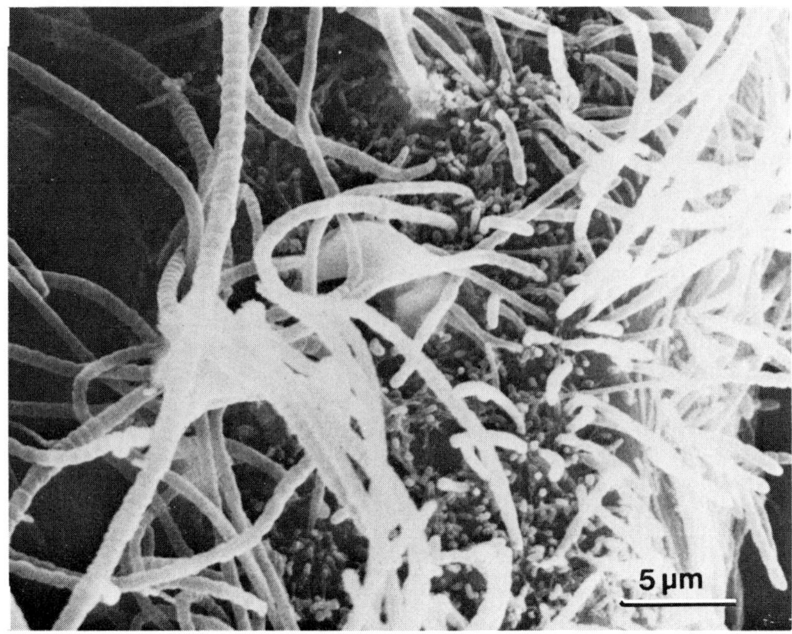

FIGURE 7B.

Table 9
SCHEME FOR NUMERICAL GRADING OF DISEASE DUE TO INFESTATIONS OF THE GILLS BY *LEUCOTHRIX MUCOR*[a,b]

Numerical grade	Status of gill examined
0	No filaments of *L. mucor* observed
0.5	A few scattered filaments present on lamellae
1	Filaments of *L. mucor* present, but not abundant (i.e., not covering or affecting more than 25% of the lamellae)
2	Filaments of *L. mucor* common, and affecting or covering at least 25% of the lamellae
3	At least 50% of the lamellae affected or covered by filaments of *L. mucor*
4	All or nearly all (75 to 100%) of the lamellar surfaces covered or affected by *L. mucor* filaments

[a] Determined by microscopic (at 50 to 100×) examination of the lamellae on a gill process biopsied from above the fifth periopod.

[b] The same system may be applied to other fouling epicommensal organisms such as *Zoothamnium* sp., diatoms, etc.

and nonliving solid substrates,[61] and in shrimp culture systems it readily attaches to the body surfaces of shrimp. In juvenile or older penaeids, it appears to especially favor attachment to the gills and accessory gill structures (Figure 7). Larval and postlarval penaeids may become so fouled by *L. mucor* filaments that respiration, feeding, locomotion, and molting may be impaired, resulting in death. *L. mucor* causes no demonstratable pathology to the surfaces on which it attaches.[62-64] Severity of disease due to *L. mucor* in juvenile or larger shrimp is determined by examination of wet mounts of the gills of affected shrimp at 100× (Table 9).

Mortalities due to *Leucothrix* infestation of the gills are thought to occur from hypoxia. Under conditions of stress due to crowding, molting, or low oxygen levels, *Leucothrix* (in grade 3 and 4 infestations, Table 9) may cause severe losses overnight. Besides sporadic

Table 10
PREFERRED METHODS OF CHEMOTHERAPY FOR VARIOUS FORMS OF GILL DISEASE AND LARVAL FOULING CAUSED BY VARIOUS TYPES OF FOULING ORGANISMS

Type of gill disease	Reported treatment(s)[a]	Concentration	Duration	Ref.
Bacterial[b] or algal[c] fouling diseases	Ponds[d], raceways, tanks	0.1 mg Cu/ℓ	24 hr, flow-through	63
	Copper (Cutrine-plus®)	0.2—0.5 mg Cu/ℓ	4—6 hr, static	22, 96
	Small volume tanks			
	Oxytetracycline	100 ppm	Indefinite	112
	Neomycin	10 ppm	Indefinite	113
	Chloramphenicol	1—10 ppm	Indefinite	30
	Streptomycin	1—4 ppm	Indefinite	25
	Malachite green oxylate	5 ppm	2-min dip	114
Ciliate[e] fouling diseases	Ponds[e], raceways, tanks			
	Formalin	25 ppm	Indefinite	67
		25—75 ppm	4—8 hr, static	22, 41
	Glutaraldehyde	2.5 ppm	6—8 hr, static	41
	Small volume tanks			
	Formalin	25 ppm	Indefinite	22
	Chloramine T	5 ppm	Indefinite	70
	Quinine bisulfate	5 ppm	Indefinite	70
	Quinacrine hydrochloride	0.6 ppm	Indefinite	70

[a] EPA or FDA approval for these uses may be required.
[b] *Leucothrix, Thiothrix, Flexibacter, Cytophaga, Vibrio,* etc.
[c] Filamentous blue-green algae and diatoms.
[d] Caution should be exercised in treatment of ponds because chemotherapeutics like algicides and formalin also may kill oxygen-producing plankton.
[e] *Zoothamnium, Epistylis, Vorticella,* etc.

but severe losses, *Leucothrix* infestations in cultured shrimp populations, if left untreated, can cause continuous low level losses.

Infestations of the gills by *L. mucor* may be managed by the use of antibiotics or the use of a seawater-soluble copper compound,[63] applied on a regular or on an "as-needed" basis (Table 10). A seawater-soluble copper compound, available commercially as Cutrine-Plus®, (Applied Biochemists, Mequon, Wis.) has been found to be effective in preventing and treating diseases due to *L. mucor;* but even better methods are needed.

Ciliate Gill Disease

Another form of gill disease that may occur alone or with *L. mucor* is due to infestation of the gills by one or more species of the peritrich protozoans, *Zoothamnium* sp., *Epistylis* sp., and *Vorticella* sp.[4,6,22,29,65] As was the case with *L. mucor*-caused gill disease, these organisms, when abundant on the surface of the gills, can cause hypoxia and death (Figure 8). In addition to impeding respiration, their abundant presence on the general body surface of larvae and postlarvae may interfere with locomotion, feeding, molting, etc. Like *L. mucor,* these peritrichs typically cause no appreciable internal damage to the gills.[4,6,22,62,66]

Formalin is reported to be effective in controlling these organisms in all forms of shrimp and crustacean culture[67-71] (Table 10).

Parasitic Protozoa

Microsporida

Microsporidans (Protozoa, Microspora)[72] cause a group of diseases in penaeids that are collectively called "cotton" or "milk shrimp disease".[2,5,22,73] At least four species of

FIGURE 8. A colony of *Zoothamnium* sp. from the gills of *P. californiensis*. (No stain.)

microsporidia are known in North American wild penaeids[5,6] (Table 11). Shrimp with these parasites have been observed in pond-cultured penaeids.[4,22,73,74] Shrimp with microsporidan infections have distinctly opaque musculature, ovaries, etc., and often have dark blue or blackish discoloration due to expansion of the cuticular chromatophores (Figure 9).

Diagnosis of this group of parasites is accomplished by demonstration of spores within the host tissues by microscopic examination of fresh squashes of affected organs or tissues. Determination of the species involved (multiple infections have been reported)[5] may also be accomplished from examination of wet mounts (Table 11). There are no proven methods of treatment for microsporidan infections in shrimp, although treatment methods using oral administration of the drug buquinolate in blue crabs, *Callinectes sapidus,* has been reported to be effective in prevention of infections by *Ameson (Nosema) michaelis.*[75,76] In one suggested control method, certain finfish must be excluded from shrimp-rearing ponds as those fish may serve as a ''conditioning intermediate'' host for microsporidan spores.[77]

Gregarines

Gregarines (Protozoa, Apicomplexa)[72] are common inhabitants of the guts of wild- and pond-reared penaeids.[5,73] Gregarines require a mollusk for completion of their life cycle and, hence, may be excluded from tank and raceway culture systems.[5,73] Even when present in very large numbers in the gut, gregarines are not thought to cause significant disease in penaeids.[73] Their presence may be demonstrated in fresh squashes of the mid- and hindgut contents. Two genera, *Nematopsis* and *Cephalolobus,* are known from penaeids[2,4,73,78] (Table 12).

NUTRITIONAL, ENVIRONMENTAL, AND TOXIC DISEASES

Ascorbic-Acid Deficiency Disease (Black Death Disease)

Only one nutritional disease syndrome of cultured penaeids has been identified. That

Table 11
MICROSPORIDAN PARASITES OF PENAEID SHRIMP

Microsporidan	Species affected	Tissue(s) affected	No. of spores per sporant	Spore dimensions (μm)	Ref.
Ameson (= *Nosema*) *nelsoni*	*Penaeus aztecus*, *P. duorarum*, *P. setiferus*	Muscle	1	2.0 × 1.2	3—6, 115, 116
Nosema sp.	*Metapenaeus monoceros*	Muscle	1	Not reported	117
Agmasoma (= *Thelohania*) *penaei*	*P. setiferus*	Blood vessels, foregut, hindgut, gonads, occasionally in muscle	8	2.0 × 5.0 and 5.0 × 8.2	3—6, 118
Thelohania duorara	*P. aztecus*, *P. duorarum*, *P. brasiliensis*	Muscle	8	5.4 × 3.6	4—6, 73, 116, 119
Pleistophora sp.	*P. aztecus*, *P. setiferus*, *P. duorarum*	Muscle, heart, stomach wall, gills, hepatopancreas	16—40+	2.6 × 2.1	4—6, 120

FIGURE 9. *P. setiferus* (top) and *P. aztecus* (middle) with "cotton shrimp" disease. The bottom shrimp (also *P. aztecus*) is normal. (Photo courtesy of S. K. Johnson, Texas A & M University, College Station, Tex.)

Table 12
GREGARINE PARASITES OF CULTURED PENAEID SHRIMP

Parasite	Species affected	Ref.
Nematopsis duorari	*P. duorarum*	6
N. penaeus	*P. aztecus, P. duorarum, P. setiferus*	2, 73
N. vannamei	*P. vannamei*	78
N. sinaloensis	*P. vannamei*	78
N. brasiliensis	*P. brasiliensis*	78
Nematopsis sp.	*P. brasiliensis, P. duorarum*	6, 78
Cephalolobus penaeus	*P. aztecus, P. duorarum, P. setiferus*	2, 73
Cephalolobus sp.	*P. vannamei, P. brasiliensis*	2, 73

disease was originally named black death disease to describe the typical, large black (melanized) lesions that occur in shrimp dying from the disease.[79,80] The disease occurs in penaeids which are reared in closed systems, aquaria, or flow-through systems in which most or all of the diet is artificial.[79,81-83] The disease has not been observed in shrimp cultured in ponds, tanks, or raceways in which there is at least some algal growth.[22,83]

Shrimp with black death (Figure 10) typically display blackened (melanized) lesions in the stomach wall, the hindgut wall, in the gills, and in the subcuticular tissues at various locations in the shrimp, especially those at the junction of body and appendage cuticular segments. The disease has been experimentally induced in *P. californiensis* and *P. stylirostris* by the feeding of diets devoid of added ascorbic acid,[80,83,84] and has been observed, although not initially recognized as being due to a deficiency of ascorbic acid, in cultured populations of other penaeid species (Table 13) including *P. japonicus*.[57,81]

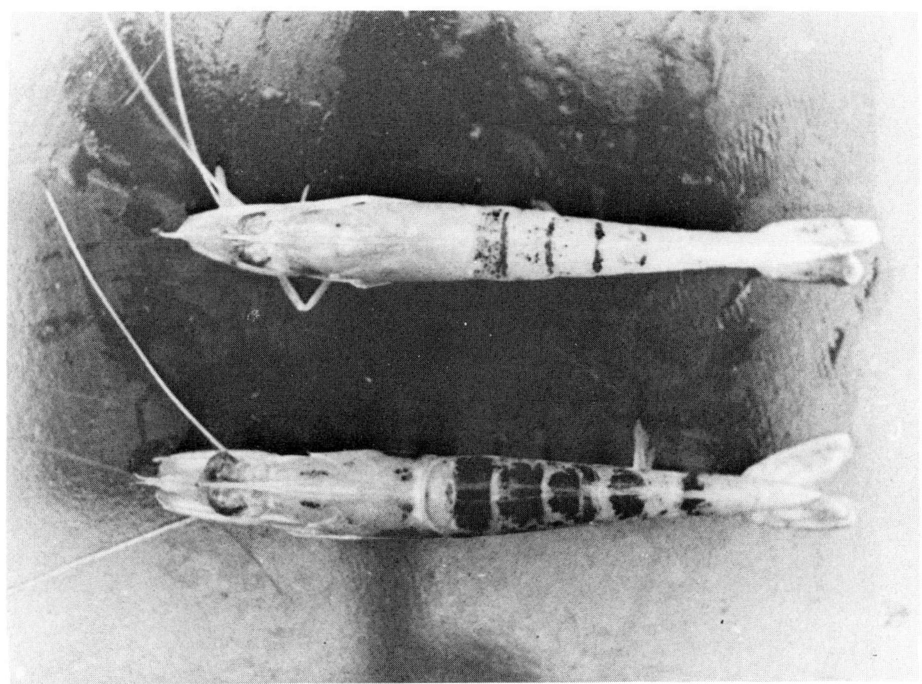

FIGURE 10. Ascorbic acid deficiency syndrome (black death disease) in juvenile *P. californiensis*. The melanized areas are under the cuticle which is not eroded.

Table 13
PENAEID SPECIES IN WHICH A DIETARY ASCORBIC ACID DEFICIENCY SYNDROME (BLACK DEATH DISEASE) HAS BEEN OBSERVED

Species	Geographic location	Ref.
Penaeus californiensis, P. stylirostris	Mexico	22, 55, 79, 80
P. aztecus	Texas	22, 55
P. japonicus	Japan	57, 81
P. japonicus	Tahiti	124

Black death lesions may be distinguished from shell disease lesions, which they superficially resemble, by the lack of erosion or damage to the cuticle in the former disease. Instead, black death lesions are melanized hemocytic, necrotic lesions present in the epithelial and subepithelial connective tissues of the stomach, gills, the general cuticle, and in the loose connective tissues in such organs as the hepatopancreas, the nerve cord, the eyestalks, etc.[83,84]

Ascorbic acid is highly unstable and is largely destroyed in the manufacture of shrimp feeds. Hence, a dietary requirement of from 2000 to 3000 mg of the vitamin per kilogram of feed has been reported as necessary,[80,81,83] while much less is actually required.

Gas-Bubble Disease

Gas-bubble disease has been reported to occur in penaeid shrimp as a result of supersaturation of atmospheric gases and oxygen.[85,86] Apparently, shrimp are similar to fish in their sensitivity to supersaturation of atmospheric gases. Oxygen-caused gas-bubble disease in

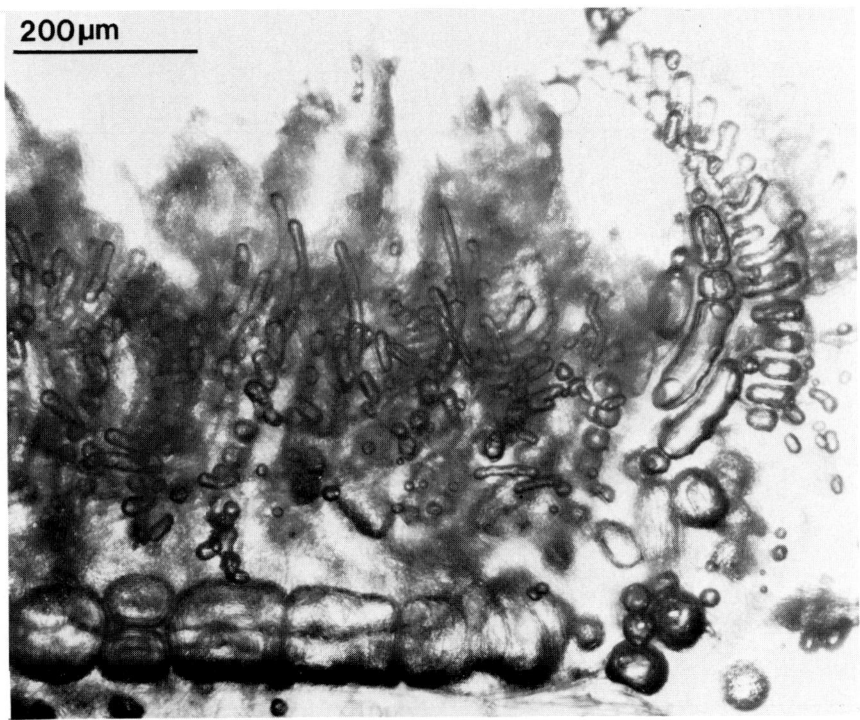

FIGURE 11. Wet mount of the gills of a juvenile *P. aztecus* with gas-bubble disease. (No stain.)

penaeids was reported to occur when dissolved oxygen levels reached or exceeded 250% of normal saturation in seawater of 24 to 26°C and 35 ppt salinity.[86] The levels of nitrogen or atmospheric gas supersaturation required to cause gas-bubble disease in penaeids are not known, but are assumed to be similar to the 118% reported for fish.[87] Regardless of the gas causing gas-bubble disease, the clinical signs are the same. The first sign of gas-bubble disease in shrimp is a rapid, erratic swimming behavior, which is soon followed by a stuporous behavior. Shrimp so affected float helplessly (in all other diseases dead or dying shrimp sink) near the water surface, with the ventral side of the cephalothorax higher than the abdomen. Examination of fresh preparations of gills or whole tissue by microscopy reveals the presence of gas bubbles (Figure 11).

Gas-bubble disease due to oxygen supersaturation was found to not necessarily be a lethal condition if corrective measures were taken to immediately lower the dissolved oxygen content of the culture tank water. Many of the affected shrimp recovered within a few hours.[86] Nitrogen- (or atmospheric gas-) caused gas-bubble disease, in contrast, is usually lethal to penaeids.[85]

Cramped Tail

This occasionally observed condition of penaeid shrimp has been reported to occur in the summer months, when both air and water temperatures are high.[22,65,88] Shrimp with cramped tails (while still alive) have a dorsal flexure of the abdomen which is rigid and cannot be straightened. The condition typical follows handling, although shrimp have been observed with partially cramped tails in undisturbed ponds.[88] The cause of the condition is unknown, but its occurrence only in summer suggests that elevated water and air temperatures, the handling of shrimp in air that is warmer than the water, and other stresses may be the cause of the syndrome.[22,65,73,88]

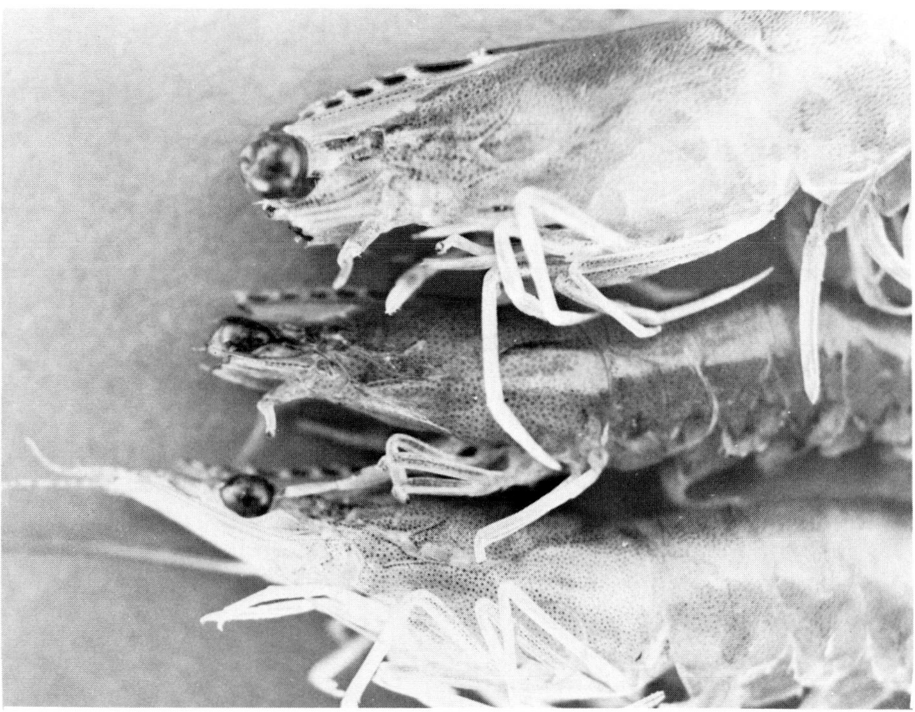

FIGURE 12. Blue-shrimp *P. stylirostris* (top two; bottom shrimp normal) with BSX disease which is of uncertain etiology, but may be related to "red tide" toxins. Blunting of the head is due to erosion of the antennae, antennules, rostrum, antennal blades, and portions of the eyes.

Muscle Necrosis (= Spontaneous Necrosis)

Muscle necrosis is the name given to a condition in all species of penaeid shrimp that is characterized by whitish opaque areas in the striated musculature, especially of the distal abdominal segments.[89] The condition follows periods of severe stress (from overcrowding, low dissolved oxygen levels, sudden temperature or salinity changes, rough handling, etc.).[90] It is reversible in its initial stages if stress factors are reduced, but it may be lethal if large areas are affected. Tail rot is the name given to the chronic and typically septic form of the disease when the distal portion of the abdomen (or appendages) becomes completely necrotic, turns red, and begins to decompose.

Toxic Diseases

Toxigenic Algae

A number of algae have been reported to cause (or are suspected to cause) mortalities in cultured penaeid shrimp. Senescent blooms of the diatom *Chaetoceros gracilis* were reported to be toxic to the larval stages of *P. stylirostris* and *P. vannamei.*[91] It was assumed that toxic substances were released by dead or dying diatom cells.

Dinoflagellate blooms (red tides) have been suspected as causing serious losses in penaeid shrimp culture in Mexico, but a cause-and-effect relationship has not been demonstrated. However, a toxicity syndrome called BSX,[41] possibly related to red tides, has been observed in cultured populations of *P. californiensis* and *P. stylirostris* in Mexico. Shrimp with this syndrome die during molting or following handling stress, and in an affected population a large percentage of the shrimp have been observed to develop "blunt heads" (Figure 12). This condition is thought to develop from repeated collisions with the walls of the culture tank, which is part of the erratic, almost convulsive behavior pattern that occurs in this

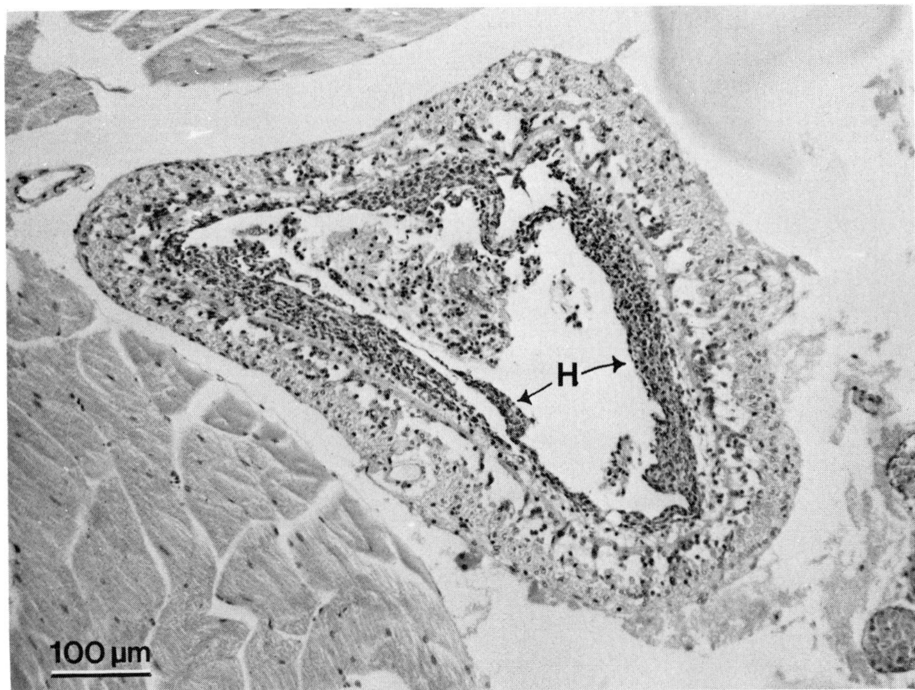

FIGURE 13. Histological cross section of the midgut of a juvenile *P. stylirostris* with blue-green algae-caused hemocytic enteritis. The gut mucosa has been sloughed and replaced by masses of hemocytes (H). (Hematoxylin and eosin.)

syndrome.[41] Dinoflagellate toxins are thought to be nontoxic to crustaceans,[92] but only short-term toxicity tests have been run on shrimp. However, during those tests, the few shrimp that molted also died.[92] That observation and the probable association of red tides and the BSX syndrome in Mexico indicates that the importance of red tide toxins to shrimp may be more significant than previously assumed.

Blue-Green Algae

Blooms of certain filamentous blue-green algae, all belonging to the family Oscillatoriaceae, have been implicated as causing the disease syndrome hemocytic enteritis (HE) in primarily young juvenile penaeids. One species of blue-green algae, confirmed by experimentation[93] to cause this syndrome, is *Schizothrix calcicola* (Agardh) Gomont.[94] That species possesses a potent endotoxin which causes gastroenteritis in man.[95] Other blue-green algae *(Spirulina subsalsa* and *Microcoleus lyngbyaceus)* are suspect, but not confirmed, as also causing HE. While HE is most typical in juvenile of 0.1- to 5-g average weight, it has been observed in 12- to 20-g *P. stylirostris*. HE apparently occurs as the result of algal toxins released in the gut from ingested algae.[96,97]

The principal lesion observed in this disease is necrosis and hemocytic inflammation of the mucosal epithelium (Figure 13) of those portions of the shrimp gastrointestinal tract that lack a chitinous lining (the midgut and the epigastric and hindgut ceca). The cause of death in shrimp with HE may be due to osmotic imbalances or to poor absorption of nutrients from the midgut due to the destruction of its mucosa, but in most instances, death appears to be due to secondary bacterial septicemias. Species of *Vibrio*, principally *V. alginolyticus*, are the organisms most commonly isolated from the hemolymph of shrimp with septic HE.[41,96,97] Mortality rates in raceway-cultured populations of *P. stylirostris* with HE have reached 85%, but usually have been less than 20% of affected populations.

Table 14
BIOLOGICAL AND CHEMICAL AGENTS REPORTED TO CAUSE BLACK GILLS IN PENAEID SHRIMP

Agent	Ref.
Vibrio spp.	20, 38
Fusarium solani	53
Apostome ciliate	6
Cadmium	6, 98, 121
Copper	22, 64, 122
Potassium permanganate	22, 64
Ascorbic acid deficiency (black death)	55, 79, 80
Ammonia and nitrate	128

Because HE typically has a septic phase, antibiotic therapy is often useful (see bacterial disease section for dose rates). The use of algicides or shading also appear promising in preventing this disease by reducing the amount of blue-green algae present in shrimp rearing tanks.

Black Gill Disease

A number of disease syndromes of cultured penaeids are accompanied by the presence of black (melanized) inflamed lesions in the gills.[22] Besides fungi and certain protozoans (discussed earlier), certain heavy metals, oil, and other chemical irritants may cause the syndrome (Table 14).

MISCELLANEOUS SHRIMP DISEASES

There exists a paucity of information for a number of important diseases of cultured penaeids. It is likely that the etiology of a number of the diseases listed in Table 15, although now of unknown (or unproven) etiology, will be determined in the coming few years. Also included in Table 15 are several conditions that have been observed in cultured penaeids, but are not believed to be potentially serious diseases of cultured penaeids.

Table 15
MISCELLANEOUS DISEASES OF PENAEID SHRIMP OF UNKNOWN OR UNCERTAIN ETIOLOGY AND DISEASES OF KNOWN ETIOLOGY, BUT OF UNKNOWN IMPORTANCE

Disease	Principal lesion(s)/sign(s)	Suggested etiology	Species affected	Geographic location	Ref.
Cramped tails	Flexed, rigid abdomen; occurs in juvenile and adult stages	Sudden increase in temperature (handling in air that is considerably warmer than culture water)	*P. aztecus, P. setiferus*	Florida	88, 73
			P. brasiliensis	Brazil	88
			P. californiensis	Mexico	22
			P. monodon	Taiwan	65
Red disease	Pale, reddish discoloration of cuticle, fouling of gills, foul odor to internal tissues, abnormally pale hepatopancreas and heart; occurs in juvenile to adult stages	Rancid diet(?)	*P. monodon, P. penecillatus*	Taiwan	65
Blue or white eye disease	Degeneration, necrosis and sloughing of eyestalk; death; occurs in juveniles	Unknown	*P. vannamei*	Ecuador	123
Blue disease	Pale blue discoloration of cuticle, thin soft cuticle, rough cuticular surface, lethargy, death; occurs in juvenile and adult stages	Probable viral etiology; nutritional cause also possible	*P. monodon*	Tahiti	19, 124
Blisters	Hemolymphomas (blisters) with colorless to melanized contents, especially in the branchiostegal region of the carapace or in ventrolateral portion of pleural plates; occurs in juvenile and adult stages	Unknown, but often seen accompanying inflammatory diseases such as fusarium and black death	*P. aztecus, P. setiferus, P. californiensis, P. stylirostris, P. vannamei*	Texas and Mexico	22, 73
Aflatoxicosis	Centronecrosis of hepatopancreas; peripheral necrosis mandibular organ cords; focal necrosis in hematopoietic organs	Feeds contaminated by aflatoxin B_1	*P. stylirostris, P. vannamei*	Mexico	125, 126

Fatty infiltration of hepatopancreas	Excessive deposits of lipid in hepatopancreatic tubule epithelium; occurs in juveniles and adults, especially in intensive culture systems	Improper dietary lipids; improper caloric-lipid balance; dietary toxins	*P. stylirostris, P. vannamei, P. californiensis, P. monodon, P. aztecus, P. japonicus*	Texas, Mexico, and Hawaii	127, 128
Tumor-like growths	Papilliform cuticular hyperplasia	Wound healing hyperplasia of cuticular hypodermis	*P. aztecus* *P. californiensis*	Texas Mexico	129 128
Hamartoma	Protrusion of muscle through abdominal pleura	Pollution; sudden and extreme reduction in salinity; trauma	*P. aztecus*	Mississippi Texas	130 128
Amebasis of larvae	Observed in protozoae stage larvae; invasion of muscle and subcuticular tissues by ameba	Unclassified ameba	*P. stylirostris, P. vannamei*	Florida and Honduras	131
White pleura disease	Lateral edges of pleural plates and branchiostegal region of carapace turn white, opaque flacid muscle, hemolymph fails to clot; occurs in juvenile and adult stages	Unknown, but *Vibrio alginolyticus* is suspect; may be related to a deficiency of ascorbic acid	*P. aztecus, P. merguiensis* *P. japonicus*	Tahiti	30 124
Golden shrimp	Distinct yellow-orange cuticular and muscle pigmentation; occurs in juvenile and adult stages	Unknown; accumulation of algae pigments (?)	*P. aztecus, P. setiferus*	Gulf of Mexico	73, 128
Gut and nerve syndrome	Extreme hypertrophy of basement membrane of anterior midgut epithelium and hyperplasia of epineurium of anterior ventral nerve cord and ganglia; occurs in juveniles	Unknown	*P. japonicus*	Hawaii	132
Multifocal opacities	Whitish deposits in subcutis, especially in rostrum; occurs in juveniles	Calcium phosphate deposits in subcutis	*P. stylirostris*	Hawaii	133
Larval encrustation	Brown to black encrusting deposits on larval penaeids	Unknown, but deposits contained iron salts	*P. stylirostris*	Costa Rica	134

REFERENCES

1. **Neal, R. A.,** Alternatives in aquacultural development: consideration of extensive versus intensive methods, *J. Fish. Res. Bd. Can.,* 30, 2218, 1973.
2. **Kruse, D. N.,** Parasites of the commercial shrimp, *Penaeus aztecus* Ives, *P. duorarum* Burkenroad and *P. setiferus* (Linnaeus), *Tulane Stud. Zool.,* 7, 123, 1959.
3. **Hutton, R. F., Sogandares-Bernal, F., Eldred, B., Ingle, R. M., and Woodburn, K. D.,** Investigations on the parasites and diseases of saltwater shrimps (Penaeidae) of sports and commercial importance to Florida, *State Fla. Bd. Conserv. Tech. Ser.,* 26, 38, 1959.
4. **Overstreet, R. M.,** Parasites of some penaeid shrimps with emphasis on reared hosts, *Aquaculture,* 2, 105, 1973.
5. **Overstreet, R. M.,** Marine maladies? Worms, germs, and other symbionts from the Northern Gulf of Mexico, *Mississippi-Alabama Sea Grant Consortium,* MASGP-78-021, 1978.
6. **Couch, J. A.,** Diseases, parasites, and toxic responses of commercial penaeid shrimps of the Gulf of Mexico and South Atlantic Coasts of North America, *Fish. Bull.,* 76, 1, 1978.
7. **Couch, J. A.,** An enzootic nuclear polyhedrosis virus of pink shrimp: ultrastructure, prevalence, and enhancement, *J. Invertebr. Pathol.,* 24, 311, 1974.
8. **Lightner, D. V. and Redman, R. M.,** A baculovirus-caused disease of the penaeid shrimp, *Penaeus monodon, J. Invertebr. Pathol.,* 38, 299, 1981.
9. **Sano, T., Nishimura, T., Oguma, K., Momoyama, K., and Takeno, N.,** Baculovirus infection of cultured Kuruma shrimp, *Penaeus japonicus,* in Japan, *Fish Pathol.,* 15, 185, 1981.
10. **Lightner, D. V., Redman, R. M., and Bell, T. A.,** Observations on the geographic distribution, pathogenesis and morphology of the baculovirus from *Penaeus monodon,* Fabricius, *Aquaculture,* 31, 1983.
11. **Bonami, J. R.,** Viruses from crustaceans and annelids: our state of knowledge, in *Proc. 1st Int. Colloquium on Invertebrate Pathology,* Queen's University Press, Kingston, Canada, 1976, 20.
12. **Johnson, P. T.,** Virus diseases of the blue crab, *Callinectes sapidus, Mar. Fish. Rev.,* 40, 13, 1978.
13. **Johnson, P. T. and Farley, C. A.,** A new enveloped helical virus from the blue crab, *Callinectes sapidus, J. Invertebr. Pathol.,* 35, 90, 1980.
14. **Couch, J. A.,** Viral diseases of invertebrates other than insects, in *Pathogenesis of Invertebrate Microbial Diseases,* Davidson, E. W., Ed., Allanheld, Osmun, Totowa, N.J., 1981, 127.
15. **Federici, B. A. and Hazard, E. I.,** Iridovirus and cytoplasmic polyhedrosis virus diseases in the fresh water daphnid *Simocephalus expinosus, Nature (London),* 254, 327, 1975.
16. **Bonami, J. R. and Pappalardo, R.,** Rickettsial infection in marine crustacea, *Experientia,* 36, 180, 1980.
17. **Federici, B. A., Hazard, E. I., and Anthony, D. W.,** Rickettsia-like organism causing disease in a crangonid amphipod from Florida, *Appl. Microbiol.,* 28, 885, 1974.
18. **Sparks, A. K. and Morado, F.,** A Chlamydial-Like Organism from the Dungeness Crab, *Cancer magister,* oral presentation to Society for Invertebrate Pathology Annual Meeting, Bozeman, Montana, August 1981.
19. **Lightner, D. V., Redman, R. M., and Bell, T. A.,** Infectious hypodermal and hemotopoietic necrosis (IHHN), a newly recognized virus disease of penaeid shrimp, *J. Invertebr. Pathol.,* in press.
20. **Cook, D. W. and Lofton, S. R.,** Chitinoclastic bacteria associated with shell disease in *Penaeus* shrimp and the blue crab *(Callinectes sapidus), J. Wildl. Dis.,* 19, 154, 1973.
21. **Nickelson, R. and Vanderzant, C.,** *Vibrio parahaemolyticus* — review, *J. Milk Food Technol.,* 34, 447, 1971.
22. **Lightner, D. V.,** Shrimp diseases, in *Disease Diagnosis and Control in North American Marine Aquaculture,* Vol. 6, Sindermann, C. J., Ed., Elsevier, New York, 1977, 10.
23. **Snieszko, S. F. and Taylor, C. C.,** A bacterial disease of the lobster *(Homarus americanus), Science,* 105, 500, 1947.
24. **Stewart, J. E. and Rabin, H.,** Gaffkemia, a bacterial disease of lobsters (Genus *Homarus*) in *A Symp. on Diseases of Fishes and Shell Fishes,* Spec. Publ. No. 5, Snieszko, S. F., Ed., American Fisheries Society, Washington, D.C., 1970, 431.
25. **Fisher, W. S., Nilson, E. H., Steenbergen, J. F., and Lightner, D. V.,** Microbial diseases of cultured lobsters: a review, *Aquaculture,* 14, 115, 1978.
26. **Barkate, J. A.,** Preliminary studies of some shrimp diseases, *Proc. World Maricult. Soc.,* 3, 337, 1972.
27. **Lewis, D. H.,** Predominant aerobic bacteria of fish and shellfish, Sea Grant Publ. No. 401, Texas A & M University, College Station, 1973, 102.
28. **Lewis, D. H.,** Response of brown shrimp to infection with *Vibrio* sp., *Proc. World Maricul. Soc.,* 4, 333, 1973.
29. **Lightner, D. V. and Lewis, D. H.,** A septicemic bacterial disease syndrome of penaeid shrimp, diseases of crustaceans, *Mar. Fish. Rev.,* 37, 25, 1975.
30. Aquacop, Observations on diseases of crustacean cultures in Polynesia, *Proc. World Maricult. Soc.,* 8, 685, 1977.

31. **Sparks, A. K.,** Bacterial diseases of invertebrates other than insects, in *Pathogenesis of Invertebrate Microbial Diseases,* Davidson, E. W., Ed., Allanheld, Osmun, Totowa, N.J., 1982, 323.
32. **Vanderzant, C., Mroz, E., and Nickelson, R.,** Microbial flora of Gulf of Mexico and pond shrimp, *J. Food Milk Technol.,* 33, 346, 1970.
33. **Vanderzant, C., Nickelson, R., and Judkins, P. W.,** Microbial flora of pond-reared brown shrimp *(Penaeus aztecus), App. Microbiol.,* 21, 916, 1971.
34. **Hood, M. A. and Meyers, S. P.,** Microbiological and chitinoclastic activities associated with *Penaeus setiferus, J. Oceanogr. Soc. Jpn.,* 33, 235, 1977.
35. **Yasuda, K. and Kitao, T.,** Bacterial flora in the digestive tract of prawns, *Penaeus japonicus* Bate, *Aquaculture,* 19, 229, 1980.
36. **Shewan, J. M. and Veron, M.,** Genus. I.*Vibrio* Pacini 1854, in *Bergey's Manual of Determinative Bacteriology,* 8th ed., Buchanan, R. E. and Gibbons, N. E., Eds., Williams & Wilkins, Baltimore, Md., 1974, 340.
37. **Anderson, J. I. W. and Conroy, D. A.,** The significance of disease in preliminary attempts to raise crustacea in sea water, *Bull. Off. Int. Epiz.,* 69, 1239, 1968.
38. **Cipriani, G. R., Wheeler, R. S., and Sizemore, R. K.,** Characterization of brown spot disease of Gulf Coast shrimp, *J. Invertebr. Pathol.,* 36, 255, 1980.
39. **Delves-Broughton, J.,** Preliminary investigation into the suitability of a new chemotherapeutic, Furanace, for the treatment of infectious prawn diseases, *Aquaculture,* 3, 175, 1974.
40. **Chan, E. S. and Lawrence, A. L.,** The effect of antibiotics on the respiration of brown shrimp larvae and postlarvae (*Penaeus aztecus* Ives) and the bacterial populations associated with the shrimp, *Proc. World Maricul. Soc.,* 5, 99, 1974.
41. **Lightner, D. V., Redman, R. M., Danald, D. A., Williams, R. R., and Perez, L. A.,** Major diseases encountered in controlled environment culture of penaeid shrimp at Puerto Peñasco, Soñora, Mexico, in Proc. UJIVR Conf. on Aquaculture, Kyoto, Japan, May 1980.
42. **Corliss, J. P., Lightner, D., and Zein-Eldin, Z. P.,** Some effects of oral doses of oxytetracycline on growth, survival and disease in *Penaeus aztecus, Aquaculture,* 11, 355, 1977.
43. **Corliss, J. P.,** Accumulation and depletion of oxytetracycline in juvenile white shrimp *Penaeus setiferus, Aquaculture,* 16, 1, 1979.
44. **Lightner, D. V. and Fontaine, C. T.,** A new fungus disease of the white shrimp *Penaeus setiferus, J. Invertebr. Pathol.,* 22, 94, 1973.
45. **Barkate, J. A., Laramore, C. R., Hirono, Y., and Persyn, H.,** Some marine organisms related to shrimp diseases, *Proc. World Maricul. Soc.,* 5, 267, 1974.
46. **Bland, C. E.,** Fungal diseases of marine crustacea, *Proc. U.S.-Japan National Resources Program,* Symposium on Aquaculture Diseases, Tokyo, 1975, 41.
47. **Lightner, D. V.,** Fungal diseases of marine crustacea, in *Pathogenesis of Invertebrate Microbial Diseases,* Davidson, E. W., Ed., Allanheld, Osmun, Totowa, N.J., 1981, 451.
48. **Couch, J. N.,** A new fungus on crab eggs, *J. Elisha Mitchell Sci. Soc.,* 58, 158, 1942.
49. **Johnson, T. W., Jr. and Bonner, R. R., Jr.,** *Lagenidium callinectes* Couch in barnacle ova, *J. Elisha Mitchell Sci. Soc.,* 76, 147, 1960.
50. **Johnson, T. W., Jr.,** Fungi in marine crustaceans, in *A Symposium on Diseases of Fishes and Shellfishes,* Spec. Publ. No. 5, Snieszko, S. F., Ed., American Fisheries Society, Washington, D.C., 1970, 405.
51. **Armstrong, D. A., Buchanan, D. V., and Caldwell, R. S.,** A mycosis caused by *Lagenidium* sp. in laboratory-reared larvae of the Dungeness crab, *Cancer magister,* and possible chemical treatments, *J. Invertebr. Pathol.,* 28, 329, 1976.
52. **Bland, C. E., Ruch, D. G., Salser, B. R., and Lightner, D. V.,** Chemical control of *Lagenidium,* a fungal pathogen of marine crustacea, *Proc. World Maricul. Soc.,* 7, 445, 1976.
53. **Egusa, S. and Ueda, T.,** A *Fusarium* sp. associated with black gill disease of the Kuruma prawn, *Penaeus japonicus* Bate, *Bull. Jpn. Soc. Sci. Fish.,* 38, 1253, 1972.
54. **Johnson, S. K.,** *Fusarium* sp. in Laboratory-Held Pink Shrimp, Leaflet No. FDDL-2, Texas A & M University, College Station, Fish Disease Diagnostic Lab, 1974.
55. **Lightner, D. V., Moore, D., and Danald, D. A.,** A mycotic disease of cultured penaeid shrimp caused by the fungus *Fusarium solani,* in *Proc. 2nd Biennial Crustacean Health Workshop,* Sea Grant Publ. No. TAMU-SG-79-114, Lewis, D. H. and Leong, J. K., Eds., Texas A & M University, College Station, 1979, 137.
56. **Lightner, D. V., Redman, R. M., Danald, D. A., and Hose, J. E.,** Pathogenesis of the imperfect fungus *Fusarium solani* in the California brown shrimp *Penaeus californiensis, J. Invertebr. Pathol.,* submitted.
57. **Shigueno, K.,** *Shrimp Culture in Japan,* Association for International Technical Promotion, Tokyo, 1975, 153.
58. **Hatai, K., Nakajima, K., and Egusa, S.,** Effects of various fungicides on the black gill disease of the Kuruma prawn *(Penaeus japonicus)* caused by *Fusarium* sp., *Fish Pathol.,* 8, 156, 1974.

59. **Johnson, P. W., Sieburth, J. M., Sastry, A., Arnold, C. R., and Doty, M. S.,** *Leucothrix mucor* infestation of benthic crustacea, fish eggs, and tropical algae, *Limnol. Oceanogr.*, 16, 962, 1971.
60. **Brock, T. D.,** Family. IV. Leucotrichaceae Buchanan, in *Bergey's Manual of Determinative Bacteriology*, 8th ed., Buchanan, R. E., and Gibbons, N. E., Eds., Williams & Wilkins, Baltimore, 1974, 118.
61. **Sieburth, J. M.,** *Microbial Seascapes. A Pictorial Essay on Marine Microorganisms and their Environments,* University Park Press, Baltimore, 1975.
62. **Lightner, D. V., Fontaine, C. T., and Hanks, K.,** Some forms of gill disease in penaeid shrimp, *Proc. World Maricul. Soc.*, 6, 347, 1975.
63. **Lightner, D. V. and Supplee, V. C.,** A possible chemical control method for filamentous gill, *Proc. World Maricul. Soc.*, 7, 473, 1976.
64. **Lightner, D. V.,** Gill disease: a disease of wild and cultured penaeid shrimp, *Proc. 66th Meeting Int. Council for the Exploration of the Sea,* 1978, F:24.
65. **Liao, I. C., Yang, F., and Lou, S.,** Preliminary report on some diseases of cultured prawn and their control methods, in *Reports on Fish Disease Research (I),* JCRR Fisheries Series, Taipei, Taiwan, 1977, 28.
66. **Foster, C. A., Sarphie, T. G., and Hawkins, W. E.,** Fine structure of the peritrichous ectocommensal *Zoothamnium* sp. with emphasis on its mode of attachment to penaeid shrimp, *J. Fish Dis.*, 1, 321, 1978.
67. **Johnson, S. K., Parker, J. C., and Holcomb, H. W.,** Control of *Zoothamnium* sp. on penaeid shrimp, in *Proc. World Maricul. Soc.*, 4, 321, 1973.
68. **Johnson, S. K.,** Ectocommensals and Parasites of Shrimp from Texas Rearing Ponds, Sea Grant Publ. No. TAMU-SG-74-207, Texas A & M University, College Station, 1974, 20.
69. **Johnson, S. K.,** Toxicity of Several Management Chemicals to Penaeid Shrimp, Leaflet No. FDDL-53, Texas A & M University, College Station, Fish Disease Diagnostic Laboratory, 1974.
70. **Johnson, S. K.,** Chemical Control of Peritrichous Ciliates on Young Penaeid Shrimp, Leaflet No. FDDL-57, Texas A & M University, College Station, Fish Disease Diagnostic Laboratory, 1976.
71. **Schnick, R. A., Meyer, F. P., Marking, L. L., and Bills, T. D.,** Candidate chemicals for crustacean culture, in *Proc. 2nd Biennial Crustacean Health Workshop,* Sea Grant Publ. No. TAMU-SG-79-114, Lewis, D. H. and Leong, J. K., Eds., Texas A & M University, College Station, 1979, 245.
72. **Sprague, V. and Couch, J.,** An annotated list of protozoan parasites, hyperparasites, and commensals of decapod crustacea, *J. Protozool.*, 18, 526, 1971.
73. **Johnson, S. K.,** *Handbook of Shrimp Diseases,* Sea Grant Publ. No. TAMU-SG-75-603, Texas A & M University, College Station, 1978, 23.
74. **Villella, J. B., Iversen, E. S., and Sindermann, C. J.,** Comparison of the parasites of pond-reared and wild pink shrimp (*Penaeus duorarum* Burkenroad) in South Florida, *Trans. Am. Fish. Soc.*, 99, 789, 1970.
75. **Overstreet, R. M.,** Buquinolate as a preventive drug to control microsporidosis in the blue crab, *J. Invertebr. Pathol.*, 26, 213, 1975.
76. **Overstreet, R.M. and Whatley, E. C., Jr.,** Prevention of microsporidosis in the blue crab, with notes on natural infections, *Proc. World Maricul. Soc.*, 6, 335, 1975.
77. **Iversen, E.S. and Kelly, J. F.,** Microsporidiosis successfully transmitted experimentally in pink shrimp, *J. Invertebr. Pathol.*, 27, 407, 1976.
78. **Feigenbaum, D. L.,** Parasites of the commercial shrimp *Penaeus vannamei* Boone and *Penaeus brasiliensis* Latreille, *Bull. Mar. Sci.*, 25, 491, 1975.
79. **Lightner, D. V., Colvin, L. B., Brand, C., and Danald, D. A.,** Black Death, a disease syndrome related to a dietary deficiency of ascorbic acid, *Proc. World Maricul. Soc.*, 8, 611, 1977.
80. **Magarelli, P. C., Jr., Hunter, B., Lightner, D. V., and Colvin, L. B.,** Black Death: an ascorbic acid deficiency disease in penaeid shrimp, *Comp. Biochem. Physiol.*, 63A, 103, 1979.
81. **Deshimaru, O. and Kuroki, K.,** Studies on a purified diet for prawn. VII. Adequate dietary levels of ascorbic acid and inositol, *Bull. Jpn. Soc. Sci. Fish.*, 42, 571, 1976.
82. **Guary, M., Kanazawa, A., Tanaka, N., and Ceccaldi, H.J.,** Nutritional requirements of prawn. VI. Requirement for ascorbic acid, *Mem. Fac. Fish. (Kogoshima Univ.),* 25, 53, 1976.
83. **Lightner, D. V., Hunter, B., Magarelli, P. C., Jr., and Colvin, L. B.,** Ascorbic acid: nutritional requirement and role in wound repair in penaeid shrimp, *Proc. World Maricul. Soc.*, 10, 513, 1979.
84. **Hunter, B., Magarelli, P. C., Jr., Lightner, D. V., and Colvin, L. B.,** Ascorbic acid-dependent collagen formation in penaeid shrimp, *Comp. Biochem. Physiol.*, 64B, 381, 1979.
85. **Lightner, D. V., Salser, B. R., and Wheeler, R. S.,** Gas-bubble disease in the brown shrimp *(Penaeus aztecus), Aquaculture,* 4, 81, 1974.
86. **Supplee, V. C. and Lightner, D.V.,** Gas-bubble disease due to oxygen supersaturation in raceway-reared California brown shrimp, *Prog. Fish Cult.*, 38, 158, 1976.
87. **Rucker, R. R.,** Gas-bubble disease of salmonids: a critical review, *U.S. Fish. Wild. Serv. Tech. Pap.*, 48, 11, 1972.
88. **Johnson, S. K.,** Cramped Condition in Pond-Reared Shrimp, Leaflet No. FDDL-56, Texas A & M University, College Station, Fish Disease Diagnostic Laboratory, 1975, 2.

89. **Rigdon, R.H. and Baxter, K. N.,** Spontaneous necrosis in muscles of brown shrimp, *Penaeus aztecus* Ives, *Trans. Am. Fish. Soc.,* 99, 583, 1970.
90. **Lakshmi, G. J., Venkataramiah, A., and Howse, H. D.,** Effect of salinity and temperature changes on spontaneous muscle necrosis in *Penaeus aztecus* Ives, *Aquaculture,* 13, 35, 1978.
91. **Simon, C.,** The culture of the diatom *Chaetoceros gracilis* and its use as a food for penaeid protozoeal larvae, *Aquaculture,* 14, 105, 1978.
92. **Sievers, A. M.,** Comparative toxicity of *Gonyaulax monilata* and *Gymnodinium breve* to annelids, crustaceans, molluscs and a fish, *J. Protozool.,* 16, 401, 1969.
93. **McKee, C.,** The Toxic Effect of Five Strains of Blue-Green Algae on *Penaeus stylirostris* Stimpson, M. S. Thesis, School of Renewable Natural Resources, University of Arizona, Tucson, 1981.
94. **Drouet, F.,** Revision of the classification of Oscillatoriaceae, in *Monograph 15 of The Academy of Natural Sciences of Philadelphia,* Fulton Press, Lancaster, Pa., 1968, 370.
95. **Keleti, G., Sykora, J. L., Lippy, E. C., and Shapiro, M. A.,** Composition and biological properties of lipopolysaccharides isolated from *Schizothrix calcicola* (Ag.) Gomont (Cyanobacteria), *Appl. Environ. Microbiol.,* 38, 471, 1979.
96. **Lightner, D. V.,** Possible toxic effects of the marine blue-green alga, *Spirulina subsalsa,* on the blue shrimp, *Penaeus stylirostris, J. Invertebr. Pathol.,* 32, 139, 1978.
97. **Lightner, D., Danald, D., Redman, R., Brand, C., Salser, B., and Reprieta, J.,** Suspected blue-green algal poisoning in the blue shrimp *(Penaeus stylirostris), Proc. World Maricul. Soc.,* 447, 1978.
98. **Nimmo, D. R., Lightner, D. V., and Bahner, L. H.,** Effects of cadmium on the shrimps, *Penaeus duorarum, Palaemonetes pugio* and *Palaemonetes vulgaris,* in *Physiological Responses of Marine Biota to Pollutants,* Vernberg, F. J., Calabrese, A., Thurberg, F. P., and Vernberg, W. B., Eds., Academic Press, New York, 1977, 131.
99. **Laramore, R.,** Shrimp Disease Studies in Panama, oral presentation to Texas A & M University Symp. on Research Priorities in Crustacean Disease Research, College Station, November 17 and 18, 1977.
100. **Currie, D. J.,** personal communication, 1981.
101. **LePennec, M. and Prieur, D.,** Les antibiotiques dans les elevages de larves de bivalves marins, *Aquaculture,* 12, 15, 1977.
102. **Gacutan, R. Q., Llobrera, A. T., and Baticados, C. L.,** Effects of Furanace on the development of larval stages of *Penaeus monodon* Fabricius, in *Proc. 2nd Biennial Crustacean Health Workshop,* Sea Grant Publ. No. TAMU-SG-79-114, Lewis, D. H. and Leong, J. K., Eds., Texas A & M University, College Station, 1979, 231.
103. **Fisher, W. S. and Nelson, R. T.,** Application of antibiotics in the cultivation of Dungeness crab, *Cancer magister, J. Fish. Res. Bd. Can.,* 35, 1343, 1978.
104. **Cook, H. L.,** Fungi parasitic on shrimp, in *FAO Aquaculture Bulletin,* Vol. 3, Pillay, T. V. R., Ed., 1971, 13.
105. **Delves-Broughton, J. and Poupard, C. W.,** Disease problems of prawns in recirculation systems in the U. K., *Aquaculture,* 7, 201, 1976.
106. **Gopalan, U. K., Meenakshikunjamma, P. P., and Purushan, K. S.,** Fungal infection in the tiger prawn *(Penaeus monodon)* and in other crustaceans from the Cochin backwaters, *Mahasagar Bull. Nat. Inst. Oceanogr.,* 13, 359, 1980.
107. **Hubschaman, J. H. and Schmitt, J. A.,** Primary mycosis in shrimp larvae, *J. Invertebr. Pathol.,* 13, 351, 1969.
108. **Shah, K. L., Jha, B. C., and Jhingran, A. G.,** Observations on some aquatic phycomycetes pathogenic to eggs and fry of freshwater fish and prawn, *Aquaculture,* 12, 141, 1977.
109. **Tharp, T. P. and Bland, C. E.,** Biology and host range of *Haliphthoros milfordensis* Vishniac, *Can. J. Bot.,* 55, 2936, 1977.
110. **Guary, J. C., Guary, M., and Egusa, S.,** Infections bacteriennes et fongiques de crustaces peneides (*Penaeus japonicus* Bate) en elevage, *Colloque Sur L'Aquaculture,* Actes de Colloques, No. 1, CNEXO Ed., 125, 1974.
111. **Overstreet, R. M. and Safford, S.,** Diatoms in the gills of the commercial white shrimp, *Gulf Res. Rep.,* in press.
112. **Solangi, M. A., Overstreet, R. M., and Gannam, A. L.,** A filamentous bacterium on the brine shrimp and its control, *Gulf Res. Rep.,* 6, 275, 1979.
113. **Steenbergen, J. F. and Schapiro, H. C.,** Filamentous bacterial infestations of lobster and shrimp gills, *Am. Zool.,* 15, 816, 1976.
114. **Fisher, W. S.,** Microbial epibionts of lobsters, in *Disease Diagnosis and Control in North American Marine Aquaculture, Developments in Aquaculture and Fisheries Science,* Vol. 6, Sindermann, C. J., Ed., Elsevier, New York, 1977, 163.
115. **Sprague, V.,** Notes on three microsporidian parasites of decapod Crustacea of Louisiana coastal waters, *Occ. Pap. Mar. Lab. (Louisiana State University),* 5, 1, 1950.

116. **Sprague, V.,** Some protozoan parasites and hyperparasites in marine decapod crustacea in *Symp. Diseases of Fishes and Shellfishes,* Spec. Publ. No. 5, Snieszko, S. F., Ed., American Fisheries Society, Washington, D.C., 1970, 416.
117. **Subrahmanyam, M.,** Incidence of microsporidosis in the prawn, *Metapenaeus monoceros* (Fabr.), *Indian J. Mar. Sci.,* 3, 182, 1974.
118. **Rigdon, R. H., Baxter, K. N., and Benton, R. C.,** Hermaphroditic white shrimp, *Penaeus setiferus,* parasitized by *Thelohania* sp., *Trans. Am. Fish. Soc.,* 104, 292, 1975.
119. **Iversen, E. S. and Manning, R. B.,** A new microsporidan parasite from the pink shrimp *(Penaeus duorarum), Trans. Am. Fish. Soc.,* 88, 130, 1959.
120. **Baxter, K. N., Rigdon, R. H., and Hanna, C.,** *Pleistophora* sp. (Microsporidia: Nosematidae): a new parasite of shrimp, *J. Invertebr. Pathol.,* 16, 289, 1970.
121. **Couch, J. A.,** Ultrastructural study of lesions in gills of a marine shrimp exposed to cadmium, *J. Invertr. Pathol.,* 29, 267, 1977.
122. **Williams, R. R., Hose, J. E., and Lightner, D. V.,** Toxicity and residue studies on cultured shrimp, *Penaeus stylirostris,* treated with the algicide Cutrine-Plus, *Prog. Fish Cult.,* 44, 196, 1982.
123. **LeBitoux, J. F. and Emerson, C.,** personal communication, 1981.
124. **Breuil, G.,** personal communication, 1981.
125. **Lightner, D. V., Redman, R. M., Wiseman, M. O., and Price, R. L.,** Histopathology of aflatoxicosis in the marine shrimp *Penaeus stylirostris* and *P. vannamei, J. Invertebr. Pathol.,* 40, 279, 1982.
126. **Wiseman, M. O., Price, R., Lightner, D. V., and Williams, R. R.,** Toxicity of aflatoxin B_1 to penaeid shrimp, *Appl. Environ. Microbiol.,* 44, 1479, 1982.
127. **Salser, B., Mahler, L., Lightner, D., Ure, J., Danald, D., Brand, C., Stamp, N., Moore, D., and Colvin, B.,** Controlled environment aquaculture of penaeids, in *Drugs and Food from the Sea. Myth or Reality?,* Kaul, P. N. and Sindermann, C. J., Eds., University of Oklahoma Press, Norman, 1978, 345.
128. **Lightner, D. V.,** unpublished data, Diagnostic Laboratory, Environmental Research Laboratory, University of Arizona, Tucson.
129. **Sparks, A.K. and Lightner, D. V.,** A tumorlike papilliform growth in the brown shrimp *(Penaeus aztecus), J. Invertebr. Pathol.,* 22, 203, 1973.
130. **Overstreet, R. M. and VanDevender, T.,** Implication of an environmentally induced harmartoma in commercial shrimps, *J. Invertebr. Pathol.,* 31, 234, 1978.
131. **Laramore, C. R. and Barkate, J. A.,** Mortalities Produced in the Protozoae Stages of Penaeid shrimp by an Unspeciated Amoeba, Leaflet No. FDDL-512, Texas A & M University, College Station, Fish Disease Diagnostic Laboratory, 1979, 7.
132. **Lightner, D. V., Redman, R. M., Bell, T. A., and Brock, J. A.,** An idiopathic proliferative disease syndrome of the midgut and ventral nerve in the Karuma prawn, *Penaeus japonicus* Bate cultured in Hawaii, *J. Invertebr. Pathol.,* submitted.
133. **Brock, J.,** personal communication, 1981.
134. **Simon, C.,** personal communication, 1976.

A BIOASSAY SYSTEM FOR STUDYING DISEASES IN JUVENILE PENAEID SHRIMP*

Jorge K. Leong

INTRODUCTION

In shrimp disease studies, laboratory bioassay is often an essential step to uncovering many kinds of important baseline information that are prerequisites to the development and establishment of reliable and valid diagnostic, treatment, and prophylactic methods. Examples of such needed information are the precise etiology of the disease in question, the virulence of a pathogen, the tolerance of the shrimp to drugs and chemicals having potential therapeutic values, and the range of effective doses of a chemotherapeutant.

In 1975, a survey of the literature showed that there was inadequate description of a convenient, replicable bioassay system that could be easily installed in a relatively small laboratory for the study of diseases of juvenile shrimp. Conventional laboratory bioassay work on juvenile penaeid shrimp generally relied on the use of 38- to 57-ℓ (10- to 15-gal) rectangular glass aquarium tanks. I found that there were many problems in adopting such a bioassay system in small laboratories. For example, if five different doses of each of five drugs plus an untreated control were to be tested in triplicates, the total number of aquarium tanks required would be 90. Considerable laboratory space was needed to accommodate 90 38- to 57-ℓ aquaria. In cases of assaying infectious diseases, there was the need to disinfect or decontaminate the aquaria and seawater to provide quality control of the experiments. The size and physical structure of those aquaria, especially when filled with seawater, made the decontamination process very difficult, if not impossible, to perform through steam-autoclaving in a normal-sized laboratory sterilizer. Therefore, the use of conventional aquarium tanks in bioassaying juvenile penaeid shrimp could be highly demanding in terms of cost, manpower, and space, and could be prohibitive to some small laboratories. In view of the above problems, our laboratory, in 1975, began examining alternative bioassay systems and succeeded in developing a practical and economical model system that could be easily installed by most laboratories for studying diseases in juvenile penaeid shrimp.

THE BIOASSAY SYSTEM

The bioassay system consists of a 3.8 ℓ (1-gal) wide-mouth glass jar (120 mm at mouth opening, and available from American Scientific Products**) fitted with a cover. This cover is converted from a hemispherical but flat-bottomed, plastic food-serving bowl which measures approximately 125 mm in diameter and is normally available in local super-markets (Figure 1J). A 100-mℓ polypropylene Tri-Pour® beaker (Sherwood No. 8889-20G200) (Figure 1I) with a perforated bottom is seated in a 5.5-cm hole made in the bottom of the bowl. Fiberglass wool (Figure 1H) is placed inside the beaker to serve as a filter for recycled seawater entering the top of the jar.

The two vertical plastic tubes (9.5 and 3.2 mm in diameter) (Figures 1C and 1E) of a Eureka Baby-Saver Spongė filter assemblage (Eureka Products Co.) are each passed through one of two round holes drilled in the bottom of the plastic bowl, so that the original sponged end (sponge removed) (Figures 1F and 1A) is in the jar while the other end protrudes outside the jar cover. A combination of elbow-shaped plastic connectors and sections of short plastic

* Contribution Number 82-31G, Southeast Fisheries Center, National Marine Fisheries Service, NOAA, Galveston, Tex. 77550.

** Mention of trade name or company's name does not constitute endorsement by the National Marine Fisheries Services, NOAA.

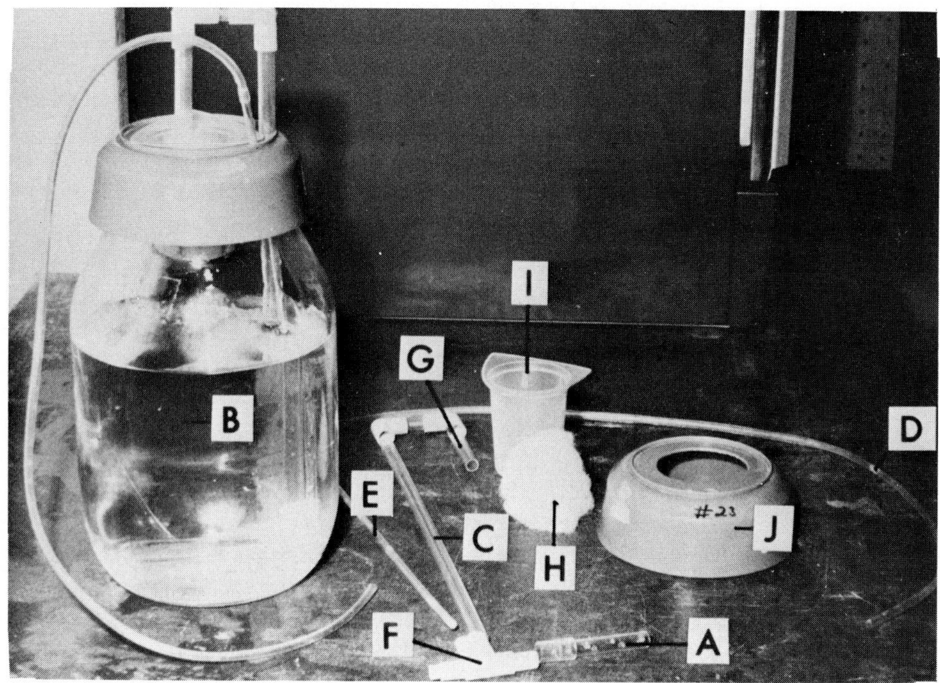

FIGURE 1. Bioassay jar and components. Sponge end (A) (sponge removed) of a Eureka sponge-filter unit is placed in aquarium jar (B), through which the aquarium water is taken in and moved by compressed air upward into plastic stem (C). (The compressed air comes in through Tygon hose (D) and the attached plastic stem (E) which is inserted into connector head (F).) The migrating water is discharged from plastic stem (G) and filtered through fiberglass wool (H) in Tri-Pour® beaker (I), which has a perforated bottom and is seated in jar cover (J) before returning to the jar.

tubes extends the top of the 9.5-mm tube (Figure 1D) to facing downward (Figure 1G) to about 2 cm above the fiberglass filter pad (Figure 1H) inside the Tri-Pour® beaker.

When compressed air is introduced from the top into the 3.2-mm plastic tube, the seawater inside the jar is forced to move up the 9.5-mm tube and reenter the jar after passing through air and the fiberglass filter. The seawater is, thus, recycled continuously. When a test material, such as a specific drug, which is known to be nonabsorbable by activated charcoal is placed in the seawater in the jar, activated charcoal flakes may be sandwiched between two pads of fiberglass wool in the beaker to provide better recycled water quality. In case of uncertainty, the activated charcoal should be omitted.

STERILIZATION

In experiments such as testing the pathogenicity of bacteria, where microbial contamination is a concern, the bioassay jar and its accessories should be sterilized in the following manner:

Fill a bioassay jar with 2.5 ℓ of seawater (prepassaged through 5 and 1-μm pore-sized cartridge filters), loosely close jar with a screw-cap, and then steam-autoclave jar (121°C, 15 psi, 15 min). (Autoclaving an empty jar is not recommended as the glass may crack.) After the water temperature has cooled, finger-tighten the lid and set the jar aside until ready for use. All the unautoclavable parts and accessories — plastic tubes, plastic beaker, plastic bowl, etc. — of the system can be disinfected by immersion for 30 min in a 2.6% sodium hypochlorite solution*, followed by thorough rinsing in autoclaved distilled water. Assemble the accessories with the sterile jars before use, taking care not to recontaminate them.

To insure quality control, take a 0.1-mℓ sample of seawater from each sterile jar for bacteriological examination after the seawater has been recycled overnight. Culture the water sample for bacterial growth on either saltwater yeast extract (SWYE) agar medium or tryptic soy agar medium (Difco) supplemented with a 3-salt solution. The formulas for these culture media are as follows:

3-Salt Solution

Sodium chloride (NaCl)	23.4 g**
Magnesium sulfate ($MgSO_40.7\ H_2O$)	6.94 g
Potassium chloride (KCl)	0.75 g

Dissolve the salts in distilled water to a final volume of 1 ℓ.

Saltwater Yeast Extract Agar Medium

Proteose peptone	10.0 g
Yeast extract	3.0 g
Bacto-agar	10.0 g
3-Salt solution	1 ℓ

Adjust pH to 7.4 to 7.6 with 1 *N* sodium hydroxide before autoclaving.

Tryptic Soy Agar Medium

Tryptic soy agar (Difco)	40.0 g
3-Salt solution	1 ℓ

After 20-hr incubation at 26°C, examine the agar plates for abundance of bacterial colonies. If excessive bacterial growth occurs (5 CFU or more per milliliter), the corresponding bioassay jar should be replaced with one that has passed the test.

When a jar with an overall diameter of 175 mm is used, ideally the size of individual juvenile shrimp should be no greater than 110 mm long, measuring from the tip of the rostrum to the end of the uropods. Under unusual circumstances, larger shrimp may be used, but the maximum body length should be about 130 mm, otherwise the shrimp will have difficulty moving around in the jar.

While there has been no formal experiment to determine the number of shrimp that can be maintained in each bioassay jar, it is suggested that one 35-mm shrimp be used for each liter of water. Above this size, only one shrimp is used per bioassay jar, disregarding water volume.

The system is not recommended for penaeid shrimp larvae before the late mysis stage. The feeding habits of younger shrimp larvae are incompatible with the aeration and seawater recycling system designed for the bioassay jar. Recently, another design suitable for naupliar and protozoeal larvae has been developed in our laboratory and will be described in a separate publication.

FOOD

Either commercial synthetic feed or cooked shrimp meat is used to feed the test shrimp.

* Prepared by mixing equal parts of Chlorox® (containing 5.25% sodium hypochlorite and available from Chlorox Co.) and distilled water.

** For preparation of tryptic soy (TS) agar medium where the TS agar already contains 5% NaCl, use only 18.4 g NaCl.

Juvenile penaeid shrimp, ranging from 50 to 130 mm long, are offered a total daily ration equal to 5% of their body weight. Computation of body weight based on body length is done according to descriptions by Fontaine[1] and Fontaine and Neal.[2] Feeding is done twice a day, once about 8 a.m. and once about 2 p.m. At the end of the work day, any unconsumed food is removed to prevent fouling of the water.

In bacteriological experiments where external sources of microbial contaminants are a concern, the shrimp feed is steam-autoclaved (121°C, 15 psi, 15 min) before being offered. Although our team has successfully maintained the test shrimp in good conditions for up to 2 weeks or more, we have been aware of the possibility that some of the heat-sensitive vitamins, such as vitamin C, may be destroyed in the process of steam-autoclaving. Resupplementation of the autoclaved food with the lost vitamins is a problem remaining to be solved. Untested suggestions have included filter-sterilization of the vitamins which are then used to marinate the autoclaved food or added to a cooling, autoclaved agar solution containing the sterile feed in a minced form. Other possibilities are to sterilize the food by gamma radiation or with ethylene dioxide.

CONTROL OF ENVIRONMENTAL CONDITIONS

Environmental conditions such as temperature, salinity, pH, and light may affect the growth, metabolism, and survival of juvenile penaeid shrimp. They may also influence the effects produced on shrimp by chemicals and microorganisms that are tested in the bioassay. In order to eliminate undesirable complications contributed by environmental factors, certain designs are incorporated into the bioassay system to maintain standard conditions of temperature, salinity, pH, and light.

Temperature

To maintain a constant water temperature in the experimental jars, ideally a walk-in, environmentally controlled chamber is used to accommodate the jars throughout the whole experiment. Alternatively, a large temperature-regulated water bath may be used.

In the Galveston Laboratory, we placed the aquarium jars in a leak-proof rectangular wooden tank (approximately 0.91 × 2.44 × 0.36 m deep internal dimensions) constructed of waterproof 2.5-cm-thick boards. Freshwater was added to the tank. The temperature of this freshwater was regulated by means of two to three submersible thermostatically controlled heating elements. A small submersible water pump was placed in the wooden tank to provide circulation and even distribution of heat. With this system, we were able to maintain a constant seawater temperature in the jars at 26 ± 1°C and up to at least 30°C.

A piece of wood (0.05 × 0.1 × 2.49 m long) was centered longitudinally across the top of the tank and fastened to the two end boards of the tank. Plastic aquarium multiple air-valve units were attached along the center crosspiece and serially connected with flexible Tygon tubing to a prefiltered compressed air source. Each outlet of the valve was in turn connected by a Tygon tubing to the 3.2-mm air tube in a bioassay jar to supply air to the seawater (Figure 2). Provided there was adequate pressure from the compressed air source, the wooden water bath could accommodate 60 bioassay jars.

Salinity

The salinity of natural seawater may vary according to the site of origin, e.g., deep sea or nearshore, and the season. A salinity refractometer, such as model AO-10440 (American Optical Co.), is commonly employed to monitor salinity. Salinity may be adjusted to the desired level by the addition of either distilled (or deionized) water or Instant Ocean Salt (Aquarium System).

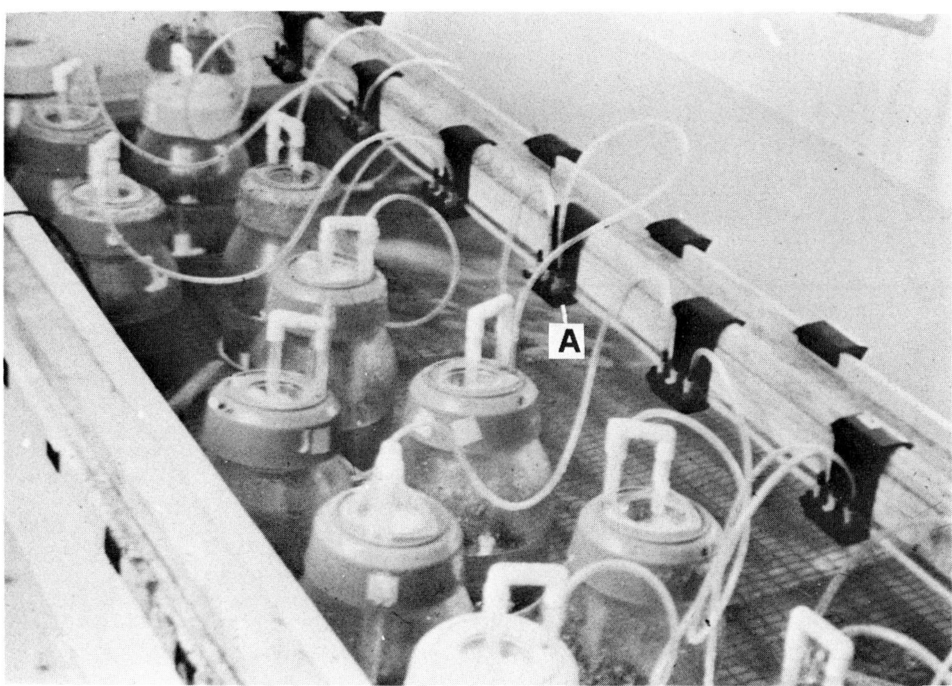

FIGURE 2. Portion of a water bath showing connections of bioassay jars. (A) Multiple air-valve unit. (Courtesy of R. Pylant.)

pH

Natural seawater usually has a pH value of about 8.0 to 8.5. For the adjustment of pH, either sodium hydroxide or hydrochloric acid may be used. If the seawater pH deviates too much from its normal range, such as below 7.5 or above 9, something may have gone wrong. In such cases, it is advisable not to use the seawater until the cause of the deviation is ascertained or until evidence shows that the abnormal seawater is not harmful to the survival or normal development of the shrimp.

Light

The optimal intensity of light for bioassay of juvenile shrimp has not been defined. In experimental work on adult *Penaeus setiferus* and *P. stylirostris,* a photoperiod of 14 to 16 hr/cmday using standard fluorescent lighting was successful in the induction of ovary maturation in adult female shrimp.[3,4] In bioassay for juvenile shrimp, I suggest that a similar photoperiod be provided.

APPLICATIONS

Some applications of the bioassay system are recommended as follows.

Chemical Bioassay

Two kinds of chemical bioassay can be done. The first one determines the tolerance level of the shrimp to specific drugs or chemicals, and evaluates nonlethal side effects of those chemicals on normal shrimp. The second type evaluates the therapeutic efficacy of specific drugs or chemicals on diseased shrimp.

When conducting chemical bioassay tests, the test chemical can be either added to the seawater in the bioassay jar or administered through intramuscular injection by means of a sterile hypodermic needle (26 to 27 gauge). The volume of inoculum at each injection varies according to the size of the test shrimp, but should not exceed 0.05 mℓ per injection. Ideally, a precision syringe is used; e.g., a Hamilton syringe can deliver repeatable, microliter-sized liquid volume. In addition, a more uniform dose delivery may be accomplished by the use of an automatic, precision syringe pump (such as a Sage 341, manufactured by Sage Instruments) in combination with the syringe.

The lateral intersection between the fourth and fifth body segments is the recommended site of injection. Just before inoculation, the shrimp is removed from the seawater and the site of injection is blotted dry with a cotton swab followed by rubbing with 70% ethyl alcohol for surface sterilization. After the alcohol is dried, the injection is made by pushing the needle at an oblique angle through the intersegmental membrane into the muscle.

In order to yield statistically sound data, a minimum of six replicates is recommended for each experimental variable. Circumstances permitting, 10 to 15 replicates are desirable.

Microbial Assay

Microbial assay is done to evaluate host-pathogen relationships such as pathogenicity and virulence in shrimp of a candidate causative microorganism. In this kind of study, the microorganism, such as a bacterium, will be used to challenge the shrimp host. Introduction of bacteria into the shrimp may be accomplished by either of several ways, such as addition of bacteria from a pure culture into the seawater in the bioassay jar, injection into the host, incorporation into shrimp feed, or introduction through an artificial wound.

The injection process of bacteria into the shrimp is similar to that described for chemical bioassay in the earlier section. In a large experimental system that involves many injections and, therefore, many hours to finish, bacterial suspensions to be used as sources of inoculum should be maintained at a low temperature such as 15°C to prevent excessive post-titration multiplication. Rapid post-titration multiplication may occur at higher temperatures over a period of time and will affect the accuracy of the dosages. We have successfully used the bioassay system and the injection method to assess the comparative virulence of four species of *Vibrio* bacteria *(Vibrio parahemolyticus, V. anguillarum, V. alginolyticus,* and *V. algosus)* in juvenile white shrimp, *Penaeus setiferus.*[5]

A method was devised by Leong and Fontaine to introduce bacteria into shrimp through the oral route.[5] Fresh shrimp meat was diced (approximately 5 × 5 × 5 mm) and the pieces placed without touching each other on a piece of wetted Whatman filter paper in a set of glass petri dishes (100 × 15 mm), and steam-autoclaved (121°C, 15 min, 15 psi). After being cooled, a small drop (0.01 mℓ) of a suspension of the candidate bacterium with a predetermined titer was placed on the upper surface of each piece of meat. After incubation overnight at 26°C, the meat dice that bore bacterial growth were transferred to the bioassay jars containing the test shrimp that had not been fed for 24 hr.

If required, the titer of the bacteria on each piece of meat may be estimated by the following method. At least one piece of meat from each set of petri dishes is placed in sterile seawater for about 1 min before it is transferred to a sterile test tube containing 5 mℓ sterile saline. The meat is then stroked with a sterile, round-ended glass rod to dislodge the bacterial growth from the surface. The suspension is next vortexed and pipetted up and down to break up lumps for even distribution of the bacteria. A standard tenfold serial titration is performed. The resulting data are used to estimate the bacterial titer on each piece of meat from the same set. This bacterial concentration is assumed to be the one ingested by the shrimp that devours the whole piece of meat within 1 hr. Any replicate in which the meat is not totally eaten within 1 hr is to be excluded from the experiment.

Pylant[6] used the bioassay jars to determine the causative organism of brown-spot disease in *P. setiferus.* He deliberately inflicted surface wounds on the integument of some of the animals by means of abrasion with a strip of carborundum paper under sterile conditions, and then he applied different species of bacteria to different wounds. The results showed that *Vibrio alginolyticus,* a chitinoclastic bacterium, could produce a brown-spot lesion on the abraded area, but not on an unabraded area. The experiment indicated that in brown-spot disease, an infection process could be initiated only through an injured integument in the shrimp.

In conclusion, a practical and relatively economical bioassay system is available for the study of diseases of juvenile penaeid shrimp. The system is particularly beneficial to smaller laboratories where manpower and space may be less readily available than larger laboratories. The system has been shown to be workable in actual laboratory experiments.

ACKNOWLEDGMENTS

I thank Daniel Patlan and Dickie Revera, Galveston Laboratory, National Marine Fisheries Service, Southeast Fisheries Center, for assisting in photography; Pete Sheridan and William Jackson, Galveston Laboratory, and James P. McVey, Panama City Laboratory, for reviewing and providing constructive criticisms to the manuscript; and Beatrice Richardson, Galveston Laboratory, for typing the manuscript.

REFERENCES

1. **Fontaine, C.T.,** Conversion Tables for Commercially Important Penaeid Shrimp of the Gulf of Mexico, Department of Commerce, NOAA, National Marine Fisheries Service, Data Report 70, 1971.
2. **Fontaine, C.T. and Neal, R.A.,** Length-weight relations for three commercially important penaeid shrimp of the Gulf of Mexico, *Trans. Am. Fish. Soc.,* 100, 584, 1971.
3. **Brown, A., Jr., McVey, J.P., Middleditch, B.S., and Lawrence, A.L.,** Maturation of white shrimp *(Penaeus setiferus)* in captivity, in *Proc. World Maricul. Soc.,* Vol. 10, Avault, J. W., Ed., 1979, 489.
4. **Brown, A., Jr., McVey, J. P., Scott, B. M., Williams, T. D., Middleditch, B. S., and Lawrence, A.L.,** The maturation and spawning of *Penaeus stylirostris* under controlled laboratory conditions, in *Proc. World Maricul. Soc.,* Vol. 11, Avault, J.W., Ed., Louisiana State University, Baton Rouge, 1980, 488.
5. **Leong, J.K. and Fontaine, C.T.,** Experimental assessment of the virulence of four species of *Vibro* bacteria in penaeid shrimp, in *Proc. 2nd Biennial Crustacean Hlth. Workshop,* Lewis, D.H. and Leong, J.K., Eds., Sea Grant Program, Texas A & M University, (TAMU-SG-79-114), College Station, 1979, 109.
6. **Pylant, R.,** A Study of Brown-Spot Disease and the Histology of the Integument in the White Shrimp, *Penaeus setiferus,* Master's thesis, University of Houston, Houston, Tex., 1980.

DISEASES (INFECTIOUS AND NONINFECTIOUS), METAZOAN PARASITES, PREDATORS, AND PUBLIC HEALTH CONSIDERATIONS IN *MACROBRACHIUM* CULTURE AND FISHERIES

James A. Brock

INTRODUCTION

Commercial aquaculture of *Macrobrachium* spp. (primarily *M. rosenbergii*) is becoming established. *Macrobrachium* farming operations can be found in areas of the world where climatic conditions permit the aquaculture of freshwater prawns. Prawn fisheries have existed for many years in Asia, Africa, South America, several Pacific Islands, and Australia. Prawns harvested from streams and rivers contribute to total fisheries yield, and *Macrobrachium* spp. are an important food item in these areas.

Attendant to human manipulated husbandry as well as being present in feral populations, a number of diseases of *Macrobrachium* have been reported or are known to occur. The purpose of this chapter is to briefly review the state-of-the-art knowledge of the diseases (infectious and noninfectious) of this genus, to discuss the nondisease-related causes of prawn losses, and to present information concerning public health considerations associated with the culture and consumption of prawns.

The term "disease", as used here, is defined as "a definite morbid process having a characteristic train of symptoms; it may affect the whole body or any of its parts, and the etiology, pathology, and prognosis may be known or unknown."[1]

Reports on *Macrobrachium* diseases from the late 70s indicated that prior to that time disease was not a major problem in *Macrobrachium* culture.[2,3] This situation is apparently changing. Within the last year in Hawaii a serious larval disease (midcycle disease) has been recognized and is under study. Epizootics have resulted in 90 to 95% sustained decline of postlarval production in hatcheries where the disease has occurred. The etiology for midcycle disease has not been determined.

An understanding of the normal is a prerequisite to recognizing the abnormal. There are few studies which have dealt with documentation of normal anatomy, histology, and defense mechanisms in *Macrobrachium*. Blewett et al.[4] have classified hemocytes in *M. rosenbergii* based on cytoplasmic granule characteristics. Huang[5] reported on the occurrence of serum agglutinins and showed them to be present in low titers in *Macrobrachium* hemolymph. Huang attempted to immunize prawns but was unable to demonstrate a rise in agglutinin titer or protection to subsequent bacterial challenge, and concluded *Macrobrachium* agglutinins were noninducible. His studies indicated that *M. rosenbergii* has a natural resistance to experimental (inoculation) infection with virulent strains of *Vibrio anguillarium*.

Young's[6] description of the gross anatomy for *Penaeus setiferus* can serve as a general guide to the gross anatomy of shrimp including *Macrobrachium*. The normal histology of the blue crab, *Callinectes sapidus*, has been detailed in a recently published text by Johnson.[7] Procedures for tissue fixations, processing, and staining which apply to the study of tissues in other decapods are outlined in this text. In general, the histology of *Macrobrachium* is similar in many aspects to that found in the blue crab. General diagnostic procedures for bacterial, mycotic, and parasitic infections appropriate for the isolation and identification of these agents from *Macrobrachium* can be found in Balows et al.[8]

Poor husbandry, overcrowding, unsuitable water quality or container conditions, and dietary imbalances are the major causes of disease in *Macrobrachium*. Unsuitable environmental conditions or nutrition probably predispose prawns to those infectious diseases (excluding metazoan parasites) that are known to affect members in this genus. All of the microbial disease agents reported are considered enzootic and opportunistic rather than obligate pathogens. Koch's postulates have not been fulfilled to confirm a causal relationship for any of these microorganisms as agents of disease in *Macrobrachium* spp.

A number of the diseases mentioned in this chapter are of undetermined etiology. Even in cases where the etiology is known, our understanding of the host-agent-environment interaction is incomplete for all of the diseases discussed. Confirmed reports of intracellular parasites (Viruses, Rickettsia, Chlamydia, and Sporozoa) or neoplasia affecting *Macrobrachium* were not found in the course of the literature search for this article. A list of the differential etiologic diagnosis for gross and microscopic lesions of the diseases of *Macrobrachium* is presented in Table 1.

DISEASES

Midcycle Larval Disease

Midcycle larval disease (MCD) of *M. rosenbergii* has been partially characterized on the basis of epizootiology, clinical signs, and microscopic pathology. MCD displays a typical mortality pattern. Dead larvae become noticeable in tanks during the middle third (day 12 to 24) of the production cycle. Although the larval mortality continues through the remainder of the cycle it is less noticeable. Epizootics of MCD have resulted in a marked reduction of postlarvae production with all production tanks being affected. In Hawaii where MCD has occurred in three hatcheries, these epizootics reduced PL production to 1 to 2 PLs per liter where previously realized production ranged 10 to 25 PLs per liter. The etiology of this disease has not been determined.

The results from experimental epizootiologic studies[9] have indicated that MCD larval mortality begins to rise on day 14 to 18 of the rearing cycle. In addition, stage-one through stage-five larvae have been shown to be susceptible, but late stage larvae have been refractory to the disease. The stocking of stage-one larvae into a system which shared common water with a larvae population in which MCD epizootic was occurring resulted in onset of typical mortality in the introduced population within 4 to 5 days suggesting the incubation period for the disease to be about 5 days.[9]

The clinical signs associated with epizootics of MCD have included weak, spiraling swimming behavior and reduced larval consumption of *Artemia*. Larval death from this disease has not been found to be associated with a particular stage in the molt cycle.

Larvae examined to date from populations in the chronic stages of MCD have had characteristic microscopic lesions. The hepatopancreas and anteriodorsal hepatic diverticular epithelium have shown a progressive reduction in size and degree of vacuolation. This has been accompanied by dilatation of the lumina in these organs (Figure 1B, Figure 1A, normal hepatopancreas supplied for comparison). Occasionally, colonies of coccobacilli have been found within the digestive tract lumina. In addition to the digestive tract lesions, striated musculature has been found to progressively shrink. Affected larvae have been noted to lose normal pigmentation and to appear bluish-gray. The epidermal tissues of the appendages characteristically have had a "cobblestone" appearance. Microbial epibiont fouling has been a sporadically observed problem late in the course of the disease.

Microscopically, the most pronounced and consistent pathology recognized in larvae affected with MCD has been hepatopancreas epithelial atrophy and degeneration. Boyd[10] lists the following as causes of atrophy: aging, deficient nourishment, disuse, toxins, pressure, and disruption of nerve supply. Deficient nourishment and toxins are potentially the most likely general cause of the larval hepatopancreas lesions in MCD.

The colonies of bacteria sometimes observed in histopathology sections of affected larvae have almost exclusively been limited to the luminal spaces of the digestive tract. Bacteremia has been an infrequent histopathologic finding in these larvae. The identity of the intraluminal

Table 1
DIFFERENTIAL ETIOLOGIC DIAGNOSIS IN *MACROBRACHIUM* SPP. BY ORGAN SYSTEM, GROSS OR MICROSCOPIC LESIONS, AND LIFE STAGE

Organ system	Gross or microscopic lesions	Life stage	Differential etiologic diagnosis	Ref.
Cuticular				
Exoskeleton	Epicuticular microbial fouling (cephalothorax, abdomen, gills, appendages, and eggs)	Larvae, juveniles, or adults	Filamentous bacteria: *Leucothrix* sp., other genera	16,22,26
		Larvae	Rod-shaped bacteria: miscellaneous genera	11,12
		Larvae, juvenile, or adults	Petritrich protozoa	
			Zoothamnium sp.	17, 22, 24
			Vorticella sp.	17, 26
			Epistylis sp.	17, 81
			Corthunia sp.	17, 22
			Vaginicola sp.	17, 26
		Juveniles or adults	Suctorian protozoa	
			Acineta sp.	17, 26
			Podophyra sp.	17
			Tokophyra sp.	17
			Algae	
			Oedogonium crassiusculum	80
			Lyngbya sp.	80
			Fungi	
			Aphanomyces sp.	3
			Achlya sp.	3
	Epicuticular metazoan ectocommensals or parasites	Juveniles or adults	Turbellaria	
			Temnocephalids	26, 58
			Mollusca	
			Barnacles	26
			Arthropoda	
			Isopoda	
			Bopyrid isopods	26
			Corallanid isopods	60
			Insecta	
			Ramphocorixa acuminata (eggs)	80
	Exuvia entrapment	Stage X—XI larvae, early PL	Etiology undetermined; nutritional deficiency, water quality, or toxins suspected	11, 34
	Melanization and necrosis with or without ulceration of any body surface or area of appendages	Larvae	Etiology unconfirmed; miscellaneous bacteria or fungi suggested	12, 37
		Juveniles or adults	Mechanical trauma; miscellaneous bacteria or fungi (*Fusarium* sp.)	27, 32, 33
	Ulceration, necrosis, and melanization principally involving the uropods	Juveniles or adults	Mechanical trauma with miscellaneous bacteria or fungi (*Fusarium* sp.)	32
	Melanization of medial surface of branchiostegites	Juveniles or adults	Etiology undetermined; precipitating chemicals and nitrogenous waste products suggested	34

Table 1 (continued)
DIFFERENTIAL ETIOLOGIC DIAGNOSIS IN *MACROBRACHIUM* SPP. BY ORGAN SYSTEM, GROSS OR MICROSCOPIC LESIONS, AND LIFE STAGE

Organ system	Gross or microscopic lesions	Life stage	Differential etiologic diagnosis	Ref.
	Focal to multifocal degenerative lesions within the deeper layers of the exoskeleton; exoskeleton in affected areas has reduced rigidity and discoloration; melanization not present	Adults, usually females	Etiology undetermined; nutritional deficiency or stress suggested	11, 34
Hypodermis	Multifocal black nodules (hemocytic nodules)	Juveniles or adults	Etiology undetermined; miscellaneous bacteria or avitaminosis C suggested	27, 48
Alimentary Tract				
Cardiac stomach	Nematode larvae	Juveniles or adults	*Angiostrongylus cantonensis* (third-stage larvae)	77
Hepatopancreas	Reduction in size; diffuse atrophy and degeneration of epithelium	Stage V—VIII larvae	Etiology undetermined; toxin suspected	9, 11
	Saturated fat preservation	Juveniles and adults	Etiology unconfirmed; organophosphate toxicosis suggested	51
Midgut	Hemocytic infiltration into intestinal mucosa, submucosa, and muscularis	Juveniles	Etiology unconfirmed in *Macrobrachium;* in penaeids lesion associated with consumption of blue-green algae	50
Muscular	Focal, multifocal to confluent areas of muscle opacity and necrosis	Larvae, juveniles, or adults	Etiology undetermined; variations in environmental conditions (stressors) associated with occurrence	44, 45
	Focal to multifocal opaque areas within the abdominal muscle	Juveniles and adults	Trematode metacercaria: *Carneophallus choanophallus*	26
Systemic	Diffuse muscle opacity, mushy muscle texture, atrophy of the hepatopancreas, and frequently generalized epibiotic fouling	Adults, usually males	Etiology undetermined	11
	Chromatophore atrophy, intermolt exoskeleton flexible, diffuse muscle opacity, mushy muscle texture, hepatopancreas atrophy	Adults, usually females	Etiology undetermined; nutritional deficiency suspected	11

bacterial has not been established. However, *Vibrio* spp. have been the most frequently isolated genera from larvae during MCD epizootics.[11] It has not been determined if the intraluminal bacteria play a role in the pathogenesis of MCD although bacterial toxemia has been suggested as a possible etiologic agent for the disease.[11]

Some other etiologic agents which have been considered for MCD include heavy metals,

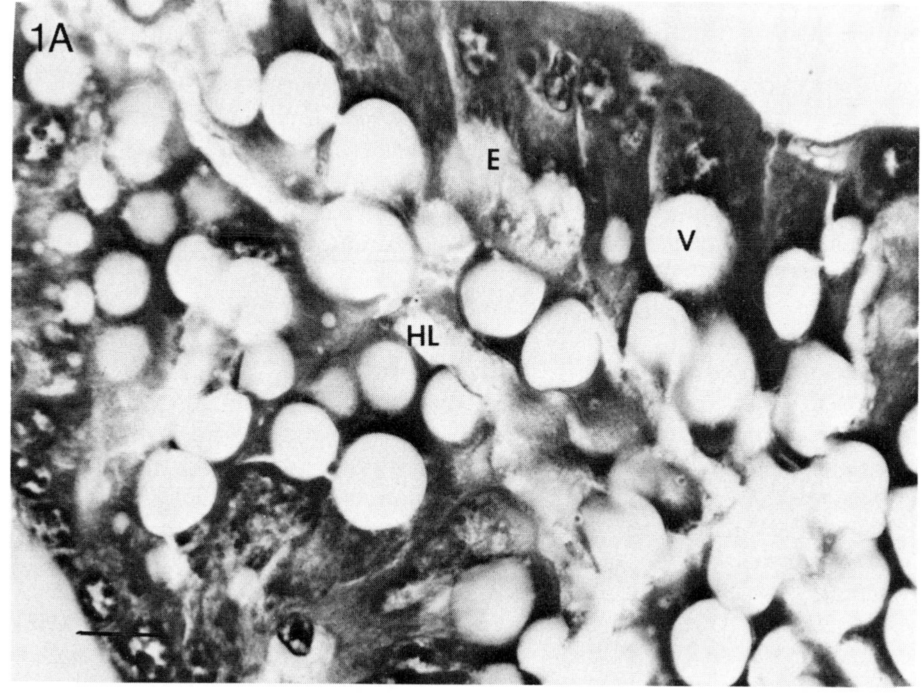

A

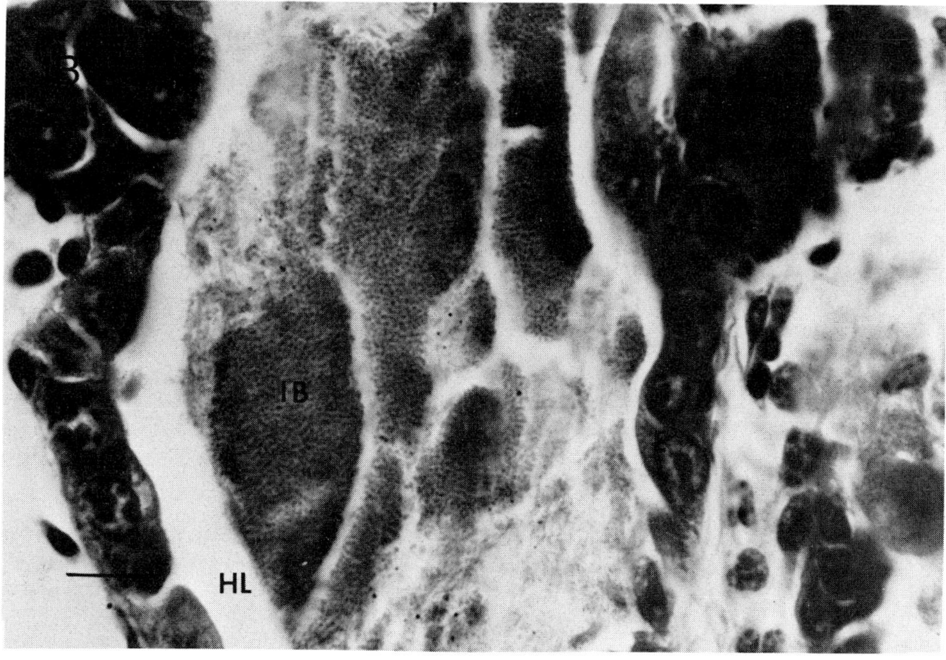

B

FIGURE 1. Larval *M. rosenbergii* (stage-VII) hepatopancreas. (A) Normal hepatopancreas in actively feeding larvae, epithelial cells columnar with abundant vacuolation; (B) hepatopancreas from larvae effected with midcycle disease; atrophy of epithelial cells and intraluminal bacteria. HL, hepatopancreas lumen; E, epithelial cell; V, vacuole; IB, intraluminal bacteria. Line = 10 μm.

synthetic organic compounds (pesticides), other biotoxins, or nutritional deficiencies. To date the etiologic role for any of these has not been established.

A variety of chemotherapeutics have been tested as treatments for MCD, but none have been beneficial in terms of showing an improvement in PL production. Depopulation and disinfection in three hatcheries in Hawaii have provided encouraging results with PL production rising to 10 to 28 PLs per liter following clean-up procedures.

The clinical signs and wet mount preparation lesions reported by Aquacop[12] for bacterial necrosis in *M. rosenbergii* larvae are indeed similar to findings for the chronic phase of MCD. However, a notable exception between bacterial necrosis and MCD is the lack of larval response in MCD to antibiotic treatment. It is unclear if MCD and bacterial necrosis have similar or different etiologies since a definitive etiologic diagnosis has not been established for either disease.

A possible control method for MCD is depopulation of all animal stocks in the hatchery, complete clean-up, and disinfection of the hatchery site. Following clean-up and disinfection, larval rearing should be conducted in distinct cycles with clean-up and disinfection between each rearing cycle. Midcycle larval disease is known only from Hawaii.

Bacterial Necrosis

Aquacop[12] has reported a disease of *Macrobrachium* larvae cultured in Tahiti called bacterial necrosis. Affected populations were reported to display a reduction of feed consumption with increased cannibalism. Moribund larvae turned bluish in color and had empty intestinal tracts.[12] Microbial fouling and focal areas of melanization on the body and appendages (Figure 2) were reported to be commonly observed on moribund larvae.[12] Deformed appendages and mortality during the molt were reported. Bacterial necrosis caused high mortality (100%) in larvae stage four to five, but older larvae and PLs were more resistant. Treatment with antibiotics was reported to effectively arrest this disease, and prophylactic use to prevent it. The authors[12] considered the etiology of this disease to be bacterial in nature. Response to antibiotic treatment, microscopic evidence of melanized lesions, microbial fouling of the appendages, and similarities to what was considered a disease of bacterial nature in penaeid larvae was the basis for their etiologic diagnosis. However, confirmatory experimental data to demonstrate the etiologic role of bacteria was not reported. Prevention and control of bacterial necrosis was reported to be possible through bath treatment with antibiotics (bipenicillin-streptomycin 2 ppm, furanace 0.1 ppm, or erythromycin phosphate 0.65 to 1 ppm). Prophylaxis entailed dosage at the above levels every third day while daily treatments were applied during outbreaks of the disease.

Bacterial necrosis has been reported from Tahiti. This or similar disease(s) are also known from Hawaii.

Exuvia Entrapment Disease

Exuvia entrapment disease of *M. rosenbergii* has been partially characterized on the basis of epizootiology and wet mount microscopic findings. This is a disease primarily of stage 11 larvae and early postlarvae (PLs) with death usually occurring at the time of the metamorphosis molt. Affected animals die during ecdysis and are not able to free the pereiopods, anterior appendages, eyes, or rostrum from the exuvia. Alternatively, some individuals shed the exuvia but die soon after molting. These animals have had malformed appendages. An earlier onset has been occasionally observed with stage-ten larvae being affected.[11] The mortality rate from exuvial entrapment disease has been variable but may reach 20 to 30% in some hatchery tanks. Exuvia entrapment disease (EED) has been diagnosed in all *Macrobrachium* hatcheries in Hawaii. The etiology for this disease has not been determined.

The microscopic lesions associated with EED have been limited to entrapment in the exuvia or postmolt appendage deformity (Figure 3). The histopathological findings in affected

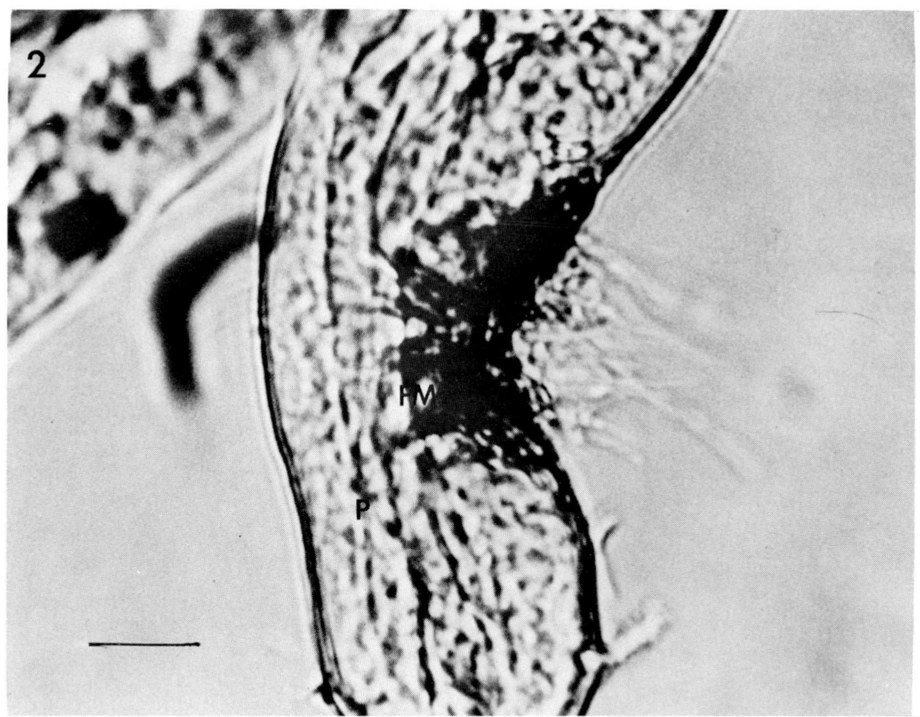

FIGURE 2. *M. rosenbergii*, stage-VII larval pleopod with focal melanization (brown spot) and associated filamentous bacterial epibionts. P, pleopod; FM, focal melanized lesion. Line = 100 μm.

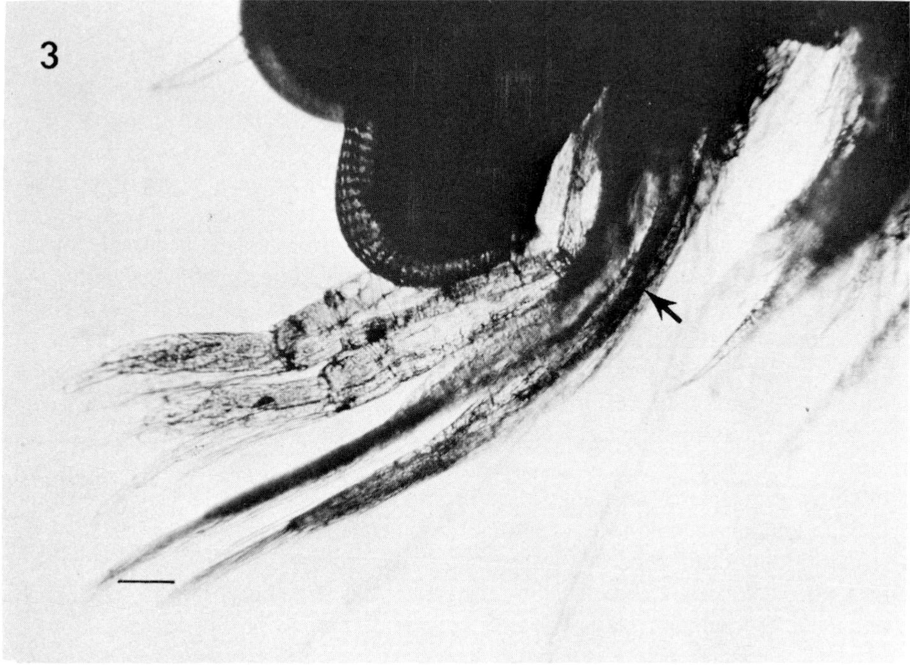

FIGURE 3. *M. rosenbergii* late stage larvae with postmolt deformed anterior appendages (arrow = abnormal dorsal flexure of antennal scales). Line = 1 mm.

larvae have been unremarkable.[11] Ultrastructure studies on affected larvae have not been undertaken.

A diagnosis of EED can be made based on larval stage and microscopic appearance of dead animals. Affected populations will be near the end of the larval cycle and some metamorphosis will be occurring. Of the dead late-stage larvae or PLs, 85% or more will show either exuvia entrapment or appendage deformity.[11]

The population mortality pattern may vary in this disease and can occur at a high acute or low persistent rate. This mortality pattern seems to reflect the rate at which the larval population undergoes metamorphosis.[11]

Exuvia entrapment disease has been found to reduce expected PL production by as much as 30%. This loss is particularly troublesome because it occurs at the end of the larval cycle, a time when maximum investment in the larvae has occurred, and is a significant disease in *Macrobrachium* hatcheries in Hawaii.

In a study[11] conducted at a commercial Hawaiian *Macrobrachium* hatchery, comparison of prevalence of EED in production tanks fed different brands of *Artemia* nauplii showed no reproducible difference between brands. When trials were conducted with and without "green water", larvae reared in clear water treatments showed a 20 to 30% increase in EED-associated mortality at the time of metamorphosis. The reason(s) why the prevalence of EED was lower in the "green water" treatments was not determined, but was thought to be related to nutritional or water quality factors. When measured, total ammonia nitrogen (NH_4-N) levels were found to be higher in the clear water treatments (clear — mean of the two treatments = 0.91 mg/ℓ; green — mean of the two treatments = 0.03 mg/ℓ). Nitrite nitrogen (NO_2-N) levels were less than 0.02 mg/ℓ in both treatment groups.

Maddox et al.[13] investigated the value of algal supplements in rearing *M. rosenbergii* larvae. These authors reported that algal supplements increased larval survival and production of postlarvae and decreased time to metamorphosis. During this study the ammonia, nitrite, and nitrate levels were often higher in green water than in clear water treatments.[13] These authors indicated that the mechanisms for algal benefit to *M. rosenbergii* larvae were unknown, but suggest possible mechanisms to be direct nutrition, indirect nutrition through *A. salina,* or water quality aspects. Microscopic appearance of the larvae and postlarvae was not reported. It is unknown if the decrease in survival of larvae and postlarvae was associated with exuvia entrapment or some other problem.

Wickins[14] has described an apparently similar condition in *Palaemon serratus* that resulted in exuvia entrapment during the metamorphosis molt in this species. This disease was associated with the use of Utah *Artemia*. However, if the *Artemia* were incubated with algae prior to addition to *Palaemon* culture or if *Palaemon* were cultured with algae in the water this allowed complete larval development to postlarvae. In addition, the disease did not occur when *Artemia* from San Francisco Bay were used. Wickins[14] postulated that the Utah *Artemia* were deficient in essential nutrient(s) required by the *Palaemon* larvae. Although a number of analyses were performed on *Artemia* eggs from Utah and San Francisco, no major difference was found which could explain the apparent poor food value of the Utah *Artemia*.

Bowser et al.[15] reported addition of lecithin to the diet decreased the incidence of molt-associated mortalities (molt-death syndrome) in juvenile lobsters (*Homarus* sp.). Animals affected with molt-death syndrome either died in the process of molting or completed the molt but were deformed and died shortly thereafter. Calcium deposits on the inner surface of the exuvial exoskeleton were associated with the occurrence of this disease.

In Hawaii similar appearing exuvia exoskeleton deposits (Figure 4) have been found to occur in *M. rosenbergii* late stage larvae, postlarvae, juveniles, and adults.[11] Examination with scanning electron microscopy and Energy Dispersive Analysis of X-rays (EDAX) suggests these deposits to be composed of calcium.[11] Preliminary results indicate a positive

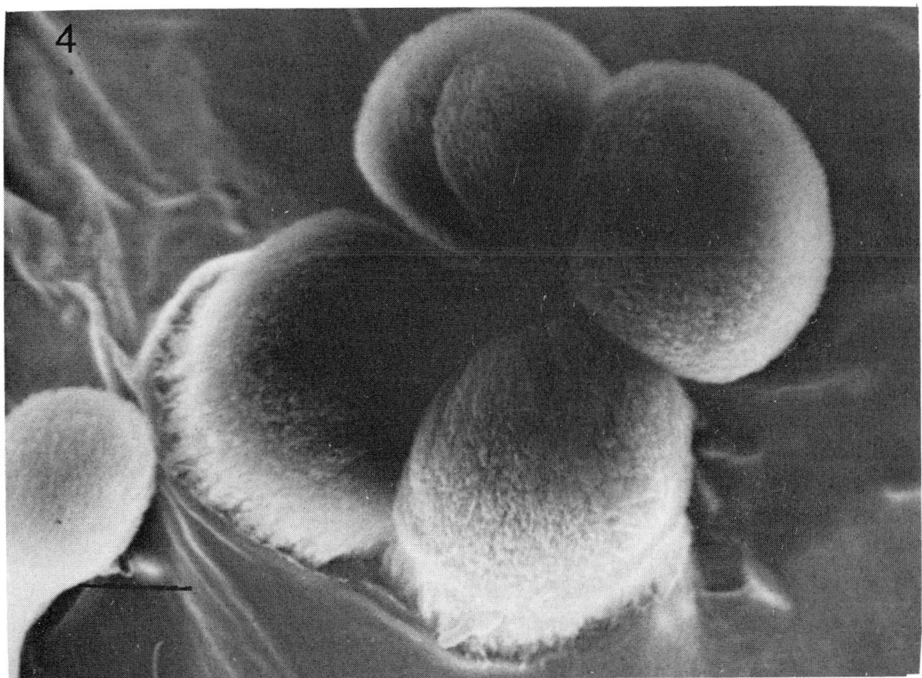

FIGURE 4. Endocuticular calcium deposits from *M. rosenbergii* postlarval exuvia. Line = 20 μm.

correlation between the prevalence of exuvia deposits and the occurrence of exuvia entrapment disease in larval *Macrobrachium*.

Several feeding trials in which soy bean curd was incorporated into larval diets fed the last week of the cycle have indicated a beneficial effect in terms of reduction of the incidence of exuvia entrapment disease. However, additional work along these lines needs to be conducted to confirm these results. The etiology of exuvia entrapment disease in *Macrobrachium* larvae and early PLs remains to be established.

Exuvia entrapment disease is known from *Macrobrachium* hatcheries in Hawaii. Until the etiology of this disease is established, prevention and control procedures are limited to use of algae in the larval culture media and maintenance of good water quality conditions within the hatchery tanks. Supplementing late stage larval diets with lecithin or products containing lecithin may be helpful in the control of this disease.

Microbial Epibiont Disease

Microbial epibionts are commonly observed disease agents in *Macrobrachium* spp. as well as other cultured crustaceans. Epibiont fouling organisms include filamentous and nonfilamentous bacteria, algae, or protozoa common to the aquatic environment. Although a variety of microorganisms are involved, sufficient similarities exist to consider microbial epibiotic fouling as a single disease syndrome. This syndrome has been characterized on the basis of gross and microscopic pathology. The lesion which heralds this disease has been the attachment of microbial agents to epicuticular surfaces with no destruction of host tissue and little or no host inflammatory response at the site of attachment. These microbial agents have a world-wide distribution and are ubiquitous in aquatic environments. In general, the occurrence of epibiotic fouling disease in culture systems has been considered as gross evidence of existing poor water quality conditions within the system[16,17] and/or an underlying abnormal condition of the host.[16] In *Macrobrachium* spp. epibiotic fouling may indicate an extended intermolt period suggesting the occurrence of an underlying chronic disease of nutritional or other etiology. Hall[17] states that no evidence exists that protozoan epibionts are pathogenic or parasitic to *Macrobrachium* spp. Koch's postulates have not been fulfilled for any of the epibiotic organisms reported in *Macrobrachium,* possibly reflecting the secondary nature of epibiont microbial disease in this genera.

Physical examination of affected individuals is sufficient to make a diagnosis of microbial epibiotic disease in *Macrobrachium.* Unless fouling is severe, larval stages require examination with a dissecting or light microscope. Presumptive etiologic diagnosis usually requires wet mount microscopic examination of representative microbial epibionts. Filamentous and nonfilamentous bacteria can be distinguished on the basis of size. Filamentous bacterial epibionts of *Macrobrachium* have tentatively been identified to the genus *Leucothrix.*[16] To my knowledge this identity has not been confirmed. Figure 5 lists a simple identification key to the protozoan epibionts of *Macrobrachium.*

Bacterial epibionts have been found on all epicuticular surfaces. In my experience with *Macrobrachium* larvae, bacterial epibionts have initially appeared attached to the setae and setaeules (distal portions of the appendages) of the pereiopods or other appendicular structures (Figures 6 and 7). Bacterial fouling of the proximal appendages, gills, or body surface of larvae has indicated a severe, late manifestation; and larval mortality has usually occurred associated with this type of fouling. As indicated by Fisher[19] for other crustacea, nonfilamentous bacterial forms have been more directly associated with mortality in *Macrobrachium* larvae than filamentous bacteria. Antibiotic treatment has been reported to decrease crustacean egg and larval mortalities associated with nonfilamentous bacterial fouling.[19] Cutrine-plus has been reported to effectively reduce penaeid shrimp mortality from filamentous bacterial gill disease.[20] These findings suggest that bacterial epibiotic fouling does result in mortality, particularly, if large numbers of bacteria have attached to epicuticular surfaces of the gills.

The protozoan epibiont *Zoothamnium* has been reported to preferentially locate on the gill lamallae of *Macrobrachium,* while other protozoan epibionts do not show site specificity.[17] Protozoan attachment of *Macrobrachium* larvae has been rarely found on the distal portions of pereiopods, and usually occurs on the eyestalks, body, antennal scales, or uropods. Egg masses on adult female *Macrobrachium* have been reported as common sites of attachment.[17] In Hawaiian production ponds terminal growth prawns or those that are unthrifty have frequently been found to be affected with generalized epibiotic fouling. In these cases fouling by bacteria, protozoa, and algae has occurred simultaneously. The prevalence of affected prawns has usually been low. However, epizootics of *Epistylis* sp. fouling on otherwise healthy appearing subadult and adult prawns in which up to 40% of the pond population were affected have infrequently occurred in Hawaiian prawn ponds.[11] In these cases attachment of *Epistylis* was on ventral surfaces initially, and spread dorsally to the base of the rostrum, eyestalks, and sometimes spread to the third abdominal segment (Figure 8). These epizootics were associated with extremely poor water quality conditions possibly resulting from lack of adequate pond water turnover. Increased water flow corrected the problem. Prawn mortalities were not observed, but marketing of affected individuals had to be postponed until *Epistylis* fouling abated. Hard water with high calcium carbonate content has been reported to be associated with an increase in *Epistylis* infestation of prawns.[2]

Several mechanisms to describe the pathophysiologic effect of microbial epibionts on the host have been advanced, but few hard data exist to demonstrate the validity of these hypotheses. Hall[17] indicated that ectocommensal protozoa may adversely affect *Macrobrachium* larvae by reducing motility and feeding sufficient enough to cause mortality through malnutrition. In addition, Hall[17] suggested suctorian protozoans because of their feeding on other protozoans, may actually be parasitic on *Macrobrachium* eggs. Several authors have indicated that epibionts may restrict gas exchange by accumulating on egg and gill surfaces in susceptible crustacea, thus causing mortality through asphyxiation,[19,20] with increased

Class - Peritrichea
Order - Sessilidae
Family - Vaginicolidae

1. Vaginicola
   Lorica vase-like with basal attachment to substrate; color clear or yellow to brown.

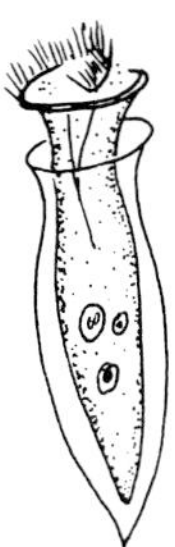

Family - Lagenophryidae

1. Lagenophrys
   Lorica opening with infolding collar; body of animal tapered or curved with posterior end attached near lorica.

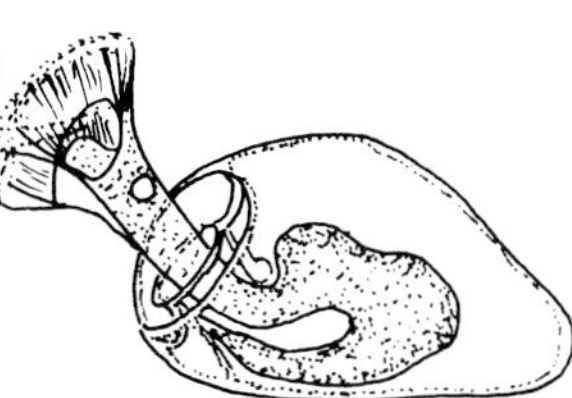

2. Corthunia
   Lorica clear or yellow to brown with short attachment stalk; lorica shape variable; lorica may adhere together to form colonies.

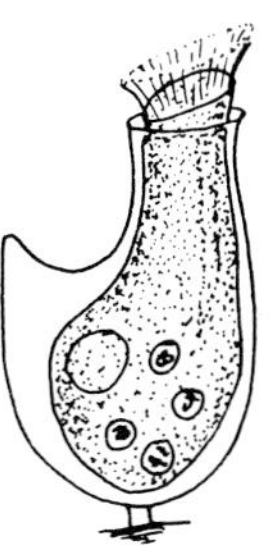

Suborder - Aloricina
Family - Vorticellidae

1. Vorticella
   Bell-shaped body attached to stalk with a contractile myoneme. Stalk is non-branching and may coil like a spring.

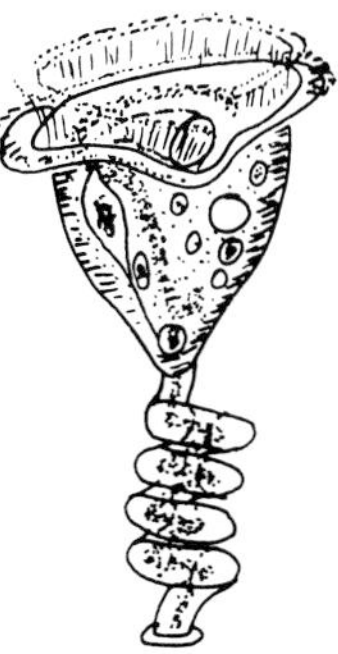

FIGURE 5. Simple identification key to the epibiont protozoan genera reported from *M. rosenbergii*.

Family - Epistylididae

1. Zoothamnium
   Like Vorticella but myoneme in branching stalk contracts colony as a unit.

2. Epistylis
   Stalk braching dichotomously and lacking contractile myoneme. Colony has no contract.

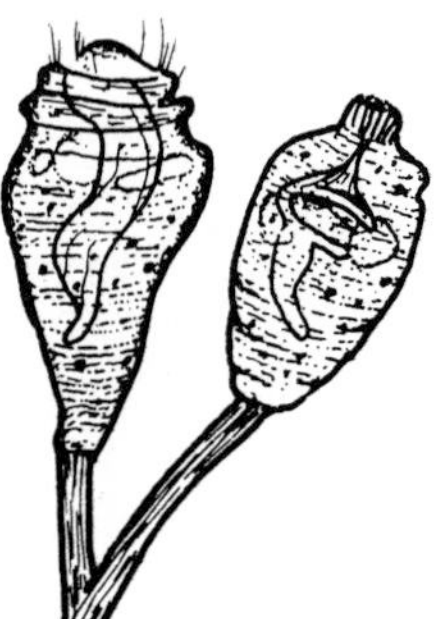

Class - Suctoria
Order - Exogenida
Family - Podophyridae

1. Podophyridae
   A usually round body on a long stalk. Tenacles radiate from around the body.

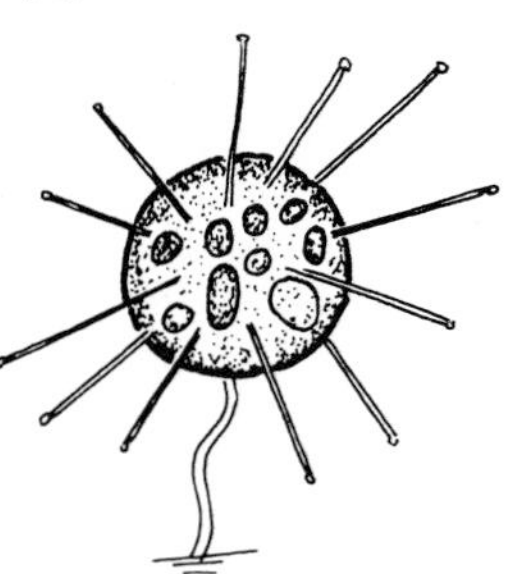

Order - Endogenidae
Family - Acinetidae

1. Acineta
   Tenacles protrude from body in clusters; body may be enclosed in a lorica.

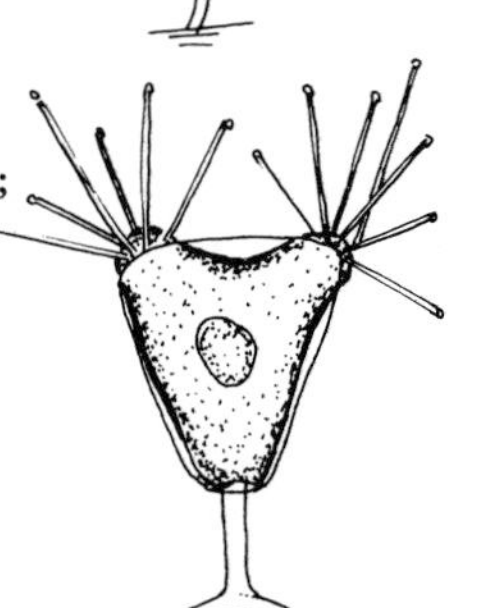

Family - Dendrosomatidae

1. Tokophrya
   Tenacles clustered in four groups.

FIGURE 5(continued)

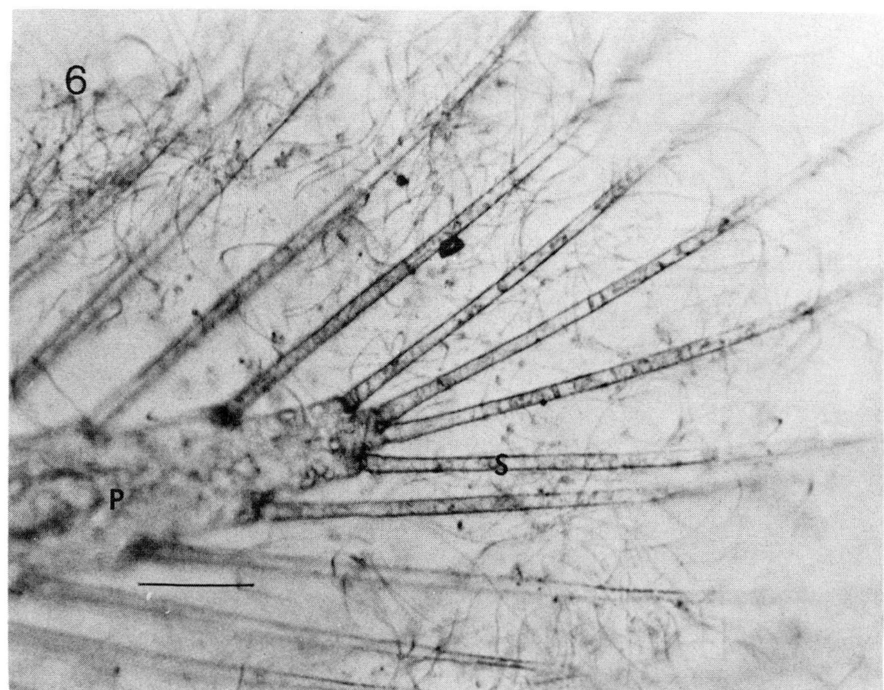

FIGURE 6. *M. rosenbergii* larvae with filamentous bacterial fouling of pereiopod setae. P, pereiopod; S, setae. Line = 20 μm.

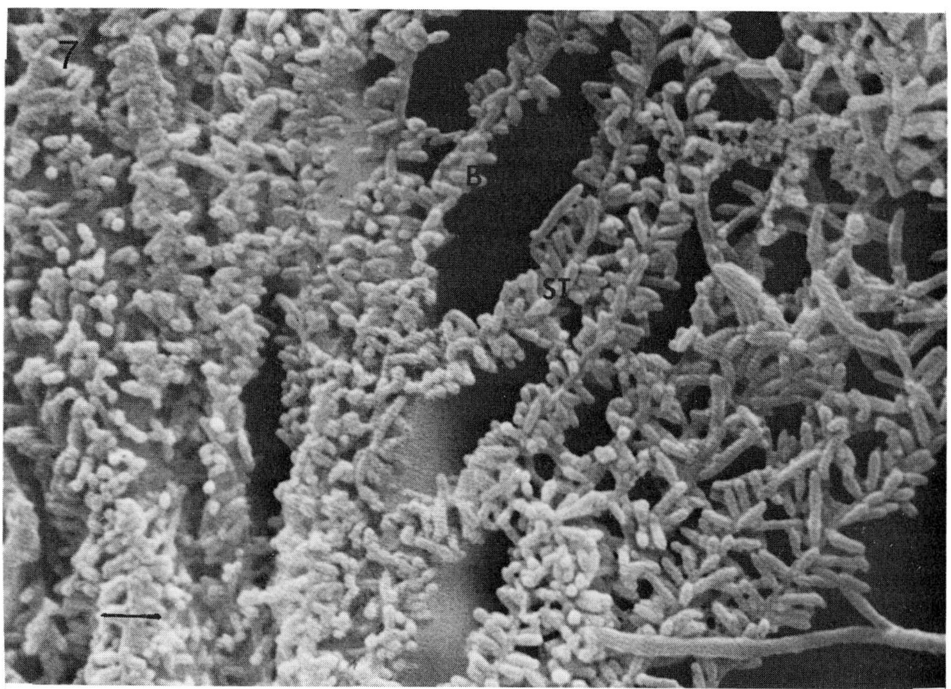

FIGURE 7. *M. rosenbergii* larvae with rod-shaped bacterial fouling of pereiopod setae and setaeules. ST, setaeules; B, bacterial rod. Line = 1.4 μm.

FIGURE 8. *Epistylis* fouling of adult *M. rosenbergii*. Arrows indicate colonies of *Epistylis*.

susceptibility during ecdysis.[19] Scheer et al.[21] found in *Leander* oxygen uptake increased markedly just prior to exuviation. This could explain why mortality in animals affected with microbial epibiont fouling occurs more frequently at the time of molting.

As microbial epibiont disease seems to indicate adverse biological conditions within a culture system or reflects a primary underlying illness in the host, control methods should be aimed at identifying and correcting the primary problems that exist. Improved water quality by reduction of nutrients and/or bacterial density,[17] decreased animal density with increased attention to container sanitation,[16,22] increased water exchange,[16,22] improved nutrition,[19] and prophylactic and therapeutic use of chemicals[16,17,19,22-24] have all been suggested for the prevention and control of epibiont microbial diseases. Measures such as selection of specific pathogen-free animals or disinfection of culture water with ultraviolet light or ozone have met with little or no success in the control of microbial epibiont disease in crustacea.[19] This probably reflects the fact that the microorganisms involved in this disease syndrome are faculative, widely distributed in aquatic environments, bloom when environmental conditions are conducive to promote their growth, and reservoir in the aquatic medium or sediments. In this regard aquaculture settings are simply ideal locations and probably provide a selective advantage for the proliferation of these types of organisms. Selected chemotherapeutants used to control microbial epibiont organisms are listed in Table 2.

Brown Spot Disease

Brown spot disease (shell disease, black spot disease) is a commonly observed disease in both cultured and feral crustaceans including *Macrobrachium*. Within this genus brown spot disease has been reported in *M. rosenbergii* (larvae and adults),[12,25-27] and adult *M. lar*,[28] *M. acanthurus*,[29] *M. carcinus*,[29] *M. ohione*,[29] and *M. vollenhovenii*.[30] Geographically, brown spot disease has been reported from Hawaii,[25,28] Florida,[25,29] Tahiti,[12] and Liberia.[30] Undoubtedly its occurrence is world-wide.

Brown spot disease can be recognized by the presence of brown to black, ulcerative to raised lesions affecting any surface of the body or appendages and varying in size from tiny

Table 2
SELECTED CHEMICAL CONTROL METHODS FOR MICROBIAL EPIBIOTIC FOULING DISEASE

Microbial agent	Chemical	Dosage/treatment duration	Species	Life stage	Ref.
Peritrich protozoa					
Epistylis sp.	Formalin	25 ppm/not stated	*M. rosenbergii*	J[a], A[b]	82
	Formalin	75 ppm/6—8 hr SB[c]	*Penaeus* spp.	J, A	82
	Formalin	25 ppm/24 hr FT[d]	*Penaeus* spp.	J, A	83
	Chloramine-T				
	Quinine bisulfate and quinine sulfate	5 ppm/72 hr SB	*P. stylirostris*	J	23
	Quinine hydrochloride	0.6 ppm/72 hr SB	*P. stylirostris*	J	23
	Acetic acid	2.0 ppm/1-min dip	*M. rosenbergii*	L[e]	81
Zoothamnium sp.	Formalin	20 ppm/24 hr	*M. acanthurus*	L	24
Filamentous and nonfilamentous bacteria					
Leucothrix sp.	Cutrine-plus	0.5 ppm/2—4 hr SB	*P. californiensis*	J	20
Miscellaneous		0.1 ppm/24 hr FT	*P. californiensis*	J	20
genera of filamentous	Malachite green	5 ppm/5- min dip	*Homarus* spp.	L	84
and rod-shaped forms	Furanace	1 ppm/not stated	*M. rosenbergii*	L	

[a] J = juvenile.
[b] A = adult.
[c] SB = static bath.
[d] FT = flow through.
[e] L = larvae.

to large. The distribution of this lesion may be focal or multifocal. In brown spot disease the external surface of the cuticle is always involved. Other cuticular and deeper tissue structures may or may not be affected.

The brown to black pigment (melanin) in brown spot lesions of crustaceans is the result of a host response to insult and does not by itself suggest a particular etiology. A variety of causes have been suggested for brown spot disease. These include bacterial species which produce extracellular lipases,[31] proteases,[31] and chitinases;[25,27,31] fungi,[32-34] mechanical trauma,[25,27,32] precipitating chemicals,[34,35] nitrogenous waste products,[34] or nutritional or developmental abnormalities which result in damage to the epicuticular layer of the exoskeleton.[36] It is important to note that the occurrence of brown spot disease seems to depend on an initial insult which results in disruption of the epicuticular layer. This is then followed by invasion by microbial agents. The epicuticle is considered to be the primary line of defense in the crustacean exoskeleton.[7] Cipriani et al.[31] were unable to experimentally produce brown spot disease in penaeids with bacterial species isolated from clinical cases of the disease until inoculation was preceded by mechanical disruption of the epicuticular layer. Burns et al.[32] were unsuccessful in attempts to infect healthy *M. rosenbergii* with *Fusarium* sp. isolated from necrotic, melanized lesions. These authors noted that mycosis only occurred in prawns which had developed prior cuticular damage.

Based on preliminary observations, I have roughly categorized brown spot lesions of *Macrobrachium* into one of three distribution patterns. These patterns are branchiostegite melanization, focal to multifocal melanization of the body or appendages, and uropod ulceration and melanization. These lesion patterns may occur simultaneously.

Branchiostegite melanization (Figure 9) has been observed to occur as confluent, spreading melanization of the medial surface of the branchiostegites. Branchiostegite melanization has

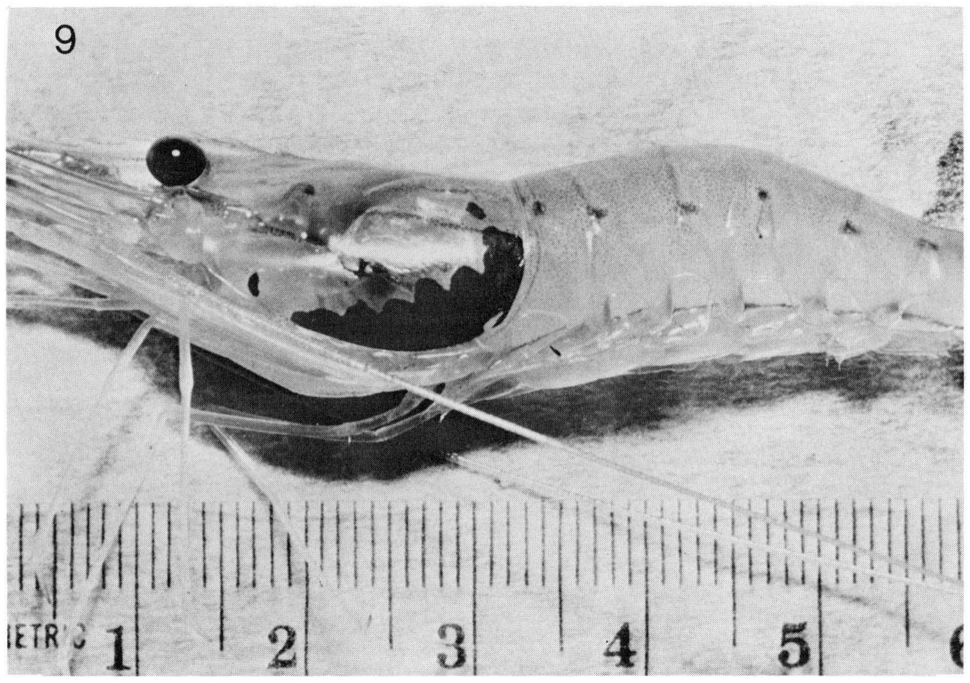

FIGURE 9. Branchiostegite melanization; juvenile *M. rosenbergii*.

been commonly noted as a nearly bilaterally symmetrical lesion on freshwater prawns.[11] This melanized lesion pattern has frequently been observed on otherwise healthy appearing adult prawns in production ponds in Hawaii. Microscopically, the lesion has consisted primarily of a melanized layer, lining the medial cuticular layer with ulceration of the cuticle, an infrequent finding. Johnson[34] reported gill carapace melanization was associated with high levels of nitrogenous substances. Figure 10 shows an area of branchiostegite melanization with fungal mycelia. However, it was not determined if the fungi were primary or opportunistic in this case.

A second frequently observed lesion pattern has been one in which melanized areas occur either focally or multifocally on any surface of the body or appendages. These lesions have been variable in size, usually small, and have not been bilaterally symmetrical in distribution.

Otherwise normal appearing prawns in Hawaiian production pond systems have frequently been observed to have these focal to multifocal melanized lesions.[11] Kubota[28] reported 17.4% (178 out of 1024) of feral *M. lar* examined in Kahana stream on Oahu were affected with this type of melanized lesion. In another study,[30] 91% (137 out of 150) of *M. vollenhovenii* examined from rivers in Liberia were found to be affected. In both cases lesions occurred more often on the larger individuals. Bacteria and/or fungi have been implicated as a cause of these lesions.[25-27,29,32]

Larval stages have also been observed to develop focal to multifocal melanized lesions primarily on the appendages (see section on bacterial necrosis). On occasion fungal hyphae have been noted on histopathologic examination within brown spot lesions of *Macrobrachium* larvae from Hawaii.[37]

A third lesion pattern that has been observed is ulceration and melanization of the uropods (Figure 11). These lesions have not been bilaterally symmetrical in distribution and seem to be common to prawns maintained on hard artificial substrate, particularly if detritus accumulates in the container. Burns et al.[32] isolated *Fusarium* sp. from necrotic uropod lesions in tank reared *M. rosenbergii*. Amborski et al.[38] identified several genera of bacteria and implicated them as the cause of this type of ulcerative lesion in crayfish.

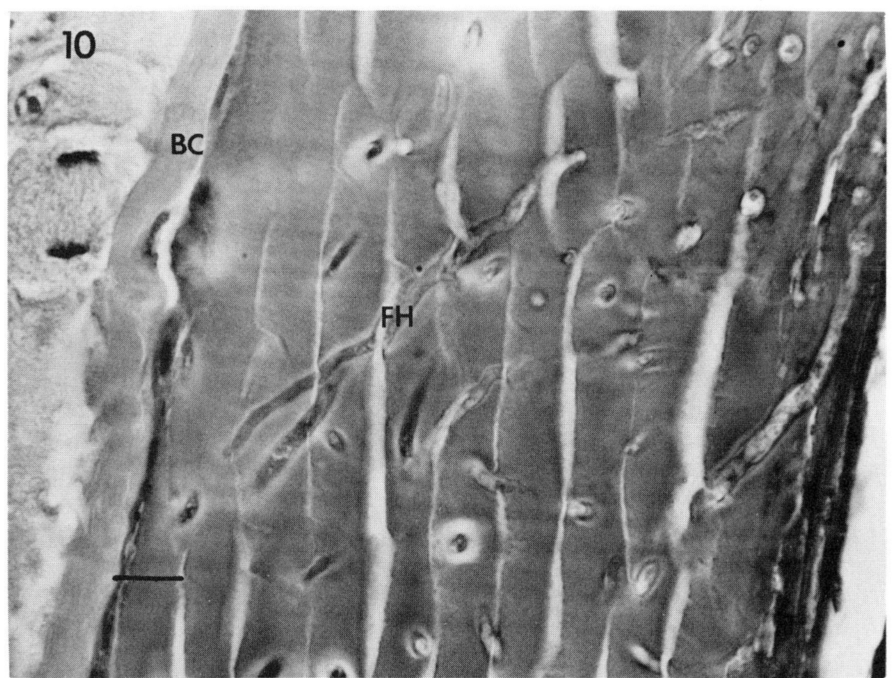

FIGURE 10. Phycomycetous fungi in branchiostegite melanized lesion of *M. rosenbergii*. FH, fungal hyphae; BC, branchiostegite cuticle (medial surface). Line = 10 μm.

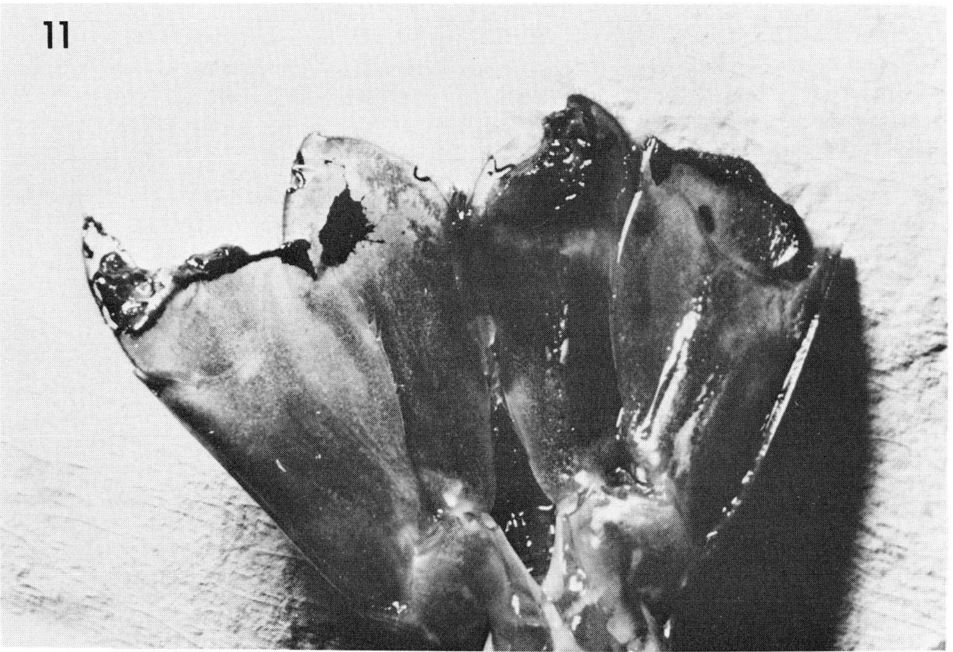

FIGURE 11. Ulceration and melanization of uropods, adult *M. rosenbergii*.

A diagnosis of brown spot disease in *Macrobrachium* can be made on the basis of gross lesion appearance. The establishment of an etiologic diagnosis requires laboratory tests. Routine culture of diseased tissues on media selective for bacteria and fungi as well as histopathology evaluation of lesions have all been employed. Appropriate diagnostic procedures can be found in Balows et al.[8] Determination of initiating or accessory etiologic factors which apparently play a role in the pathogenesis of brown spot disease may be quite difficult (mechanical trauma is probably an exception to this).

The control of brown spot disease in *Macrobrachium* involves improvement of culture conditions through increased attention to husbandry and nutrition. Providing adequate substrate, shelter, and space to minimize intraspecies aggression as well as insuring optimal water quality conditions are all recommended. As suggested by Delves-Broughton et al.,[27] "Shell disease does not lend itself to prophylactic or chemotherapeutic control." The use of chemicals are generally not recommended in the control of this disease. An exception to this exists, however, antibiotics have been reported to successfully control bacterial necrosis (I consider this a form of brown spot disease) in *M. rosenbergii* larvae cultured in Tahiti.[12]

Idiopathic Muscle Necrosis

Idiopathic muscle necrosis (spontaneous muscle necrosis, muscle opacity) has been partially characterized on the basis of epizootiology, gross, and microscopic lesions. This disease has been reported from *Penaeus* spp.[39-42] and *M. rosenbergii.*[43-45] Idiopathic muscle necrosis (IMN) has been associated with exposure to environmental stressors such as salinity and temperature fluctuations,[39-42,44] hypoxia,[39,41,42] hyperactivity associated with handling,[42,45,46] overcrowding,[40-42,44] and narcotizing with quinaldine.[46] The etiology and pathogenesis of IMN are not known. In one case study a 40% prawn mortality was reported associated with the occurrence of IMN.[45] IMN has been observed in larvae, juvenile, and adult *M. rosenbergii.*[11,12,43-45,47]

The gross lesions in *M. rosenbergii* have been focal, multifocal, to confluent opacity of the striated muscle (Figure 12). Microscopically, IMN lesions have been reported as areas of necrotic muscle cells with slight to moderate hemocyte infiltration (Figure 13).[39,42,45] Sarcolemmal nuclear proliferation has been a common finding in older, regenerating lesions.[39,45]

A tentative diagnosis of IMN in freshwater prawns can be made based on gross lesions (focal to multifocal muscle opacity) and a history of recent exposure to environmental stressors. This diagnosis can be confirmed by demonstration of typical lesions by histopathology.

The prevalence of postlarvae (PLs) with IMN has been described by Sarver et al.[47] as a good indicator of the general fitness of postlarvae leaving the hatchery. These authors suggested that monitoring PL batches for IMN lesions may prove useful as a management tool to assess the condition of PLs prior to stocking in dirt ponds.

Prevention of IMN is through reduction of environmental stressors during routine culturing and particularly at times of handling and transfer. This disease has been reported from South Carolina,[44] the U.K.,[27] and Hawaii.[43-45,47]

Black Nodule Disease

Black nodule disease has been partially characterized by microbiology results, gross, and microscopic lesions. The disease has only been reported to occur in laboratory cultured *M. rosenbergii* in the U.K.[27] The gross lesions observed were black nodules located within the epidermis and hypodermis of the cephalothoracic cuticle. There was no ulceration of the overlying cuticle.[27] Microscopically, these nodules were foci of necrotic melanized tissue debris surrounded by numerous hemocytes (hemocytic nodules). Bacteria were isolated from the tissues of prawns with the black nodule lesions.[27]

The authors reported it was probable that these lesions were caused by bacteria as furanace

FIGURE 12. Multifocal idiopathic muscle necrosis of abdominal muscles of juvenile *M. rosenbergii*. (From Akiyama, et al.[45])

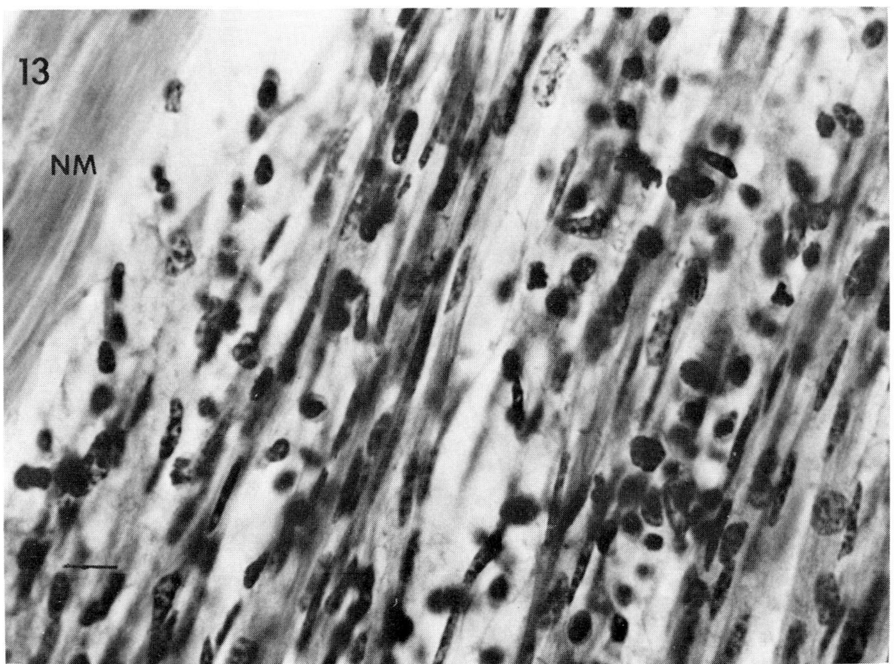

FIGURE 13. Idiopathic muscle necrosis lesion of juvenile *M. rosenbergii* with moderate hemocyte infiltration and sarcolemmal nuclear proliferation. NM, normal muscle tissue. Line = 10 μm. (From Akiyama, et al.[45])

treatment (0.09 ppm) arrested the disease.[27] However, bacteria were not observed in the lesions or elsewhere in the tissues on histopathology; infectivity studies failed to demonstrate Koch's postulates for the microorganisms isolated; and controls were not maintained during the furanace treatments. In my opinion an etiologic diagnosis of bacterial septicemia was not well supported by the information presented by the authors and other etiologies should be considered for the pathology reported.

Lightner[48] has pointed out the similarities in clinical history, gross, and microscopic pathology between black nodule disease and black death, a disease of *Penaeid* spp. Penaeids affected with black death disease develop melanized hemocytic nodules in the subcuticular tissues of the general body surface, foregut and hindgut, and gills.[48,49] Black death disease has been shown to be caused by a deficiency in dietary vitamin C. It is possible that black nodule disease has a similar etiology, but experimental studies to confirm this hypothesis need to be conducted. A specific vitamin C deficiency syndrome in *Macrobrachium* has not been reported.

Hemocytic Enteritis

Hemocytic enteritis (HE) is a disease that has been characterized by epizootiology, gross, and microscopic pathology in *Penaeus stylirostris.*[50] This disease has been associated with high, acute, or low chronic mortality in susceptible-sized shrimp populations.[50] Shrimp affected with this disease have been found to develop necrosis of the mid- and hindgut epithelium with moderate to marked accumulation of hemocytes in the gut lumen, and all layers of the midgut wall in affected areas.[50] This disease has been reported to be caused by ingestion of blue-green algae, primarily *Spirulina subsalsa,* and possibly other blue-green species.[50] Septicemia bacterial disease has often accompanied epizootics of hemocytic enteritis in penaeids, with *Vibrio* spp. being the most common genera isolated.[50]

Juvenile and subadult pond-cultured *M. rosenbergii* in Hawaii have sporadically been found affected with midgut lesions (Figure 14) consistent in appearance with those described for hemocytic enteritis in *P. stylirostris*. These midgut lesions have been presumed to result from ingestion of blue-green algae (*Oscillatoria* spp.).

Preliminary survey results using histopathology examination of subadult prawns from two farms have suggested a low and/or sporadic occurrence of this disease in ponds at these locations.[11] Additional studies of this nature to substantiate these findings and experimental feeding trials to show that ingestion of blue-green algae will induce the HE lesions in prawns should be undertaken.

The effect of hemocytic enteritis on prawn production has not been established and epizootics similar to those described in *P. stylirostris* have not been observed or diagnosed as such in *M. rosenbergii*.

Control methods are limited to keeping adequate algae density in prawn ponds to prevent significant growth of filamentous blue-green benthic forms. Application of antibiotics in feed has been utilized as a treatment for secondary bacterial infections in penaeids affected with hemocytic enteritis.[50] Without more information about this disease in *Macrobrachium,* the use of antibiotics as a treatment is not recommended. In *Macrobrachium* culture, hemocytic enteritis is only known from Hawaii.

Hepatopancreas Saturated Fat Preservation

Hepatopancreas saturated fat preservation (hepatopancreas mummification) has been recognized as a disease of pond-cultured *M. rosenbergii* and has been characterized by the occurrence of grossly intact hepatopancrea (Figure 15) recovered from prawn ponds during routine sampling or harvesting.[51] Once preserved these hepatopancrea appear resistant to normal degenerative processes. Specimens have been kept for more than 1 year without the use of fixatives.[51]

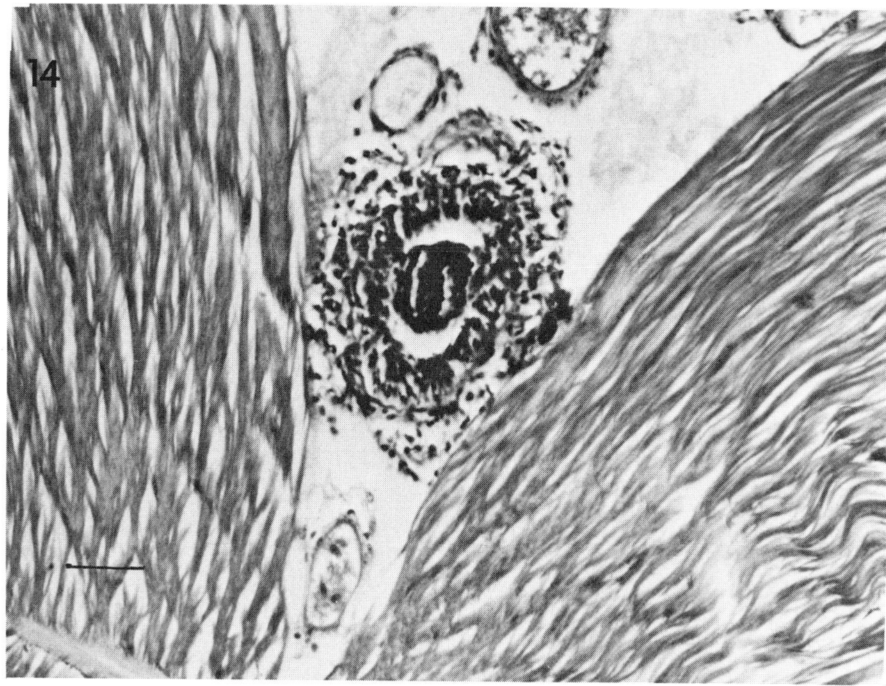

FIGURE 14. Abdominal midgut of juvenile *M. rosenbergii* with necrosis of mucosal epithelial and marked hemocyte infiltration (hemocytic enteritis) Line = 50 μm.

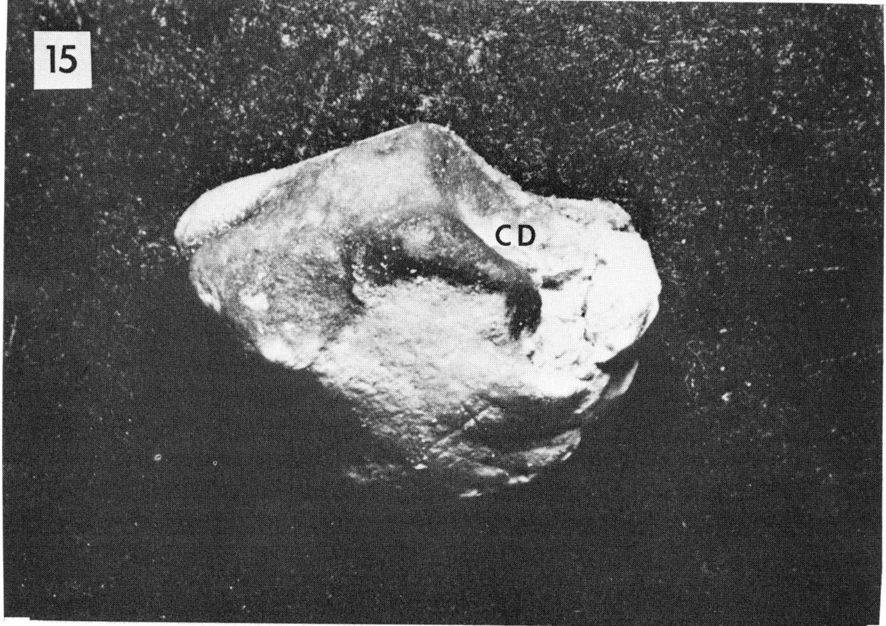

FIGURE 15. Preserved *M. rosenbergii* hepatopancreas recovered from a prawn pond. CD, cardiac stomach depression on anterior dorsal surface.

Histopathology examination of preserved glands has shown the absence of normal hepatopancreas tissue structure, with occasional bacteria and protozoa being present. When heated to 80°C these glands have turned into a liquid that solidified into a waxy substance once cooled.[11]

Results of fatty acid analysis have found the preserved glands to be composed of increased amounts of saturated fatty acids, principally palmitic acid.[51] Analysis for pesticides in these glands have in some cases been positive for the organophosphates malathion (0.49 ppm), thimet (0.03 ppm), parathion (0.06 ppm), and methyl parathion (0.03 ppm).[51] Other types of pesticides were not detected. Heavy metal levels were not found to be significant.[51]

The recovery of organophosphate pesticides from preserved glands suggest these as potential etiologic agents for this disease. Saturated fat accumulation has been postulated to result from abnormal lipid metabolism due to chemical blockage of normal lipase activity in the hepatopancreas.[51] The etiologic role of organophosphates and their proposed mode of action needs to be confirmed. Studies should be undertaken to accomplish this. The glands apparently resist degradation due to their high saturated fat content (45%, dry weight basis).[52]

Clinically, preserved hepatopancrea have been found in ponds on three farms in Hawaii. In one of these cases the occurrence of large numbers of preserved glands was associated with over a 90% reduction of prawn biomass in the affected pond.[51] Interestingly, moribund or dead prawns were not observed by the farm personnel although what was graded as a minor oxygen kill (20 to 30 lb of prawns died) was recorded in the daily log book approximately 1 month prior to discovering that few prawns remained in the pond. Examination of the pond bottom revealed large numbers of preserved hepatopancrea. *Gambusia* sp. were also noted to be greatly reduced in number and the water color of the pond was reported to be dark brown. Surviving prawns in this pond appeared grossly normal. No other ponds on the farm were found to be affected. Preserved hepatopancrea have been sporadically recovered from only two other farms and have not been associated with a high, unexplained loss of prawns.[11]

Preliminary studies have revealed malathion at 20- to 500-ppb levels in feed ingredients of plant origin used in prawn feed formulations.[52] However, it is not established that these low pesticide levels constitute a health hazard to prawns through oral- or water-borne exposure. Studies to determine the safe levels of pesticides in crustacean feeds need to be undertaken.

Hepatopancreas saturated fat preservation is only known from Hawaii. At the present level of understanding of this disease, prevention is limited to applying feeds to ponds with low or nondetectable levels of organophosphates.

Endocuticular Degeneration

Endocuticular degeneration (exoskeleton spotting) is a poorly understood disease of adult *M. rosenbergii* which has been partially characterized by epizootiology, gross, and microscopic lesions. Johnson[22] reported this or a similar disease in larger cultured *Macrobrachium* from the southern U.S. The etiology of endocuticular degeneration (ED) has not been determined.

In Hawaii endocuticular degeneration has been noted from only one farm and almost exclusively from one pond on that farm. Preliminary study data[11] from this pond showed only female prawns 10.5 cm or larger affected with ED lesions. The prevalence of affected prawns in this pond irrespective of sex and size was 1.0% (N = 420). The prevalence of affected female prawns 10.0 cm or larger was 38% (N = 21). The farm manager reported no observed mortality associated with the occurrence of ED in this pond. The effect of this disease on prawns is unknown. Prawns with ED lesions held in the laboratory have been observed to successfully complete a molt with no apparent change in gross appearance of the ED lesions.[11] A spreading or enlargement of these lesions was not observed in these captive prawns.

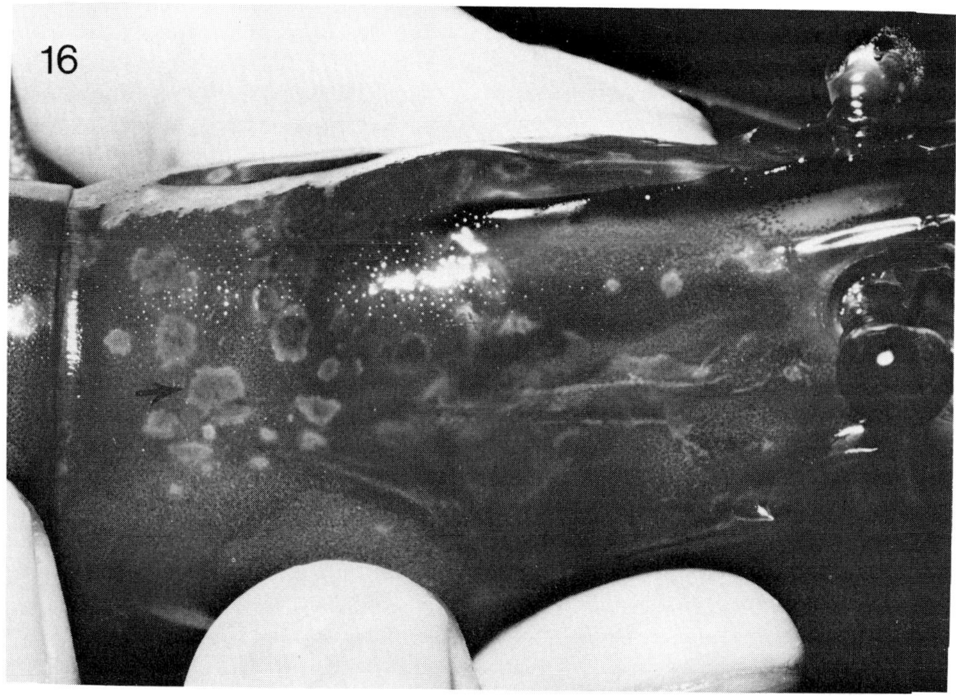

FIGURE 16. Endocuticular degeneration from adult *M. rosenbergii*. Arrow indicates a lesion.

The gross lesions noted in examination of these prawns were focal to multifocal, irregular shaped, brown to orange-brown discolored areas involving the deeper layers of the cuticle (Figure 16). Lesions were distributed in a nonbilaterally symmetrical fashion on the cephalothorax and abdomen. The cuticle was typically not ulcerated, but was in some locations more flexible in affected areas. Gross lesions in other organ systems were not observed.

Microscopically, ED lesions appeared as patchy areas of degeneration within the membranous and endocuticular layers of the cuticle. Subcuticular amorphous deposits were frequently present, and underlying epidermis appeared disorganized with fewer epidermal cells than were found in regions supporting normal cuticle (Figure 17). Microscopic lesions have not been observed in other organ systems examined.

Johnson[22] reported attempts to isolate infectious agents from ED lesions were negative. Microscopically, I have not observed bacteria, fungi, protozoan parasites, or evidence of virus (inclusion bodies) in ED lesions. Ultrastructure studies have not been conducted.

Endocuticular degeneration (exoskeleton spotting) has been reported from Texas and is known to occur in one location in Hawaii. In view of the lack of a known etiology for this disease, prevention and control methods are not recommended.

Terminal Growth (Enzootic Cachexia)

Terminal growth (TG) is a poorly understood disease of undetermined etiology that has been partially characterized by epizootiology, clinical signs, gross postmortem, and microscopic lesions. Reports concerning terminal growth have been limited to a brief comment by Kato[44] at the second *Macrobrachium* workshop in which he described a condition in large male prawns which had "mushy flesh and separation of the carapace and abdomen before death." However, the disease is a common enzootic problem in all pond culture systems in Hawaii.

Terminal growth prawns have been known to occur in pond culture systems for as long

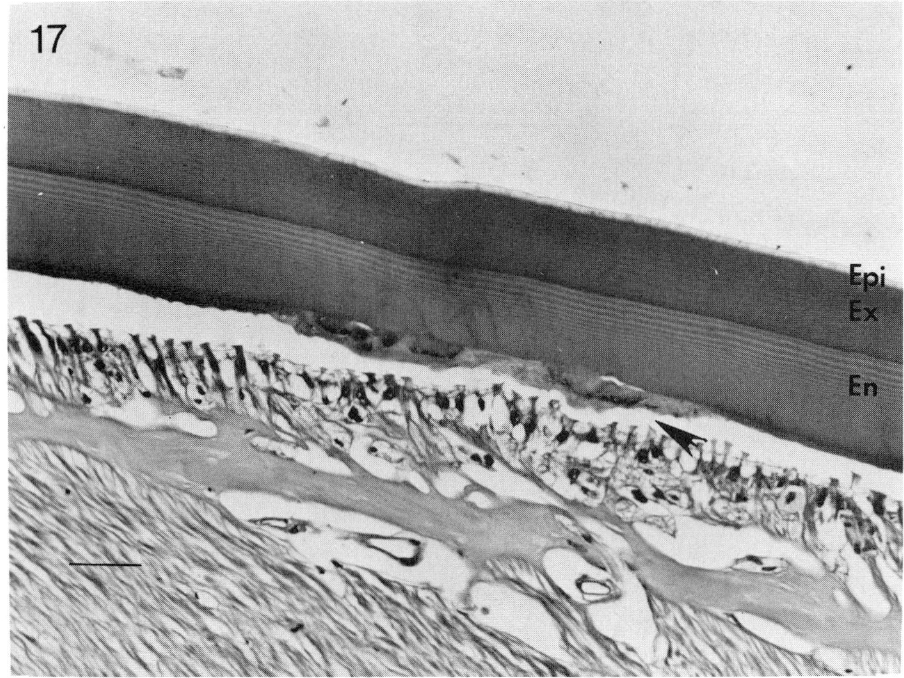

FIGURE 17. Endocuticular amorphous deposits in *M. rosenbergii*. Note normal appearance of epicuticle, exocuticle, and exterior layers of the endocuticle. Arrow indicates amorphous deposits. Epi, epicuticle; Ex, exocuticle; En, endocuticle. Line = 50 μm.

as prawn culture has been present in Hawaii.[54] The findings of a preliminary study[11] indicated the following. The disease was found to occur almost exclusively in large, blue claw, male prawns, with female animals rarely being affected. The prevalence of TG prawns in ponds has been in the order of 0 to 6% of the harvest. Higher prevalence have been found in ponds that were infrequently or inadequately harvested. The disease was found to be present at all times of the year. Severely affected prawns have been weak as evidenced by sluggish movements, not being able to raise their chela up level with the body, and death within 1 to 2 hr after capture. Less severely affected animals have been more active but less so than normal appearing animals. Terminal growth prawns examined have been in molt stage C. The presence of stomach contents or feces in the abdominal intestine has been a variable finding in TG prawns.

The gross lesions (Figure 18) associated with this disease have been diffuse opacity of the entire striated musculature of the body (in severely affected individuals the muscle will have a mushy texture); marked reduction in the size (atrophy) of the hepatopancreas; engorgement of epigastric sac (bladder); and mild to marked epibiotic fouling involving the ventral surfaces of the body initially, and later spreading to other areas. Chromatophore atrophy or thinning of the exoskeleton has not been noted. Focal to multifocal brown spot lesions have been a frequent but inconsistent finding in TG prawns examined.

When TG prawns were boiled, drained, and opened, the muscle mass was found to have shrunken away from the exoskeleton (Figure 19). Muscle tissue from affected prawns has been found to be approximately 10% higher in moisture content than found in normal appearing animals.[11]

The results of clinical pathology tests were variable for those prawns graded as mild to moderately affected by the disease. However, severely affected prawns consistently were hemocytopenic, hypoproteinemic, hypoglycemic, and had a prolonged hemolymph clotting time. These prawns frequently were bacteremic at the time of postmortem examination.[11]

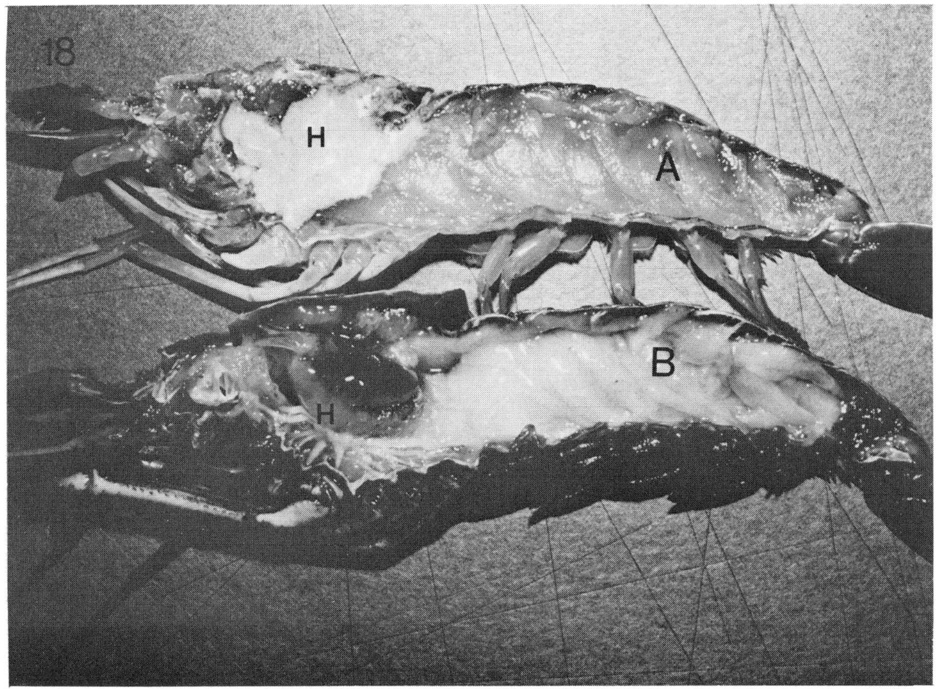

FIGURE 18. Midsagittal section through normal, molt C, adult, male *M. rosenbergii* (A), and terminal growth (TG) prawn (B). Note diffuse opacity of the muscle and marked hepatopancreas atrophy of the TG prawn. M, muscle; H, hepatopancreas.

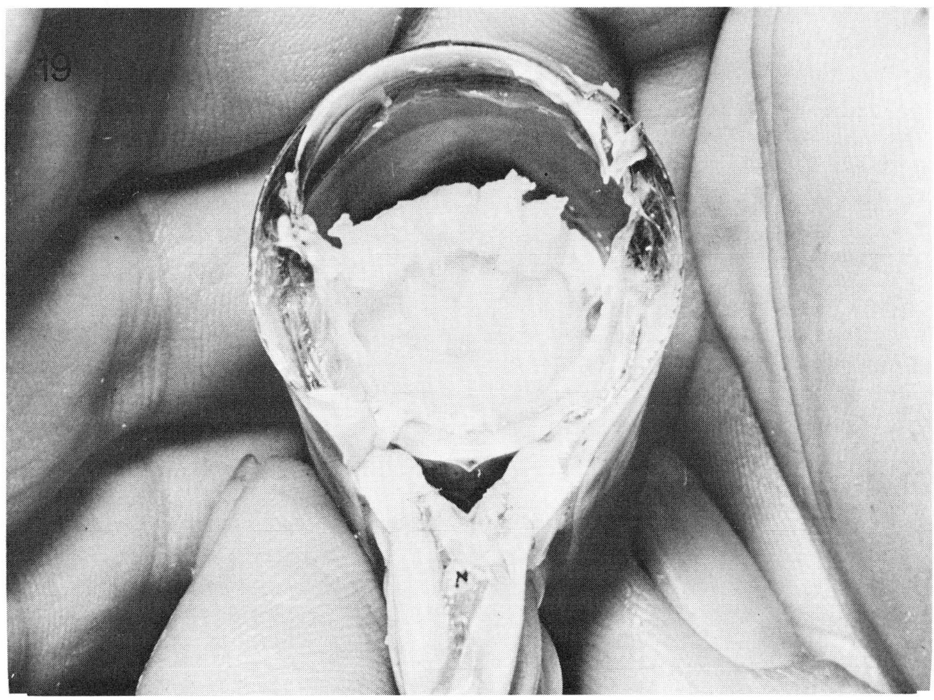

FIGURE 19. Shrinkage of abdominal muscle tissues following boiling of terminal growth (TG), *M. rosenbergii*.

The microscopic lesions that have been noted include atrophy and edema of striated muscle (Figure 20B, Figure 20A, normal striated muscle supplied for comparison) and reduction of secretory and absorptive vacuoles in hepatopancreas epithelium (Figure 21B, Figure 21A, normal hepatopancreas epithelium supplied for comparison). The severity of the above lesions have varied from slight to marked probably reflecting the severity of the disease in a particular animal. Epibiotic fouling has been a frequently observed finding, but severity of fouling has not correlated well with severity of muscle and hepatopancreas lesions. Broken and partial loss of setae and discoloration and pitting of the exoskeleton surface have been common findings. Other lesions noted in the TG prawns (N = 9) examined were multifocal hemocytic nodules of varying size in the heart (67%), antennal gland (43%), hepatopancreas (33%), gills (88%), muscle (22%), and rarely in other organ systems. Similar appearing hemocytic nodules have also been observed on histopathology examination of grossly normal appearing prawns (N = 14) (heart [64%], antennal gland [29%], hepatopancreas [21%], gill [7%], and muscle [7%]). In a few cases bacteria have been found associated with these lesions, but in the majority of nodules infectious agents were not noted. Focal to multifocal areas of idiopathic muscle necrosis have also occasionally been identified in both TG and control prawns.[11]

The etiology and pathogenesis for this disease are undefined. The principal lesions (atrophy of the hepatopancreas and edema and atrophy of the striated musculature) are suggestive of generalized cachexia (wasting) in the animals. The pitted condition of the exoskeleton and broken, fouled setae imply that affected prawns are in an extended intermolt period. However, low vacuolation of the hepatopancreas indicates that proper storage of nutritional reserves is not taking place in these animals. The hepatopancreas should be full of nutrient reserves in the later stages of intermolt to provide the necessary supply for the animal when feeding ceases during late premolt, ecdysis, and early postmolt.[7,55] In *Cancer* hepatopancreas lipid increases over sevenfold from molt stage C_1 through D_1.[55] Terminal growth prawns are apparently feeding as gut contents have been found even in severely affected animals.[11] However, no assessment of nutrient digestion, absorption, or assimilation have been made so that functional starvation could be occurring even though affected animals are eating.

Microscopic evaluation of hepatopancreas epithelium, ventral nerve cord neurosecretory cells, antennal gland podocytes, antennal gland labyrinthal epithelium, branchial lamellar, and stem cellular elements have not revealed cellular lesions which are suggestive of a particular etiology for this disease. However, ultrastructure studies have not been conducted and higher magnification may reveal lesions of diagnostic significance in one or more of these organ systems. The Y-organ has not been evaluated microscopically and its morphologic state in relation to TG disease is undefined.

The occurrence of focal to multifocal hemocytic nodules has been a variable finding in terminal growth prawns examined. In all cases the gill lesions were limited to focal areas of cuticle ulceration and hemocyte nodule formation. These lesions probably reflect the extended intermolt period and possibly attack by low virulence bacterial or fungal organisms infecting a host compromised by an underlying chronic disease problem. However, additional study should be aimed at understanding the etiology of these lesions. These microscopic lesions have also been recognized in grossly normal appearing prawns harvested from ponds.[11]

A terminal molt stage (C_4T) in which no further molting occurs has been recognized in certain crustaceans.[7] The term "terminal growth" has been applied as a descriptive name for the TG condition in *M. rosenbergii*. However, it is not clear that the TG condition is actually a C_4T state because crustaceans in the C_4T stage have not been reported to be suffering from hepatopancreas atrophy and generalized cachexia which occurs in TG prawns; the TG condition is not restricted to only the larger individuals as intermediate size adult males have also been found to be affected; and while TG prawns are lethargic other crustaceans in C_4T are reported to be fully active.[7]

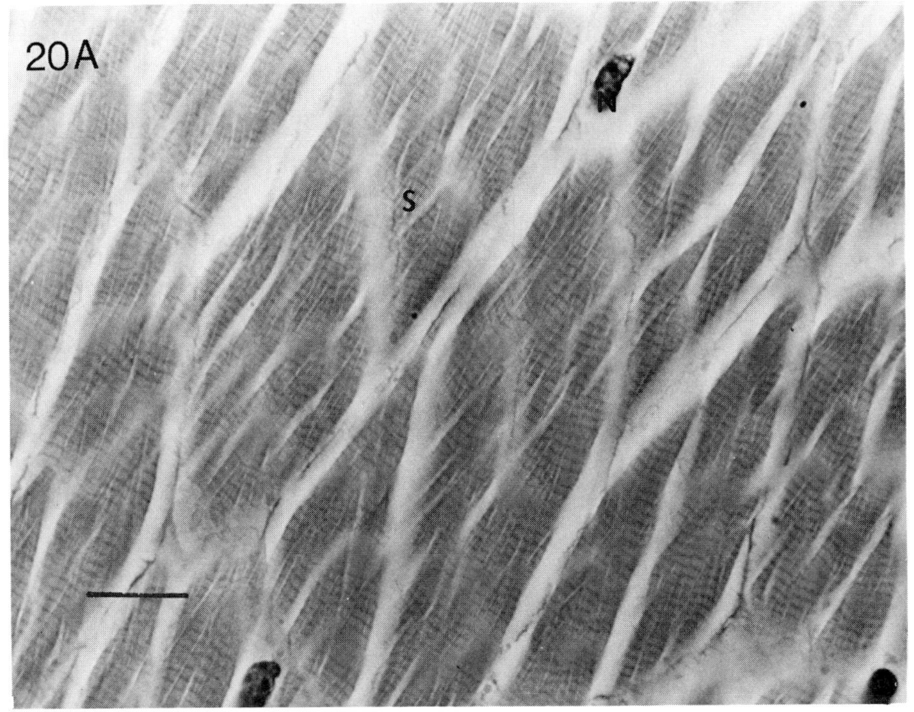

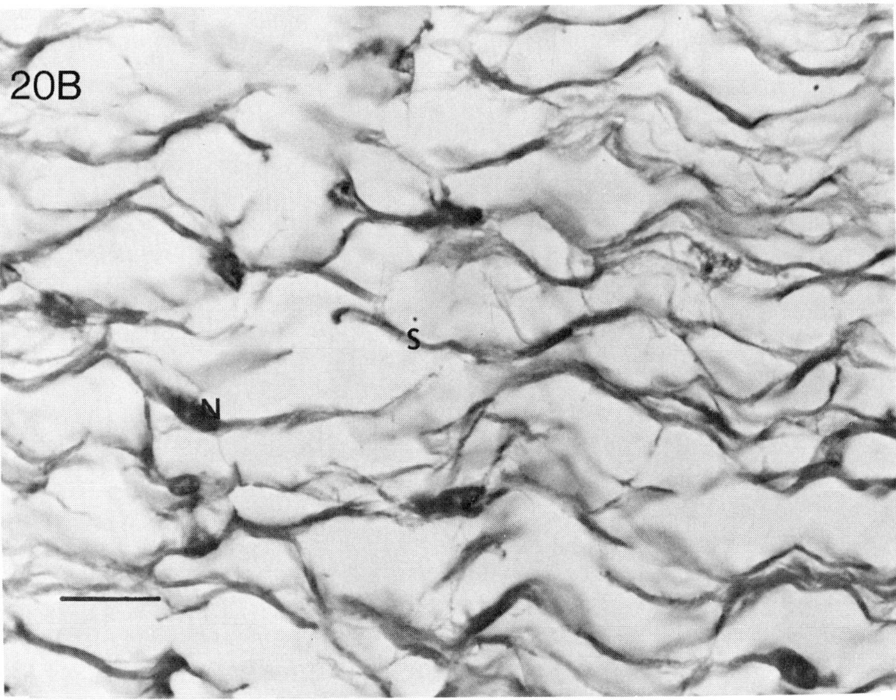

FIGURE 20. (A) Normal abdominal muscle *M. rosenbergii;* (B) severe sarcomere atrophy and edema of muscle from terminal growth prawn *(M. rosenbergii)*. N, nucleus; S, sarcomere. Line = 20 μm.

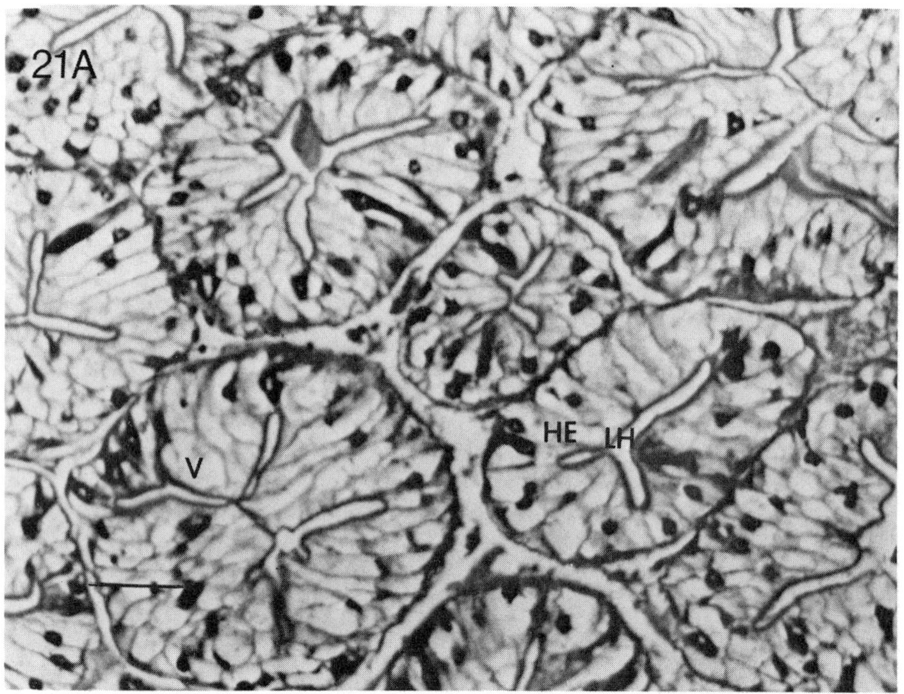

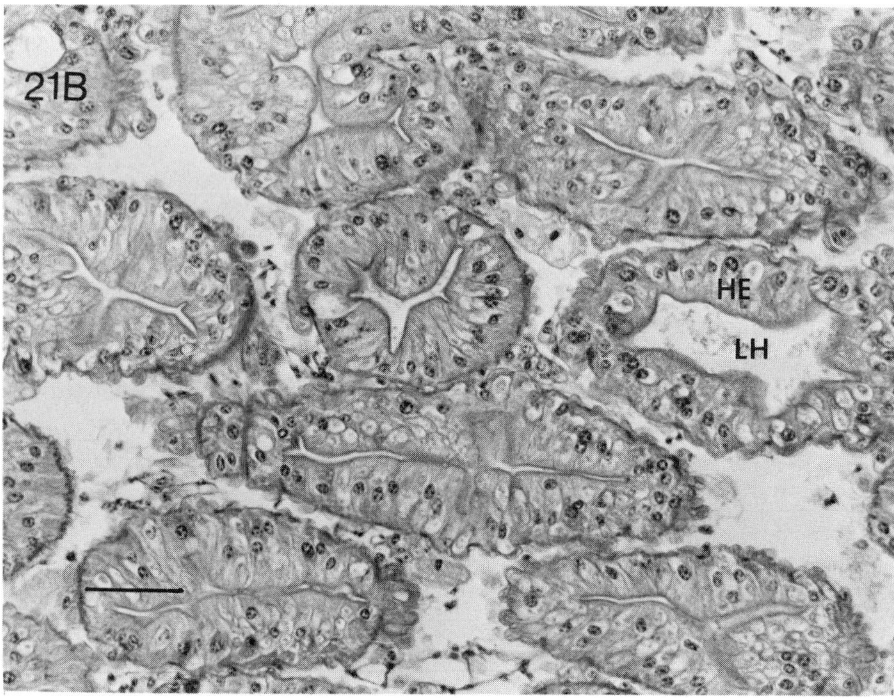

FIGURE 21. (A) Normal, adult *M. rosenbergii* hepatopancreas (molt C) (note abundant vacuolation of the cytoplasm); (B) terminal growth prawn hepatopancreas (molt C) with sparse epithelial vacuolation. HE, hepatopancreas epithelium; V, vacuole; LH, lumen of hepatopancreas diverticula. Line = 10 μm.

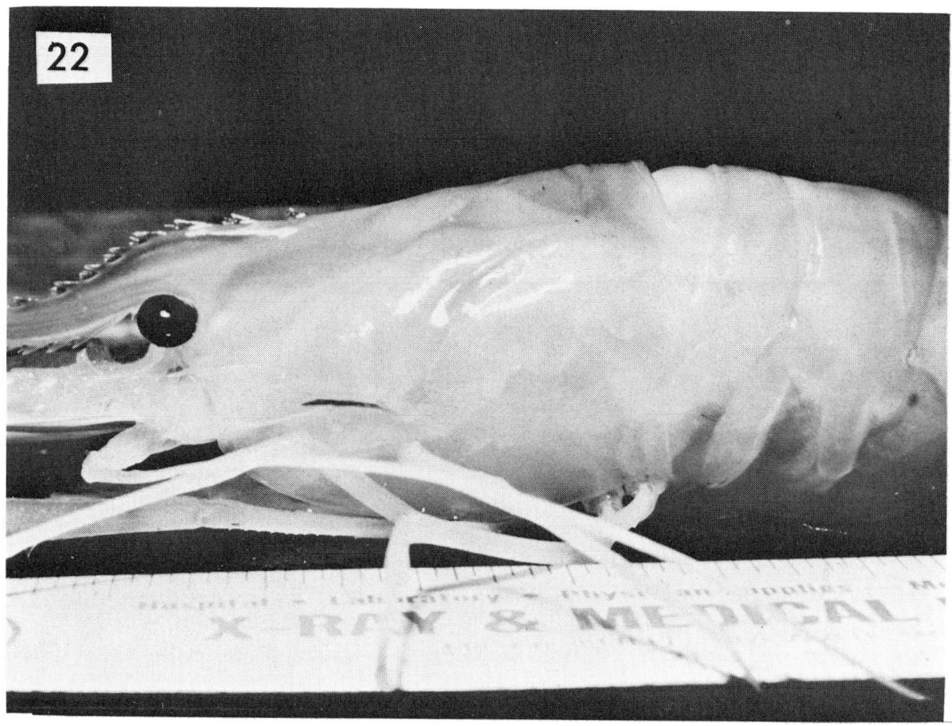

FIGURE 22. Adult, female *M. rosenbergii* with diffuse loss of hypodermal pigmentation and generalized white coloration. Background color of ruler in the foreground is white.

Although the Y-organ has not been microscopically evaluated in TG prawns, it is not likely that a primary dysfunction of this organ results in the systemic lesions which occur in this condition. Y-organ extirpation blocks the molting cycle in C_4 but it does not influence metabolic events leading up to proecdysis.[55] The molting hormone is only needed to initiate normal molting during molt stages D_0 through D_1.[55]

With the presently available information regarding terminal growth in prawns, a diagnosis of this disease can be made based on clinical signs and gross lesions. Affected animals will be weak, have diffuse muscle opacity, and a small hepatopancreas. Epibiotic fouling may or may not be present and the exoskeleton will have the expected rigidity and epidermal pigmentation will be present.

It is important that prawn farmers recognize and remove TG prawns from their harvest. The flesh quality of these animals can be extremely poor and unsuitable as a food item. If TG prawns enter commercial channels they have a marked, negative impact on product acceptance because of their poor quality.

Control methods for TG disease are limited to culling of affected animals recovered during the harvest. TG prawns should be removed and not returned to the pond. This disease is known from Hawaii, but probably occurs wherever *M. rosenbergii* is cultured in ponds.

White Prawn Disease

White prawn disease (white syndrome, haole prawn) is a disease of *Macrobrachium* that has been partially characterized by epizootiology, gross, and microscopic pathology. White prawn disease (WPD) has been found to be a chronic, progressive disease of adult prawns fed an artificial diet and maintained in a prawn culture system lacking exposure to direct sunlight. Affected prawns slowly loose normal pigmentation (Figure 22) presumably due to atrophy of epidermal chromatophores and take on a white coloration. In addition to loss of pigmentation, a progressive softening of the cuticle occurs. This begins as a prolongation in time needed for exoskeleton hardening to occur following ecdysis, and progresses over a period of 6 months or longer until the affected animal's exoskeleton fails to harden properly through the entire intermolt period. In addition, diffuse opacity of the striated muscle and hepatopancreas atrophy has been consistently observed in prawns affected with this disease. The etiology and pathogenesis of WPD of freshwater prawns is unknown.

In one study[11] 12 adult, female prawns (mean orbital length 13.1 cm and mean total weight 55.8 g) with WPD were examined (males only rarely develop this condition). All these animals had been held in artificial culture conditions for 6 months or longer before loss of pigmentation became grossly evident. This pigmentation loss was slowly progressive but usually by 1 year affected animals were totally white. Interestingly, not all animals in the system developed pigmentation loss. In a prevalence study 19% of the female prawns in the system for 1 year or longer were found to be white.[11] Increased flexibility of the exoskeleton occurred late in the course of the disease with progressively longer periods postecdysis for the exoskeleton to harden. White prawns were observed to eat applied feed as well as the other normally pigmented animals. The mean intermolt period for these prawns was 56 days with no apparent lengthening of the intermolt period occurring with increased time in the system. Those animals that died prior to examination did so in late premolt or within 24 hr following ecdysis. Of the white prawns, 75% were 2 years or older at the time of death (due to natural causes or sacrificed for postmortem examination).

The gross lesions noted (N = 12) included atrophy of epidermal chromatophores (100%), diffuse muscle opacity (100%), epibiont fouling gills (40%), body (20%), cuticular ulceration and melanization (18%), reduced size of hepatopancreas (100%), and focal to multifocal melanization of the ovary (25%) and green gland (17%). Clinical pathology tests conducted on five of these prawns showed all were hemocytopenic with prolonged hemolymph clotting time.

Microscopic lesions noted (N = 12) on histopathology included muscle atrophy and edema (100%), reduction of hepatopancreas epithelial vacuolation (100%), focal to multifocal hemocytic nodules in the ovary (100%), heart (80%), green gland (38%), muscle (38%), hepatopancreas (33%), and gills (29%). Bacteria were observed in these lesions in only one animal. Other microbial agents were not noted. The muscle and hepatopancreas lesions were microscopically similar to those observed in terminal growth prawns. Focal areas of idiopathic muscle necrosis were occasionally noted but could not account for the diffuse opacity observed on gross examination. The microscopic appearance of the focal lesions was typical for that described for idiopathic muscle necrosis. Microscopic sections of cuticle and underlying epidermis in some affected prawns showed an apparent decrease in endocuticle thickness with epicuticle, exocuticle, and underlying epithelium normal in appearance.

At time of postmortem bacteria were isolated from 50% of hemolymph specimen cultured. All cases revealed mixed infections. *Citrobacter* sp., *Klebsiella* sp., *Alpha Streptococcus, Bacillus* sp., and *Sarcina* sp. were isolated. Attempts to culture bacteria from ovary or other organ tissues were not made.

The prolonged course and occurrence of white prawn disease in Hawaii in a single population of prawns maintained in an area without exposure to direct sunlight and fed a formulated diet suggest that this disease has a nutritional or environmental related etiology. Additional studies are required to determine if dietary or other factors are causing this disease.

White prawn disease shares similarities with terminal growth disease. Both are chronic, progressive diseases. Hepatopancreas atrophy and diffuse sarcomere atrophy and edema are lesions common to both conditions. It is reasonable to speculate that these lesions have a similar etiologic basis in these two diseases.

However, differences between white prawn and terminal growth disease are also apparent. Dramatic chromatophore atrophy and loss of exoskeleton rigidity do not occur in terminal growth disease. Also, terminal growth occurs almost exclusively in male prawns while white prawn disease is largely limited to females.

Other authors have reported the occurrence of white discoloration or loss of pigmentation in *Macrobrachium*. Johnson[26] reported yellow discoloration in *Macrobrachium* sp., but was uncertain if this represented a normal color variation or a pathological condition. This yellow color was found throughout the muscle tissues of affected prawns. A condition characterized by white discoloration involving the tissues below the exoskeleton has been reported.[26] This disease was noted in feral *M. ohione* adults after collection from the wild. The condition was progressive and examination of affected tissues failed to reveal infectious agents.

Delves-Broughton et al.[27] reported a disease called white syndrome affecting laboratory cultured *M. rosenbergii* and *P. serratus*. This disease was characterized on the basis of gross and microscopic muscle lesions. Grossly, the affected animals were dense white in color and the exoskeleton was softer than normal. In addition, growth impairment and high mortality in affected animals were reported. Microscopically, muscle lesions in both species were described as areas of muscle necrosis with substantial infiltration of hemocytes into the affected areas. Areas within the muscle which stained basophilic were also noted. These authors suggested the muscle opacity resulted from tissue reaction to a foreign body or was a response to stress. No control methods were reported for this disease.[27]

The diffuse muscle opacity, softening of the exoskeleton, and history of culture in a covered environment reported for white syndrome in *M. rosenbergii* are similar to the findings for white prawn disease. However, the microscopic pathology described is different and these two diseases may not have the same etiology. Additional studies need to be conducted on WPD to determine the etiology and pathogenesis of this condition.

White prawn disease is a disease of prawns held over extended periods of time under artificial lighting conditions and fed formulated diets. This disease is not known to be a problem in pond culture settings. In the absence of understanding the etiology of white prawn disease control methods are limited to addition of natural food items to the diet of prawns held under artificial conditions.

METAZOAN PARASITES

Four metazoan parasites have been reported from feral *Macrobrachium* spp. None of these have been found to present problems in culture settings.

Trematoda

Carneophallus choanophallus

Freshwater shrimp including *Macrobrachium* spp. in the Gulf Coast area of the U.S. are commonly parasitized by the fluke *C. choanophallus*.[26] Prawns serve as the second intermediate host in the life cycle of this fluke. *C. choanophallus* metacercaria encyst in the abdominal muscle of prawns. Mammals such as raccoons or rats serve as the final host and snails as the first intermediate host. Apparently, this fluke is of minor consequence to the survival of prawns.[26] A diagnosis of infection can be established by demonstration of fluke metacercaria in wet mounts of prawn muscle.

Turbellaria

Temnocephalids

Temnocephala (Turbellaria: Rhabdocoela) are a group of flatworms reported as ectocommensals on freshwater animals, primarily crustaceans.[56] Temnocephalids are characterized by having four, five, or six anterior simple tentacles, a single pair of eyes and posterior sucker, and usually two pairs of testis[56] (Figure 23). These flatworms have been reported from South, Central, and North America, Australia, New Zealand, New Guinea, Indonesia,

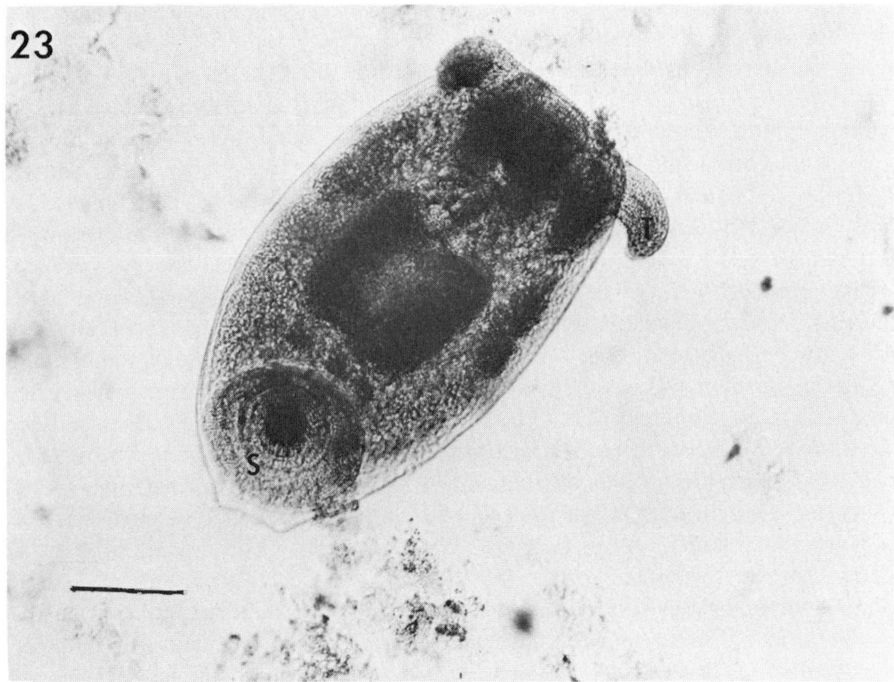

FIGURE 23. Temnocephalid recovered from branchial chamber of *Macrobrachium* sp. from Australia. T, tenacle; S, posterior sucker. Line = 100 μm.

and several Pacific Islands,[56] but not Hawaii. *T. brenesia* has been reported to be associated with *M. americanum* in Costa Rica.[57] *Temnocephala* sp. has been observed from *Macrobrachium* sp. in Australia.[58]

Adult temnocephalid flatworms principally locate within the branchial chamber in *Macrobrachium.* These flatworms have a direct life cycle and cement oval eggs on the exoskeleton. Temnocephalids feed on small animals or algae on the host's surface. Although considered commensal rather than parasitic, large numbers within the branchial chamber have been presumed to reduce water flow through the gills and decrease respiratory efficiency.[58]

Control methods for temnocephalids have not been reported but formalin 50 to 100 ppm would probably be effective. The introduction of temnocephalids into pond culture systems and the extension of their geographic range through transfer on *Macrobrachium* or other hosts should be avoided.

Isopoda

Bopyrid Isopods

Bopyrid isopods have been reported from feral *Macrobrachium* in the Gulf Coast area of the U.S.[26] Similar bopyrid isopods have been found to infect feral *M. rosenbergii* in Malaysia.[59] These are large parasites in relation to the size of the host. Bopyrid isopods locate on the medial surface of the branchiostegite within the branchial (gill) chamber. Infected prawns may show swelling and discoloration in the area of parasite attachment.

Bopyrid isopods have an indirect life cycle. Shrimp are the definitive hosts and copepods serve as intermediate hosts.

Diagnosis can be made by demonstration of the isopod on the carapace using gross examination procedures. Two female bopyrid isopods are pictured in Figure 24. Males are much smaller in size than the females.

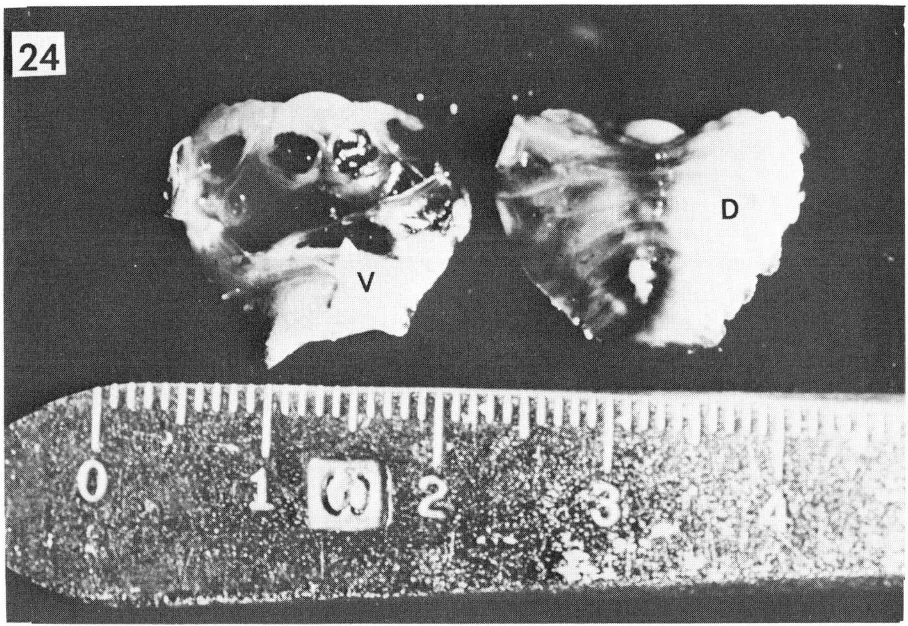

FIGURE 24. Two female bopyrid isopods recovered from the branchial chamber of adult *M. rosenbergii* from Malaysia. D, dorsal surface; V, ventral surface.

Control methods are limited to physical removal of parasites from female prawns sourced from the wild and preventing the extension of the range of these parasites into new geographical areas.

Corallanid Isopods

Corallanid isopods (*Austoargathona* spp.) have been reported as parasitic on feral *Macrobrachium* sp. and other freshwater shrimp from inland New South Wales and Queensland, Australia.[60] Corallanid isopods parasitic on *Macrobrachium* attach to cephalothorax and are clearly visible (size 10 to 12 mm) on gross examination. The life cycle and control methods for corallanids are similar to those for bopyrid isopods.

PREDATION AND CANNIBALISM

Both predation and cannibalism have been reported as causes of prawn losses in *Macrobrachium* culture. A 40 to 60% reduction of stocked animals is a feature common to both the hatchery and pond grow-out phases. The usual reasons given for this population decline have been cannibalism (intraspecific killing)[34,61,62] and, in some cases, predation.[63] An absence of animal carcasses has been a consistent feature. This supports the idea that the losses result from some form of killing followed by consumption or removal of the animal from the system.

Several authors[64,65] have reported the incidence of cannibalism to increase when *M. rosenbergii* is cultured at high density. Peebles[62] and Segal and Roe[64] reported prawns undergoing molt were more susceptible to cannibalism. Although observational data suggest that cannibalism causes losses, there is little actual data to prove the occurrence of intraspecific killing in hatchery or pond production systems. However, laboratory study results strongly imply that cannibalism accounts for the majority of these losses of prawns.

Cannibalism in juvenile prawns can be reduced by providing additional substrate in culture or holding tanks, individual segregation, or by low stocking density.[65] The latter two measures are likely to be economically prohibitive for commercial prawn culture.[65] Additional substrate in ponds while improving prawn survival has the disadvantage of needing to be moved during harvests. In Hawaii the use of additional substrate in ponds varies from farm to farm. However, use of these substrates seems to have a distinctly beneficial effect in promoting prawn survival in nursery ponds.[59]

Interspecific predation refers to predation by a different species. Hydrozoans (several types) have been reported to result in losses of *Macrobrachium* larvae following inadvertent introduction into the culture system.[66] Losses in younger stages resulted from direct predation by Medusae. The hydrozoans also competed with *Macrobrachium* larvae for *Artemia* nauplii. Hydrozoans were also found to be able to kill late stage prawn larvae through the action of their nematocyst. These authors[66] reported 250 ppm formalin as an effective control method but recommended prevention of introduction of hydrozoans into hatchery water systems as the preferred means of control.

Bird predation has been implicated as a cause of losses in prawn ponds. Smith et al.[63] reported green herons (*Butorides virescens*), herons, egrets, and kingfishers to frequent experimental ponds in South Carolina, and losses attributable to bird predation were significant in one location. In Hawaii night herons and cattle egrets are commonly observed around prawn ponds and are considered predaceous on prawns by farmers, but hard data concerning their impact on prawn production are lacking. However, Beynon et al.[67] reported that bird predation can result in up to 75% loss of stocks in penaeid shrimp culture. Presumably, substantial numbers of birds must be present in order to significantly reduce crustacean populations.

Bullfrogs (*Rana* sp.) are also known to feed on prawns in Hawaiian prawn ponds, and mosquito fish (*Gambusia* sp.) are thought to prey on PLs newly stocked in ponds.[54] Fujimura[61] reported dragonfly nymphs to be predaceous on early postlarval prawns stocked into shallow dirt ponds in Hawaii.

Interspecific predator control methods employed in *Macrobrachium* pond culture in Hawaii are limited to reduction or removal of predatory species from pond areas. In Hawaii bullfrogs are captured during harvests and are usually given to one of the employees. Dragonfly nymph control is through predation by *Gambusia* sp. and larger prawns. *Gambusia* sp. pond populations are periodically reduced in order to minimize their impact on newly stocked PLs. Some Hawaiian farms do not have *Gambusia* sp. stocked in ponds. Bird control procedures are generally not employed on prawn farms in Hawaii.

ENVIRONMENTAL AND WATER QUALITY PARAMETERS AS ETIOLOGIC FACTORS IN PRAWN DISEASES

Environment and water quality parameters may be direct or accessory etiologic agents of prawn diseases. General environmental conditions, container and substrate characteristics, and physical, chemical, and biological components of the culture medium are included in this broad category. It is generally regarded as common knowledge that environment and/or water quality factors are often required as accessory etiologic agents (stressors) for many infectious diseases to develop into clinical problems in aquatic animal populations. These factors may also affect populations through growth reduction or serve as specific disease agents.

Several water quality parameters (oxygen, ammonia, nitrite, nitrate, and calcium carbonate) have been studied to establish toxic or growth inhibitory levels to *M. rosenbergii* larvae and juveniles. Reports citing water quality parameters as primary or accessory disease agents in prawn culture have appeared,[2,17,27,34] but with the exception of low dissolved oxygen, actual clinical cases of prawn disease from these agents have not been demonstrated.

Dissolved Oxygen

Hatchery

Water hypoxia has not been reported as a cause of *Macrobrachium* larval mortality, probably because this problem does not arise due to standard culture methodology with exuberant mixing of the culture medium with compressed air.

As a general rule of thumb, dissolved oxygen levels should not drop below 5 ppm in hatchery tank water to insure an adequate supply for prawn larvae.

Acute Pond Hypoxia

Acute pond hypoxia has been recognized as the most significant cause of prawn mortality in *Macrobrachium* pond culture. Thirty prawn kills in ponds on Oahu were reported to the Anuenue Fisheries Research Center, Sand Island, Oahu from January 1977 through September 1978. In all cases oxygen depletion was thought to cause the prawn mortalities. Estimated losses in these cases ranged from 4.5 to 1000 lb of prawns. Green et al.[68] reported a 45% loss of prawns stocked in a 0.19-ha pond in Malaya. Early morning dissolved oxygen (DO) levels reported were 0.5 ppm during the die-off. Oxygen shortage has also been reported as a common problem in crawfish pond culture.[22]

In Hawaii acute hypoxia prawn kills typically have occurred but have not been limited to the months of July through November. This corresponds to periods when light to variable southerly winds commonly occur more and beneficial stirring action in ponds from the northeasterly (trade) winds have been reduced. In some cases pond (DO) crises followed several days of overcast weather with low sunlight levels. The early morning (6:00 am) DO readings have frequently been 3 ppm or less for several consecutive days with secchi disc readings usually less than 20 cm. In this precrisis period, prawns have often been noticed gathering around the pond edges (stress behavior) during the early morning critically low oxygen periods. Prawn mortality has occurred during the precrises period but was usually limited to those animals that have been molting or were unthrifty.

Ponds in the precrisis period that were unattended have progressed to the point where a DO prawn kill occurred. Prawns dying from hypoxia have been found dead primarily distributed along the edges of the pond or accumulated near the water inlet. All sizes of prawns have been found to be susceptible. Freshly dead or prawns dying from anoxia have shown diffuse muscle opacity, but were otherwise grossly unremarkable. Prawns that have been dead 24 hr or longer were in advanced stages of deterioration (red coloration, separation of the cephalothorax, and abdomen and putrid smelling).

A tentative diagnosis of pond hypoxia as a cause of an acute prawn mortality can be made on epizootiology (acute onset of mortality, all size classes of prawns affected); clinical signs (lethargy, weak or dead prawns gathered along the edges of the pond or near the water inlet), and gross lesions (diffuse muscle opacity). A diagnosis of acute pond hypoxia can be confirmed by measuring early morning dissolved oxygen levels in the pond. To be of diagnostic significance DO levels should be measured just prior to sunrise as later readings may be above known lethal levels for *M. rosenbergii*. DO levels less than 2.0 ppm support a diagnosis of hypoxia as the cause of an acute prawn kill.

Prawn pond hypoxia usually follows a phytoplankton die-off with a resultant marked increase in BOD of the pond water and reduced oxygen production from photosynthesis during periods of sunlight. Significant pond hypoxia may also occur when phytoplankton densities are extremely high (secchi less than 15 cm) due to respiratory activity (oxygen consumption) of phytoplankton during periods of darkness.

Laboratory studies have shown *M. rosenbergii* to be a metabolic regulator at oxygen levels greater than its critical level and a conformer at oxygen tensions below this level.[69] This critical value increases with rising temperature in a linear manner and is 2.08 ppm DO at 23°C, 2.9 ppm at 28°C, and 4.65 ppm at 33°C for a 0.2-g dry weight juvenile prawn.[69] Prawn weight, if more than 0.05 g dry weight, has only a small effect on QO_2.[69]

These laboratory results relate well to empirical observation in production ponds. When DO levels approach the critical point, prawns display "stress" behavior (gather along the edges of the pond and show increased activity levels). When DO levels fall below the critical point prawns stop moving and lay on the bottom displaying behavior typical for an oxygen conformer at low levels of ambient O_2.

By accumulating along the pond edges prawns may decrease their chances for survival. This gathering in high density probably drives the oxygen levels down considerably in these areas.

Prevention of acute pond hypoxia is related to good pond husbandry practices. In my opinion all ponds should be checked by the farm manager or a reliable assistant each morning at daybreak. In addition to visual inspection, pond DO and secchi disc measurements should be made and the results recorded. If secchi disc readings fall below 25 cm (indicating phytoplankton densities are above beneficial levels for O_2 production) or if morning DO levels decline below 3 ppm, ponds should be flushed with clear water and reduction of feed levels be considered. Quantitative aspects of these changes will vary in each particular case. In some cases algacides may be applied but care must be taken regarding the possible detrimental effects these may have on prawns.

Once a hypoxic crisis has occurred control methods include dropping the water level of the pond and flushing it at as high a rate as possible to remove dead phytoplankton and organic debris which contribute to the high BOD; installation of portable aerolaters or running a small boat with an outboard motor in the pond to put air into the water; removal of all dead prawns from the pond; and, once the pond is well flushed (24 to 48 hr), reinoculation with algae from an adjacent "healthy" appearing pond. The addition of potassium permanganate to the pond is not recommended as this has been shown not to increase DO levels in the water.[70]

Acute pond hypoxia has been noted in *Macrobrachium* culture in South Carolina, Malaysia (experimental pond), and Hawaii; undoubtedly this problem occurs wherever *Macrobrachium* are cultured in ponds.

Ammonia, Nitrite, and Nitrate

As in many other commercially important aquatic species ammonia, nitrite, and nitrate have been studied to determine the concentrations toxic to *Macrobrachium* larvae and juveniles. Selected toxicity data for these substances are listed in Table 3.

Total ammonia nitrogen includes nitrogen in both ionic states (ionized [NH_4^+] and unionized [NH_3]). Unionized ammonia is considered the principal toxic form to aquatic species.[71,72] Ionized and unionized ammonia exist in an equilibrium state ($NH_4^+ \leftrightarrows NH_3$) in water. Factors which influence shifts between these two ionic states include temperature, salinity, and pH.[71] The effect of temperature and salinity is relatively minor in comparison to the effect of pH.[72] A shift of one pH unit (7 to 8) will increase NH_3 levels tenfold.[72] Lower pH shifts the equilibrium in favor of NH_4^+, thus, decreasing toxicity to aquatic species. Armstrong[72] has presented some evidence that NH_4^+ can also be toxic to *M. rosenbergii* larvae.

Analysis of total ammonia nitrogen has been commonly used to determine ammonia concentrations in culture water. The unionized ammonia level can be calculated if the total ammonia nitrogen and pH is known. Conversion tables are available.[71,73]

The levels of total ammonia nitrogen necessary for the unionized ammonia nitrogen to reach lethal levels to *Macrobrachium* larvae and juveniles are of sufficient magnitude to probably not be commonly found in any functioning *Macrobrachium* culture system (see Table 3). However, levels at which sublethal effects to juveniles (i.e., growth suppression) may occur are sufficiently low to be considered a possible problem in culture situations (Table 3).

Table 3
SELECTED TOTAL AMMONIA NITROGEN (NH_4-N), NITRITE NITROGEN (NO_2-N), AND NITRATE NITROGEN (NO_3-N) TOXICITY DATA FOR *M. ROSENBERGII* LARVAE AND JUVENILES

Life stage (weight)	Toxin	Exposure: Concentration (mg/ℓ)	Exposure: Duration	Measured effect on test population	Temp (°C)	Salinity (‰)	pH	Ref.
Juveniles (0.09 g)	NH_4-N	192	30 min	LT_{50}[a]	29.2	3	7.0	71
Juveniles (0.6—1 g)	NH_4-N	>3.6—0.7 (>0.1 mg NH_3-N/L)	6 weeks	30—40% Reduction of growth when compared to controls	28.0	0	7.6—8.4	71
Larvae	NH_4-N	112.9	24 hr	LC_{50}[b]	28.0	12	7.6	72
Larvae	NH_4-N	43.2	144 hr	LC_{50}	28.0	12	7.6	72
Larvae	NH_4-N	32	168 hr	GR[c]	28.0	12	7.6	72
Juveniles (0.04 g)	NO_2-N	419	6 hr	LC_{50}	22.2	3	7.4	71
Juveniles (1—1.4 g)	NO_2-N	15.4	4 weeks	LC_{50}	—	—	—	71
Larvae	NO_2-N	5.1—6.4	144 hr	LC_{50}	28	12	7.9	74
Larvae	NO_2-N	1.8	192 hr	GR	28	12	7.9	74
Juveniles (0.15 g)	NO_3-N	3869	15 hr	LT_{50}	23.4	1	7.1	71
Juveniles (0.1—0.2 g)	NO_3-N	160	3—4 weeks	LC_{50}	—	—	—	71

[a] LT_{50} = estimated time for 50% of the population to be killed at a given concentration of toxin.
[b] LC_{50} = estimated toxin concentration in which 50% of population was killed for the given exposure time.
[c] GR = concentration that caused reduction in growth of exposed population when compared to controls.

M. rosenbergii larvae are sensitive (mortality and growth suppression) to nitrite levels at low milligram-per-liter concentrations (Table 3). Although clinical cases of nitrite toxicity in prawns have not been reported, larval sensitivity to nitrite is sufficiently high to indicate that nitrite toxicosis may be a problem in some *Macrobrachium* hatcheries.

The concentrations of nitrate which have been shown to be toxic to *M. rosenbergii* juveniles are high (Table 3) and clinical nitrate toxicosis probably does not pose a problem in *Macrobrachium* culture.

Information is unavailable regarding the signs, gross and microscopic lesions of ammonia or nitrite toxicosis in *Macrobrachium*. Diagnosis of disease caused by these agents can tentatively be established based on demonstration of the substance at known toxic levels. Additional criteria such as signs and lesions of ammonia and nitrate toxicosis need to be determined to improve the understanding of the impact these substances have on prawn culture.

Calcium Carbonate

Experimentally, hard water (total hardness of $CaCO_3$/200 ppm) has been shown to cause reduction in growth rate of juvenile *M. rosenbergii*.[75] The mechanism of action on prawns of high $CaCO_3$ levels has not been determined. Water high in $CaCO_3$ has also been suggested to enhance the growth of *Epistylis* on prawns,[2] but the relationship between *Epistylis* and water hardness has not been firmly established.

PUBLIC HEALTH CONSIDERATIONS

Diseases That May Affect Consumers

Angiostrongylosis

Angiostrongylus cantonensis, the rat lungworm, is a common parasitic infection of rats in many tropical areas of the world. This nematode parasite is the etiologic agent of angiostrongylosis (eosinophilic meningitis), an occasionally fatal central nervous system disease of man. Human cases of eosinophilic meningitis in Tahiti have been associated with the consumption of raw *M. lar.*[76] *M. lar* has been shown to harbor *A. cantonensis* in Tahiti, Western Samoa, and other Pacific Islands.[77] The geographic range of *A. cantonensis* includes Madagascar, Mauritius, South and East Asia, Australia, and most of the Pacific Islands including Hawaii. The parasite has not been found in Africa, the American or European continents, or New Zealand.[77]

Macrobrachium sp. is not required as an intermediate host in the life cycle of *A. cantonensis*. However, prawns can serve as transport (paratenic) hosts probably picking up the third-stage nematode larvae through ingestion of infected snails. Rats serve as definitive hosts. In the rat the lungworm larvae migrate via the vasculature through the brain and into the pulmonary arteries where they develop into adults. If third-stage, *A. cantonensis,* infective larvae are ingested by a mammal other than a rat, the larvae go through their migration pattern but lodge in the brain where an intense inflammatory response to the parasite occurs (hence the name eosinophilic meningitis).

Although *A. cantonensis* occurs in Hawaii, no cases of eosinophilic meningitis in the state have been reported associated with the consumption of *Macrobrachium* spp. Preliminary surveys of pond cultured *M. rosenbergii* have failed to show these animals to be infected, although rodents trapped around prawns ponds do harbor the parasite.[11]

Control methods for *A. cantonensis* on prawn farms include reduction of rodent and mollusk populations and prevention of introduction of infected intermediate hosts (snails) into prawn ponds. Species of mollusks that can harbor *A. cantonensis* should never be fed to prawns. *Macrobrachium* spp. cultured or harvested from streams or rivers in areas where *A. cantonensis* is endemic should always be thoroughly cooked prior to human consumption.

Bacterial Infections and Intoxications

Although there seems to be no reports implicating *Macrobrachium* spp. as a vector of transmission of human bacterial infections or intoxications, the possibility exists that prawns can serve to transmit these diseases to man. Therefore, every effort should be made to prevent introduction of known human enteric bacterial pathogens into prawn ponds, prawn fishery areas, or processing channels for prawns.

Occupational Diseases

Leptospirosis

Leptospirosis is a bacterial disease of domestic and wild animals, and man caused by *Leptospira* spp. This disease is an endemic disease of world-wide distribution (temperate and tropical regions). Leptospirosis is classified as a direct anthropozoonosis.[78] Leptospires are transmitted from one infected mammal to another via direct contact with urine or through contact with a contaminated vehicle such as soil, water, etc. Man is considered a deadend host for leptospiral dissemination.

Leptospira may enter the body through mucous membranes of the conjunctivae, nose, mouth, or genital tract, or through skin abrasions or lacerations. It has been demonstrated that leptospira can survive several weeks in moist soil, stagnant ponds, or slow moving streams which are neutral or slightly alkaline. Leptospires do not survive in brackish or saltwater or when exposed to environmental temperatures less than 20°C.

The incubation period for leptospirosis in man is usually 10 days.[78] Signs or symptoms of leptospirosis in man are variable and include inapparent infections through serious disease involving major organ systems that can be fatal.

Leptospirosis is considered an occupational disease for animal handlers, field workers, or any group that comes in contact with infected animals, soil, or water. Human leptospirosis is usually a disease of sporadic occurrence. However, exceptions exist (rice field workers in Italy), and because the symptoms of this disease are nonspecific and diagnosis requires serologic confirmation, many cases of leptospirosis probably are not correctly identified.

A recent study[79] conducted by the Hawaii State Department of Health found 22.9% of prawn farmers in Hawaii to be seropositive to leptospira (indicates prior exposure to the bacteria). This antibody prevalence was higher than that found in the control group (2.3%) suggesting a higher risk level to prawn farmers than individuals engaged in saltwater aquaculture (the control group). Results of the study did not show a particular relationship with exposure to prawn pond water and prevalence of antibody, and left unanswered the question of route of exposure for prawn farmers. Rodents on farms were frequently found to be shedding pathogenic leptospira and the data suggested rodents to be leptospira reservoirs in areas where prawn farms are located in Hawaii.[79]

When a correct diagnosis is made, human cases of leptospirosis usually can be successfully treated with antibiotics. It is important that persons engaged in many types of agriculture including prawn farming or living in leptospira endemic areas be aware of the disease and promptly seek medical attention for any illness with symptoms of high fever and general malaise.

Schistosomiasis (Bilharziasis)

Schistosomiasis is a widespread, serious, parasitic disease of mammals and birds caused by flukes which reside in the veins of the host. Three species of these trematodes are known to affect man *(S. japonicum, S. mansoni,* and *S. haematobium).*[78] The life cycle of these parasites involves man and possibly other mammals as definitive hosts and species of snails as intermediate hosts.

Transmission to man occurs through skin contact with cercariae-infected water. Infective stages can penetrate the intact skin of man. Prevention is through snail control or avoidance of contact with contaminated freshwater. While effective treatments are available, every effort should be made to prevent any human infection by this parasite. Schistosomiasis is endemic to many areas of South America, Africa, and Asia. Individuals establishing freshwater prawn farms in these areas should consult with local public health officials on prevention and control procedures for schistosomiasis in their area.

REFERENCES

1. *Dorland's Illustrated Medical Dictionary,* 24th ed., W.B. Saunders, Philadelphia, 1960, 428.
2. **Sindermann, C.J.,** Disease and disease control in *Macrobrachium* culture, in *Freshwater Prawn Farming in the Western Hemisphere,* Hanson, J.A. and Goodwin, H.L., Eds., Dowden, Hutchinson and Ross, Stroudsburg, Pa., 1977, 210.
3. **Sindermann, C.J.,** Freshwater shrimp *(Macrobrachium)* diseases, in *Disease Diagnosis and Control in North American Marine Aquaculture,* Sindermann, C.J., Ed., Elsevier, New York, 1977, chap. 3.2.
4. **Blewett, C. and Eble, A.F.,** Cytology and cytochemistry of hemocytes from the freshwater prawn *Macrobrachium rosenbergii, Proc. 2nd Bienn. Crus. Health Workshop,* TAMU-SG-79-114, College Station, Tex., 38.

5. **Huang, M.,** Defense Mechanisms of the Freshwater Prawn *Macrobrachium rosenbergii* (de Man): Serum Agglutinins and the Host Response to Vibrio Infection, M. S. thesis, University of Rhode Island, Kingston, 1979.
6. **Young, J.,** Morphology of the White Shrimp *Penaeus setiferus* (Linnaeus 1758), Fishery Bulletin 145, Vol. 59, United States Government Printing Office, Washington, D.C., 1959.
7. **Johnson, P.T.,***Histology of the Blue Crab, Callinectes sapidus, a model for the Decapoda,* Praeger, New York, 1980.
8. **Balows, A. and Hausler, W.J., Jr., Eds.,** *Diagnostic Procedures for: Bacterial, Mycotic and Parasitic Infections,* 6th ed., American Public Health Association, Washington, D.C., 1981.
9. **Akita, G., Nakamura, R., Brock, J., Miyamoto, G., Fujimoto, M., Oishi, F., and Sumikawa, D.,** Epizootiologic study of mid-cycle disease of larval *Macrobrachium rosenbergii,* Presented at World Mariculture Society Meeting, Charleston, S. C., 1982.
10. **Boyd, W.,** Growth and its disorders, in *A Textbook of Pathology, Structure and Function in Disease,* 8th ed., Lea & Febiger, Philadelphia, 1979, 304.
11. **Brock, J.A.,** unpublished data, 1981.
12. Aquacop, Observations on disease of crustacean culture in Polynesia, *Proc. World Maricul. Soc.,* 8, 685, 1977.
13. **Maddox, M.B. and Manzi, J.J.,** The effects of algal supplements on static system culture of *Macrobrachium rosenbergii* (de Man) larvae, *Proc. World Maricul. Soc.,* 7, 677, 1976.
14. **Wickens, J.F.,** The food value of brine shrimp, *Artemia salina* L., to larvae of the prawn, *Palaemon serratus* Pennant, *J. Exp. Mar. Biol. Ecol.,* 10, 151, 1972.
15. **Bowser, P.R. and Rosemark, R.,** Mortalities of cultured lobsters, *Homarus,* associated with a molt death syndrome, *Aquaculture,* 23, 11, 1981.
16. **Sindermann, C.J.,** Filamentous bacterial infestation (*Leucothrix*), in *Disease Diagnosis and Control in North American Marine Aquaculture,* Sindermann, C.J., Ed., Elsevier, New York, 1977, chap. 3.2.2.
17. **Hall, T.J.,** Ectocommensals of the freshwater shrimp, *Macrobrachium rosenbergii* in culture facilities at Homestead, Florida, *Proc. 2nd Bienn. Crus. Health Workshop,* TAMU-SG-79-114, College Station, Tex., 214.
18. **Jahn, L.T., Bovee, E.C., and Jahn, F.F.,***How to Know the Protozoa,* 2nd ed., Wm. C. Brown, Dubuque, Iowa, 1979.
19. **Fisher, W.S.,** Epibiotic microbial infestations of cultured crustaceans, *Proc. World Maricul. Soc.,* 8, 673, 1977.
20. **Lightner, D.V. and Supplee, V.C.,** A possible chemical control method for filamentous gill disease, *Proc. World Maricul. Soc.* 7, 473, 1976.
21. **Scheer, B.T. and Scheer, M.A.R.,** The hormonal control of metabolism in crustaceans. VIII. Oxygen consumption, *Leander serratus, Pubbl. Staz. Zool. Napoli.,* 25, 419, 1954.
22. **Johnson, S.K.,** Some Disease Problems in Crawfish and Freshwater Shrimp Culture, FDDL-S11., Texas A & M University, College Station, 1978.
23. **Johnson, S.K.,** Chemical Control of Peritrichous Ciliates on Young Penaeid Shrimp, FDDL-S7, Texas A & M University, College Station, 1976.
24. **Roegge, M.A., Rutledge, W.P., and Guest, W.C.,** Chemical control of *Zoothamnium* sp. on larval *Macrobrachium acanthurus, Proc. 2nd Bienn. Crustacean Health Workshop,* p. 295, 1979.
25. **Sindermann, C.J.,** Black-spot disease of freshwater shrimps, in *Disease Diagnosis and Control in North American Marine Aquaculture,* Sindermann, C.J., Ed., Elsevier, New York, 1977, chap. 3.2.1.
26. **Johnson, S.K.,** Handbook of Crawfish and Freshwater Shrimp Diseases, TAMU-SG-77-605, College Station, Tex., 1977.
27. **Delves-Broughton, J. and Poupard, C.W.,** Disease problems of prawns in recirculation systems in the U.K., *Aquaculture,* 5, 201, 1976.
28. **Kubota, W.T.,** The Biology of an Introduced Prawn *Macrobrachium lar* (Fabricius) in Kahana Stream, M.S. thesis, University of Hawaii, Honolulu, 1972.
29. **Dugan, C.C. and Frakes, T.A.,** Culture of brackish-freshwater shrimp, *Macrobrachium acanthurus, M. carcinus,* and *M. ohione, Proc. World Maricul. Soc.,* 3, 185, 1973.
30. **Miller, G.C.,** Commercial fishery and biology of the fresh-water shrimp *Macrobrachium* in the lower St. Paul River, Liberia, 1952—55, N.O.A.A., N.M.F.S., Spec. Sci. Rept., Fish. No. 626, U.S. Department of Commerce, Washington, D.C., 1971.
31. **Cipriani, G.R., Wheeler, R.S., and Sizemore, R.D.,** Characterization of brown spot disease of Gulf Coast shrimp, *J. Invert. Pathol.,* 36, 255, 1980.
32. **Burns, C.D., Berrigan, M.E., and Henderson, G.E.,** *Fusarium* sp. infections in the freshwater prawn *Macrobrachium rosenbergii* (de Man), *Aquaculture,* 16, 193, 1979.
33. **Dugan, C.C., Hagood, R.W., and Frakes, T.A.,** Development of spawning and mass larval rearing techniques for brackish-freshwater shrimps of the genus *Macrobrachium* (Decapoda, Palemonidae), *Fla. Mar. Res. Publ.,* 12, 5, 1975.

34. **Johnson, S.K.,** Diseases of *Macrobrachium,* presented at Giant Prawn 1980 Conf., Bangkok, June 15 to 20, 1980.
35. **Nimmo, D.W.R., Lightner, D.V., and Bahner, L.H.,** Effects of cadmium on shrimps, *Penaeus duorarum, Palaemonetes pugio* and *Palaemonetes vulgaris,* in *Physiological Responses of Marine Biota to Pollutants,* Vernberg, F.J., Calabrese, A., Thurberg, F.T., and Vernberg, W.B., Eds., Academic Press, New York, 1977, 131.
36. **Fisher, W.S., Rosemark, R., and Wilson, E.H.,** The susceptibility of cultured American lobsters to a chitinolytic bacterium, *Proc. World Maricul. Soc.,* 7, 511, 1976.
37. **Lightner, D.V.,** personal communication, 1981.
38. **Amborski, R.K., LoPiccolo, G., Amborski, G.F., and Huner, J.,** A disease affecting the shell and soft tissues of Louisiana crayfish, *Procambarus clarkii,* in Freshwater Crayfish, Papers from the 2nd Int. Symp. on Freshwater Crayfish, Avault, J., Ed., Louisiana State University, Baton Rouge, La., 1976, 299.
39. **Rigdon, R.H. and Bacter, K.N.,** Spontaneous necrosis in muscles of brown shrimp, *Penaeus aztecus* Ives, *Trans. Am. Fish. Soc.,* 99, 583, 1970.
40. **Venkataramiah, A.,** Necrosis in shrimp, *F.A.O. Aquaculture Bull.,* 3, 11, 1970.
41. **Lightner, D.V.,** Muscle necrosis of shrimps, in *Disease Diagnosis and Control in North American Marine Aquaculture,* Sindermann, C.J., Ed., Elsevier, New York, 1977, 75.
42. **Lakshmi, G.T., Venkataramiah, A., and House, H.D.,** Effect of salinity and temperature changes on spontaneous muscle necrosis in *Penaeus aztecus* Ives, *Aquaculture,* 13, 35, 1978.
43. **Fujimura, T. and Okamoto, H.,** Notes on progress made in developing a mass culturing technique for *Macrobrachium rosenbergii* in Hawaii, *Indo-Pac. Fish. Counc. Proc.,* 14th Session, Bangkok, Thailand, Symp., 53, 17, 1970.
44. **Sindermann, C.J.,** Muscle opacity and necrosis, in *Disease Diagnosis and Control in North American Marine Aquaculture,* Sindermann, C.J., Ed., Elsevier, New York, 1977, 95.
45. **Akiyama, D., Brock, J.A., and Haley, S.R.,** Idiopathic muscle necrosis in the cultured freshwater prawn, *Macrobrachium rosenbergii,* VM/SAC, 1119, 1982.
46. **Johnson, S.K.,** Use of Quinaldine with Penaeid Shrimp, FDDL–54, Texas A & M University,College Station, 1977.
47. **Sarver, D., Malecha, S., and Onizuka, D.,** Possible Sources of Variability in Stocking Mortality in Post Larval *Macrobrachium rosenbergii,* presented at Giant Prawn 1980 Conf., Bangkok, June 15 to 20, 1980.
48. **Lightner, D.V.,** "Black death" disease of shrimps, in *Disease Diagnosis and Control in North American Marine Aquaculture,* Sindermann, C.J., Ed., Elsevier, New York, 1977, 65.
49. **Lightner, D.V., Colvin, L.B., Brand, C., and Danald, D.A.,** Black death, a disease syndrome of penaeid shrimp related to a dietary deficiency of ascorbic acid, *Proc. World Maricul. Soc.,* 8, 611, 1977.
50. **Lightner, D.V., Danald, D.A., Redman, R.M., Brand, C., Salser, B.R., and Rerpieta, J.,** Suspected blue-green algae poisoning in the blue shrimp, *Proc. World Maricul. Soc.,* 9, 447, 1978.
51. **Laramore, C.R. and Brock, J.A.,** Investigation into the "mummification" of the digestive gland of the freshwater shrimp *Macrobrachim rosenbergii,* in preparation, 1982.
52. **Laramore, C.R.,** personal communication, 1981.
53. **Peebles, B.,** personal communication, 1982.
54. **Fujimoto, M.,** personal communication, 1982.
55. **Passano, L.M.,** Molting and its control, in *The Physiology of Crustacea,* Vol. 1, Waterman, T.H., Ed., Academic Press, New York, 1960, 473.
56. **Hyman, L.H.,** The Acoelomate Bilateria—phylum Platyhelminthes, in *The Invertebrates: Platyhelminthes and Rhynchocoela The Acoelomate Bilateria,* Vol. 2, McGraw-Hill, New York, 1951, 132.
57. **Jennings, J.B.,** A new Temnocephalid flatworm from Costa Rica, *J. Nat. Hist.,* 2, 117, 1968.
58. **Marum, B.,** personal communication, 1980.
59. **Murphy, J.,** personal communication, 1981.
60. **Riek, E.F.,** A new Corallanid Isopod parasitic on Australian freshwater prawns, *Proc. Linn. Soc. N.S.W.,* 91, 176, 1966.
61. **Fujimura, T.,** Development of a Prawn Industry in Hawaii, N.O.A.A. Job Completion Report, Sub-Project No. H-14-D, 1974.
62. **Peebles, B.,** Molting and mortality in *Macrobrachium rosenbergii, Proc. World Maricul. Soc.,* 9, 39, 1978.
63. **Smith, T.I.J., Sandifer, P.A., and Trimble, W.C.,** Pond culture of the Malaysian prawn, *Macrobrachium rosenbergii* (de Man), in South Carolina, 1974—1975, *Proc. World Maricul. Soc.,* 7, 625, 1976.
64. **Segal, E. and Roe, A.,** Growth and behavior of post juvenile *Macrobrachium rosenbergii* (de Man), in close confinement, *Proc. World Maricul. Soc.,* 6, 67, 1975.
65. **Smith, T.I.J. and Sandifer, P.A.,** Increased production of tank-reared *Macrobrachium rosenbergii* through use of artificial substrate, *Proc. World Maricul. Soc.,* 6, 55, 1975.
66. **Sandifer, P.A., Smith, T.I.J., and Calder, D.R.,** Hydrozoans as pests in closed-system culture of larval decapod crustaceans, *Aquaculture,* 4, 55, 1974.

67. **Beynon, J.L., Hutchins, D.L., Lawrence, A.L., and Chapman, B.R.,** Nocturnal activity of birds on shrimp mariculture ponds, Presented at World Mariculture Society Meeting, Seattle, Washington, 1981.
68. **Green, J.P., Richards, T.L., and Singh, T.,** A massive kill of pond-reared *Macrobrachium rosenbergii, Aquaculture,* 11, 263, 1977.
69. **Sharp, J.,** Effects of Dissolved Oxygen, Temperature, and Weight on Respiration of *Macrobrachium rosenbergii,* Department of Water Science and Engineering, University of California, Davis, 1976, 5.
70. **Boyd, C.E.,** Chemical treatment, in *Water Quality in Warmwater Fish Ponds,* Aquacultural Experiment Station, Auburn University, Craftmaster Printers, Opelika, 1979, 169.
71. **Wickins,J.F.,** The tolerance of warmwater prawns to recirculated water, *Aquaculture,* 9, 19, 1976.
72. **Armstrong, D.A.,Chippendale, D., Knight, A.W., and Colt, J.E.,** Interaction of ionized and un-ionized ammonia on short-term survival and growth of prawn larvae, *Macrobrachium rosenbergii, Biol. Bull.,* 154, 15, 1978.
73. **Trussell, R.P.,** The percent un-ionized ammonia in aqueous ammonia solutions at different pH levels and temperatures, *J. Fish Res. Board Can.,* 29, 1505, 1972.
74. **Armstrong, D.A.,**Acute Toxicity of Nitrite to Larval Stages of *Macrobrachium rosenbergii,* Department of Water Science and Engineering, University of California, Davis, 1976, 24.
75. **Cripps, M.C.,** The Growth Inhibitory Effect of Calcium Carbonate Water Hardness on Adult Malaysian Prawns *(Macrobrachium rosenbergii),* M.S. thesis, Animal Science Department, University of Hawaii, Honolulu, 1976.
76. **Alicata, J.E.,** The Parasitology Laboratory of the Hawaii Agricultural Experiment Station: a Brief Review of Its History and Contributions (1935—1970), Research Report 193, Hawaii Agricultural Experiment Station, University of Hawaii, Honolulu, 1970, 20.
77. **Alicata, J.E.,** Present status of *Angiostrongylus cantonesis* infection in man and animals in the tropics, *J. Trop. Med. Hyg.,* (3), 53, 1969.
78. **Benenson, A.S., Ed.,** *Control of Communicable Diseases in Man,* American Public Health Association, Washington, D.C., 1975, 282.
79. **Anderson, B.S., Brock, J.A., Higa, H.H., Gooch, J.M., Wiebenga, N.H., Palumbo, N.E., Perri, S., and Sato, V.T.,** A Study of the Epidemiology of Leptospirosis on Aquaculture Farms in Hawaii, State of Hawaii, Department of Health, Honolulu, 1982.
80. **Smith, T.I.J., Sandifer, P.A., and Manzi, J.J.,** Epibionts of pond-reared adult Malaysian prawns, *Macrobrachium rosenbergii* (de Man), in South Carolina, *Aquaculture,* 16, 299, 1979.
81. **Sindermann, C.J.,** Ciliate infestation *(Epistylis)* in *Disease Diagnosis and Control in North American Marine Aquaculture,* Sindermann, C.J., Ed., Elsevier, New York, 1977, chap. 3.2.4.
82. **Liao, I.C., Yang, F.R., and Lou, S.W.,** Preliminary report on some diseases of cultured prawn and their control methods, *JCRR Fish. Ser.,* No. 29, 1977.
83. **Lightner, D.V.,** Ciliate disease of shrimps, in *Disease Diagnosis and Control in North American Marine Aquaculture,* Sindermann, C.J., Ed., Elsevier, New York, 1977, chap. 3.1.9.
84. **Fisher, W.S.,** Microbial epibionts of lobsters, in *Disease Diagnosis and Control in North American Marine Aquaculture,* Sindermann, C.J., Ed., Elsevier, New York, 1977, chap. 3.5.3.

LOBSTER PATHOLOGY AND TREATMENTS

Renée Rosemark and Douglas E. Conklin

INTRODUCTION

Diseases are always a potential threat to the success of any animal production operation. This remains true at the University of California Bodega Marine Laboratory even though significant improvements have been made with respect to diseases which were once the major hindrance to lobster aquaculture research. While improvements in general lobster husbandry have certainly contributed to the lower incidence of disease, presently the principle means of disease control at the laboratory is early recognition and subsequent elimination or treatment of infected groups of lobsters. The following review outlines the various diseases of lobsters, symptoms of infected animals, and treatments which have been conducted at the Bodega Marine Laboratory and by other investigators.

BACTERIAL DISEASES

Gaffkaemia

Gaffkaemia or "red tail" caused by the Gram-positive bacterium *Aerococcus viridans* var. *homari*, has been most recently reviewed by Stewart.[1] It is most prevalent in overcrowded conditions that usually exist in lobster pounds. This bacterium can cause extensive losses in *Homarus* sp. and mortalities can increase sharply if the water temperature exceeds 15°C.[2]

Once infected, the host usually stops eating, becomes progressively weaker, often drops one or both chelipeds, and may exhibit a pinkish color underneath the abdomen. In advanced infections, the hemolymph fails to clot and microscopic examinations reveal large numbers of tetrad-forming bacteria visible under oil immersion.

Since *A. viridans* is ubiquitous and enters the host through lesions, it is important to minimize injuries. Thus, any necessary immobilization of the chelipeds on experimental animals should be done by banding rather than pegging. It is also important to maintain a high standard of water quality. Calcium hypochlorite treatment has been recommended for disinfecting holding facilities.[3] For therapeutic treatment, Fisher et al.[4] suggest separating infected animals, holding at temperatures of 5 to 10°C, and injection of 40,000 units of penicillin per kilogram of body weight. Vancomycin, an antibiotic, was found to be effective (25 mg/kg body weight) if injections were begun in early stages of infection.[5] Induced immunity as a means of prophylaxis was proposed by Stewart and Zwicker.[6] Other infection and immunization studies with gaffkaemia have been conducted on the spiny lobster, *Panulirus interruptus*.[7]

Gaffkaemia has not been a major problem at the Bodega Marine Laboratory as imported lobsters are held in quarantine for a minimum of 3 weeks at ambient temperatures (12 to 15°C) before being introduced into the regular holding systems. Additionally, individual holding facilities in the laboratory minimize wounds from aggressive behavior which might allow invasion of the bacterium into uninfected animals.

Shell Disease

Shell disease is thought to be caused primarily by Gram-negative chitinolytic bacteria and was first reported on *Homarus americanus* by Hess.[8] This disease affecting *Homarus* sp. has been reviewed by Rosen,[9] Sindermann,[2] Fisher et al.,[4] and more recently by Stewart.[1] Shell disease affects all stages of the lobster and is thought to be somewhat contagious. Progress of the disease in infected animals is temperature dependent.[10]

Gross signs include pitting and marring of the host exoskeleton which in advanced stages may result in necrotic lesions. These lesions can become a portal of entry for secondary invaders such as *A. viridans* and *Vibrio* sp. Lobsters can overcome minor shell disease by molting,[11] but death may occur at ecdysis due to cohesion between the exoskeleton and subskeleton at the lesion site.[12] Gill filaments can also be affected by shell disease which can interfere with gas exchange.[13] Chitinolytic bacteria isolated from lesions can be tested on chitin media as described by Bauman et al.[14]

Shell disease can be controlled to a large extent by proper nutrition,[15] removal of diseased exuvia waste, and infected animal isolation, and the use of filtered, ultraviolet irradiated seawater. *H. americanus* larvae were successfully reared despite the presence of shell disease by dipping the animals in 20-ppm malachite green for 8 min every other day during the larval period.[16]

Shell disease at the Bodega Marine Laboratory is a particularly annoying problem for broodstock animals. Many of these adult lobsters, which are probably space limited and, thus, molting infrequently, show continuing signs of shell erosion. Although advanced cases which result in death are infrequent, the value of these animals for genetic analysis and eventual broodstock improvement heightens the concern.

Vibrio sp.

Bowser et al.[17] have reported *Vibrio* sp. to be pathogenic to *H. americanus*. The bacterium was isolated from the hemolymph of both moribund juvenile and adult *H. americanus* and more recently isolated from juvenile *H. grammarus* × *H. americanus* hybrids at the Bodega Marine Laboratory.

Infected animals become lethargic, followed by death. *Vibrio* sp. can be isolated on TCBS (Difco) or brain-heart infusion (BHI) agar (Difco) enriched with 3% NaCl. Yellow colonies on TCBS indicate *Vibrio* sp. and colonies on BHI can be characterized by biochemical tests (Rosemark, unpublished manuscript).

Since most *Vibrio*-infected animals observed at the Bodega Marine Laboratory were on inadequate diets or in advanced stages of shell disease, good animal husbandry is an important deterrent for vibriosis.

Sporocytophaga sp.

A *Sporocytophaga* sp. was isolated from moribund fifth-stage *H. americanus* lobsters at the Bodega Marine Laboratory (Rosemark, unpublished data). The culture batch exhibited a 98% mortality within approximately 1 week which was attributed to this bacterium.

All diseased animals had light orange lesions at the base of missing appendages. Some animals had orange lesions on the telson and uropods. Microscopic examination of the lesions (400×) showed numerous long, thin, nonmotile rods.

No treatment was initiated.

FUNGAL DISEASES

Lagenidium

The pathogenic *Lagenidium* sp. is responsible for causing mortalities of *H. americanus* eggs, larvae, and occasionally postlarvae.

Nilson et al.[18] reported that the fungus can destroy animal tissue within 24 to 48 hr and cause a 90% or more mortality in larval batches. It has also been observed on *H. gammarus* larvae at the Bodega Marine Laboratory.

Dead larvae appear opaque or white. With the use of a dissection microscope, the fungal mycelium can often be seen replacing the host's tissue. The mycelium is thin walled, highly branched, sparingly septate, filled with globules, and 10 to 14 μm in diameter.[19] The spore cysts, which measures 9.0 to 10.5 μm, germinate usually into a single germ tube.[20] The mycelium can be isolated on corn meal extract agar supplemented with both streptomycin sulfate and penicillin at concentrations of 50 mg/mℓ each.[18] Different isolates of fungi thought to be *L. callinectes* varied greatly as to their need for sodium and other nutrients suggesting a reevaluation of the taxonomic characterization of this genus.[21]

Bland et al.[22] tested 12 potentially fungitoxic compounds on *L. callinectes* and found only two, malachite green and DS 9073 (an experimental fungicide), that could possibly be used as disease deterrents in mariculture. Lobster larvae in systems infected with Lagenidium responded well to 2-min baths of 5-ppm malachite green, two to three times per week.[23] Therapeutic levels of the fungicides trifluralin and captan were found to suppress zoospore viability.[24] It should be noted that antibiotic treatment for bacteria may optimize conditions for fungal growth.[19]

Haliphthoros

Homarus americanus and *H. gammarus* juveniles have been found to be susceptible to the fungus, *Haliphthoros milfordensis*.[25] These researchers conducted a bioassay with this fungus which resulted in a 46% mortality in *Homarus americanus* juveniles in a 22-day period.

The mycosis causes dark red-brown melanized areas, usually found at the base of the walking legs. Other infection sites include the gills, swimmerettes, and the walls of the branchial chamber. Fungal hyphae can be seen in the diseased tissue with a compound microscope (100×). The mycelium is branched, nonseptate with a diameter of 12 to 16 μm. It can be isolated on cornmeal extract agar made with seawater and the addition of 50 mg/mℓ each of streptomycin and penicillin to inhibit bacterial growth. The fungus requires sodium[25,26] and will grow satisfactorily on several media.[4,27]

Abrahams and Brown[28] evaluated 22 potentially fungitoxic compounds and found the most effective to be malachite green at 0.25 ppm and Furanace at 2.5 ppm.

Fusarium

Lightner and Fontaine[29] isolated *Fusarium* sp. from diseased lobsters at one culture facility and stated a 35% loss in a year was attributable to this fungus. It more recently was implicated in an erosive exoskeleton pathology of cultured lobsters, *H. gammarus (vulgaris)*.[30] It is known to infect juvenile and adult lobsters but has not been reported to infect larval stages. This disease is often referred to as ''black spot''.

The fungus infects dead or damaged tissue and is visible as expanding cuticular or subcuticular lesions that usually darken with melanin.[4] Gills exhibit a brownish discoloration. *Fusarium* diagnosis is made by microscopic observation (250×) of the canoe-shaped macroconidia in the necrotic tissue. Macroconidia may be three to five celled but usually four celled, 31 to 49 μm in length and 3.8 to 4.5 μm in width.[29] Sabourad's dextrose agar (Difco) can be used to isolate *Fusarium* which will form a brown to purplish-brown diffusable pigment.

Various fungicides have been tested[31] but none have been effective in treating established *Fusarium* infections under culture conditions.[4]

PARASITES

Parasites have not been a problem in culturing lobsters at the Bodega Marine Laboratory although a number are known to infect lobsters of the genus *Homarus* (see Table 1 modified from Stewart[1]).

Table 1
PARASITES OF *HOMARUS*

Parasite	Host	Tissue	Ref.
Helmintha			
Stichocotyle nephropis (immature)	*H. americanus*	Stomach and intestinal wall	32—36
Ascarophis sp. (larva)	*H. americanus*	Rectal wall Gills	37[a]—39
Acanthocephalan (*Corynosoma* sp.)	*H. americanus*	Intestinal wall, heart, and body muscle (occasionally)	35
Histriobdella homari[b]	*H. gammarus*, *H. americanus*	Exterior of eggs, larvae, and adults	39,40[a],41[a],42
Nemertean	*H. americanus*	Eggs	43
Pseudocarcino-nemertes homari	*H. americanus*	Eggs, gills	44
Copepoda			
Nicothoe astaci	*H. gammarus*	Gills	2,45—47
Choniostomatidae			
Unicaleuthes	*H. gammarus*, *H. americanus*	Gills Exoskeleton	2 2
Protozoa			
Ephelota gemmipara[b]	*H. gammarus*	Eggs	48
Porospora gigantea	*H. gammarus*, *H. americanus*	Digestive tract	35,39,49
Anophrys sp.[c]	*H. americanus*	Hemolymph	50

[a] Cited by Sindermann, 1970.
[b] On occasion held responsible for the poor larval production in hatcheries.
[c] Reported once as cause of unusually high mortalities in lobsters held in captivity.

OTHER

Epibiotic Fouling

The bacterium, *Leucothrix mucor,* and a variety of other filamentous microorganisms are thought to be responsible for occasional episodes of high mortalities in laboratory reared *H. americanus* larvae.[51] At the Bodega Marine Laboratory, epibiotic fouling has been observed on lobster *(H. americanus* and *H. gammarus)* eggs, larvae, and juveniles with the highest mortalities affecting the larval stages.

The epibionts include filamentous and nonfilamentous bacteria, blue-green algae, and stalked protozoans. In advanced stages of epibiotic fouling, there is a fuzzy appearance on the eggs or exoskeleton which can be seen with the dissecting microscope (12×). The filamentous algae are 10 to 15 μm thick and may be more than 1 mm long; the *L. mucor* trichomes are 2 μm thick and range from 14 to 50 μm in length.[4] The stalked protozoans can be easily identified by use of a protozoan key.[52] The long algal filaments mechanically impair ecdysis by entangling setae and appendages[53] and may cause asphyxia by the occlusion of egg membranes or gills.[4] It is thought that filamentous epiphytes may also trap zoospores which may initiate fungal infections.[51]

Fisher[53] recommends control of epibionts by bathing larvae in 5-ppm malachite green for 2 min every other day. Quinacrine hydrochloride at 0.6 mg/ℓ was found effective against the stalked protozoan *Epistylis* on young penaeid shrimp.[54] Steenbergen and Shapiro[55] controlled *L. mucor* infestations on lobsters by using 10 mg neomycin per liter. By successfully molting, the animal can shed the fouled exoskeleton but may soon be reinfected. Epibionts can be controlled by lowering animal density, water temperature, and nutrients in the seawater.

Molt Death Syndrome

Molt death syndrome (MDS) was reported to cause up to a 100% mortality in juvenile *H. americanus* and *H. americanus* × *H. gammarus* hybrids.[56] The mortalities were related to the amount of soy lecithin in purified diets. (For more complete discussion see Lobster Nutrition, p. 413.)

Most animals suffering from MDS die in the process of molting. Calcium deposits are found embedded on and in the inner surface of the exuvial exoskeleton. The deposits are pink to colorless, range up to 0.2 mm in diameter, and are visible with a dissecting microscope (6 ×). Most deposits occurred on the exuvial carapace and claws but in severe cases were visible on all exuvial parts.

Increasing soy lecithin to 7.5% of the dry weight of the diet significantly decreased mortality.[56] Although MDS appears to be mainly a result of an inadequate diet, other factors such as physiological stress may be involved since it has been observed in other rearing situations at the Bodega Marine Laboratory.

CONCLUSIONS

Although diseases are presently not a major problem for lobster aquaculture research, this does not indicate it will be an insignificant factor for commercial culture. As indicated, for most diseases of lobsters therapeutic treatments have not been devised to prevent epizootics. Thus, the commercial culturist will have to depend heavily on prophylactic treatments which are presently being used in the laboratory. The requirement for multiple isolated subsystems to limit the extent of disease transmission along with the expense of appropriate monitoring efforts to detect the presence of disease before a real problem develops will undoubtedly add to the costs of lobster production. Unfortunately, the magnitude of these costs are presently difficult to anticipate because of the lack of experience with large-scale culture operations.

ACKNOWLEDGMENTS

This work is a result of research sponsored in part by NOAA, Office of Sea Grant, Department of Commerce, under Grant No. 04-8-M01-189 R/A 28. The U.S. government is authorized to produce and distribute reprints for governmental purposes notwithstanding any copyright notation that may appear hereon.

REFERENCES

1. **Stewart, J. E.,** Diseases, in *The Biology and Management of Lobsters,* Vol. 1, Cobb, J. S. and Phillips, B. F., Eds., Academic Press, New York, 1980, 301.
2. **Sindermann, C. J.,** Diseases of shellfish, in *Principal Diseases of Marine Fish and Shellfish,* Academic Press, New York, 1970, 106.
3. **Goggins, P. L. and Hurst, J. W.,** Progress Report on Lobster Gaffkyaremia (Red Tail), Department Sea and Shore Fisheries, Augusta, Maine, unpublished mimeo report, 1960.
4. **Fisher, W. S., Nilson, E. H., Steenbergen, J. F., and Lightner, D. V.,** Microbial diseases of cultured lobsters: a review, *Aquaculture,* 14, 115, 1978.

5. **Stewart, J. E. and Aire, B.,** Effectiveness of vancomycin against gaffkemia, the bacterial disease of lobsters (genus *Homarus*), *J. Fish Res. Board Can.*, 31, 1873, 1974.
6. **Stewart, J. E. and Zwicker, B. M.,** Comparison of various vaccines for inducing resistance in the lobster *Homarus americanus* to the bacterial infection, gaffkemia, *J. Fish. Res. Board Can.*, 31, 1887, 1974.
7. **Schapiro, H. C., Mathewson, J. H., Steenbergen, J. F., Kellogg, S., Ingram, C., Nierengarten, G., and Rabin, H.,** Gaffkemia in the California spiny lobster, *Panulirus interruptus:* infection and immunity, *Aquaculture,* 3, 403, 1974.
8. **Hess, E.,** A shell disease in lobsters *(Homarus americanus)* caused by chitinovorous bacteria, *J. Biol. Board Can.*, 3, 358, 1937.
9. **Rosen, B.,** Shell disease of aquatic crustaceans, in *A Symp. on Diseases of Fishes and Shellfishes,* Special Publication No. 5, Snieszko, S. F., Ed., American Fisheries Society, Washington, D.C., 1970, 409.
10. **Taylor, C. C.,** Shell disease as a mortality factor in the lobster *(H. americanus)*, Department Sea and Shore Fisheries, Augusta, Maine, *Fish. Circ.*, 4, 1948.
11. **McLeese, D. W.,** Lesions on the abdominal membrane of lobsters, *J. Fish. Res. Board Can.*, 22(2), 639, 1965.
12. **Fisher, W. S.,** Shell disease of lobsters, in *Disease Diagnosis and Control, in North American Marine Aquaculture, Developments in Aquaculture and Fisheries Science,* Vol. 6, Sindermann, C. J., Ed., Elsevier, Amsterdam, 1977, 158.
13. **Sawyer, W. H., Jr. and Taylor, C. C.,** The effect of shell disease on the gills and chitin of the lobster *(Homarus americanus), Maine Dept. Sea Shore Fish. Res. Bull.*, 1, 1949.
14. **Baumann, P., Baumann, L., and Mandel, M.,** Taxonomy of marine bacteria: the genus *Beneckea, J. Bacteriol.*, 107(1), 268, 1971.
15. **Fisher, W. S., Rosemark, T. R., and Nilson, E. H.,** The susceptibility of cultured American lobsters to a chitinolytic bacterium, *Proc. World Maricul. Soc.*, 7, 511, 1976.
16. **Fisher, W. S., Rosemark, T. R., and Shleser, R. A.,** Toxicity of malachite green to cultured American lobster larvae, *Aquaculture,* 8, 151, 1976.
17. **Bowser, P. R., Rosemark, R., and Reiner, C. R.,** A preliminary report of vibriosis in cultured American lobsters, *Homarus americanus, J. Invertebr. Pathol.*, 37, 80, 1981.
18. **Nilson, E. H., Fisher, W. S., and Shleser, R. A.,** A new mycosis of larval lobster *(Homarus americanus), J. Invertebr. Pathol.*, 27, 177, 1976.
19. **Nilson, E. H. and Fisher, W. S.,** Lagenidium disease of lobsters, in *Disease Diagnosis and Control in North American Marine Aquaculture, Developments in Aquaculture and Fisheries Science,* Vol. 6, Sindermann, C. J., Ed., Elsevier, Amsterdam, 1977, 168.
20. **Bland, C. E. and Amerson, H. V.,** Observations on *Lagenidium callinectes:* isolation and sporangial development, *Mycologia,* 65, 310, 1973.
21. **Bahnweg, G. and Bland, C. E.,** Comparative physiology and nutrition of *Lagenidium callinectes* and *Haliphthoros milfordenis,* fungal parasites of marine crustaceans, *Bot. Mar.*, 23, 689, 1980.
22. **Bland, C. E., Ruch, D. G., Salser, B. R., and Lightner, D. V.,** Chemical Control of *Lagenidium,* a Fungal Pathogen of Marine Crustacea, Sea Grant Publ. UNC-SG-76-02, 1976.
23. **Fisher, W. S., Nilson, E. H., Follett, L. F., and Shleser, R. A.,** Hatching and rearing lobster larvae *(Homarus americanus)* in a disease situation, *Aquaculture,* 7, 75, 1976.
24. **Armstrong, D. A., Buchanan, D. V., and Caldwell, R. S.,** A mycosis caused by *Lagenidium* sp. in laboratory reared larvae of Dungeness Crab, *Cancer magister,* and possible chemical treatments, *J. Invertebr. Pathol.*, 28, 329, 1976.
25. **Fisher, W. S., Nilson, E. H., and Shleser, R. A.,** Effect of the fungus *Haliphthoros milfordensis* on the juvenile stages of the American lobster *Homarus americanus, J. Invertebr. Pathol.*, 26, 41, 1975.
26. **Vishniac, H.,** A new marine phycomycete, *Mycologia,* 50, 66, 1958.
27. **Bahnweg, G.,** Phospholipid and steroid requirements of *Haliphthoros milfordensis,* a parasite of marine crustaceans, and *Phytophthora epistomium,* a facultative parasite of marine fungi, *Bot. Mar.*, 23, 209, 1980.
28. **Abrahams, D. and Brown, W. D.,** Toxicity to juvenile European lobster *(Homarus gammarus)* of several anti-fungal agents used to control *Haliphthoros milfordensis, Aquaculture,* 12, 31, 1977.
29. **Lightner, D. V. and Fontaine, C. T.,** A mycosis of the American lobster, *Homarus americanus,* caused by *Fusarium* sp., *J. Invertebr. Pathol.*, 25, 239, 1975.
30. **Alderman, D. J.,** *Fusarium solani* causing an exoskeleton pathology in cultured lobsters, *Homarus vulgaris, Trans. Br. Mycol. Soc.*, 76(1), 25, 1981.
31. **Hatai, K., Nakajima, K., and Egusa, S.,** Effects of various fungicides on the black gill disease of the Kuruma prawn *(Penaeus japonicus)* caused by *Fusarium* sp., *Fish Pathol.*, 8(2), 156, 1974.
32. **Nickerson, W. S.,** On *Stichocotyle nephropis* Cunningham, a parasite of the American lobster, *Zool. Jahr. Abt. Anat. Ontog. Tiere,* 8, 447, 1894.
33. **Herrick, F. H.,** Natural history of the American lobster, *Bull. U.S. Bur. Fish.*, 29, 149, 1909.

34. **Odhner, T.,** *Stichocotyle nephropis* J. T. Cunningham ein aberranter Trematode det Digenenfamilie Aspidogastridae, *K. Sven. Vetenskapsakad. Handl.,* 45(3), 1910.
35. **Montreuil, P.,** Parasitological investigations, *Rapp. Annu. Stn. Biol. Mar. Dept. Pech. Quebec,* Contrib. No. 50, Append. 5, 69, 1954.
36. **MacKenzie, K.,** *Stichocotyle nephropis* Cunningham, 1887 (Trematoda) in Scottish waters, *Ann. Mag. Nat. Hist.,* 6(13), 505, 1963.
37. **Anonymous,** Boothbay studies parasite of ocean lobster, *U.S., Fish Wildl. Serv. Fish Wildl. Rep.,* p. 21, 1966.
38. **Uzmann, J. R.,** Juvenile *Ascarophis* (Nematoda: Spiruroidea) in the American lobster, *Homarus americanus, J. Parasitol.,* 53, 218, 1967.
39. **Boghen, A. D.,** A parasitological survey of the American lobster *Homarus americanus* from the Northumberland Strait, *Can. J. Zool.,* 56, 2460, 1978.
40. **Havinga, B.,** Rapport over de kreeften-visserij in Zeeland en de kunstmatige kreeftenteelt, *Meded. Versl. Vissch. Insp.,* 30, 1921.
41. **Sund, O.,** Beretning om anlaeg av statens hummeravlsstation og drriften: 1913, *Arsberet VedKomm. Nor. Fisk.,* 4, 525, 1914.
42. **Uzmann, J. R.,** *Histriobdella homari* (Annelida: Polychaeta) in the American lobster, *Homarus americanus, J. Parasitol.,* 53, 210, 1976.
43. **Aiken, D. E., Waddy, S. L., Whazy, L. S., and Campbell, A.,** A Nemertean destructive to the eggs of the lobster, *Homarus americanus, Int. Counc. Explor. Sea, No. 4,* Special Meeting on Diseases of Commercially Important Marine Fish and Shellfish, 1980.
44. **Fleming, L. C. and Gibson, R.,** A new genus and species of monostiliferous hoplonemerteans, ectohabitant on lobsters, *J. Exp. Mar. Biol. Ecol.,* 52, 79, 1981.
45. **Gibson, F. A.,** Gaffkaemia in stored lobsters, *Int. Counc. Explor. Sea,* Shellfish Comm. No. 58, 1961.
46. **Gibson, F. A. and Francis, C.,** Pathological conditions in lobsters *(H. gammarus), Int. Counc. Explor. Sea,* Shellfish Comm., K, 2, 1972.
47. **Mason, J.,** The biology of *Nicothoe astaci* Audoin and Milne Edwards, *J. Mar. Biol. Assoc. U.K.,* 38, 3, 1959.
48. **Dannevig, A.,** Beretning om Flodevigens utkleknimgsanstalt for 1926—1927, *Arsberet. Vedkomm. Nor. Fisk.,* p. 150, 1928.
49. **Hatt, P.,** L'evolution de la gregarine du homard *(Porospora gigantea* E. V. Bened.) Chez les mollusques, *C. R. Seances Soc. Biol. Ses. Fil.,* 98, 647, 1928.
50. **Aiken, D. E., Sochasky, J. B., and Wells, P. G.,** Ciliate infestation of the blood of the lobster *Homarus americanus, Int. Counc. Explor. Sea, Shellfish Comm.,* K, 46, 1973.
51. **Nilson, E. H., Fisher, W. S., and Shleser, R. A.,** Filamentous infestations observed on eggs and larvae of cultured crustaceans, *Proc. 6th Annu. Meet. World Maricul. Soc.,* p. 367, 1975.
52. **Jahn, T. L. and Jahn, F. F.,** *How to Know the Protozoa,* Wm. C. Brown, Dubuque, Iowa, 1949.
53. **Fisher, W. S.,** Microbial epibionts of lobsters, in *North American Marine Aquaculture, Developments in Aquaculture and Fisheries Science,* Vol. 6, Sindermann, C. J., Ed., Elsevier, Amsterdam, 1977, 163.
54. **Johnson, S. K.,** Chemical Control of Peritrichous Cilates on Young Penaeid Shrimp, FDDL-S7, Texas A & M University, College Station, 1976.
55. **Steenbergen, J. F. and Schapiro, H. C.,** Filamentous bacterial infestations of lobster and shrimp gills, *Am. Zool.,* 15, 816, 1976.
56. **Bowser, P. R. and Rosemark, T. R.,** Mortalities of cultured lobsters, *Homarus,* associated with a molt death syndrome, *Aquaculture,* 23, 11, 1981.

Section IV
Crustacean Nutrition

DIETARY AND NUTRIENT REQUIREMENTS FOR CULTURE OF THE ASIAN PRAWN, *MACROBRACHIUM ROSENBERGII*

Lowell V. Sick and Mark R. Millikin

INTRODUCTION

Initial attempts to use nonliving natural foods or formula feeds for the culture of the Asian prawn, *Macrobrachium rosenbergii,* originated in Thailand.[1] Although the culture of *M. rosenbergii* began in the 1950s, a continuous effort, including hatchery support, did not become established in Thailand until the late 1960s and early 1970s.[2] Grow-out of juveniles within the culture methodology developed in Thailand has used dried and chopped animal matter (largely chicken remains from the poultry industry), rice bran, calf bones, and broiler chicken starter pellets.[3] Larval foods used in Thailand hatcheries include live *Artemia salina,* chopped fish, chopped egg, and soybean curd.[1] Since the deployment of natural foods and broiler starter feeds, advances in formula feed technology for *M. rosenbergii* have included evaluation of binders and palatability of feeds for juvenile stages as well as development of formula feeds suitable for larval stages. However, such attempts have been based largely on empirical knowledge of growth requirements rather than on a quantitative understanding of nutritional-metabolic requirements.

This section reviews nutritional investigations to date concerning the Asian prawn *Macrobrachium* sp. In addition, recommendations are proposed for continued research that is considered prerequisite for the development of improved feeds for the efficient culture of this species.

FORMULA FEEDS USED TO CULTURE *MACROBRACHIUM ROSENBERGII*

Juvenile Diets

A variety of materials have been used in attempts to bind formula (i.e., manufactured entirely from raw ingredients) or semiformula feeds (i.e., manufactured partially from protein meals or chopped animal and plant matter) in a form that would be readily ingested by *Macrobrachium rosenbergii* juveniles (Table 1). Wheat gluten added to *M. rosenbergii* diets may, at least partially, satisfy a metabolic requirement, as well as serve to physically bind feed ingredients.[4-6] Sick and Beaty[7] found that semiformulated extruded feeds isocalorically balanced and bound with amylose starch produced significantly faster growth rates than similar feeds bound with collagen. Using a repelleting of crushed Ralston Purina® marine ration (25% crude protein), Farmanfarmaian and Lauterio[8] reported that more efficient feed conversions and higher growth rates were obtained using algin rather than carboxymethylcellulose as a binder. Heinen[9] compared 11 binding agents and found agar and sodium alginates yielded greatest stability for Ralston Purina® trout chow placed in an aquatic environment (Table 1).

The potential dissolution of nutrients from formula feeds used in aqueous systems has obvious importance in evaluating the nutritional significances of various diets. Goldblatt et al.[10] reported that amino acids, in general, were dissolved from formula feeds which were submerged. Recently, Fortner and Sick[11] reported that concentrations of selected individual amino acids decreased in the processing and storage of formula feeds designed for culture of *M. rosenbergii*. Furthermore, they reported that due to selected dissolution of nutrients other than amino acids, upon submersion of formula diets, selected amino acids remained in some aquatic diets at higher concentrations than contained in the original formulations. However, if diets having greater than 40% total protein were submerged for 1 hr, loss of

Table 1
NATURAL FOODS AND TYPES OF FORMULA FEEDS USED TO CULTURE JUVENILE *MACROBRACHIUM* SP.

General description	Processing technique	Type of binder	Ref.
Chopped animal and plant matter (i.e., chopped fish, rice bran, calf bones)	Often without processing or occasionally cooked	Gelatination by cooking natural composition of starches	2
Elongated pellet with protein meal	Pelletization	Wheat gluten	5,6
Elongated pellet with protein meal	Steam extrusion	Amylose starch vs. collagen	7
Crushed pellet from repelletized Ralston Purina Marine Ration	Pelletization	Algin vs. carboxymethylcellulose	8
Various binders added to Purina trout chow and repelleted	Pelletization	Comparisons of cornstarch, carboxymethylcellulose, chitosan, collagen, Guar gum, low viscosity chitosan, sodium carrageenin, gum mixtures, agar, sodium alginate, and sodium hexametaphosphate	9

several amino acids occurred.[11] In the case of dissolved individual amino acids, the ratio of concentrations among individual amino acids becomes as nutritionally significant as absolute concentrations of individual amino acids.

Larval Diets

Nutritional sources for culturing *Macrobrachium* sp. larval stages have traditionally consisted of live zooplankton, chopped fish flesh, or combinations of live food and chopped animal matter (Table 2). Ling[3] first successfully reared *M. rosenbergii* through all the larval stages using *ad libitum* additions of live zooplankton (i.e., rotifers, cyclops, copepods, and insect larvae) and chopped fish, shellfish, and steamed chicken egg yolk. Similarly, Minamizawa and Morizane[12] successfully cultured all larval stages of *M. rosenbergii* using live, newly hatched brine shrimp *A. salina*, chopped fish, and chopped short neck clam. Using live protozoans and relatively finely chopped fish flesh, Lewis[13] was able to culture larval stages of *M. carcinus* only through the second metamorphosis. Murai and Andrews[14] compared larval development using either live *A. salina* nauplii with freeze-dried oyster meat, freeze-dried catfish, or commercial trout chow. The most rapid rate of development was achieved using *A. salina* plus freeze-dried oyster meat and slowest rates occurred when larvae were fed oyster meat or trout chow without live *A. salina* nauplii.

The potential role of unicellular algae in the successful culture of *Macrobrachium* sp. has been discussed by several investigators, but no cause and effect relation between algae and larval nutrition has ever been established. Several culturists have reported that additions of various species or combinations of species of unicellular algae enhanced survival rates among cultures of metamorphosing *M. rosenbergii* larvae.[15-17] Maddox and Manzi[18] studied effects of seven species of unicellular algae on rates of survival and metamorphosis and concluded that all algal species studied enhanced successful culture of larvae. Subsequently, Manzi and Maddox[19] reported that a diatom, *Phaeodactylum tricornutum,* was the most effective species of algae investigated when used in either static or recirculating culture systems. However, while Manzi and Maddox[19] reported enhanced larval growth in association with ingestion of algae by larvae, Sick and Beaty[7] could demonstrate no nutritional value of phytoplankton added to larval culture systems. Furthermore, Joseph[20] was not able to demonstrate any fatty acid assimilation from algal populations used to culture *M. rosenbergii* larvae. Cohen et al.[21] reported that ^{14}C-tagged algae were not ingested by *M. rosenbergii* larvae. Manzi and Maddox[19] also demonstrated that algae were not effective in controlling

Table 2
NATURAL FOODS AND FORMULA FEEDS USED TO CULTURE LARVAL STAGES OF *M. ROSENBERGII*

General description	Processing technique	Composition	Ref.
Live food	None	Zooplankton consisting of rotifers, cyclops, copepods, and insect larvae	3
Live food and chopped animal flesh	None	Newly hatched brine shrimp, *Artemia salina,* and chopped fish	7,12,15
Live food and chopped animal flesh[a]	None	Line protozoan and finely chopped fish	13
Live food, chopped animal matter, and live algae	None	Newly hatched brine shrimp, *A. salina,* chopped fish, and unicellular phytoplankton (primarily diatoms)	7,15,18,19
Live food plus freeze-dried oyster meat, freeze-dried cat fish meat, or trout chow vs. freeze-dried food without live food	None	*A. salina* (live) vs. freeze-dried animal matter	14
Formula feed	Alginate-gelatin cubes measuring 3 mm on a side	Blocks impregnated with ovalbumin as an attractant plus fish, soybean, and shrimp meals	22

[a] Larvae successfully reared through only two metamorphic stages.

ammonia concentrations of culture systems used to culture *M. rosenbergii* larvae. Therefore, unicellular algae may have nutritional contributions or may assist in maintaining acceptable water quality conditions in the culture of *M. rosenbergii* larvae, but those specific contributions have not been detected in investigations conducted to date.

Sick[22] compared several amine substrates added to a diet consisting of fish, soybean, and shrimp meals in an attempt to develop a formula diet that would be acceptable by and support growth and development of *M. rosenbergii* larvae. Although Fujimura[15] reported that size of particles was critical and specific for each larval stage, Sick[22] found that porous, pliable gels cut into cubes 3 mm on a side were readily acceptable to all larval stages. Such acceptance was enhanced by the additions of selected amines and glucosamines as appropriate attractants. Results from these investigations of selected attractants indicated that ovalbumin was significantly more potent when added to larval diets than avidin, betaine, ovoglobulin, or ovomucin. Among several amino acids added to larval diets containing ovalbumin,[22] diets containing either supplements of proline or a combination of proline, arginine, glycine, and taurine resulted in significantly higher rates of ingestion than control diets.

NUTRITIONAL REQUIREMENTS FOR THE CULTURE OF *M. ROSENBERGII* JUVENILES

While requirements for appropriate types of foods, both natural and formulated, or semiformulated, have received most of the nutritional research emphasis given to this species, some investigations have considered qualitative and quantitative requirements for specific nutrients. Most of the research effort to date has concerned determining requirements for dietary protein and energy concentrations necessary to achieve maximum rates of growth. Much less emphasis has been given to determining specific metabolic requirements for other macronutrients (i.e., carbohydrates and lipids) or micronutrients (e.g., individual amino acids, fatty acids, vitamins, and minerals).

Table 3
ESTIMATES OF DIETARY REQUIREMENTS FOR *MACROBRACHIUM* SP. JUVENILES BASED ON ANALYSES OF DIGESTIVE ENZYMES

Indicated dietary requirements	Supporting evidence	Ref.
Broad range of proteins	Broad range of proteolytic enzymes, carboxypeptidases, and specific amino acid peptidases	26, 27
An omnivorous, rather than carnivorous, feeding habit	Quantities and relative activities of proteases and amylases	26
Available evidence suggests limited role of carbohydrates in dietary requirements	Other than cellulase, amylase, no evidence for carbohydrases in digestive system	26, 28
Lipids	Nonspecific esterases and lipases	26

Digestive Enzymes

Analyses of digestive enzymes in *Macrobrachium* sp. have been used to suggest potential nutritional requirements. Although the measured activities of respective digestive enzymes may have relatively large individual organismic and phylogenetic variations,[23-25] Lee et al.[26] recently suggested that enzyme activities measured in the digestive system of various species of *Macrobrachium* sp. are indicative of food habits and empirical estimators of macronutrient requirements. Based on comparable activities and quantities of proteases and amylases, Lee et al.[26] proposed that *Macrobrachium* sp. is a true omnivore. Murthy,[27] furthermore, reported that the wide range in types of digestive enzymes measured in both *M. lamerei* and *M. rosenbergii* is indicative of a capability to digest a relatively large range of nutrient substrates, including various classes of proteins, carbohydrates, and lipids. Lee et al.[26] concluded, however, that large activities of proteases found in *Macrobrachium* sp. are probably not indicative of a carnivorous food habit, but rather indicate a surplus production of protease relative to low dietary concentrations of protein. The ability to utilize ingested starch and cellulose due to the presence of active amylase and cellulase has also been demonstrated for *M. rosenbergii*.[26,28]

Specific nutritional requirements for *Macrobrachium* sp. have also been suggested based on analyses of digestive enzymes (Table 3). Among a wide variety of proteases, as discussed above, trypsin, aminotripeptidase, leucine amino-peptidase, and L-glycine-leucine dipeptidase have specifically been found. The variety of carbohydrases found in *Macrobrachium* sp. has suggested that complex carbohydrates are readily digested.[29] Relatively high amylase activity found in the digestive system of *M. dayanum*[30] indicated that starch may make a significant contribution to the metabolic energy requirements for this species. Although both lipases and esterases are present in *Macrobrachium* sp.,[27] most dietary lipids are assumed to be hydrolyzed by esterases.[26]

Protein-Energy Requirements

While several growth studies have indicated that a dietary protein concentration of 15 to 35% may be adequate to maintain maximum rates of growth,[6,7] more recent evidence has indicated that concentrations of 40% or higher may be required for optimum growth.[31] Both dietary concentrations of proteins as well as dietary carbohydrate-to-lipid ratios have been correlated with significant differences in growth rates of *Macrobrachium* sp.

A study with 0.1-g prawns indicated that over a 35-week period, the highest protein concentrations examined (35%) provided better growth of *M. rosenbergii* than 15 and 25% protein.[6] Prawn with an initial size of 0.15 g fed either 23, 32, 40, or 49% protein grew significantly better when fed diets with 40 or 49% total protein concentrations after 10 weeks.[31] However, prawns greater than 1.0-g weight had better feed conversion rates and protein efficiency ratios and grew significantly better on a 40% protein diet than individuals fed 49% protein over the next 4-week period, suggesting excessive protein catabolism in the prawns fed the 49% protein diet during this period.[31] Prawn fed diets having a protein concentration of 25% and 1:3 or 1:4 ratios of dietary lipids to carbohydrates catabolized less protein than prawns fed 25% protein diets having higher ratios of lipids to carbohydrates.[32,33] However, in a long-term feeding study, Millikin et al.[31] reported greater growth and better feed conversion in *M. rosenbergii* fed 40% protein plus a 1:2 lipid-to-carbohydrate ratio than in prawn fed 32% protein plus a 1:3 lipid-to-carbohydrate ratio or 23% protein plus a 1:4 lipid-to-carbohydrate ratio. The apparent discrepancy between the two studies is at least partially explained by the difference in prawn size between the studies. Most of the prawn used by Clifford and Brick[32] were appreciably larger on a wet weight (live) basis than those used by Millikin et al.;[31] therefore, a lower protein requirement for the larger size prawn is predictable. The lower protein requirement for larger *M. rosenbergii* was further substantiated in a 20-week feeding study in which prawn with initial mean weight of 3.9 g grew slightly, but not significantly better when fed the 40% protein diet compared to 25% protein (Ralston Purina® Marine Ration) plus 8 or 13% lipid (T. Smith, personal communication). In conclusion, early juvenile *M. rosenbergii* (0.1 to 3.0 g) apparently require about 40% protein, whereas larger prawn (4.0 to 20 g) require approximately 25 to 30% protein.

Amino Acid Requirements

In addition to interaction with other macronutrients (i.e., carbohydrates and lipids), dietary availability of specific amino acids may have an effect on total dietary protein required for maximum rates of growth. Several investigations concerning specific amino acid requirements have generally indicated that such requirements for *Macrobrachium* sp. are qualitatively similar to most other organisms.[34,29] However, despite several studies conducted with *Macrobrachium* sp., both specific qualitative and quantitative estimates of dietary amino acid requirements are not known.

Among the many studies conducted with *Macrobrachium* sp., there is considerable disagreement among investigators regarding both quantitative and qualitative amino acid requirements (Table 4). Although not measuring tryptophan, Miyajima et al.[35] reported that nonessential and essential amino acid requirements for this genus were generally the same as other animals, including mammals, except for a requirement for tyrosine. Watanabe[36] did not assay for threonine or tryptophan but found a requirement for all essential amino acids plus tyrosine while not finding a requirement for lysine. Stahl and Ahearn[37] attempted to verify, through short-term growth studies, the previous conclusions of Miyajima et al.[35] and Watanabe.[36] They were not able to qualitatively establish requirements previously demonstrated for arginine, methionine, and tryptophan, nor could they demonstrate a requirement for lysine. In a recent investigation, Fair and Sick[38] reported that some amino acids previously reported as nonessential for *Macrobrachium* sp. are metabolically required, although they may not necessarily be dietarily essential. They reported that appreciable reductions of proline, aspartic acid, and serine removed from hemolymph during 5 days of starvation were indicative of high rates of metabolic demand, and, therefore, rendered these amino acids metabolically essential. Miyajima et al.[35] suggested that for *Macrobrachium* sp. tyrosine is synthesized from dietary concentrations of phenylalanine. Undoubtedly, the myriad of methods and experimental approaches used to study or estimate amino acid requirements in this species has contributed to the failure of various investigators to verify each other's conclusions. In the case of estimating amino acid requirements from selected dietary supplementations,[37-39] gut flora, particularly bacteria, could excrete amino acids and contribute to variable growth results. Also, dissolution of crystalline amino acids frequently supplemented in aquatic formula feeds[11] may explain variable results among independently conducted investigations. This variability among investigations in itself may be responsible for failure to achieve verifiable results among many investigators.

Table 4
SUMMARY OF AMINO ACID REQUIREMENTS DETERMINED FOR *MACROBRACHIUM* SP.

Amino acids required	Species investigated	Methods	Ref.
All amino acids considered essential for mammals plus tyrosine (tryptophan not evaluated)	*M. ohione*	Metabolic labeling of essential amino acids from a ^{14}C-glucose substrate	35
All essential amino acids including tyrosine but excluding lysine (tryptophan and threonine not evaluated)	*M. rosenbergii*	Metabolic labeling of essential amino acids from a ^{14}C glucose substrate	36
No growth requirement found based on investigations of lysine, arginine, methionine, and tryptophan	*M. rosenbergii*	Growth study (4 to 12 weeks) using diets having supplements of selected amino acids	37
Growth requirement for arginine, phenylalanine, leucine, and isoleucine; but no requirement determined for lysine, histidine, methionine, or threonine	*M. rosenbergii*	Supplementation of repelletized commercial feed with selected amino acids and a feed conversion ratio measured	8
Growth requirement for lysine, arginine, tryptophan, leucine, and isoleucine	*M. rosenbergii*	Supplementation of repelletized commercial feed with selected amino acids and a feed conversion ratio measured	39
Metabolic requirement for proline, glutamic acid, methionine, phenylalanine, and leucine	*M. rosenbergii*	Selected utilization of amino acids from hemolymph during starvation	38

Physiological adaptations of crustaceans such as *Macrobrachium* sp. may alter classic concepts of "essential" and "nonessential" amino acids. Although metabolic processes at the cellular level dictate that amino acid requirements will be similar for all organisms, adaptations to a brackish water environment and growth by molting may dictate specific amino acid requirements that are unique for crustaceans and possibly even unique for *Macrobrachium* sp. The adaptation to brackish conditions, for example, renders *Macrobrachium* sp. uniquely adapted to an environment that is different from those that either truly marine or freshwater crustaceans inhabit. Amino acids have traditionally been classified as dispensable (nonessential) and indispensable (essential). Indispensable amino acids cannot be cellularly synthesized in sufficient quantities while dispensable amino acids can be metabolically produced.[40] Perhaps if the rate of metabolic demand is considered against rate of metabolic use, as suggested by Maynard and Loosli[40] and as proposed for *Macrobrachium* sp. by Fair and Sick,[38] a category of truly "semiessential" amino acids would be an appropriate category to consider for this genus.

Carbohydrate Requirements

Requirements for specific carbohydrates have not been determined for *Macrobrachium* sp. While the contribution of various carbohydrates to energy requirements and specific needs such as carbohydrate skeletal components of a myriad of metabolic pathways are undoubtedly significant in this genus, such contributions can only be estimated from investigations conducted with other genera of crustaceans.[34] As cited above, cellulose may contribute either directly or indirectly to a nutrition requirement for *Macrobrachium* sp. Fair et al.[28] speculated that the presence of cellulase activity may be indicative of cellulose assimilation. In addition, they speculated that in contrast to most mammalian systems, dietary cellulose ingested by *M. rosenbergii* may facilitate the absorption of nutrients from the intestine. The mechanism accounting for increased nutrient absorption may be a decreased rate of food passage through the digestive system due to the presence of cellulose.[28]

Lipid Requirements

Like dietary requirements for carbohydrates, little is known about specific lipid and fatty acid requirements for *Macrobrachium* sp. Joseph and Williams[41] and Sandifer and Joseph[42] reported that dietary shrimp *(Penaeus setiferus)* head oils added at a concentration of 3% supplementation to 7% total dietary lipid (Marine Ration 25) and having relatively high concentrations of ω3 fatty acids resulted in relatively high rates of growth for postlarval *M. rosenbergii*. Sandifer and Joseph[42] further reported that ω6 fatty acids tended to inhibit prawn growth. The optimum ratio of ω3 to ω6 fatty acids in *Macrobrachium* sp. diets has not been investigated.

Mineral and Vitamin Requirements

Dietary requirements for vitamins and minerals have virtually been ignored in research investigations with *Macrobrachium* sp., as reviewed by Biddle.[29] Not only is there a paucity of information concerning micronutrient requirements, but many micronutrient additions to formula feeds used for *Macrobrachium* sp. and crustacean culture by virtue of *ad libitum* supplementation may actually be deleterious for maximum rates of growth. *Ad libitum* addition of various vitamins, for example, have ranged from dietary concentrations considered deficient to concentrations as high as 500 times as high as those considered optimum for normal metabolism and growth in terrestrial animals and cultured finfish.[34]

RECOMMENDATIONS

1. In conjunction with cost-benefit analyses, for use of given feeds, further development through research of readily acceptable formula feeds is required. Specific aspects of such research and development that should receive research priority are (1) experimentation with various binders, processing techniques (e.g., pelletization vs. steam extrusion), or encapsulation; (2) attempt to increase rates of formula feed ingestion through comparisons among dietary substrates that may serve as attractants; and (3) further development of formula feeds that will be physically and nutritionally acceptable forms of feed for larval stages.
2. Obviously, the further development of formula feeds as suggested above will be integrally associated with having more information concerning dietary requirements for specific macro- and micronutrients than presently exists. While there is relatively adequate knowledge of nutritional requirements for dietary concentrations of proteins, qualitative requirements for proteins, carbohydrates, and lipids are not known for *Macrobrachium* sp. Furthermore, an understanding of even general requirements concerning micronutrient (i.e., vitamins and minerals) nutritional factors, which are undoubtedly more severely growth limiting than macronutrients (i.e., proteins, carbohydrates, and lipids), are totally unknown for this genus of crustacean.
3. Designs for future nutrient requirement experiments should include not only growth data, but approaches that include metabolic investigations either conducted in series and separate, or combined with growth data. Understanding of nutritional requirements for specific macronutrients (i.e., amino acids, monosaccharides, and fatty acids) and, particularly, micronutrients has not yet been evaluated. Such information can be appropriately gained using traditional nutritional, physiological, and biochemical techniques including radioassay, in vitro tissue and cell culture, and molecular separatory procedures. Although such techniques have been predominantly used in mammalian or terrestrial invertebrate research, their application, with perhaps some modifications, is appropriate and recommended for studying nutritional-metabolic requirements for *Macrobrachium* sp.

REFERENCES

1. **Singholka, S., New, M. B., and Vorasayan, P.,** The status of *Macrobrachium* farming in Thailand, *Proc. World Maricul. Soc.,* 11, 60, 1980.
2. **Ling, S. W.,** Aquaculture in Southeast Asia: A Historical Overview, Washington Sea Grant Publication, University of Washington Press, Seattle, 1977, 94.
3. **Ling, S. W.,** The general biology and development of *Macrobrachium rosenbergii,* in Proc. World Conf. on Shrimps and Prawns (Mexico), 1967, FAO Ref. Pap. E30, FAO, Rome, 1969, 589.
4. **Balazs, G. H., Ross, E., and Brooks, C. C.,** Preliminary studies on the preparation and feeding of crustacean diets, *Aquaculture,* 2, 369, 1973.
5. **Balazs, G. H., Ross, E., Brooks, C. C., and Fujimura, T.,** Effect of Protein Source and Level on Growth of the Captive Freshwater Prawn, *Macrobrachium rosenbergii,* Hawaii Institute of Marine Biology Contribution No. 000, 1974.
6. **Balazs, G. H. and Ross, E.,** Effect of protein source and level on growth and performance of the captive freshwater prawn, *Macrobrachium rosenbergii, Aquaculture,* 7, 299, 1976.
7. **Sick, L. V. and Beaty, H.,** Development of formula foods designed for *Macrobrachium rosenbergii* larval and juvenile shrimp, *Proc. World Maricul. Soc.,* 6, 89, 1975.
8. **Farmanfarmaian, A. and Lauterio, T.,** Amino acid supplementation of feed pellets of the giant shrimp *(Macrobrachium rosenbergii) Proc. World Maricul. Soc.,* 10, 674, 1979.
9. **Heinen, J. M.,** Evaluation of some binding agents for crustacean diets, *Prog. Fish. Cult.,* 43(3), 142, 1981.
10. **Goldblatt, M. J., Conklin, D. E., and Brown, W. D.,** Nutrient leaching from pelleted rations, in Proc. Symp. on Finfish Nutrition Fish Feed Technology, Hamburg 1978, 28; Vol. 2, 1979, 118.
11. **Fortner, A. R. and Sick, L. V.,** Effects of feed processing and submersion on concentrations of selected amino acids in formula feeds designed for aquatic organisms, NOAA Tech. Rep. (submitted).
12. **Minamizawa, A. and Morizane, T.,** Report on a Study about Cultivation Techniques for Freshwater Shrimp, Ehime Prefecture Laboratory, Bureau of Commercial Fisheries, Office of Foreign Fisheries, U.S. Department of Interior, Washington, D.C., 1970.
13. **Lewis, J. B.,** Preliminary experiments on the rearing of the freshwater shrimp, *Macrobrachium carcinus,* in Proc. 14th Annual Session, University of Miami Institute of Marine Science, Gulf and Caribbean Fisheries Institute, Miami Beach, Fla., November 1961, 1962.
14. **Murai, T. and Andrews, J. W.,** Comparison of feeds for larval stages of the giant prawn *(Macrobrachium rosenbergii), Proc. World Maricul. Soc.,* 9, 189, 1978.
15. **Fujimura, T.,** Notes on the development of a practical mass culturing technique for the giant prawn, *Macrobrachium rosenbergii,* in Proc. Indo-Pacific Fisheries Council, 12th Session, IPFC/C66/WP47, Honolulu, 1966.
16. **Fujimura, T. and Okamoto, H.,** Notes on progress made in developing a mass culturing technique for *Macrobrachium rosenbergii* in Hawaii, in *Coastal Aquaculture in the Indo-Pacific Region,* Pillay, T. V. R., Ed., Fishing News Books, Farnham, Surrey, England, 1972, 313.
17. **Wickins, J. F.,** Experiments on the culture of the spot prawn, *Pandalus platycerus* Brandt, and the giant freshwater prawn, *Macrobrachium rosenbergii, Fish. Invest. Ser.,* 27(5), 23, 1972.
18. **Maddox, M. B. and Manzi, J. J.,** The effects of algal supplements on static system culture of *Macrobrachium rosenbergii* larvae, *Proc. World Maricul. Soc.,* 7, 677, 1976.
19. **Manzi, J. J. and Maddox, M. B.,** Algal supplement enhancement of static and recirculating system culture of *Macrobrachium rosenbergii* larvae, *Helgolander Wiss. Meeresuntersuchungen,* 28, 447, 1976.
20. **Joseph, J. D.,** Assessment of the nutritional role of algae in the culture of larval prawns *(Macrobrachium rosenbergii), Proc. World Maricul. Soc.,* 8, 853, 1977.
21. **Cohen, D., Finkel, A., and Sussmann, M.,** On the role of algae in larviculture of *Macrobrachium rosenbergii, Aquaculture,* 8, 199, 1976.
22. **Sick, L. V.,** Selected studies of protein and amino acid requirements for *Macrobrachium rosenbergii* larvae fed neutral density formula diets, in *Proc. 1st Int. Conf. Aquaculture Nutr.,* 1976, 215.
23. **Sova, V. V., Elyakova, L. A., and Vaskovsky, V. E.,** The distribution of laminarianases in marine invertebrates, *Comp. Biochem. Physiol.,* 32, 459, 1970.
24. **Elyakova, L. A.,** Distribution of cellulases and chitinases in marine invertebrates, *Comp. Biochem. Physiol.,* 43B, 67, 1972.
25. **Vonk, H. J.,** Digestion and metabolism, in *Physiology of Crustacea,* Waterman, T. H., Ed., Academic Press, New York, 1960, 291.
26. **Lee, P. G., Blake, J. J., and Rodrick, G. E.,** A quantitative analysis of digestive enzymes for the freshwater prawn *Macrobrachium rosenbergii, Proc. World Maricul. Soc.,* 11, 392, 1980.
27. **Murthy, R. C.,** Study of proteases and esterases in the digestive system of *Macrobrachium lamarrei,* Crustacea: Decapoda, *J. Anim. Morphol. Physiol.,* 24, 211, 1977.

28. **Fair, P. H., Fortner, A. R., Millikin, M. R., and Sick, L. V.,** Effects of dietary fiber on growth, assimilation, and cellulase activity of the prawn, *(Macrobrachium rosenbergii), Proc. World Maricul. Soc.,* 11, 369, 1980.
29. **Biddle, G. N.,** The nutrition of freshwater prawns, in *Shrimp and Prawn Farming in the Western Hemisphere,* Hanson, J. A. and Goodwin, H. L., Eds., Dowden, Hutchinson and Ross, Stroudsburg, Pa., 1977, 272.
30. **Tyagi, A. P. and Prakash, A.,** A study on the physiology of digestion in freshwater prawn, *Macrobrachium dayanum, J. Zool. Soc. India,* 19, 77, 1967.
31. **Millikin, M. R., Fortner, A. R., Fair, P. H., and Sick, L. V.,** Influence of dietary protein concentration on growth, feed conversion, and general metabolism of juvenile prawn *(Macrobrachium rosenbergii), Proc. World Maricul. Soc.,* 11, 382, 1980.
32. **Clifford, H. C. and Brick, R. W.,** Protein utilization in freshwater shrimp, *Macrobrachium rosenbergii, Proc. World Maricul. Soc.,* 9, 195, 1978.
33. **Clifford, H. C. and Brick, R. W.,** A physiological approach to the study of growth and bioenergetics in the freshwater shrimp, *Macrobrachium rosenbergii, Proc. World Maricul. Soc.,* 10, 701, 1979.
34. **New, M. B.,** A review of dietary studies with shrimp and prawns, *Aquaculture,* 9, 101, 1976.
35. **Miyajima, L. S., Broderick, G. A., and Reimer, R. O.,** Identification of the essential amino acids of the freshwater shrimp, *Macrobrachium ohione, Proc. World Maricul. Soc.,* 7, 699, 1976.
36. **Watanabe, W. O.,** Identification of Essential Amino Acids of the Fresh Water Prawn *Macrobrachium rosenbergii,* M.S. thesis, Department of Zoology, University of Hawaii, 1975.
37. **Stahl, M. S. and Ahearn, G. A.,** Amino acid studies with juvenile *Macrobrachium rosenbergii, Proc. World Maricul. Soc.,* 9, 209, 1978.
38. **Fair, P. H. and Sick, L. V.,** Serum amino acid concentrations during starvation in the Asian prawn, *Macrobrachium rosenbergii,* as an indicator of metabolic requirements, *Comp. Biochem. Physiol.,* 73(B) 2, 195, 1982.
39. **Farmanfarmaian, A. and Lauterio, T.,** Amino acid composition of tail muscle of *Macrobrachium rosenbergii* — comparison to amino acid patterns of supplemental commercial feed pellets, *Proc. World Maricul. Soc.,* 4, 454, 1980.
40. **Maynard, L. A. and Loosli, J. K.,** *Animal Nutrition,* 6th ed., McGraw-Hill, New York, 1969, 613.
41. **Joseph, J. D. and Williams, J. E.,** Shrimp head oil: a potential feed additive for mariculture, *Proc. World Maricul. Soc.,* 6, 147, 1975.
42. **Sandifer, P. A. and Joseph, J. D.,** Growth responses and fatty acid composition of juvenile prawns *(Macrobrachium rosenbergii)* fed a prepared ration augmented with shrimp head oil, *Aquaculture,* 8, 129, 1976.

FEEDING PRACTICES AND NUTRITIONAL CONSIDERATIONS FOR *MACROBRACHIUM ROSENBERGII* CULTURE IN HAWAII

John S. Corbin,
Michael M. Fujimoto, and
Thomas Y. Iwai, Jr.

INTRODUCTION

Experimental culture of the giant Malaysian prawn, *Macrobrachium rosenbergii,* began in the U.S. in Hawaii.[1] Work by the State's Department of Land and Natural Resources developed viable mass-culture techniques for larval and juvenile prawns over a period of years.[2-5] Commercial scale testing of hatchery and grow-out technology was carried out and a concerted effort was undertaken by the state to demonstrate to the private sector that prawn farming was a technologically and economically viable investment.[6-9] Once workable techniques were developed and transferred to the private sector, state research and development efforts focused on reducing the costs of production and improving yields from production systems.[10]

In general, feed management and nutrition are major areas of concern for aquaculturists because quality and quantity of compounded feeds or feed organisms are major determinants in the profitability of commercial farms.[11,12] Nutritional considerations in culturing commercially valuable shrimps and prawns have similarly been of great interest due to private and public sector attempts at fostering an industry in the U.S., Japan, and elsewhere.[13] In recent years, this pivotal area has been the subject of several comprehensive literature reviews.[14-17] Extensive research by Japanese scientists into the macro- and micronutrient requirements of the marine shrimp, *Penaeus japonicus*, is particularly noteworthy,[17,18] though these results may not be applicable to *Macrobrachium.*[14,16] For further background on the subject of aquatic animal nutrition, the reader may wish to consult basic references in fish nutrition, albeit mostly trout, salmon, and catfish.[19-23]

Feeding Practices and Nutritional Considerations in Prawn Farming

Macrobrachium farming can be made up of four component phases: broodstock maintenance, hatchery, nursery, and grow-out. Each phase has its own set of nutritional considerations, feeds, and feeding practices, which, in broad terms, should provide the most efficient and cost-effective means of satisfying the prawn macro- and micronutrient requirements in order to optimize production. Consideration of nutritional requirements involves the quality and quantity of the commercially formulated rations consumed by the prawn, as well as any natural foods which may be presented or are available during each culture phase. Current feeds and feeding practices utilized for *Macrobrachium* culture in Hawaii are presented (Table 1) and available nutritional information is described.

Broodstock Maintenance

Macrobrachium broodstock are brought to the state hatchery located at the Anuenue Fisheries Research Center (AFRC), Sand Island, Oahu, for two purposes: one, production of postlarval prawns for distribution to commercial farmers; and, two, genetic experimentation and domestication through intraspecific breeding. *Macrobrachium* are known for the ease in which they mature, mate, and spawn in the laboratory.

Feeding Practices

The broodstock utilized for the production of postlarval prawns for distribution to com-

Table 1
PRINCIPAL SOURCES OF NUTRITION AND AVAILABLE SUPPLEMENTS FOR VARIOUS PHASES OF PRAWN FARMING IN HAWAII

Phase	Principal source(s)	Available supplements
Broodstock	Mixed commercial feeds	Mosquito fish (*Gambusia*), aquatic invertebrates, aquatic plants, bacterial aggregates, detritus, and vegetable matter (e.g., bean sprouts)
Hatchery	Brine shrimp nauplii (*Artemia*), fish flesh (tuna fillets and pollack surimi)	Phytoplankton, protozoans
Nursery	Commercial feed	Aquatic invertebrates, aquatic plants, bacterial aggregates, and detritus
Grow-out	Commercial feed	Mosquito fish (*Gambusia*), aquatic invertebrates, aquatic plants, bacterial aggregates, and detritus

mercial farms are gravid females brought to the hatchery from commercial production ponds. The number of females held at any one time depends upon the postlarval requirements of the hatchery. Incubation time ranges from 2 to 25 days, depending on the stage of egg development. During this period, up to 75 animals are maintained in 1600-ℓ tanks and are fed a daily mid-afternoon ration of Waldron's Prawn Pellet No. 1. Uneaten food and feces are removed daily.

Broodstock maintained over long periods of time for genetic selection experiments are either Anuenue stock or imported stock from various countries within the geographic range of *M. rosenbergii*.[24] Populations of each ecotype of animal are maintained at the center by reproduction of mature stock and subsequent maturation, mating, and spawning of offspring. *Macrobrachium* females are known to spawn as many as four times a year in the laboratory.[25]

Two holding systems are employed for maturing ecotypes of *Macrobrachium:* one, outdoor pools which simulate pond environments; and, two, covered, compartmentalized tanks. In general, the procedure is to hold animals in mixed sex groups in the outdoor pools at low densities until they are mature. Mature males and females are then transferred to the compartmentalized tanks, where they are held individually for genetic selection experiments. Feeding regimes for the pools and the tanks differ slightly. Both groups are fed commercially available feeds (e.g., Clark's trout pellets) supplemented by vegetable material (e.g., bean sprouts), live snails, *Daphnia,* and duckweed. Outside pools are fed once a day and inside tanks are fed twice a day. Amounts are adjusted from observations of leftover food. Natural productivity such as live snails, duckweed, insect larvae, mosquito fish, and water hyacinths is maintained in the pools and probably forms a significant portion of the prawns' diet. Some natural food organisms (e.g., mosquito fish and live snails) are added from time to time to the tanks as supplemental food.

Nutritional Requirements

Little is known about the nutritional requirements of *Macrobrachium* broodstock and the feeds and feeding practices utilized are the result of pragmatic observation and empirical testing. Prawns readily mature in commercial earthen grow-out ponds and, with the continuous harvest method, adequate numbers of gravid females are readily available for the Anuenue hatchery from commercial farms. Since incubation time in the production hatchery for gravid animals is short, relative to pond residence time, their feeding history while in the pond is probably of greater nutritional significance to laboratory spawning success and survivorship of larvae. *Macrobrachium* females collected from a typical harvest in Hawaii on the average produce 10,000 larvae per 40- to 50-g female.

Broodstock maintained in the laboratory over long periods of time for breeding and genetic selection must have a nutritionally adequate diet for successive maturation, mating, and spawning to occur. Natural food organisms are made available to these animals along with pelleted feeds. Specific qualitative and quantitative requirements for *Macrobrachium* broodstock are not known. The feeding regime utilized at the AFRC appears adequate judging from the success in maintaining stocks, though variable spawning success has been observed.[26] Experimental evidence from penaeid research suggests that diet, particularly fatty acid components, may be an important factor in maturation and spawning in artificial environments.[27] In general, the Hawaii strategy is to provide a broad spectrum diet such that nutritional adequacy is reasonably assured.

Hatchery Production

Commercial-scale larval rearing techniques for *Macrobrachium* are of two basic types: the Anuenue "green water" method, developed in Hawaii, and the Aquacop "clear water" method, developed by the French.[28] The Anuenue method is utilized at the state hatchery, as well as three private commercial hatcheries in Hawaii. At temperatures of about 28°C, the normal larval cycle is between 30 and 40 days.

Feeding Practices

The "green water" method involves rearing *Macrobrachium* larvae in a medium of mixed plankton, which is prepared by "aging" brackish water (15‰), in tanks stocked with fish (*Tilapia* spp.). Fish densities are held at approximately 1.4 kg/1000 ℓ and the fish are fed a daily ad libitum ration of prawn pellets. Phytoplankton densities of 1×10^5 to 1×10^6 cells per milliliter are maintained as the culture medium. Usually the chlorophytes, *Chlorella* or *Palmellococcus*, or the cyanophyte *Nannochloris* are the dominant species; however, other phytoplankton species have been used in successful larval rearing (Table 2). Microorganisms such as various ciliated protozoans and metazoans have also been noted in both the "green water" and the larval production tanks and are consistently correlated with healthy larvae and good survival (Table 3).

In practice, newly hatched, free-swimming zoea are placed into 9500-ℓ larval rearing tanks. Densities of about 1×10^6 (105 larvae per liter) are initially stocked in each tank for ease of feeding and management. This amount is split into groups of 300,000 (31 larvae per liter) during the second week of the cycle. Of this amount about 190,000 or 63% are expected to reach postlarvae. Phytoplankton-rich water is added to the larval tanks regularly (between days 6 to 25, approximately 3700 ℓ are exchanged every 1 to 2 days).

Introduction of fish flesh and 2-day-old *Artemia salina* nauplii begins after the first or second day of stocking in the larval tanks. Fish flesh is prepared by forcing pollack surimi (*Theragra calcogramma*)[29] or skipjack tuna fillets (*Katsuwonus pelamis*) through stainless steel screens with high pressure water. Screen openings of 850 μm (20 mesh), 425 μm (40 mesh), or 250 μm (60 mesh) are used. The smaller 425-μm and the 250-μm screens are used to process smaller fish particles for the first 10 to 12 days of the cycle, while the larger screen is used to process larger fish particles for the older larvae.

Application rates for the fish flesh are adjusted ad libitum to ensure that a build-up of uneaten feed does not occur. At the AFRC, newly stocked tanks (stocking density of 105 larvae per liter) are fed four times a day at 3-hr intervals with approximately 100 mℓ of slurry (50% water by volume) per feeding. The amount is adjusted daily until a level of 400 to 500 mℓ is reached by the end of the second week of the cycle. At this time, larval stocking densities are reduced to approximately 31 larvae per liter (i.e., the larvae are removed, counted, and stocked into clean tanks) and feeding levels are readjusted to about 100 mℓ

Table 2
DOMINANT PHYTOPLANKTON OBSERVED IN THE GREEN WATER MEDIA USED IN THE CULTURING OF PRAWNS AT THE ANUENUE FISHERIES RESEARCH CENTER, OAHU, HAWAII

Types of phytoplankton
Chlorella ellipsoida[a]
C. vulgaris[a]
Palmellococcus minutes[a]
P. protothecoides[a]
Amphiprora paludosa
Schroederia setigera
Chaetoceros sp.
Thalassiosira sp.
Nannochloris sp.[a]
Oscillatoria sp.
Navicula sp.
Nitzschia sp.

[a] Most frequently utilized.

Table 3
REPRESENTATIVE MACROSCOPIC AND MICROSCOPIC FLORA AND FAUNA OBSERVED IN PRAWN LARVAL REARING TANKS AT THE ANUENUE FISHERIES RESEARCH CENTER, OAHU, HAWAII

Flora	Fauna
Achnanthes	*Anisonema*[a]
Amphidinium	*Epistylis*
Amphiprora	*Mayorella*[a]
Chloramoeba[a]	*Mesodinium*[a]
Chlorella	*Stylonchia*[a]
Cymbella	*Vorticella*
Gymnodinium	*Zoothamnium*
Gyrodinium	
Melosira	Amoeba
Microcystis	Copepods
Nannochloris	Nematodes
Navicula	Planaria
Nitzschia	Radiolaria
Oscillatoria	Rotifers
Palmellococcus	Suctoria
Schizothrix	
Schroederia	Misc. flagellates
Synechocystis[a]	Misc. ciliates

[a] Sato.[126]

of slurry per feeding. This feeding level is again increased as the larvae grow until a maximum of 400 to 500 mℓ of slurry per feeding is reached.

At AFRC, *Artemia* cysts are routinely chlorinated before hatching. This soaking procedure utilizes approximately 70 ppm of 5.25% active sodium hypochlorite (Chlorox®) for a period of 20 to 25 min. The cysts are then rinsed with fresh water before being placed in the hatching containers. After 48 hr, the live nauplii are separated from the empty cysts and distributed among the rearing tanks. Decapsulation and washing reportedly reduce the bacterial load associated with feeding *Artemia* nauplii.[30]

Larvae are fed the *Artemia* nauplii once a day, usually following the last application of the fish slurry. First-stage larvae can survive off residual yolk;[31] however, feeding of *Artemia* begins on the second day. Feeding rates begin at 60 g (dry cyst weight) per tank gradually increasing to 120 g per tank by the end of the second week. When the larval density is reduced to 31 larvae per liter, the amount of nauplii fed is also adjusted to 60 g per tank. The feeding level is then gradually increased to 150 to 180 g per tank. Amounts are adjusted according to *Artemia* hatching success, which usually is 60%.

Towards the end of the cycle, the number of metamorphosed larvae (postlarvae) increases in the rearing tanks and the feeding level of nauplii and fish slurry is reduced. A mixed diet of ground trout pellets and Marine Ration #25 is then provided for the postlarvae until they are removed for transport to nursery or grow-out ponds. During a normal larval rearing period of 30 to 40 days, survival averages around 60%.

In their natural environment, larvae would feed on zooplankton including rotifers, copepods, and other minute crustacea, small worms, and larval stages of various aquatic invertebrates. When living food is not presented, organic detritus of animal and plant origin may be consumed.[32] In culture situations around the world, *Artemia* nauplii are the principal food during the first week of larval rearing. Subsequently, a variety of specially prepared and locally-available low cost foods may be substituted, e.g., fish roe, aquatic worms, minced mussles, clams and squid, and steamed egg and soy custards.[1]

Nutritional Requirements

Little is known about the qualitative and quantitative nutritional requirements of larval prawns, though successful feeding regimes have been developed. At AFRC, fish flesh (tuna fillets and pollack surimi) and 2-day-old *Artemia* nauplii are the principal sources of nutrition available in the larval culture tanks. Several local commercial hatcheries are supplementing the standard brine shrimp and fish flesh diet with mullet roe (*Mugil* sp.) and/or chicken egg custard in the later stages of the production cycle. Recently, one commercial hatchery has developed a workable formulated larval feed to replace the fish flesh.[33]

The fish flesh utilized is predominantly protein (Table 4) and probably is the principal source of protein, by weight, consumed by the developing larvae. Tuna fillets, in particular, are known to have a balanced representation of the essential amino acids.[34] The usage by the AFRC of one or the other fish is dependent on availability and cost.

Artemia nauplii are routinely used for the larval culture of many aquatic species. Though biochemical composition and nutritional value are known to vary with the source of cysts, *Artemia* are generally considered a convenient and nutritious source of food.[35] Newly hatched *Artemia* are generally high in protein, lipid, and energy (Table 5); however, it may be important to use the nauplii as soon after hatching as possible, i.e., day two or less.[36] The levels of essential amino acids present in nauplii vary with the source but are generally considered nutritionally adequate.[37] Essential fatty acids from various strains of *Artemia* vary to greater degree, but also generally appear adequate to promote good growth and survival. However, it has been determined that *Artemia* strains can be classified into two groups: one contains predominantly 18:3ω3 and 18:4ω3, while the other contains chiefly 20:5ω3.[38,39] Such variation of these essential fatty acids may be large enough to affect the

Table 4
REPRESENTATIVE BIOCHEMICAL COMPOSITION OF FISH FLESH UTILIZED IN THE HATCHERY PHASE OF PRAWN PRODUCTION IN HAWAII

	Analysis	Tuna fillet	Pollack surimi
Fresh[a]			
	Dry matter	28.6	25.1
	Moisture	71.4	—
Dry matter[b]			
	Crude protein	97.2	69.2
	Crude fat	0.7	0.5
	Crude fiber	0.1	0.1
	Ash	6.4	2.9
	Phosphorus	0.96	0.52
	Potassium	1.75	0.63
	Calcium	0.0006	0.04
	Magnesium	0.15	0.10
	Sodium	0.14	0.52

[a] Percentage in fresh material.
[b] Percentage in dried material.

nutritional value of the nauplii.[40] Probably, the *Artemia* nauplii are the principal source of nonprotein dietary requirements for the prawns, though microorganisms and phytoplankton also present in the culture medium may furnish significant growth factors.

Moller[41] reports ingestion rates by previously fed stage-IV larvae of 0.65 nauplii per hour, but starved larvae can exceed this rate by as much as 12 times, for a period of 1 hr. Manzi and Maddox[42] found that early larval stages (I to V) can be fed on a concentration as low as four *Artemia* nauplii per milliliter per day (larvae stocked at 75/ℓ) and not be food limited. They also found that a concentration of five nauplii per milliliter allocated on an every-other-day schedule was suitable for later larval stages when a daily supplement of an acceptable fish roe was incorporated in the feeding regime. A survey of *Macrobrachium* hatchery operators found feeding at rates of 5 to 15 nauplii per milliliter, depending on the stage of the larvae, with concentrations often being decreased in the latter half of the larval cycle.[28]

Experimental feeding of a variety of supplemental fresh animal and plant materials in the latter portion of the cycle has been reported. Care must be exercised to make the fresh and the dry foods in appropriate particle size to achieve a homogeneous distribution in suspension. Dugan and associates[43] found foods of animal origin (e.g., *Artemia* nauplii, copepods, ground fish, ground beef heart) were better accepted than foods of plant origin (e.g., chopped rice, cornmeal, chicken mash, chopped beans). Moreover, fresh and frozen foods were better accepted than dried foods. Experimental observations at AFRC support these results. Of the numerous larval diets tried, the most successful was live *Artemia* nauplii, later supplemented with ground fish and beef heart. Other foods which have been used experimentally to supplement live *Artemia* diet include ground squid, fish eggs (*Mugil* sp.), scrambled chicken eggs, and frozen adult *Artemia;* [44] mixtures of eggs, ground fish flesh, and ground clam;[45] and freeze-dried oyster meat and freeze-dried catfish.[46]

Sick and Beaty[47] reported on the development of larvae using only formula foods. Six diets based on different combinations of fish, soybean, and shrimp meal were used. Cellulose was included as a filler and vitamin and mineral supplements were added. Two of the rations

Table 5
REPRESENTATIVE BIOCHEMICAL AND NUTRITIONAL INFORMATION FOR NEWLY HATCHED *ARTEMIA SALINA* NAUPLII

Component	Measurement
Selected Biochemical Composition[a]	
Individual dry wt (mg)	1.48
Ash wt (% dry wt)	11.28
Total lipid (% dry wt)	13.7
Fatty acids (% dry wt)	10.9
Caloric content per gram ash-free dry wt (cal)	5.503
Individual caloric content (μcal)	7.30
Selected Biochemical Composition[b]	
Water (% wet wt)	90.85
Dry matter (% wet wt)	9.15
Carbon (% dry wt)	27.5
Nitrogen (% dry wt)	8.09
Phosphorus (% dry wt)	1.24
Range of Certain Essential Amino Acids between Sources (g/100 g Protein)[c]	
Threonine	4.8— 6.0
Valine	3.1— 5.5
Methionine	2.2— 3.7
Isoleucine	4.9— 6.8
Leucine	7.9—10.1
Phenylalanine	5.1—10.4
Histidine	2.7— 4.9
Lysine	8.7—11.7
Arginine	9.7—11.5
Range of Selected Fatty Acids between Sources (mg/g dry wt Lipid)[d]	
18:0	2.79— 6.83
18:1ω9	26.97—31.2
18:2ω6	3.69— 9.59
18:3ω3	4.87—33.59
18:4ω3	0.96— 4.88
20:1ω9	0.35— 0.52
20:2ω6/ω9	0.06— 0.24[e]
20:3ω6	0.05— 2.76
20:3ω3/20:4ω6	1.48— 2.69[e]
20:5ω3	1.68—13.63
22:6ω3	0.06— 0.26[e]

[a] Benijts et al.[36]
[b] Reference 127.
[c] Seidel et al.[37]
[d] Schauer et al.[39]
[e] May be totally absent from certain sources.

had 15 or 30% *Artemia* meat added, while two others contained 5 and 15% albumin. Freeze-dried, gel, and dry flake forms were fed. Regardless of form, formula diets having *Artemia* meat were assimilated at higher rates than the other diets and supported net growth. Larvae fed a diet of *Artemia* nauplii and freeze-dried catfish produced stage-VIII larvae in significantly less time than the formula feeds.

Sick,[48] in other experiments with *Macrobrachium* larvae and formulated feeds, found larvae had oxygen-to-nitrogen ratios (O:N) of 8 or lower for diets 20% or less in total protein content. O:N ratios of this magnitude generally indicate that protein is being utilized as the primary substrate for catabolism[49] and perhaps further indicate that these diets are inadequate in total protein (starvation) or have a poor amino acid balance. Experimental evidence from feeding studies with juvenile prawns and shrimps, which indicate protein requirements for earlier life stages are likely to be higher than later stages, may suggest that prawn larval protein requirements may be higher than 20%.[50,51]

While it is certain that the presence of phytoplankton in the larval production tanks enhances larval growth and development,[4,52,53] the mechanism of enhancement remains unclear.[54] Evidence to date suggests that certain phytoplankton species, while ingested directly and indirectly (through fed *Artemia*), do not contribute to larval nutrition.[55] Cohen and associates[56] suggest that the presence of phytoplankton may facilitate larval growth through the removal of ammonia and other toxic substances. Joseph,[57] however, notes that the possibility that larval prawns metabolize unusual plant fatty acids cannot be discounted and, furthermore, it is possible that algal supplements provide nutritionally essential water-soluble trace compounds not available from primary and secondary feeds.

Several laboratory studies of daily caloric requirements of larvae have been carried out, providing limited information on larval energy requirements. Sick and Beaty[58] report that respiration rates for stage-I larvae averaged 0.053 cal/mg dry wt/hr and for stage-VII and -VIII larvae, 0.03 cal/mg dry wt/hr. Stephenson and Knight[59] demonstrate comparable larval caloric requirements which vary from 0.762 cal/mg dry wt/day (0.032 cal/mg dry wt/hr) for stage-I larvae to 0.576 cal/mg dry wt/day (0.024 cal/mg dry wt/hr) for stage-VIII larvae. Energy expended in respiration and growth and the energy assimilated by larvae were similar during each stage of development. Moreover, net growth efficiency varied between 21 and 36% and estimated assimilation efficiency ranged from 60 to 70%.

Again, as with broodstock, satisfactory feeds for prawn larval production have evolved from pragmatic testing and empirical observation. Improved understanding and control of the feed components of the AFRC "green water" system would be desirable. Optimization of production and least-cost feed formulation await comprehensive research on qualitative and quantitative larval nutrition and development of satisfactory techniques for nutrient delivery such as microencapsulation.[60]

Nursery Production

The use of nursery ponds in Hawaii is at the experimental stage. Notably, much work has been carried out by the prawn research group at the University of South Carolina on intensive indoor nursery systems.[61-63] An important difference between the two approaches is that the nursery phase in Hawaii is carried out in earthen ponds where natural productivity, with the exception of mosquito fish, is encouraged. Cycles generally are carried out for 60 to 90 days.

Feeding Practices

Nursery pond experiments are being conducted by the Hawaii Prawn Aquaculture Program as part of a test for a multirotational prawn stocking and harvesting system.[64] Postlarvae are stocked at between 500 to 800 individuals per square meter in 148-m^2 earthen ponds.[65] Prior to stocking, experimental habitats are placed in the ponds and natural productivity, such as phytoplankton, zooplankton, and various aquatic invertebrates, is encouraged. Postlarvae are fed once a day in the afternoon using varying amounts of commercial feed (e.g., Waldron's No. 1 pellet, Tables 6 and 7). Amounts, which usually approximate 5% biomass per day, are adjusted based on observations of leftover feed. Prawns grow during the 60- to 90-day residence time in the experimental ponds to between 2.5 and 3.5 cm in length or approximately 2 g in wet weight.

Table 6
INGREDIENT COMPOSITION OF VARIOUS FEEDS BEING UTILIZED BY HAWAII PRAWN FARMERS

Ingredients	Feeds (%)			
	Waldron's Broiler Starter	Waldron's Game Cock Pellets	Waldron's Prawn No. 1	Waldron's Prawn No. 2
Alfalfa	—	—	4.0	4.0
Corn	53.25	50.25	56.75	56.75
Cottonseed meal	10.00	15.75	—	—
Soybean meal	24.25	20.50	27.00	25.00
Meat and bone meal	7.00	7.00	11.00	8.00
Tuna meal	—	—	—	5.00
Vitamin mix	1.25	1.25	1.25	1.25
Mineral mix	1.25	1.25	—	—
Molasses	3.00	4.00	—	—

From Waldron's Feed Mill, personal communication, 1981.

Table 7
PROXIMATE COMPOSITION OF COMMERCIAL PRAWN FEEDS BEING UTILIZED IN HAWAII

Brand name	Guaranteed analysis (%)				
	Crude protein (min)	Crude fat (min)	Crude fiber (max)	Ash (max)	Minerals (max)
Carnation® Special Prawn Feed[a]	27.0	4.5	5.0	8.0	NA[b]
Trojan Fresh Water Prawn Feed[c]	38.5	3.1	2.0	12.0	3.0
Waldron's Prawn Pellet No. 2,[d,e]	24.0	3.8	5.1	6.5	NA
Waldron's Broiler Starter[e]	24.0	8.0	4.0	9.0	NA
Waldron's Game Cock Pellets[e]	23.8	3.1	3.9	5.7	NA
Waldron's Prawn Pellet No. 1[e]	24.0	3.0	3.0	6.0	NA
Clarks New Age Trout Pellets (3/32nd)	38.0	5.5	7.0	15.0	3.0
Purina® Experimental Marine Ration 25	25.0	10.0	5.0	NA	NA

[a] Source, Carnation Co., Honolulu, Hawaii.
[b] Not available.
[c] Source, Western Consumers Inc.
[d] Average proximate analysis performed on prawn feed samples taken from three farms.
[e] Source, Waldron's Milling Co., Honolulu, Hawaii.

In a commercial-scale test of nursery ponds at a local farm, postlarvae are stocked at approximately 70 to 90/m^2 in 1619- to 3238-m^2 earthen ponds. Natural productivity is encouraged, but no mosquito fish are stocked. Feeding is carried out once a day in the afternoon using available experimental feeds or Carnation® Special Prawn Feed (Table 7). Rates begin at 15% of the biomass per day and are gradually stepped down to 3% of the biomass per day by the end of the 90-day cycle. Conversion ratios have ranged between 1.8:1 and 2.8:1 and animals ranged between 4 and 5 cm in length at the end of a cycle. Natural productivity typical of pond environments (Tables 9 to 11) is observed with the exception of fish species.[33]

In their natural environment newly metamorphosed juveniles consume mainly insect larvae, worms, and small crustacean animal detritus.[32] In Southeast Asia, farmed juvenile prawns are fed a variety of supplemental materials, including small pieces of fish, shrimp, chironomid larvae, squid, mussel, cockle, worms, grains, and beans.[1]

Best results in the South Carolina tank experiments were obtained by feeding a commercially available marine shrimp ration (Purina® Marine Ration 25) regularly supplemented with a mixture of cooked chicken eggs, fish roe (mullet, *Mugil* sp. and sea trout, *Cynoscion* spp.), squid, and spinach.[66] The daily ration was apportioned into several feedings (three to four times a day at approximately 6 to 10% total body weight per day) to achieve better feed utilization and to prevent the accumulation of uneaten food. Other studies have varied feeding rates during the first month following metamorphosis from 20% of the biomass daily, decreasing to 15% the second week and 10% the third and fourth weeks.[67] Recent measurements of juvenile prawn activity patterns lend support for the need for multiple feeding, since observations indicated prawns were active in both light and dark photophases.[68]

Nutritional Requirements

Qualitative and quantitative nutritional requirements for nursery stage juveniles are contained in the discussion of grow-out systems.

Grow-Out Production

The grow-out strategy most frequently used in Hawaii is the continuous harvest system. It consists of regular stockings of earthen ponds with newly metamorphosed postlarval prawns and regular, selective harvesting of marketable animals.[69] Postlarvae of about 1 cm are stocked at approximately 16 animals per square meter (range 10 to 20/m^2) of pond bottom. Commercial pond sizes vary somewhat, but most frequently are around 0.4 ha. Approximately 7 to 9 months after stocking, harvesting of market-size animals (greater than 11 cm and 30 g) begins with a seine and a given pond is harvested every 3 weeks. In Hawaii, prawns are generally marketed at between 13 and 22 whole animals per kilogram. Restocking of postlarval prawns occurs once or twice a year. Mortality during the growth cycle averages around 70%. Populations of prawns grown in ponds are characterized by both sexual dimorphic and heterogenous individual growth.[5,70]

Feeding Practices

For newly stocked ponds in their first year of production prawns are fed daily, usually in the afternoon at approximately 2.5 to 7.4 kg of feed per hectare per day. Farmers are instructed to increase feeding if the food disappears within 24 hr or decrease the amount fed if there are leftovers. This feeding rate stabilizes at between 27.2 and 44.5 kg/ha/day. A recently determined Hawaiian industry average is 34.6 kg/ha/day for producing ponds. In addition, a few Hawaiian farmers irregularly feed supplemental trashfish (*Tilapia* spp.) when it is available. Feed conversion ratios generally range between 2:1 and 4:1 for the commercial diets utilized and growth rates average around 1 to 2 cm/month.

Most farmers apply feeds by hand either by walking or driving along the pond bank. Several larger farms utilize automatic blower feeders. Feed application is a function of pond size and shape. The larger farms, using automatic feeders, broadcast feed towards the center of the pond to achieve an "even" distribution. Farmers broadcasting manually or from a motorized vehicle feed along either the windward or leeward berm in a horseshoe pattern. In the case of windy days, feed is broadcast with the wind for better distribution and less wastage. Some operations recommend alternating sides of the pond.

In the Hawaiian earthen pond grow-out system, the ecosystem is managed to encourage the presence of natural foods. Farmers try to maintain a rich bloom of phytoplankton (500,000 to 2,000,000 cells per milliliter) in each pond. Various grasses are planted along the banks

Table 8
VITAMIN AND MINERAL MIXES USED IN CERTAIN PRAWN FEEDS MANUFACTURED IN HAWAII[a]

Vitamin mix ingredient	Amount/kg diet[b]	Mineral mix ingredient	Amount/kg diet[c](mg)
Vitamin A	5500.0 IU	Zinc oxide	55.1
Vitamin D	1237.0 IU	Ferrous sulfate and carbonate	59.5
Vitamin E	4.1 IU	Manganous oxide	56.0
Vitamin K	0.8 mg	Copper oxide	4.5
Vitamin B_2	3.3 mg	Ethylenediaminedihydroiodide	0.25
Pantothenic acid	4.9 mg	Cobalt sulfate	0.50
Niacin	24.7 mg	Sodium selenite	0.10
Choline chloride	67.1 mg	Sodium chloride	2646.0
Vitamin B_{12}	8.2 mg		
Folic acid	0.3 mg		

[a] Source, Waldron's Milling Co., Honolulu, Hawaii.
[b] Vitamin mix is added to the prepared diet at 0.025%.
[c] Mineral mix is added to the prepared diet at 0.29%.

for stabilization and refuge habitat for molting prawns and to foster natural productivity (Table 9). Mosquito fish (*Gambusia* sp.) are stocked in new ponds to control dragonfly nymphs and to serve as a supplemental food source for the growing prawns. However, since *Gambusia* are observed to consume prawn feed and postlarvae, their population in the pond should be reduced prior to stocking. Currently, one commercial farm is conducting a large-scale demonstration of nursery and grow-out systems without stocking fish.[33]

Adult *Macrobrachium* are nocturnal, omnivorous benthic feeders. Their natural diet consists mainly of aquatic worms, insects, mollusks, crustacea, fish, aquatic plants, and detritus.[32] Quantitative analysis of prawn digestive enzymes indicates that *Macrobrachium* can be classified as an omnivore.[71] In the earthen-pond grow-out systems in Hawaii, the major sources of prawn nutrition available are the prawn feed (Tables 6 to 8) and the natural productivity of the ponds, i.e., mosquito fish, various invertebrate planktonic and benthic fauna, macro- and microalgae, and detritus and associated bacteria (Tables 9 to 11). *Macrobrachium* in the size range between 5.0 and 75.0 g are known to consume these various sources.[72]

The composition of various types of commercial feeds utilized in Hawaii is shown in Tables 6 and 7. Waldrons Prawn Pellet No. 1 or No. 2 is used by the majority of the farms. Historically, chicken broiler starter has been the major feed utilized in *Macrobrachium* culture and the specially developed prawn feeds are modifications of the broiler starter formula. Reportedly, broiler starter has been shown to be nutritionally deficient when no natural productivity or other supplemental feedstuffs are available.[73]

Most currently used prawn feeds in Hawaii leach nutrients and disintegrate soon after being placed in water. Concern has been expressed by farmers as to whether prawns actually consume pellets in significant quantities, or whether most pellets simply serve as a relatively expensive fertilizer for the pond.[72] Further contributing to potential loss of nutrients, *Macrobrachium* do not possess a gastric mill like some penaeids, but rather manipulate and masticate their food outside the buccal cavity with their anterior appendages.[74,75] Unless food is well bound, particles become dislodged and are swept away by the prawn's exhalent gill current. In addition, prawns may delay consuming feeds due to failure to locate the ration or lack of appetite, thus allowing feeds to disintegrate. Hawaii prawn feeds currently have no special binders added (Table 6), such as those tested by Forester.[76]

Table 9
TYPES OF BANK STABILIZATION VEGETATION USED IN POND CULTURE OF PRAWNS IN HAWAII

Scientific name	Common name
Cynodon dactylon	Bermuda grass, manienie
Commelina diffusa	Honohono, day flower, wandering Jew
Pennisetum clandestinum	Kikuyu grass
Paspalum vaginatum	Seashore paspalum
Brachiaria mutica	California grass, para grass, tall panicum

From Iwai, T., unpublished data, 1981.

Table 10
PHYTOPLANKTON GENERA COMMONLY OBSERVED IN PRAWN PONDS IN HAWAII

Cyanophyta (bluegreen)
- Anabaena
- Anabaenopsis
- Aphanocapsa
- Chroococcus
- Coelosphaerium
- Cylindrospermum
- Gloeocapsa
- Gomphosphaeria
- Merismopedia
- Microcystis/Anacystis/Polycystis
- Oscillatoria

Euglenophyta (euglenoids)
- Euglena
- Phacus

Pyrrhophyta (dinoflagellates)
- Amphidinium
- Ceratium
- Glenodinium
- Gonyaulax
- Gymnodinium
- Gyrodinium
- Peridinium
- Urococcus

Chrysophyta (yellow-green)
- Amphora
- Asterionella
- Chaetoceros
- Chromulina
- Cyclotella
- Cymbella
- Diceras
- Fragilaria
- Gomphonema
- Melosira
- Navicula
- Nitzschia
- Pinnularia
- Surirella
- Synedra
- Tabellaria
- Thalassiosira

Chlorophyta (green)
- Actinastrum
- Allorgeia
- Ankistrodesmus
- Chlamydomonas
- Chlorella
- Chlorogonium
- Chlosteriopsis
- Closterium
- Coelastrum
- Conochaete
- Cosmarium
- Crucigenia
- Desmococcus
- Dictyosphaerium
- Dimorphococcus
- Golenkinia
- Golenkiniopsis
- Kirchneriella
- Chodatella
- Micractinium
- Nephrocytium
- Oedogonium
- Oocystis
- Pediastrum
- Planktosphaeria
- Polyedriopsis
- Radiococcus
- Rhizoclonium
- Scenedesmus
- Schroederia
- Selenastrum
- Sphaerocystis
- Spirogyra
- Staurastrum
- Synura
- Tetraedron
- Tetrallantos
- Treubaria
- Ulothrix
- Westella

From Iwai, T., unpublished data, 1981.

Table 11
FAUNA COMMONLY OBSERVED IN PRAWN PONDS IN HAWAII

Class	Order (family)	Common name	Scientific name
Crustacean	Amphipoda	Scuds	*Gammarus* sp.
	Cladocera	Water fleas	*Daphnia* sp.
	Eucopepoda	Copepods	
	Podocopa[a]	Seed shrimps	
Insecta	Coleoptera	Beetles	
	Diptera[a]	Midge flies, mosquitoes	*Chironomus* sp.
	Hemiptera	Water strider	
	Odonata	Dragonflies, damsel flies	
Mollusca	Mollusca	Snails	*Fossaria ollula*
		Red ramshorn snail	*Planorbis cornrus*
Annelida	Oligochaeta[a]	Red worms	*Tubifex* sp.
Osteichthyes	Cyprinodontiformes	Moons, platys	*Xiphophorus* sp.
	(Poeciliidae)	Mollies	*Poecilia* sp.
		Swordtails	*Xiphophorus helleri*
		Guppies	*P. reticulata*
		Mosquito fish[a]	*Gambusia affinis*
Amphibia	Anura (Ranidae)	Common bullfrog	*Rana catesbeiana*
		Wrinkled bullfrog	*R. rugosa*
	Anura (Bufonidae)	Toad	*Bufo marinas*

[a] Predominant orders observed.

Nutritional Requirements

The state-of-knowledge of optimal dietary requirements for commercial grow-out of *Macrobrachium* is largely based on empirical and economical approaches to feed formulation. Least-cost prawn feed formulations[77] are not possible due to lack of some of the basic information, i.e., nutrient requirements of the animal and availability of nutrients to the animal from various feed ingredients.

Utilizable qualitative and quantitative experimental information is scarce, despite numerous studies identifying the area as a major research priority.[14,16,17] Moreover, differences in experimental design make comparisons between studies difficult.[14] Important considerations include qualitative and quantitative requirements for proteins and amino acids, fats and fatty acids, carbohydrates, energy, vitamins, minerals, and fiber. For earthenpond culture of *Macrobrachium*, the role of natural productivity is also important and largely undefined.

Protein and Amino Acid Requirements

The qualitative protein requirements for juvenile and adult prawns have been investigated utilizing feeds made from conventional feedstuff ingredients. The source, dietary level, and amino acid composition of the proteins in relation to prawn nutrition have received the most attention. Several protein sources have been tested; these include squid, soybean meal, shrimp meal, and several types of fish meal.[78-80] Other research has tested materials indigenous to tropical islands, such as koa haole (*Leucaena luecocephala*) and copra meal with satisfactory results.[81]

According to New,[14] a consensus of opinion indicates that the optimal level of dietary protein for shrimp and prawns may lie between 27 and 35%. Research results suggest that the requirement for juvenile *Macrobrachium* may be somewhat higher than for older animals,[50] due to the animal being in a high-growth, low-maintenance phase. Millikin and associates[82] found that after 14 weeks, prawns fed a 40% protein diet produced a significantly higher cumulative weight gain than prawns fed diets containing 49, 32, and 23% protein. However, they noted that optimum protein requirements will vary considerably with alterations of the dietary amino acid profile, variation in dietary supplementation of macro- and micronutrients other than protein, and changes in selected environmental factors.

Current feeds utilized by the Hawaiian prawn industry reflect available research results to a limited extent (Tables 6 and 7). Soybean and meat and bone meal are the major contributors of pellet protein. Ground corn, which is approximately 9% protein,[78] makes up the bulk of the prawn feed.

Similarly, limited information is available on amino acid requirements for juvenile and adult prawns. Qualitatively, essential dietary amino acids for *Macrobrachium* appear similar to other animals.[83,84] Observed variation in this pronounced pattern noted by Watanabe,[83] i.e., apparent ability to synthesize lysine, may be due to artifacts in the experimental design.[85]

Quantitative amino acid requirements for *Macrobrachium* remain undefined; however, of interest is the work of Farmanfarmaian and Lauterio.[86] These researchers tested the amino acid balance of a commercially available pellet (Purina® Marine Ration 25) and found that several essential amino acids were in insufficient supply in the ration. Ration performance was appreciably improved by addition of an algin binder and fortification by 1% arginine, phenylalanine, leucine, and isoleucine.

New[14] provides a summary comment that while it is unlikely that any multiingredient shrimp ration will be qualitatively deficient in any of the 11 essential amino acids, it is still impossible to formulate rations with a balanced amino acid profile. Current diets are certain to be quantitatively deficient in some amino acids while providing excess of others. Some aquatic animal nutritionists tentatively suggest that amino acid composition of formulated diets should closely mimic the amino acid pattern of the cultured species.[87,89] Amino acid composition of *Macrobrachium* can be found by Farmanfarmaian and Lauterio.[89]

Fats and Fatty Acid Requirements

Fats in animal diets are used as energy and stored as depot lipid or incorporated into phospholipids in vital tissues. Very little has been published on the qualitative and quantitative lipid and fatty acid requirements of juvenile and adult *Macrobrachium*. Generally, it is believed that crustaceans cannot tolerate high levels (in excess of 10% of the diet) of dietary fat, when added in the form of fish oils (menhaden and cod), vegetable oil (corn), and beef tallow.[90-92] Moreover, most commercial aquaculture diets do not exceed 8% lipid largely because of manufacturing problems encountered when high levels of lipid supplementation are attempted.[12] Existing commercial diet formulations for *Macrobrachium* culture in Hawaii generally reflect these suppositions (Table 7).

Joseph and Williams[93] and Sandifer and Joseph[94] demonstrated superior growth in juvenile prawns fed diets (either a commercial shrimp food or a semidefined diet) supplemented with 3% shrimp head oil, which is a rich source of linolenic fatty acids. Results suggest that dietary ω3 fatty acids may be retained while ω6 fatty acids are metabolized for energy production. Additional observations indicated that dietary fatty acids of the linoleic family (principally 18:2ω6) inhibited growth. Research on two marine crustacea, *Penaeus japonicus* and *Homarus americanus,* supports the conclusion drawn for *Macrobrachium,* as these species have dietary requirements for both 18:2ω6 and 18:3ω3 fatty acids.[95] No animal has demonstrated the ability to convert through biosynthesis one family of fatty acids to another, e.g., oleic to linoleic,[12] therefore, these components must be included in the diet.

Cholesterol Requirement

Macrobrachium, like other crustacea, may have a dietary requirement for cholesterol; however, there are no species-specific data available. Cholesterol is thought to be a precursor of important steroid, brain, and molting hormones and of vitamin D in shrimp. Kanazawa et al.[96] found a dietary cholesterol requirement for *P. japonicus* of about 0.5% of the diet; but Deshimaru and Kuroki[97] found that a level of 2.1% in the diet achieved the best growth for the same species. Dietary levels of 0.5% are generally not difficult to achieve in multiingredient diets.[14]

Carbohydrate Requirement

Based on data gathered using penaeid shrimps, crustaceans appear to be able to utilize complex polysaccharides (starches) more efficiently than simple sugars.[92,98] Differing levels and types of dietary carbohydrate were found to affect body composition. Clifford and Brick[99] present convincing evidence for juvenile *Macrobrachium* that a dietary ratio for fat to carbohydrate of 1:3 to 1:4 results in more efficient utilization of dietary protein than ratios of 1:1 and 1:2. Balazs and Ross[80] have conducted dietary experiments with juvenile prawns using a high amylose starch binder with acceptable survival and conversion ratios, though growth rates were inferior to other diets.

Energy Requirement

The energy value of a diet affects the partitioning and utilization of the protein, lipid, and carbohydrate components of a feed. Protein-to-energy ratios, protein and energy densities, and available nonprotein energy (lipid and carbohydrate) sources are major dietary factors in achieving maximum tissue growth and maximum efficiencies of dietary utilization by crustacea. Insufficient nonprotein components in the diet can lead to metabolism of dietary proteins for energy.[100,101]

Limited data are available on physiological energetics and energy requirements of juvenile and adult *Macrobrachium*.[102-104] Sze,[105] reporting a 15.8% (dry weight) body lipid level in juvenile *Macrobrachium,* referred to the contention of Neiland and Scheer[106] that protein, rather than fat or carbohydrate, was the primary energy source for crustacea. Clifford[107] observed the metabolic responses of juvenile *M. rosenbergii* to various levels of dietary protein, lipid, and carbohydrate and concluded that higher carbohydrate levels resulted in a greater efficiency of protein utilization. He further concluded that the maximum protein-sparing effect of nonprotein energy sources is achieved at a lipid-to-carbohydrate ratio of 1:4 (25% protein diet), and optimum growth was achieved at a protein-to-energy ratio of 97.4 mg protein per kilocalorie.

Mineral, Vitamin, and Fiber Requirements

Dietary mineral and vitamin requirements have not been determined for juvenile and adult *Macrobrachium*. Sze[105] and Iwai[108] determined the ash content of prawns to be as high as 15.9 and 21.3% of the dry weight, respectively, suggesting that mineral nutrition may be important to overall animal health and well being. Penaeid shrimp experimental data indicate that the ratio of phosphorus to calcium in the diet may be significant.[87,109] Deshimaru and Yone[110] discuss the mineral requirements for *P. japonicus*. Since pond waters in Hawaii may be relatively low in mineral content,[111] dietary mineral consumption may be of greater significance than for brackish water or marine crustacea.

Currently utilized *Macrobrachium* feeds contain ingredients with various inherent concentrations of minerals. For example, ground corn may have 1.3% and cottonseed meal 6.7% ash content on a dry weight basis.[21] Moreover, fish and meat and bone meals also contain substantial quantities of minerals, though the availability of all these sources to the prawns is unknown. Certain feeds used as prawn diets utilize supplemental mineral premixes (Table 6). The composition of premixes used can be found in Table 8. Other premix formulas for experimental shrimp diets are found by New.[14] Differences between the shrimp and the prawn formulations are apparent with the *Macrobrachium* premix having no calcium, phosphorus, and potassium components. These dietary minerals can probably be supplied by the other feed ingredients and from natural productivity.

No qualitative or quantitative studies have been carried out on the vitamin requirements of *Macrobrachium*. Limited qualitative data are available for crustacea. Reportedly, B group, C, and E vitamins are required in crustacean diets.[17] Dietary needs for vitamin C have conclusively been demonstrated for a penaeid shrimp.[113,114] Vitamin D may be partly ingested in the diet but also can be synthesized. Vitamin K may be antagonistic to some species of crustacea.[17] The importance of vitamin A in shrimp and prawn diets is suggested by the importance of its precursor substances, the carotenoids, to pigmentation.[95]

As in the case of minerals, dietary vitamins are included in shrimp diets as premixes. New[14] presents the composition of several premixes. The composition of the Hawaii prawn premix can be found in Table 8. Again various meal components of the prawn feed provide significant levels of vitamins,[21] though their availability to the animal is unknown. It is important to note that commercially, prawn vitamin premixes have been formulated according to specific requirements established for domesticated terrestrial species and may not only be wasteful but counterproductive.[14]

In addition to mineral and vitamin requirements for *Macrobrachium*, another neglected area of research has been the role of dietary fiber in prawn nutrition.[85] Cellulose is often incorporated into experimental crustacean rations as a dietary filler, e.g., see Clifford.[107] A recent study on *Macrobrachium* concluded that: (1) replacement of up to 30% precooked starch with fiber cellulose resulted in no detriment to growth, (2) diets of up to 20% cellulose may stimulate growth in relatively mature prawns, (3) an endogenous source of cellulase activity has been suggested with the highest levels being in adult prawns, and (4) dietary cellulose could successfully be included in nutritionally adequate and more cost-effective diets for juveniles.[115] Crude fiber values for Hawaii prawn diets are 5% or less (Table 7) suggesting an area for possible feed improvement and cost reduction.

Role of Natural Pond Productivity

The natural productivity in earthen grow-out ponds is an additional source of nutrition for juvenile and adult prawns. Weidenbach[72] concluded from stomach content analysis that: (1) prawns ingest commercial pellets when available, (2) natural foods (natural vegetation and aquatic animal remains) are frequent dietary components of prawns, regardless of the presence of commercial pellets, and (3) prawns adjust to an absence of pellets with increased consumption of available vegetation. AFRC extension specialists report that prawn ponds which consistently produce higher yields also have greater availability of natural foods.[116]

Currently used prawn feeds in Hawaii leach nutrients and break down rapidly in the pond environment, and how much nutrient is actually consumed and utilized,[72,117] directly or indirectly by the prawn for maintenance and growth, is not known. Unconsumed particulate and/or dissolved feeds contribute significant amounts of organic carbon and nitrogen and inorganic phosphorus to the food web in the pond and, in particular, to the microbial/detrital component.[117,118] Excessive nutrient loading from dissolved feed inputs may contribute to periodic phytoplankton blooms, water quality problems, and the need for increased pond flushing rates.[119]

Qualitative experimental evidence supplemented by casual observations suggests that the natural productivity could significantly contribute to prawn production. Mosquito fish (*Gambusia* sp.) can be caught and consumed by juvenile and adult prawns.[120] Various benthic epifauna and infauna present in ponds (Table 11) are periodically consumed and could be significant dietary components. Weidenbach[72] found that prawn pieces and mosquito fish remains occurred more frequently in the stomachs of large prawns (20.0 to 75.0 g wet weight) while chironomid larvae and terrestrial ants occurred more frequently in the stomachs of small prawns (5.0 to 19.9 g wet weight). Bacteria-rich aggregates of detritus prevalent in eutrophic pond environments are available for consumption.[117,121] Experimental evidence suggests that juvenile prawns grow poorly on solely aquatic plant diets[122] or in simulated natural environments fertilized by cow manure or pulverized feed.[118] However, prawns have also been successfully pond cultured in polyculture with Chinese carps with no supplemental feeding but with enrichment with swine manure.[123]

Experimental results suggest that a major portion of macronutrients can be provided by applied feeds and that natural or stimulated productivity may only supplement "limiting factor" micronutrients.[118] This conclusion is supported by results from warm water, omnivorous fish culture.[124,125] Lovell[125] concludes that for species that feed lower on the food chain, the nutritional benefits of natural food will be greater and the opportunity for altering the quality and the quantity of the supplemental feed to achieve economic optima will be correspondingly greater.

SUMMARY

In summary, feeds and feeding practices for prawn farming in Hawaii involve a combination of live, fresh, and pelleted feeds. Larval production is dependent on live *Artemia* nauplii and prepared fish flesh, which are fed to larvae grown in tanks containing a mixture of phytoplankton and various microorganisms. Broodstock maintenance, experimental nursery production, and grow-out production utilize a combination of commercial feeds and available natural productivity. Commercial feed formulations used in Hawaii appear to be based on practical and on economic considerations, rather than the limited nutritional information that is available. Applied feeds are known to break down rapidly in water and the amount which is consumed directly by prawns is unknown. Also, the nutritional role of natural productivity is not well defined and, hence, management of desirable components is generally not possible.

Refinements in feeds and feeding practices should address several major areas. Low-cost nutritious feeds which are reasonably water stable, palatable, and digestible are needed. More knowledge of the diel patterns of food consumption of various life stages in commercial ponds is needed, particularly in the continuous culture system which may have several age groups present at any one time. Identification of those components of the natural productivity which have major nutritional significance and can be managed to reduce the amounts of applied feeds needed appear highly desirable for pond grow-out systems.

Knowledge of the qualitative and quantitative nutritional requirements for *M. rosenbergii* at various life stages is limited. Experiments with juvenile prawns have provided the greatest amount of information on macronutrient requirements. Prawns are omnivores and supplemental natural productivity is available to some extent at each phase of production, perhaps contributing to the observation that research efforts have not emphasized defining prawn macro- and micronutrient requirements. The Hawaii approach may be described as providing a broad spectrum of dietary components such that nutritional adequacy is reasonably assured.

Feeding and nutritional information which could greatly assist commercial prawn farming in Hawaii should focus on the areas of (1) defining the daily requirements of macronutrients, (2) determining the digestibility of nutrients in conventional feedstuffs processed in different ways, (3) establishing the qualitative protein, total lipid, and essential fatty acid requirements, as well as the total energy requirements for various life stages of prawns, (4) determining the optimum protein-to-energy ratios in diets for various life stages, and (5) establishing the nutritional value of the natural productivity consumed by the prawns, including the development of methods of selectively managing desirable components. Results in these areas may allow development of a least-cost feeding approach which combines applied pelleted rations and available natural supplements. Intensification of existing earthen pond production systems will require greater attention to development of nutritionally complete applied feeds.

REFERENCES

1. **Ling, S.-W. and Costello, T. J.,** Review of culture of freshwater prawns, *FAO Tech. Conf. on Aquaculture,* FIR:AQ/Conf./76/R, Kyoto, Japan, 1976, 29.
2. **Fujimura, T.,** Notes on progress made in culturing technique for *Macrobrachium rosenbergii,* in *Proc. Indo Pac. Fish. Counc.,* 12th Session, 1966.
3. **Fujimura, T. and Okamoto, H.,** Notes on progress made in developing a mass culturing technique for *Macrobrachium rosenbergii* in Hawaii, in *Proc. Indo Pac. Fish. Counc.,* 14th Session, Bangkok, Thailand, 1970.
4. **Fujimura, T. and Okamoto, H.,** Notes on progress made in developing a mass culturing technique for *Macrobrachium rosenbergii* in Hawaii, in *Coastal Aquaculture in the Indo. Pac. Reg.,* Pillay, T. V. R., Ed., Fishing News Books, Farnham, Surrey, England, 1972, 313.
5. **Fujimura, T.,** Development of a Prawn Industry, Development of a Rearing Technique for the Giant Long-Legged Prawn *Macrobrachium rosenbergii,* Quarterly Progress Report to the National Marine Fisheries Service, 1974.
6. **Shang, Y. C.,** *Economic Feasibility of Freshwater Prawn Farming in Hawaii,* Economic Research Center, University of Hawaii, Honolulu, 1972.
7. **Shang, Y. C. and Fujimura, T.,** The production economics of freshwater prawn *(Macrobrachium rosenbergii)* farming in Hawaii, *Aquaculture,* 11, 99, 1977.
8. **Lee, S. R.,** *The Hawaiian Prawn Industry: a Profile,* Department of Planning and Economic Development, State of Hawaii, Honolulu, 1979.
9. **Shang, Y. C.,** Freshwater Prawn *(Macrobrachium rosenbergii)* Production in Hawaii: Practices and Economics, UNIHI-SEAGRANT-MR-81-07, 1981.
10. **State of Hawaii,** *Aquaculture Development for Hawaii,* Department of Planning and Economic Development, State of Hawaii, Honolulu, 1978.
11. **Bardach, J., Ryther, J. H., and McLarney, W. O.,** *Aquaculture, The Farming and Husbandry of Freshwater and Marine Organisms,* Wiley-Interscience, New York, 1972, 10.
12. **Stickney, R. R.,** *Principles of Warmwater Aquaculture,* Wiley-Interscience, New York, 1979, chap. 5.
13. **McVey, J.,** Current developments in the penaeid shrimp culture industry, *Aquaculture Mag.,* July-August, 20, 1980.
14. **New, M. B.,** A review of dietary studies with shrimp and prawns, *Aquaculture,* 9, 101, 1976.
15. **Wickens, J. F.,** Prawn biology and culture, *Oceanogr. Mar. Biol. Ann. Rev.,* 14, 435, 1976.
16. **Hanson, J. A. and Goodwin, H. L.,** *Shrimp and Prawn Farming in the Western Hemisphere,* Dowden, Hutchinson and Ross, Stroudsburg, Pa., 1977, chap. 8.
17. **New, M. N.,** A bibliography of shrimp and prawn nutrition, *Aquaculture,* 21, 101, 1980.
18. **Shigueno, K.,** *Shrimp Culture in Japan,* Association for International Tech. Promotion, Tokyo, 1975.
19. **Cowey, C. B. and Sargent, J. R.,** Fish nutrition, in *Advances in Marine Biology,* Vol. 10, Russell, F. S. and Yonge, M., Eds., Academic Press, New York, 1972, 383.
20. **Halver, J. E.,** *Fish Nutrition,* Academic Press, New York, 1972.
21. National Research Council, *Nutrient Requirements of Trout, Salmon and Catfish,* National Academy of Sciences, Washington, D.C., 1973.
22. **National Research Council,** *Nutrient Requirements of Warmwater Fishes,* National Academy of Sciences, Washington, D.C., 1977.
23. **Halver, J. E. and Tiews, K.,** *Finfish Nutrition and Fishfeed Technology,* Vols. 1 and 2, Heenemann Verlagsgesellschaft mbH, Berlin, 1980.
24. **Malecha, S. R.,** Genetics and selective breeding, in *Shrimp and Prawn Farming in the Western Hemisphere,* Hanson, J. A. and Goodwin, H. L., Eds., Dowden, Hutchinson and Ross, Stroudsburg, Pa., 1977, 328.
25. **Wickens, J. F. and Beard, T. W.,** Observations on the breeding and growth of the giant freshwater prawn *Macrobrachium rosenbergii* (de Man) in the laboratory, *Aquaculture,* 3, 159, 1974.
26. **Malecha, S. R.,** Development and general characterization of genetic stocks of *Macrobrachium rosenbergii* and their hybrids for domestication, *Sea Grant Q.,* 2, 1, 1980.
27. **Middleditch, B. S., Missler, S. R., Ward, D. G., McVey, J. P., Brown, A., and Lawrence, A. L.,** Maturation of penaeid shrimp: dietary fatty acids, *Proc. World. Maricul. Soc.,* 10, 472, 1979.
28. **Sandifer, P. A., Hopkins, J. S., and Smith, T. I. J.,** Status of Macrobrachium hatcheries, 1976, in *Shrimp and Prawn Farming in the Western Hemisphere,* Hanson, J. A. and Goodwin, H. L., Eds., Dowden, Hutchinson and Ross, Stroudsburg, Pa., 1977, 220.
29. **Miyauchi, D., Kudo, G., and Patashnik, M.,** Surimi, a semi-processed wet fish protein, *Mar. Fish. Rev.,* 35, 7, 1973.
30. **Austin, B. and Allen, A.,** Microbiology of laboratory-hatched brine shrimp *(Artemia), Aquaculture,* 26, 369, 1981.

31. **Dugan, C. C., Hagood, R. W., and Frakes, T. A.,** Development of spawning and mass larval rearing techniques for brackish water shrimps of the genus *Macrobrachium, Fla. Marine Res. Publ.,* 12, 2, 1975.
32. **Ling, S. W.,** The general biology and development of *Macrobrachium rosenbergii,* in Proc. World Conf. on Shrimps and Prawns, 1967, FAO, Rome, 1969, 607.
33. **Gibson, R. T.,** personal communication, 1982.
34. **Neilands, J. B., Sirny, R. J., Sohljell, I., Strong, F. M., and Elvelyem, C. A.,** Canned food chemical composition II. Amino acid content of fish and meat products, *J. Nutr.,* 39, 187, 1949.
35. **Persoone, G., Sorgeloos, P., Roels, O., and Jaspers, E., Eds.,** *The Brine Shrimp Artemia, Vol. 3, Ecology, Culturing, Use in Aquaculture,* Universa Press, Wetteren, Belgium, 1980.
36. **Benijts, F., Van Voorden, E., and Sorgeloos, P.,** Changes in the biochemical composition of the early larval stages of the brine shrimp, *Artemia salina* L., in *Proc. 10th Eur. Symp. on Mar. Biol.,* Vol. 1, Persoone, G. and Jaspers, E., Eds., Universa Press, Wetteren, Belgium, 1976, 1.
37. **Seidel, C. R., Kryznowek, J., and Simpoon, K. L.,** International study on *Artemia.* XI. Amino acid composition and electrophoretic protein patterns of *Artemia* from five geographical locations, in *The Brine Shrimp Artemia,* Vol. 3, Persoone, G., Sorgeloos, P., Roels, O., and Jaspers, E., Eds., Universa Press, Wetteren, Belgium, 1980, 375.
38. **Watanabe, T., Oowa, F., Kitajima, C., and Fujita, S.,** Nutritional quality of brine shrimp, *Artemia salina,* as a living feed from the viewpoint of essential fatty acids for fish, *Bull. Jpn. Soc. Sci. Fish.,* 44, 1115, 1978.
39. **Schauer, P. S., Johns, D. M., Olney, C. E., and Simpson, K. L.,** International study on *Artemia.* IX. Lipid level, energy content and fatty acid composition of the cysts and newly hatched nauplii from five geographical strains of *Artemia,* in *The Brine Shrimp Artemia,* Vol. 3, Persoone, G., Sorgeloos, P., Roels, O., and Jaspers, E., Eds., Universa Press, Wetteren, 1980, 365.
40. **Fujita, S., Watanabe, T., and Kitajima, C.,** Nutritional quality of *Artemia* from different localities as a living feed for marine fish from the viewpoint of essential fatty acids, in *The Brine Shrimp Artemia,* Vol. 3, Persoone, G., Sorgeloos, P., Roels, O., and Jaspers, E., Eds., Universa Press, Wetteren, 1980, 277.
41. **Moller, T. H.,** Feeding behaviour of larvae and post-larvae of *Macrobrachium rosenbergii* (de Man) (Crustacea: Palaemonidae), *J. Exp. Mar. Biol. Ecol.,* 35, 251, 1978.
42. **Manzi, J. J. and Maddox, M. B.,** Requirements for *Artemia* nauplii in *Macrobrachium rosenbergii* (de Man) larviculture, in *The Brine Shrimp Artemia,* Vol. 3, Persoone, G., Sorgeloos, P., Roels, O., and Jaspers, E., Eds., Universa Press, Wetteren, 1980, 313.
43. **Dugan, C. C., Hagood, R. W., and Frakes, T. A.,** Development of spawning and mass larval rearing techniques for brackish water shrimps of the genus *Macrobrachium, Fla. Mar. Res. Publ.,* 12, 3, 1975.
44. Aquacop, *Macrobrachium rosenbergii* culture in Polynesia: progress on developing a mass intensive larval rearing technique in clear water, *Proc. World Maricul. Soc.,* 8, 311, 1977.
45. **Menasveta, P. and Piyatiratitvokul, S.,** A comparative study on larviculture techniques for the giant freshwater prawn, *Macrobrachium rosenbergii* (de Man), *Aquaculture,* 20, 239, 1980.
46. **Murai, T. and Andrews, J. W.,** Comparison of feeds for larval stages of the giant prawn *(Macrobrachium rosenbergii), Proc. World Maricul. Soc.,* 9, 189, 1978.
47. **Sick, L. V. and Beaty, H.,** Development of formula foods designed for *Macrobrachium rosenbergii* larval and juvenile shrimp, *Proc. World Maricul. Soc.,* 6, 89, 1975.
48. **Sick, L. V.,** Selected studies of protein and amino acid requirements for *Macrobrachium rosenbergii* larvae fed neutral density diets, in Proc. 1st Int. Conf. on Aquaculture Nutrition, Delaware, 1976, 263.
49. **Conover, R. J. and Corner, E. D. S.,** Respiration and nitrogen excretion by some marine zooplankton in relation to their life cycles, *J. Mar. Biol. Assoc. U.K.,* 48, 49, 1968.
50. **Colvin, L. B. and Brand, C. W.,** The protein requirement of penaeid shrimp at various life-cycle stages in controlled environment systems, *Proc. World Maricul. Soc.,* 8, 821, 1977.
51. **Clifford, H. and Brick, R.,** Protein utilization in the freshwater shrimp *Macrobrachium rosenbergii, Proc. World Maricul. Soc.,* 9, 195, 1978.
52. **Maddox, M. B. and Manzi, J. J.,** The effects of algal supplements on static system culture of *Macrobrachium rosenbergii* larvae, *Proc. World Maricul. Soc.,* 7, 677, 1976.
53. **Manzi, J. J. and Maddox, M. B.,** Algal supplement enhancement of static and recirculating system culture of *Macrobrachium rosenbergii* larvae, *Helgolander Wiss. Meeresunters.,* 28, 447, 1977.
54. **Manzi, J. J., Maddox, M. B., and Sandifer, P. A.,** Algal supplement enhancement of *Macrobrachium rosenbergii* larviculture, *Proc. World Maricul. Soc.,* 8, 207, 1977.
55. **Maddox, M. B. and Manzi, J. J.,** The effects of algal supplements on static system culture of *Macrobrachium rosenbergii* (de Man) larvae, *Proc. World Maricul. Soc.,* 7, 677, 1976.
56. **Cohen, D., Finkel, A., and Sussman, M.,** On the role of algae in the larviculture of *Macrobrachium rosenbergii, Aquaculture,* 8, 199, 1976.
57. **Joseph, J. D.,** Assessment of the nutritional role of algae in the culture of larval prawns, *Macrobrachium rosenbergii, Proc. World Maricul. Soc.,* 8, 853, 1977.

58. **Sick, L. V. and Beaty, H.,** Culture techniques and nutrition studies for larval stages of the giant prawn, *Macrobrachium rosenbergii, Ga. Mar. Sci. Cont. Tech. Rep. Ser.*, 74-75, 1974.
59. **Stephenson, M. J. and Knight, A. W.,** Growth respiration and caloric content of larvae of the prawn *Macrobrachium rosenbergii, Comp. Biochem. Physiol.*, 66A, 385, 1980.
60. **Jones, D. A., Kanazawa, A., and Rahman, S. A.,** Studies on the presentation of artificial diets for rearing the larvae of *Penaeus japonicus* Bate, *Aquaculture*, 17, 33, 1979.
61. **Smith, T. I. J. and Sandifer, P. A.,** Increased production of tank-reared *Macrobrachium rosenbergii* through use of artificial substrates, *Proc. World Maricul. Soc.*, 6, 55, 1975.
62. **Sandifer, P. A. and Smith, T. I. J.,** Experimental aquaculture of the Malaysian prawn, *Macrobrachium rosenbergii* (de Man) in South Carolina (U.S.A.), FAO Tech. Conf. on Aquaculture, F.I.R.:AQ/Conf./76/E, Kyoto, Japan, 1976, 3.
63. **Sandifer, P. A. and Smith, T. I. J.,** Intensive rearing of post-larval Malaysian prawns *(Macrobrachium rosenbergii)* in a closed-cycle nursery system, *Proc. World Maricul. Soc.*, 8, 225, 1977.
64. **Malecha, S. R., Polovina, J. J., and Moav, R.,** A multi-rotational stocking and harvesting system for pond culture of the freshwater prawn, *Macrobrachium rosenbergii, Aquaculture*, in press.
65. **Fujimoto, M.,** personal communication, 1982.
66. **Smith, T. I. J. and Sandifer, P. A.,** Development and potential of nursery systems in the farming of Malaysian prawns, *Macrobrachium rosenbergii, Proc. World Maricul. Soc.*, 10, 369, 1979.
67. **McSweeney, E. S.,** Intensive culture systems, in *Shrimp and Prawn Farming in the Western Hemisphere*, Hanson, J. A. and Goodwin, H. L., Eds., Dowden, Hutchison and Ross, Stroudsburg, Pa., 1977, 255.
68. **Scudder, K. M., Pasanello, E., Krafsur, J., and Ross, K.,** Analysis of locomotory activity in juvenile giant Malaysian prawns, *Macrobrachium rosenbergii* (de Man) (Decapoda, Palaemonidae), *Crustacea*, 40, 31, 1981.
69. **Fujimoto, M., Fujimura, T., and Kato, K.,** An idiot's guide to prawn ponds, in *Shrimp and Prawn Farming in the Western Hemisphere*, Hanson, J. A. and Goodwin, H. L., Eds., Dowden, Hutchinson and Ross, Stroudsburg, Pa., 1977, 237.
70. **Smith, T. I. J., Sandifer, P. A., and Smith, M. H.,** Population structure of Malaysian prawns, *Macrobrachium rosenbergii* (de Man), reared in earthen ponds in South Carolina, 1974—1976, *Proc. World Maricul. Soc.*, 9, 21, 1978.
71. **Lee, P. G., Blake, N. J., and Rodrick, G. E.,** A quantitative analysis of digestive enzymes for the freshwater prawn *Macrobrachium rosenbergii, Proc. World Maricul. Soc.*, 11, 392, 1980.
72. **Weidenbach, R. P.,** Dietary components of prawns reared in Hawaiian ponds, *Proc. of the Giant Prawn Conf. 1980*, Bangkok, International Foundation for Science, 1980, Rept. 9.
73. **Goodwin, H. L. and Hanson, J. A.,** *The Aquaculture of Freshwater Prawns (Macrobrachium species)*, The Oceanic Institute, Waimanalo, Hawaii, 1975.
74. **Patwardhan, S. S.,** On the structure and mechanism of the gastric mill in decapoda. V. The structure of the gastric mill in *Natantous macrura Caridea, Proc. Ind. Acad. Sci. B.*, 11, 693, 1935.
75. **Patwardhan, S. S.,** On the structure and mechanism of the gastric mill in decapoda. VI. The structure of the gastric mill in *Natantous macrura Penaeidea* and *Stenopidea, Proc. Ind. Acad. Sci. B.*, 12, 155, 1935.
76. **Forester, J. M. R.,** Some methods of binding prawn diets and their effects on growth and assimilation, *J. Cons. Int. Explor. Mer.*, 34, 200, 1972.
77. **Lovell, T.,** Least-cost fish feeds, *Commer. Fish Farmer Aquaculture News*, May-June, 26, 1976.
78. **Balazs, G. H., Ross, E., and Brooks, C. C.,** Preliminary studies on the preparation and feeding of crustacean diets, *Aquaculture*, 2, 269, 1973.
79. **Balazs, G. H., Ross, E., Brooks, C. C., and Fujimura, T.,** Effects of protein source and level on growth of the captive freshwater prawn *(Macrobrachium rosenbergii), Proc. World Maricul. Soc.*, 5, 1, 1974.
80. **Balazs, G. H. and Ross, E.,** Effect of protein source and level on growth and performance of the captive freshwater prawn, *M. rosenbergii, Aquaculture*, 7, 299, 1976.
81. **Glude, J.,** Nutritional consideration in the culture of tropical species, *Proc. 1st Int. Conf. on Aquaculture Nutr.*, p. 107, 1975.
82. **Millikin, M. R., Fortner, A. R., Fair, P. H., and Sick, L. V.,** Influence of dietary protein concentration on growth, feed conversion and general metabolism of juvenile prawn *(Macrobrachium rosenbergii), Proc. World Maricul. Soc.*, 11, 382, 1980.
83. **Watanabe, W. O.,** Identification of the Essential Amino Acids of the Freshwater Prawn, *Macrobrachium rosenbergii*, M.S. thesis, University of Hawaii, Honolulu, 1975.
84. **Miyajima, L. S., Broderick, G. A., and Reimer, D. R.,** Identification of the essential amino acids of the freshwater shrimp, *Macrobrachium ohione, Proc. World Maricul. Soc.*, 8, 245, 1977.
85. **Biddle, G. N.,** The nutrition of *Macrobrachium* species, in *Shrimp and Prawn Farming in the Western Hemisphere*, Hanson, J. A. and Goodwin, H. L., Eds., Dowden, Hutchinson and Ross, Stroudsburg, Pa., 1977, 272.
86. **Farmanfarmaian, A. and Lauterio, T.,** Amino acid supplementation of feed pellets of the giant shrimp *(Macrobrachium rosenbergii), Proc. World Maricul. Soc.*, 10, 674, 1979.

87. **Deshimaru, O. and Shigeno, K.,** Introduction to the artificial diet for prawn, *Penaeus japonicus, Aquaculture,* 1, 115, 1972.
88. **Colvin, P. M.,** Nutritional studies on penaeid prawn: protein requirements in compounded diets for juvenile *Penaeus indicus, Aquaculture,* 7, 315, 1976.
89. **Farmanfarmaian, A. and Lauterio, T.,** Amino acid composition of the tail muscle of *Macrobrachium rosenbergii* — comparison to amino acid patterns of supplemented commercial feed pellets, *Proc. World Maricul. Soc.,* 11, 454, 1980.
90. **Andrews, J. W., Sick, L. V., and Baptist, G. J.,** The influence of dietary protein and energy levels on growth and survival of penaeid shrimp, *Aquaculture,* 1, 341, 1972.
91. **Forster, J. R. M. and Beard, T. W.,** Growth experiments with the prawn *Palaemon serratus* fed with fresh foods, *Fish. Invest. Ser.,* 2, 27, 1973.
92. **Sick, L. V. and Andrews, J. W.,** Effects of selected dietary lipids, carbohydrates and proteins on the growth, survival, and body composition of *Penaeus duorarum, Proc. World Maricul. Soc.,* 4, 263, 1973.
93. **Joseph, J. D. and Williams, J. E.,** Shrimp head oil: a potential feed additive for mariculture, *Proc. World Maricul. Soc.,* 6, 147, 1975.
94. **Sandifer, P. A. and Joseph, J. D.,** Growth response and fatty acid composition of juvenile prawn *(Macrobrachium rosenbergii)* fed a prepared ration augmented with shrimp head oil, *Aquaculture,* 8, 129, 1976.
95. **Castell, J. D.,** Fatty acid metabolism of crustaceans, *Proc. 2nd Int. Conf. on Aquaculture Nutrition,* 1982, in press.
96. **Kanazawa, A., Tanaka, N., Teshima, S., and Kashiwada, K.,** Nutritional requirements for prawn. II. Requirement for sterols, *Bull. Jpn. Soc. Sci. Fish.,* 37, 211, 1971.
97. **Deshimaru, O. and Kuroki, K.,** Studies on a purified diet for prawn. II. Optimum contents of cholesterol and glucosamine in the diet, *Bull. Jpn. Soc. Sci. Fish.,* 40, 421, 1974.
98. **Andrews, J. W. and Sick, L. V.,** Studies on the nutritional requirements of penaeid shrimp, *Proc. World Maricul. Soc.,* 3, 403, 1972.
99. **Clifford, H. C. and Brick, R. W.,** A physiological approach to the study of growth and bioenergetics in the freshwater shrimp *(Macrobrachium rosenbergii), Proc. World Maricul. Soc.,* 10, 701, 1979.
100. **Capuzzo, J. M. and Lancaster, B. A.,** The effects of dietary carbohydrate levels on protein utilization in the American lobster, *Proc. World Maricul. Soc.,* 10, 689, 1979.
101. **Capuzzo, J. M.,** Crustacean bioenergetics, *Proc. 2nd Int. Conf. on Aquaculture Nutrition,* 1982, in press.
102. **Nelson, S. G., Knight, A. W., and Li, H. W.,** The metabolic cost of food utilization and ammonia production by juvenile *Macrobrachium rosenbergii, Comp. Biochem. Physiol.,* 58A, 67, 1977.
103. **Nelson, S. G., Li, H. W., and Knight, A. W.,** 1977, Calorie, carbon and nitrogen metabolism of juvenile *Macrobrachium rosenbergii, Comp. Biochem. Physiol.,* 58, 319, 1977.
104. **Iwai, T.,** A Preliminary Investigation on Oxygen Consumption of *Macrobrachium rosenbergii,* University of Hawaii Sea Grant Prog. Working Paper No. 31, 1978.
105. **Sze, C. P.,** The biochemical composition of juveniles of *Macrobrachium rosenbergii, Malaysian Agric. J.,* 49, 8, 1973.
106. **Neiland, K. A. and Scheer, B. T.,** The influence of fasting and of sinus gland removal on body composition of *Hemigrapsus nudus, Physiol. Comp. Oecol.,* 2, 198, 1953.
107. **Clifford, H. C.,** Bioenergetics and Protein Metabolism in the Freshwater Shrimp *Macrobrachium rosenbergii,* M.S. thesis, Texas A & M University, College Station, 1979.
108. **Iwai, T.,** Energy Transformation and Nutrient Assimilation by the Freshwater Prawn *Macrobrachium rosenbergii* under Controlled Laboratory Conditions, M.S. thesis, University of Hawaii, Honolulu, 1976.
109. **Huner, J. V. and Colvin, L. B.,** A short-term study on the effects of diets with varied calcium: phosphorus ratios on the growth of juvenile shrimp, *Penaeus californiensis:* a short communication, *Proc. World Maricul. Soc.,* 8, 775, 1977.
110. **Deshimaru, O. and Yone, Y.,** Studies on a purified diet for prawn. X. Requirement of prawn for dietary minerals, *Bull. Jpn. Soc. Sci. Fish.,* 44, 970, 1978.
111. **Iwai, T.,** A Limnological Investigation of Selected Water Quality Parameters from a Freshwater Prawn *(Macrobrachium rosenbergii)* Rearing Pond, Oahu, Hawaii, unpublished report, 1979.
112. **Fisher, L. R.,** Vitamins, in *The Physiology of Crustacea,* Vol. 1, Waterman, T. H., Ed., Academic Press, New York, 1960, 259.
113. **Lightner, D. V., Colvin, L. B., Brand, C., and Dawald, D. A.,** Black death, a disease syndrome of penaeid shrimp related to a dietary deficiency of ascorbic acid, *Proc. World Maricul. Soc.,* 8, 611, 1977.
114. **Lightner, D. V., Magarelli, P. C., Hunter, B., and Colvin, L. B.,** Ascorbic acid. II. Wound repair in ascorbic acid deficient shrimp, *Proc. World Maricul. Soc.,* 10, 513, 1979.
115. **Fair, P. H., Fortner, A. R., Millikin, M. R., and Sick, L. V.,** Effects of dietary fiber on growth, assimilation and cellulase activity of the prawn *(Macrobrachium rosenbergii), Proc. World Maricul. Soc.,* 11, 369, 1980.
116. **Iwai, T.,** personal communication, 1982.

117. **Stahl, M. S.,** The role of natural productivity and applied feeds in the growth of *Macrobrachium rosenbergii, Proc. World Maricul. Soc.,* 10, 92, 1979.
118. **Fair, P. H. and Fortner, A. R.,** The role of formula feeds and natural productivity in culture of the prawn, *Macrobrachium rosenbergii, Aquaculture,* 24, 233, 1981.
119. **Laws, E. and Malecha, S. R.,** Application of a nutrient saturated growth model to phytoplankton management in freshwater prawn *(Macrobrachium rosenbergii)* ponds in Hawaii, *Aquaculture,* 24, 91, 1981.
120. **Sukumaran, N. and Kutty, M. N.,** Vulnerability of prey to predation by freshwater prawn, *Macrobrachium malcolmsonii, Aquaculture,* 16, 363, 1979.
121. **Schroeder, G. L.,** Autotrophic and heterotrophic production of microorganisms in intensely manured fish ponds and related fish yields, *Aquaculture,* 14, 303, 1978.
122. **Knight, A. W.,** Laboratory studies on selected nutritional, physical, and chemical factors affecting the growth, survival, respiration, and bioenergetics of the giant prawn *(Macrobrachium rosenbergii),* Water Sci. and Eng. Paper No. 4501, University of California, Davis, 1976.
123. **Malecha, S. R., Buck, D. H., Baur, R. J., and Onizuka, D.,** Polyculture of the freshwater prawn, *Macrobrachium rosenbergii,* Chinese and common carp in ponds enriched with swine manure, *Aquaculture,* in press.
124. **Lovell, T.,** Estimate needed on contribution of pond organisms to fish feed, *Commer. Fish Farmer Aquaculture News,* 3, 32, 1977.
125. **Lovell, T.,** Formulating diets for aquaculture species, *Feedstuffs,* 51, 29, 1979.
126. **Sato, V.,** personal communication, 1982.
127. **Oppenheimer, C. H. and Moreira, G. S.,** Carbon, nitrogen and phosphorous content in the developmental stages of the brine shrimp *Artemia,* in *The Brine Shrimp Artemia,* Vol. 2, Persoone, G., Sorgeloos, P., Roels, O., and Jaspers, E., Eds., Universa Press, Wetteren, 1980, 609.
128. **Waldron's Feed Mill,** personal communication, 1981.
129. **Iwai, T.,** unpublished data, 1981.

LOBSTER NUTRITION

D. E. Conklin, L. R. D'Abramo, and K. Norman-Boudreau

INTRODUCTION

The primary goal of lobster nutrition research at the University of California Bodega Marine Laboratory over the last decade has been the development of information necessary to formulate rations for a commercial culture industry. It is anticipated that these rations would be used with environmentally controlled, technologically sophisticated, intensive systems for growing individually housed lobsters of the genus *Homarus*. While alternative methods of lobster culture have been suggested, such as natural impoundments, polyculture raceways, and suspended cages in productive bays, a nutritional approach to these techniques will not be specifically addressed. As available information regarding nutritional needs of lobsters is still somewhat fragmentary and limited, the following discussion should be supplemented with nutritional information concerning other arthropods. Although taxonomic proximity does not guarantee similar nutritional requirements, a recent review of insect nutrition research[1] combined with information available on shrimp[2,3] will provide a useful and comprehensive summary of arthropod nutrition in general.

COMMERCIAL FEED CONSIDERATIONS

The potential commercial lobster aquaculturist is always concerned with feed, the principal raw material input, and, in particular, wants to answer two questions. Is it nutritionally adequate? How much does it cost? Unfortunately the researcher's advice in answer to both questions is at present somewhat limited: "Feed a variety of marine invertebrates, the least expensive you can obtain; lobsters appear to do well on these in nature". There are some unavoidable problems with this apparently simplistic approach as a culturist will quickly indicate. For example, a commercial facility would require startlingly large amounts of these food items. Using a 36,364-kg (80,000-lb)/month lobster production figure and the conversion ratio of 3.3:1 (dry weight of feed to wet weight of animal) adapted by Johnson and Botsford[4] in their model, the culturist needs to locate (assuming the average dry weight per wet weight percentage of 17% for marine foodstuffs) almost 24 metric tons daily. A substantial portion of this tonnage would have to be reduced to smaller portions and presumably fed by hand. If an additional 2-weeks supply were kept on hand for those inevitable times of procurement problems, an additional 330 metric tons would need to be accommodated in frozen storage. The problem of adequate storage facilities is compounded by the lack of a dependable, cost-stable supply of natural food due to a frequent lack of alternative suppliers or foods. Moreover, the nutritional value of particular natural food items can differ depending upon harvesting season and locale. Provision of natural food items in an acceptable form would be highly labor intensive. Also, these materials could substantially contribute to the fouling of a system, particularly a recirculating one.

In order to mitigate these problems of supply, reliability, compositional consistency, and storage, all of which increase production costs, the potential lobster culturist is interested in the development of formulated or artificial diets. These diets would be derived from the knowledge of specific dietary requirements of these animals. Ideally, formulated diets in which optimum nutrient levels would be supplied by an appropriate but varying mixture of feedstuffs would be useful in reducing feed costs relative to the normal fluctuations within the commodities market. Preferably, these rations could be pelletized or otherwise adapted to automated feeding and hopefully could be stored in a nonfrozen form.

Artificial Food Development

In nature *Homarus* juveniles and adults appear to feed opportunistically on a variety of available invertebrates: crabs, polychaetes, mussels, gastropods, as well as sea urchins and starfish.[5] Out of this variety, lobsters appear to prefer crustaceans both in nature[6,7] and in the laboratory.[8,9] Mollusks, another primary constituent of their natural diet, are used extensively for laboratory culture (chopped mussels and squid).

A number of practical problems has prevented the straightforward development of artificial diets by analyzing and then mimicking the composition of natural prey items.[10] For example, the apparent requirement of juvenile lobsters for phosphatidylcholine (to be discussed later) could not be anticipated although a detailed chemical analysis would reveal brine shrimp (*Artemia* sp.), an excellent laboratory food source, to contain all the individual components, choline, phosphorus, and fatty acids, of this phospholipid compound. Additionally, the development of artificial diets is confronted with problems of achieving suitable palatability, attractiveness, stability, and digestibility. These factors, in association with the feeding habits[11] and molt cycle stage[12] of the lobster, could influence the nutrient input by altering consumption rates.

The development of a purified diet[13] which has been recently modified and improved (Table 1) represents a significant advance in the understanding of lobster nutrition. This defined diet has yielded good growth (0.05 mm/day) and survival (85 to 95%) of juvenile lobsters and should serve as the foundation for the rational development of inexpensive rations. With this diet, components are readily modifiable and specific nutritional requirements can be determined.[14] Extrapolation of the average growth which can be achieved on the diet of live adult brine shrimp is often used as a standard to compare the quality of other diets.[15] Growth rates that are 85% of this brine shrimp standard have been achieved through a modification of the newly developed purified diet using shrimp meal as a major component. Such growth rates indicate the feasibility of producing traditionally marketable sized (.5 kg) lobsters within 2.5 years at temperatures of 20°C.[16]

The first 60 days, following completion of the larval rearing phase, appear to be the most critical in the survival of juvenile lobsters. Variation in the survival rate has been noted even on the purified diet,[13] which is now routinely used as a control, from experiment to experiment. Survival during this time may be partially dependent upon maternal as well as larval nutritive history. Unfortunately, no precise research has been conducted regarding larval or maternal nutrition. This variation in average number of mortalities could be substantially reduced by initially using a brine shrimp diet which in past work has had great success.[17] Even commercially, the higher survival elicited could possibly outweigh the costs associated with the maintenance of live food for this short time period.

Most probably, effective large-scale commercial culture will entail the use of age-specific diets that would be routinely employed for (1) the first critical early juvenile growth period, (2) juvenile grow-out, and, perhaps, (3) broodstock maintenance. Presently, broodstock lobsters are fed a diet of varying fresh frozen crustaceans and mollusks. The need, if any, for special dietary requirements associated with reproduction is yet to be determined.

Physical Properties of Diets

For research and commercial usage, an ideal diet should be stable in water for at least 12 hr to minimize leaching and should be appealing to the lobster to maximize efficient consumption. Effective achievement of such characteristics simultaneously is a considerable problem. For research purposes moist and dry diets have been employed. Moist diets contain a gelling agent such as agar or gelatin.[18] This material is mixed with other dietary ingredients and hot water and then allowed to cool. Dry diets[11,13] typically contain a binder such as gluten, starch, or carboxymethycellulose. These diets usually consist of pellets that have been derived from an extruded dough and oven dried.

Table 1
BODEGA MARINE LABORATORY PURIFIED DIET 81S+

Ingredient	Dry weight (%)	Source[a]
Casein	31	(1)
Corn starch	24	(1)
Cellulose	12.1	(1)
Soy lecithin-refined	10	(1)
Lipid mix S	6	(3)
Gluten	5	(1)
Vitamin mix BML-2[b]	4	(1)
Spray dried egg white	4	(2)
Mineral mix BTm	3	(4)
Cholesterol	0.5	(1)
Vitamin E acetate 50% (500 IU/g)	0.2	(2)
Vitamin A acetate (500,000 IU/g)	0.1	(2)
Vitamin D_3 (400,000 IU/g)	0.1	(2)

Note: +Diet 81S differs from published diet 79F[13] in that the gluten has been reduced from 15 to 5%, the soy lecithin has been increased from 8 to 10%, and the corn starch has been reduced from 26.7 to 24%. Vitamins A and D_3 are now added separately (0.1% each) rather than as an A/D_3 mix (0.1%). Vitamin A activity has decreased from 650,000 to 500,000 IU/g. Vitamin D_3 activity has increased from 325,000 to 400,000 IU/g. The resulting net difference (10.6%) has been compensated via an increase in the cellulose content (1.5 to 12.1%).

[a] (1) ICN Pharmaceuticals, Inc., Cleveland, Ohio 44128; (2) Bio-serv, Inc., P.O. Box 100-B, Frenchtown, N.J. 08825; (3) Contains: cod liver oil 66% (Bio-serv), corn oil 33.8% (Bio-serv), ethoxyquin 0.2% (Monsanto); (4) Mineral Mix Bernhart — Tomarelli; modified (ICN).

[b] Vitamin mix BML-2 contains: thiamin mononitrate 0.5%, riboflavin 0.8%, nicotonic acid 2.6%, Ca-pantothenate 1.5%, pyridoxine HCl 0.3%, cobalamine 0.1%, folic acid 0.5%, biotin 0.1%, inositol 18%, ascorbic acid 12.5%, PABA 3%, cellulose 60%, BHA 0.1%.

No precise study has been completed to determine whether a diet's moisture content or binding characteristics has any significant effect on palatability and consumption in lobsters. Heinen[19] tested various binders in association with a trout feed intended for use as a crustacean diet. He found that dry pellets were less stable than moist ones. Under his experimental conditions, only agar and sodium alginate gave 24-hr stability for both types of pellets. Ideally, if two diets of different moisture content or chemical composition are to be tested, then consumption rates should be determined. Differential consumption rates could affect the relative nutritive value of particular diets, as determined by growth rates. Precise consumption measurements, however, are difficult since the feeding activity of lobsters causes the diet to be broken into minute particles which are often nonretrievable in a system where significant flow is maintained to insure proper water quality.

Nutrient leaching from artificial diets is another factor which affects the optimal provision of water-soluble nutrients. A significant reduction in the concentration of water-soluble nutrients has been demonstrated in pelleted trout feeds after only 10 sec of submergence;[20] the majority of water-soluble nutrients were leached from extruded crustacean diets after 2 hr.[21] Quantification of water-soluble nutrient requirements will be extremely difficult and it may be possible only to indirectly approximate these requirements unless some method of encapsulation can be developed. The design of the capsule should significantly reduce leaching of water solubles, yet still be amenable to disjunction by the lobster's masticatory activities. Recent developments in capsule technology for aquatic invertebrates are encouraging, however, these capsules are not effective in trapping water-soluble nutrients for any extended period of time.[22] Alternatively, the development of an effective attractant might improve growth and survival on artificial diets particularly during the early juvenile period. The attractant would increase the probability of early encounter and consumption and thereby minimize nutrient losses due to leaching.

Storage practices and processing effects, particularly those associated with the application of heat and techniques of mixing, should also be a subject of concern.[20] Changes in the chemical composition of the diet subsequent to processing should be monitored and duly compensated for. Precaution should be exercised to insure that particular ingredients do not lose potency or become toxic.

Feeding and Digestion

Lobsters have an elaborate system of chemoreception and artificial diets should be designed with this in mind. The earlier a lobster encounters and is enticed to consume its food, the less leaching will occur. The chemosensory cells of the lobsters antennae are particularly sensitive to the amino acid hydroxyproline.[23]

Baker and Gibson[24] provide the most recent study of food manipulation and digestive tract physiology. Briefly, after extensive manipulation and tearing of food by various mouthparts the food is passed into the esophagus. From the esophagus food travels to a two-chambered stomach. In the cardiac stomach, the food is triturated by a gastric mill and mixed with digestive enzymes provided by the hepatopancreas. Eventually the food is turned into fine particles and is transferred to the pyloric stomach where it eventually finds its way to the hind gut or the hepatopancreas. The hepatopancreas is both a digestive and storage organ. Gross and microscopic analysis of the hepatopancreas is often a good indicator of the nutritional state of the lobster.[25] The hepatopancreas is the major site of digestive enzyme synthesis. Enzymes reported as part of the gastric juice include various types of lipases, proteinases, and carboxyhydrases. Very little is known regarding the kinetics and specificity of particular enzymes that are part of the digestive physiology of the lobster. It is reported that complete digestion may take 12 hr and that a periodicity of enzyme activity may exist.[24]

SPECIFIC REQUIREMENTS

Protein

Successful development of artificial diets for commercial lobster culture will be dependent upon effective utilization of protein. Protein will undoubtedly be one of the major dietary components, both in terms of cost and quantity added.

In nature, protein represents the major portion of the preferred foods of the lobster. Prey tissue contains high quantities of essential amino acids.[26] An examination of the gut contents of wild lobster populations indicated that protein intake varied in amount over the molt cycle but that protein quality was similar.[27]

Early attempts to adequately define quantitative and qualitative protein requirements by use of artificial diets were plagued by other essential nutrients being either absent or present

Table 2
SUMMARY OF PROTEINS USED AND THEIR SUCCESS IN ARTIFICIAL DIETS OF JUVENILE LOBSTERS

Dry weight of protein source (%)	Protein (%)	Binder	Survival (%)	Growth[a]	Ref.
Casein 0—60	0—58	Agar	40—80	0	18[b]
Wheat gluten 8 Yeast 15 Casein (varied) 5—45	20—60	Wheat gluten	33—97	0—1	32
Egg white 5—25 Shrimp meal 1—5 Tuna meal 4—19 Yeast 4—20	11—54	Alginate	89—98	Not given	32
Casein 31 Egg white 4 Wheat gluten 5	40	Wheat gluten	89	1	13
Rice bran 20 Shrimp meal 7 Herring meal (varied) Soy meal 3 Yeast 12	16—23	Kelgin	90—97	1	29
Casein 50	44	Gelatin	25—50	1	
Whole egg 50	44	Gelatin	0—25	1	28
Feather meal 50	44	Gelatin	25—50	0	
Shrimp 50	44	Gelatin	75—100	1	
Casein 31 Wheat gluten 15 Egg white 4 Shrimp meal 25	53	Wheat gluten	97	1	16

[a] 0, either a loss or none; 1, increase.
[b] Adult animals used.

at suboptimal levels. Having developed the first artificial diets for the lobster, Castell and Budson[18] found that adult lobsters lost weight if dietary protein, casein, was reduced to below 60% (dry weight). As a result they suggested that the lobster utilizes protein as a principal source of energy in addition to tissue growth. This apparent high protein dietary requirement was undoubtedly an artifact due to other nutrient imbalances in this early diet.[10]

Table 2 summarizes past research investigating the effect of quantity and quality of protein on growth of lobsters. In general, marine animals have a high essential to dispensable amino acid ratio in their tissues.[26] Casein, which is known to be deficient in several essential amino acids, has been demonstrated to be a poor singular source of protein when added in excess of 50% (dry weight) of the diet.[18,28] In particular, arginine and sulfur amino acids (methionine and cysteine) are commonly found in low concentrations in most terrestrial sources when compared to lobster tissue proteins. However, this is not the case in protein from marine animal sources. As a consequence, shrimp meal and fish meal are good dietary supplements for lobsters on the basis of amino acid profile.

As mentioned earlier, natural lobster diets are high in protein and probably some is used as a source of energy. Crustaceans are ammoniotylic and, therefore, are spared the energy cost of producing urea or uric acid as in terrestrial animals. However, the use of protein for energy is still economically wasteful and alternative sources would be preferable in formulated diets. Capuzzo[29] utilized diets containing different levels of herring meal to vary protein levels and found protein sparing could be achieved without sacrificing growth by replacement with dietary carbohydrates. Similar results[30] have been demonstrated in several domesticated animals including fish. These results were confirmed in feeding trails which demonstrated no differences in growth for juvenile lobsters fed diets containing 30.5 to 53.0% protein from various feedstuffs.[16] In these experiments the relative proportions of the various protein sources remained constant. It is evident that a good balance of dietary protein can be achieved at low levels without sacrificing growth. While promising, this potential protein sparing effect has yet to be confirmed using diets containing purified protein sources. Also, it is yet to be determined for lobsters if protein utilization and requirements change with age or physiological state.

For comparative research purposes, a protein standard, either a single isolated protein source or mixture of purified proteins, suitable for the lobster needs to be identified. While an isolated crustacean protein would seem to be an ideal reference standard, efforts for successful incorporation are still plagued with other deficiencies in the diet.[28]

The technique of defining individual amino acid requirements through crystalline amino acid supplementation commonly used for other animals is not applicable in the lobster due to the rapid leaching of water-soluble nutrients from formulated diets. Previously, unsuccessful efforts to supplement lobster diets with pure amino acids[31] are probably related to this phenomenon. Using radioactive tracer techniques, Gallagher[32] demonstrated that ten amino acids — arginine, histidine, isoleucine, leucine, lysine, methionine, phenylalanine, threonine, tryptophan, and valine — could not be synthesized endogenously and were, therefore, considered essential for the lobster. However, the ability to synthesize particular amino acids does not preclude their possible requirement since synthetic rates may also be limiting. Thus, definitive work concerning minimal protein requirements will require greater understanding of enzymatic capacities as well as appropriate reference protein(s). Efforts to produce formulations that will yield efficient protein utilization and optimal growth must be united with studies involving other nutrients, i.e., protein metabolism is dependent upon several vitamins.

Lipids and Fat-Soluble Vitamins

In lobsters, the primary lipid pool is the hepatopancreas. The lipid content of the hepatopancreas is dependent upon dietary lipid intake but normally represents 15 to 20% of the wet weight of this tissue. In contrast, the lipid content of the remaining tissue is only 1 to 2% (wet weight). Both polar and neutral lipids are stored in the hepatopancreas. Diglycerides and triglycerides appear to be used as sources of energy while phospholipids appear to be intimately associated with lipid transport activities and reproduction.

A summary of the dietary lipids required by the lobster is presented in Table 3. D'Abramo and co-workers[33] have demonstrated that the deletion of a source of triglycerides (cod liver oil, corn oil) from an artificial diet significantly reduced growth in juvenile lobsters. A specific requirement for polyunsaturated fatty acids (PUFA) of the linolenic series has been suggested by the work of Castell and co-workers.[34,35] This requirement is also found in a wide variety of arthropods.[10] Juvenile lobster growth was enhanced when the lobsters were fed diets containing oils rich in polyunsaturated fatty acids of the linolenic series and it was found the PUFA are preferentially incorporated into the polar lipids of the tissue.[33] For arthropods, fatty acids of the linoleic series appear to have some sparing effect on the requirements of the linolenic series fatty acids. This phenomenon appears to be true for lobsters, also.[36]

Table 3
SUMMARY OF LIPID REQUIREMENTS FOR THE SURVIVAL AND OPTIMAL GROWTH OF JUVENILE LOBSTERS

Lipid	Dry weight in the diet (%)	Feeding period	Observed effect when absent	Ref.
Cholesterol	0.5—1.0	10 Months		34
Triglycerides	4 — Cod liver oil 2 — Corn oil	60 Days	Lower growth rates	36
Phosphatidylcholine	As 8—10 — refined soy lecithin	120 Days	Very poor survival	14
ω3 Polyunsaturated fatty acids	As 6 — cod liver oil	10 Months	Lower feed conversion, serum protein levels, and serum hemocyte counts	35[a]
ω3 Polyunsaturated fatty acids	4 — Tuna oil, 2 — cod liver oil, or 4 — cod liver oil, 2 — tuna oil	120 Days	Lower growth rates	36

[a] Adult animals used.

Recent work has demonstrated that juvenile lobsters fed a purified diet (Table 1) require a source of dietary lecithin for survival.[13] The active ingredient in the lecithin is phosphatidylcholine and the relative effectiveness of various forms of phosphatidylcholine in preventing high mortality appears to be related to the degree of fatty acid unsaturation.[14] A relationship between the quantities of serum phospholipids and serum cholesterol is indicated.[36] The phospholipids, primarily phosphatidylcholine, are apparently a component of high density lipoproteins which are found in the hemolymph and operate to efficiently transport cholesterol from the hepatopancreas to target tissues during specific stages of the molt cycle. The phosphatidylcholine requirement of juvenile lobsters is interesting since it has been shown that they are able to synthesize this molecule from component precursors.[37] However, the rate of synthesis may not be sufficient during a time of accelerated growth (short molt intervals). Further investigations of these various interactions among triglycerides, phosphatidylcholine, and cholesterol are needed.

Zandee[38] pinpointed the lack of cholesterol biosynthesis from ^{14}C-labeled precursors in *Homarus*. This observation is in agreement with the seemingly general inability for cholesterol synthesis in all arthropods. Castell and co-workers[34] found that growth and survival were maximized when juvenile lobsters were provided a diet containing 0.5% (dry weight) cholesterol. Depending upon the amount of cholesterol in other components of the diet such as cod liver oil, the 0.5% may be marginal and have to be increased;[35] however, 2% cholesterol resulted in a definite decrease in growth.[34] The required level of cholesterol in the diet of the lobster may be affected by the presence of phytosterols such as sitosterol, stigamosterol, and desmosterol. It has been demonstrated that several crustacean species are able to convert various of these plant-derived sterols into cholesterol.[39] However, recent experiments in our laboratory indicate that phytosterols cannot completely replace cholesterol in artificial diets fed lobsters. However, a cholesterol sparing action may still be possible with the use of a phytosterol-cholesterol mixture, similar to that described in certain species of insects.[1] As indicated earlier, the required level of cholesterol in the diet may also be affected by the quality and quantity of phospholipids.

Little is known about the requirement for fat-soluble vitamins by invertebrates. However, specific requirements for these nutrients have been noted for some insect species[40] and the possibility exists that quantifiable requirements will be identified for the lobster. As vitamin E is generally required to protect polyunsaturated fatty acids from oxidation, it can be anticipated that lobsters will also have a requirement although no definitive work regarding required levels has been accomplished. Vitamin D has also been suggested as a requirement for lobsters,[35,41] but, again, definitive results are lacking.

Water-Soluble Vitamins

It is assumed that lobsters, like other animals that have been studied, will require a complete range of dietary B vitamins: thiamin, riboflavin, pyridoxine (B_6), pantothenic acid, niacin, folacin, cyanocobalamin (B_{12}), biotin;[10] a response to individual deletion of these nutrients from a purified diet has not been conclusively demonstrated. A significant growth response by juvenile lobsters has been demonstrated in response to increases in a vitamin mix present as part of a purified diet.[42]

Other crustaceans have an apparent requirement for choline[43,44] and this nutrient is often present in artificial diets as part of the additive, phosphatidylcholine. Although a requirement for vitamin C has not been demonstrated for lobsters, it is known to be important for growth[45] and wound repair[46] in penaeid shrimp. Vitamin C has also been shown to be required by a number of insects.[40]

Defining the requirement for individual water-soluble vitamins is hampered by the rapid leaching of these nutrients from the diets. Interestingly, the use of phosphatidylcholine in artificial diets in the form of soy lecithin at the optimum level of 7 to 10% significantly reduces the leaching of water-soluble nutrients from the pellet.[47] For practical diets it may be possible to circumvent this problem through the use of appropriate feedstuffs in which these nutrients are tightly complexed, thereby minimizing the problem of leaching. However, basic research to define suitable levels is presently hindered by the lack of an appropriate delivery system. Although its technology is in its infancy, microencapsulation may be a partial answer not only for research but also for commercial culture purposes. For example, a requirement for a water-soluble vitamin at high levels, such as vitamin C which is readily lost during processing and storage, could only be supplied through supplementation. The precise definition of vitamin requirements of the lobster may also be confounded by the contribution of some of these compounds by endogenous gut bacteria.

Minerals

Very little is known about the mineral requirements of the lobster. In nature an apparent change to a high calcium diet prior to molting has been suggested.[6] The addition of calcium to artificial diets improved the mineralization of the exoskeleton but no survival or growth enhancement was observed.[31] In terrestrial animals, phosphorus influences calcium balance and bone mineralization and excess phosphorus is known to promote bone resorption. It has been suggested that an optimal Ca/P of approximately 1:2 should be maintained in lobster diets.[48] However, since all the diets in the experiment yielded poor growth and survival relative to a brine shrimp control diet this interaction is still open to question. A similar ratio (1:2) has been found to be desirable for fish.[49] While lobsters could presumably derive all their calcium needs from seawater,[50] phosphorus is less abundant in seawater than calcium and, thus, any phosphorus requirement would have to be met by the diet. Presently, mineral mixes are routinely supplied as ingredients of artificial diets.

DIET EVALUATION AND FUTURE CONSIDERATIONS

The recently developed purified diet (Table 1) for lobsters serves as an experimental tool for the development of an inexpensive ration that would be feasible for commercial production. The purified diet will hopefully be refined, possibly with a purified crustacean source protein. In the future, such a diet could become a standard reference diet for all crustacea of aquaculture importance.

The evaluation of the nutritional quality of diets for the lobster has in the past been basically dependent upon measurements of wet weight increases and survival through time. As the development of successful diets proceeds, more sophisticated and precise measurements of growth and nutritional state should be developed. Diet evaluations could possibly involve the monitoring of physiological indexes such as the activity of a particular enzyme or the ratio of two particular nutrients in the hemolymph or the synthesis of a particular organic compound as related to the stage of the molt cycle.

Although artificial diets currently in use are cost prohibitive they do yield growth rates that suggest marketable size can be achieved within 2.5 years.[16] Better growth rates are possible.[51] Given excellent culture conditions such as water quality maintenance and adequate space, growth optimization hopefully can be achieved via an inexpensive ration based upon a knowledge of quantitative and qualitative nutritional requirements. This knowledge will need to be supplemented with studies into the effective use of binders, digestibility of feedstuffs, encapsulated nutrients, attractants, diurnal or molt cycle feeding periodicities, and age-specific growth requirements.

ACKNOWLEDGMENTS

This work is a result of research sponsored in part by NOAA, Office of Sea Grant, Department of Commerce, under Grant No. 04-8-M01-189 R/A 28. The U.S. government is authorized to produce and distribute reprints for governmental purposes notwithstanding any copyright notation that may appear hereon.

REFERENCES

1. **Dadd, R. H.,** Qualitative requirements and utilization of nutrients: insects, in *CRC Handbook Series in Nutrition and Food,* Vol. 1, Rechcigl, M., Jr., Ed., CRC Press, Boca Raton, Fla., 1977, 305.
2. **New, M. B.,** A review of dietary studies with shrimp and prawns, *Aquaculture,* 9, 101, 1976.
3. **New, M. B.,** A bibliography of shrimp and prawn nutrition, *Aquaculture,* 21, 101, 1980.
4. **Johnson, W. E. and Botsford, L. W.,** Systems Analysis for Lobster Aquaculture, EIFAC/80/Symp.: E/56, symp. on new developments in the utilization of heated effluents and of recirculation systems for intensive aquaculture, Stavanger, Norway, 1980.
5. **Cooper, R. A. and Uzmann, J. R.,** Ecology of juvenile and adult *Homarus*, in *The Biology and Management of Lobsters,* Vol. 2, Cobb, J. S. and Phillips, B. F., Eds., Academic Press, New York, 1980, 97.
6. **Weiss, H. M.,** The Diet and Feeding Behavior of the Lobster, *Homarus americanus,* in Long Island Sound, Ph.D. thesis, University of Connecticut, Storrs, 1970.
7. **Ennis, G. P.,** Food, feeding, and condition of lobsters, *Homarus americanus*, throughout the seasonal cycle in Bonavista Bay, Newfoundland, *J. Fish Res. Board Can.*, 30, 1905, 1973.
8. **Evans, P. D. and Mann, K. H.,** Selection of prey by American lobsters *(Homarus americanus)* when offered a choice between sea urchins and crabs, *J. Fish. Res. Board Can.*, 34, 2203, 1977.
9. **Hirtle, R. W. and Mann, K. H.,** Distance chemoreception and vision in the selection of prey by American lobster *(Homarus americanus), J. Fish. Res. Board Can.*, 35, 1006, 1978.
10. **Conklin, D. E.,** Nutrition, in *The Biology and Management of Lobsters,* Vol. 1, Cobb, J. S. and Phillips, B. F., Eds., Academic Press, New York, 1980, 277.
11. **Bordner, C. E. and Conklin, D. E.,** Food consumption and growth of juvenile lobsters, *Aquaculture,* 24, 285, 1981.
12. **Davis, A. L.,** Importance of Palatability, Approach Time and Effort, Nutritional Adequacy and Molt Stage in Prey Choice in *Homarus americanus* (Milne-Edwards), M.Sc. thesis, University of Rhode Island, Kingston, 1979.

13. **Conklin, D. E., D'Abramo, L. R., Bordner, C. E., and Baum, N. A.,** A successful diet for the culture of juvenile lobsters: the effect of lecithin, *Aquaculture,* 21, 243, 1980.
14. **D'Abramo, L. R., Bordner, C. E., Conklin, D. E., and Baum, N. A.,** Essentiality of dietary phosphatidylcholine for the survival of juvenile lobsters, *Homarus, J. Nutr.,* 111, 425, 1981.
15. **Conklin, D. E., Devers, K., and Bordner, C.,** Development of artificial diets for the lobster *Homarus americanus, Proc. World Maricul. Soc.,* 8, 841, 1977.
16. **D'Abramo, L. R., Bordner, C. E., Conklin, D. E., Baum, N. A., and Norman-Boudreau, K. E.,** Successful artificial diets for the culture of juvenile lobsters, *Proc. World Maricul. Soc.,* 12, 325, 1981.
17. **Shleser, R. A. and Gallagher, M. L.,** Formulation of rations for the American lobster, *Homarus americanus, Proc. World Maricul. Soc.,* 5, 157, 1974.
18. **Castell, J. D. and Budson, S. D.,** Lobster nutrition: the effect on *Homarus americanus* of dietary protein level, *J. Fish. Res. Board Can.,* 31, 1363, 1974.
19. **Heinen, J. M.,** Evaluation of some binding agents for crustacean diets, *Prog. Fish Cult.,* 43, 142, 1981.
20. **Slinger, S. J., Razzaque, A., and Cho, C. Y.,** Effect of feed processing and leaching on the losses of certain vitamins in fish diets, in *Finfish Nutrition and Fishfeed Technology,* Vol. 2, Halver, J. E. and Tiews, K., Eds., Heenemann Verlags-Gesellschaft, Berlin, 1978, 425.
21. **Goldblatt, M. J., Brown, W. D., and Conklin, D. E.,** Nutrient leaching from pelleted rations, in *Finfish Nutrition and Fishfeed Technology,* Vol. 2, Halver, J. E. and Tiews, K., Eds., Heenemann Verlags-Gesellschaft, Berlin, 1978, 117.
22. **Teshima, S., Kanazawa, A., and Sakamoto, M.,** Attempt to culture rotifers with a microencapsulated diet, *Bull. Jpn. Soc. Sci. Fish.,* p. 47, 1981.
23. **Ache, B. W.,** The sensory physiology of spiny and clawed lobsters, *Circ. CSIRO, Div. Fish. Oceanogr. (Aust.),* 7, 103, 1977.
24. **Barker, P. L. and Gibson, R.,** Observations on the feeding mechanism, structure of the gut, and digestive physiology of the European lobster *Homarus gammarus* (L.) (Decapoda: Nephropidae), *J. Exp. Mar. Biol. Ecol.,* 26, 297, 1977.
25. **Rosemark, R., Bowser, P. R., and Baum, N.,** Histological observations of the hepatopancreas in juvenile lobsters subjected to dietary stress, *Proc. World Maricul. Soc.,* 11, 471, 1980.
26. **Borgstrom, G.,** Shellfish protein — nutritive aspects, in *Fish as Food,* Vol. 2, Borgstrom, C., Ed., Academic Press, New York, 1962, 115.
27. **Leavitt, D. F., Bayer, R. C., Gallagher, M. L., and Rittenburg, J. H.,** Dietary intake and nutritional characteristics in wild American lobsters, *Homarus americanus, J. Fish. Res. Board Can.,* 36, 965, 1979.
28. **Boghen, A. D. and Castell, J. D.,** Nutritional value of different dietary proteins to juvenile lobsters, *Homarus americanus, Aquaculture,* 22, 343, 1981.
29. **Capuzzo, J. M. and Lancaster, B. A.,** The effects of dietary carbohydrate levels on protein utilization in the American lobster, *Homarus americanus, Proc. World Maricul. Soc.,* 10, 689, 1979.
30. **Garling, D. L. and Wilson, R. P.,** Optimum dietary protein to energy ratio in channel catfish fingerlings, *Ictalurus punctatus, J. Nutr.,* 106, 1368, 1976.
31. **Conklin, D. E., Devers, K., and Shleser, R. A.,** Initial development of artificial diets for the lobster, *Homarus americanus, Proc. World Maricul. Soc.,* 6, 237, 1975.
32. **Gallagher, M. L.,** The Nutritional Requirements of Juvenile Lobster, *Homarus americanus,* Ph.D. thesis, University of California, Davis, 1976.
33. **D'Abramo, L. R., Bordner, C. E., Conklin, D. E., Daggett, G. R., and Baum, N. A.,** Relationships among dietary lipids, tissue lipids and growth in juvenile lobsters, *Proc. World Maricul. Soc.,* 11, 335, 1980.
34. **Castell, J. D., Mason, E. G., and Covey, J. F.,** Cholesterol requirements in the juvenile lobster *Homarus americanus, J. Fish. Res. Board Can.,* 32, 1431, 1975.
35. **Castell, J. D. and Covey, J. F.,** Dietary lipid requirements of adult lobsters *Homarus americanus, J. Nutr.,* 106, 1159, 1976.
36. **D'Abramo, L. R., Bordner, C. E., and Conklin, D. E.,** Relationship between dietary phosphatidylcholine and serum cholesterol in the lobster, *Homarus americanus, Mar. Biol.,* 67, 231, 1982.
37. **Shieh, M. S.,** The biosynthesis of phospholipids in the lobster *Homarus americanus, Comp. Biochem. Physiol.,* 39, 679, 1969.
38. **Zandee, D. I.,** Absence of cholesterol synthesis as contrasted with the presence of fatty acid synthesis in some arthropods, *Comp. Biochem. Physiol.,* 20, 811, 1967.
39. **Teshima, S. and Kanazawa, A.,** Bioconversion of dietary ergosterol to cholesterol in *Artemia salina, Comp. Biochem. Physiol.,* 38B, 603, 1971.
40. **Dadd, R. H.,** Insect nutrition: current developments, *Annu. Rev. Entomol.,* 18, 381, 1973.
41. **Stewart, J. E. and Castell, J. D.,** Various aspects of culturing the American lobster, *Homarus americanus,* in *Advances in Aquaculture,* Pillary, T. V. R. and Dill, W. A., Eds., Fishing News Books, Farnham, Surrey, England, 1979, 314.

42. **Conklin, D. E.,** Recent progress in lobster nutrition at the Bodega Marine Laboratory, in 1980 Lobster Nutrition Workshop Proc., Bayer, R. C. and D'Agostino, A., Eds., Marine Sea Grant Tech. Report #58, Walpole, 1980, 29.
43. **D'Abramo, L. R. and Baum, N. A.,** Choline requirement for the microcrustacean *Moina macrocopa:* a purified diet for continuous culture, *Biol. Bull.*, 161, 357, 1981.
44. **Kanazawa, A., Teshima, S., and Tanaka, S.,** Nutritional requirement of prawn. V. Requirements for choline and inositol, *Mem. Fish. Kagoshima Univ.*, 25, 47, 1976.
45. **Guary, M., Kanazawa, A., Tannaka, N., and Ceccaldi, H. J.,** Nutritional requirements of prawn. VI. Requirement for ascorbic acid, *Mem. Fac. Fish. Kagoshima Univ.*, 25, 53, 1976.
46. **Lightner, D. V., Colvin, L. B., Brand, C., and Donald, D. A.,** Black death — a disease syndrome of penaeid shrimp related to a dietary deficiency of ascorbic acid, *Proc. World Maricul. Soc.*, 8, 611, 1977.
47. **Castell, J. D.,** personal communication, 1981.
48. **Gallagher, M. L., Brown, W. D., Conklin, D. E., and Sifri, M.,** Effects of varying calcium phosphorus ratios in diets fed to juvenile lobsters *(Homarus americanus), Comp. Biochem. Physiol.*, 60A, 467, 1978.
49. **Sakamoto, S. and Yone, Y.,** Effects of dietary calcium on growth, feed efficiency and blood serum calcium and phosphorus levels in red sea bream, *Bull. Jpn. Soc. Sci. Fish.*, 39, 343, 1973.
50. **Deshimaru, O., Kuroki, K., Sakamato, S., and Yone, Y.,** Absorbtion of labelled calcium — ^{45}Ca by prawn from seawater, *Bull. Jpn. Soc. Sci. Fish.*, 44, 975, 1978.
51. **Hughes, J. T., Sullivan, J., and Shleser, R.,** Enhancement of lobster growth, *Science*, 177, 1110, 1972.

Index

INDEX

A

D

E

F

G

H

I

J

K

L

O

P

Q

R

T

Z